Integration of Insect-Resistant Genetically Modified Crops within IPM Programs

Progress in Biological Control

Volume 5

Published:

Volume 1
H.M.T. Hokkanen and A.E. Hajek (eds.):
Environmental Impacts of Microbial Insecticides – Need and Methods for Risk Assessment. 2004
ISBN 978-1-4020-0813-9

Volume 2
J. Eilenberg and H.M.T. Hokkanen (eds.):
An Ecological and Societal Approach to Biological Control. 2007
ISBN 978-1-4020-4320-8

Volume 3
J. Brodeur and G. Boivin (eds.):
Trophic and Guild Interactions in Biological Control. 2006
ISBN 978-1-4020-4766-4

Volume 4
J. Gould, K. Hoelmer and J. Goolsby (eds.):
Classical Biological Control of *Bemisia tabaci* in the United States. 2008
ISBN 978-1-4020-6739-6

Volume 5
J. Romeis, A. M. Shelton, and G. G. Kennedy (eds.):
Integration of Insect-Resistant Genetically Modified Crops within IPM Programs. 2008
ISBN 978-1-4020-8372-3

Forthcoming:

Use of Microbes for Control and Eradication of Invasive Arthropods
Edited by A.E. Hajek, M. O'Callaghan and T. Glare

Ecological & Evolutionary Relationships among Entomphagous Arthropods and Non-prey Foods
By J. Lundgren

Biocontrol-based Integrated Management of Oilseed Rape Pests
Edited by I.H. Williams and H.M.T. Hokkanen

Biological Control of Plant-Parasitic Nematodes: Building Coherence between Microbial Ecology and Molecular Mechanisms
Edited by Y. Spiegel and K. Davies

Egg Parasitoids in Agroecosystems with emphasis on Trichogramma
Edited by F. Consali, J. Parra, R. Zucchi

Jörg Romeis • Anthony M. Shelton
George G. Kennedy
Editors

Integration of Insect-Resistant Genetically Modified Crops within IPM Programs

Springer 2008

Editors
Jörg Romeis
Agroscope Reckenholz-Tänikon Research Station ART
Reckenholzstrasse 191
8046 Zurich
Switzerland
joerg.romeis@art.admin.ch

Anthony M. Shelton
Department of Entomology
Cornell University/NYSAES
Geneva, NY 14456
USA
ams5@cornell.edu

George G. Kennedy
Department of Entomology
North Carolina State University
Raleigh, NC 27695-7630
USA
George_Kennedy@ncsu.edu

ISBN 978-1-4020-8459-1 (PB)
ISBN 978-1-4020-8372-3 (HB)
e-ISBN 978-1-4020-8373-0 (e-book)

Library of Congress Control Number: 2008923181

Cover Illustration:
Upper left: Scouting a maize crop.
Lower left: Cotton crop.
Upper right: European corn borer, *Ostrinia nubilalis* (Lepidoptera: Crambidae), damage and fungal infection in non-*Bt* (left) maize and *Bt* maize.
Lower right: A green lacewing, *Chrysoperla rufilabris* (Neuroptera: Chrysopidae), larva preying on whitefly nymphs.
The picture in the upper right was kindly provided by Gary Munkvold (Iowa State University, IA, USA). All others are from the USDA-ARS Image Gallery.

Printed on acid-free paper

9 8 7 6 5 4 3 2 1

springer.com

Endorsements

The products of biotechnology will be essential for moving agriculture forward to help meet the food and fiber needs of the growing world population. Biotech crops (GM crops) offer tremendous advances in our ability to manage agricultural pests safely and effectively, and have been rapidly adopted by farmers worldwide. Until recently, plant breeders have been unable to develop crops that are highly resistant to many of our most serious insect pests, but this changed when plants expressing proteins from the bacterium *Bacillus thuringiensis* (*Bt*) were developed. *Bt* crops fit in well with the concept and practice of Integrated Pest Management (IPM), and are becoming the cornerstone for IPM in the world's most important crops. This comprehensive book provides valuable information and analysis by many of the world's leading experts involved with integrating transgenic insect-resistant crops into IPM.

Norman E. Borlaug - Nobel Peace Prize Laureate, 1970

Using transgenic plants for pest management requires the best of science to retain both the public's trust and the durability of the technology. This comprehensive book contains the best scientific knowledge to date about transgenic insecticidal plants and the importance of their use within an IPM context. Transgenes, especially those from *Bacillus thuringiensis*, are increasingly used to protect the world's most important crops (cotton, maize, potato and rice) from insect damage. However the durability of their effectiveness is under pressure from insect evolution, and should thus be protected by appropriate IPM practices. This book has collected the wisdom and experience of many of the leading experts on this extremely important aspect of food and fiber security and will serve as an important guide to the future of IPM in transgenic crop management for students, regulators, and a wide array of scientists in developed and developing countries.

Thomas Lumpkin, former Director General, AVRDC - The World Vegetable Center and new Director General of CIMMYT

Foreword

The Green Revolution of the 1960s, 1970s and 1980s demonstrated the potential of science and technology to contribute to agricultural development, food security and economic growth in poor and predominantly agrarian countries as well as rich industrial countries.

The benefits reached many of the world's poorest people and the proportion of the population that is undernourished in developing countries declined from 40% in 1960 to 17% in 2000. While this was a great accomplishment, further research and development clearly needs to be done to better feed those that remain undernourished. And, since agro-ecosystems are not static but rather are continually evolving, considerable research and development is needed to maintain the productivity gains already achieved and to do so through farming practices that are more sustainable and leave a much smaller environmental footprint than current practices. Research to reduce crop losses caused by insect pests and pathogens has made and will continue to make important contributions toward the necessary increases in yield, productivity and sustainability.

This book reviews the potential for integrating, and thereby strengthening, two insect pest control technologies that have each already made significant contributions to reducing both crop losses and insecticide use in many countries. Integrated pest management (IPM) was developed as an insect control strategy in part due to the failure of insecticides to keep insect pests under control. For some crops, such as cotton and rice, inordinant insecticide applications had resulted in development of insects resistant to insecticides, emergence of new pests that were worse than those being targeted, increasing crop losses and negative environmental impacts. IPM has gone a long way in solving these problems by utilizing a collection of pest monitoring and control strategies designed to maintain pest populations below levels causing economic loss. This almost always includes genetic host plant resistance combined with biological control, cultural methods, behavioral methods and farmer knowledge. Effective IPM strategies have now been developed for many crops, including those that feed the developing world, and further improvements are continually being made.

The second pest control technology reviewed utilizes crop genetic engineering. Genes from the bacterium, *Bacillus thuringiensis* (*Bt*), strains of which have long been used as microbial insecticides, are added to the genome of crop plants. There the *Bt* genes express proteins that are toxic to target agronomic pests but not to other organisms. The technology has spread rapidly and in 2007 maize and cotton crops having this new form of host plant resistance were planted on 42 million hectares in 22 countries. Control of target insects has been excellent, insecticide use has been reduced significantly and strategies designed to delay or prevent the development of insects resistant to the *Bt* proteins have so far worked successfully. Field trials of numerous other crops containing *Bt* genes have demonstrated similar efficacy. Clearly this is a powerful new pest control technology that needs to be used wisely and for the benefit of a much greater number of the world's farmers, including those who cannot afford premium priced seed.

Several chapters in this book present evidence indicating that it should be possible to integrate crop plants having host plant resistance from *Bt* genes into existing and emerging IPM strategies. Unlike insecticides, *Bt* proteins are toxic only to the specific targeted pests and only to those insects that feed on *Bt* plant tissues. They are not toxic to all the other beneficial insects and organisms that are essential for biocontrol and ecosystem balance within an effective IPM system. To achieve integration and broader adoption of these two pest control strategies, further research is needed to: (1) develop an even better understanding of the impact of *Bt* crops on the general ecology of pests populations and their natural enemies, particularly under field conditions, (2) develop *Bt* based host plant resistance in a broader range of locally adapted crop varieties, including those that are essential for food security and economic growth in developing countries, and (3) develop strategies for incorporating *Bt* varieties into IPM systems in a ways that are most compatible with all other components of the IPM systems, are durable and empower farmers to become even more competent in the management of both pests and natural resources.

This book is an excellent first step in bringing together in one volume the relevant information necessary to achieve this integration of technologies. Now it is up to the IPM specialists and the crop genetic engineers to work together more effectively than they have to date to provide farmers throughout the world with the best pest control methods science has to offer.

Gary Toenniessen
Managing Director
Rockefeller Foundation

Preface

Insect pests remain one of the main constraints to food and fiber production worldwide despite farmers deploying a range of techniques to protect their crops. Modern pest control is guided by the principles of integrated pest management (IPM), defined as "a decision support system for the selection and use of pest control tactics, singly or harmoniously coordinated into a management strategy, based on cost/benefit analyses that take into account the interests of and impacts on producers, society, and the environment" (Kogan, 1998[1]). Pest resistant germplasm should be an important part of the foundation for IPM, but traditional breeding has not been able to achieve insect-resistant germplasm to many of our most serious pests. In the past decades, molecular tools of biotechnology have become available that allow the transfer of genes that provide strong plant resistance to certain groups of pests. Products of such genetic engineering procedures have been termed "genetically modified (GM)" by the public, although we take issue with this term since all of our agriculturally important plant species have been "modified" by farmers and breeders in some way over the last 10,000 years of agriculture. However, the editors and authors use the term GM because of its common use, as well as the terms "genetically engineered", "transgenic crops", or "biotech crops".

Since 1996, when the first insect-resistant GM maize variety was commercialized in the USA, the area planted to insect-resistant maize and cotton varieties has grown to 42.1 million hectares in 22 countries in 2007. This represents the fastest adoption rate of any agricultural technology in human history. While GM varieties have proven to be a powerful tool for pest management and their use has been accompanied by dramatic economic and environmental benefits, parts of the world (including most of Europe) are still engaged in discussions about potential negative impacts of these crops on the environment. Fear about potential negative effects of GM crops has lead to the implementation of very stringent regulatory systems in several countries and regulations that are far more restrictive for GM crops than for

[1] Kogan, M., 1998. Integrated pest management: Historical perspectives and contemporary developments. Annual Review of Entomology 43: 243–270.

other agricultural technologies. This has precluded many farmers and consumers from sharing benefits these crops can provide.

In this book we focus on insect-resistant GM plants and their place in agricultural IPM systems. These plants are designed to protect the crop from specific major insect pests in a very effective manner. As such the deployment of GM varieties will affect the way farmers manage their crop and, in particular, the way they apply other pest control measures. The intent of this book is to provide an overview of the development, adoption, and impact of insect-resistant GM plants and the role they play or could potentially play in IPM in different crop systems worldwide. We hope that the book will contribute to a more rational debate about the role GM crops can play in plant protection for food and fiber production.

Jörg Romeis
Anthony M. Shelton
George G. Kennedy

Progress in Biological Control

Series Preface

Biological control of pests, weeds, and plant and animal diseases utilising their natural antagonists is a well-established and rapidly evolving field of science. Despite its stunning successes world-wide and a steadily growing number of applications, biological control has remained grossly underexploited. Its untapped potential, however, represents the best hope to providing lasting, environmentally sound, and socially acceptable pest management. Such techniques are urgently needed for the control of an increasing number of problem pests affecting agriculture and forestry, and to suppress invasive organisms which threaten natural habitats and global biodiversity.

Based on the positive features of biological control, such as its target specificity and the lack of negative impacts on humans, it is the prime candidate in the search for reducing dependency on chemical pesticides. Replacement of chemical control by biological control – even partially as in many IPM programs – has important positive but so far neglected socio-economic, humanitarian, environmental and ethical implications. Change from chemical to biological control substantially contributes to the conservation of natural resources, and results in a considerable reduction of environmental pollution. It eliminates human exposure to toxic pesticides, improves sustainability of production systems, and enhances biodiversity. Public demand for finding solutions based on biological control is the main driving force in the increasing utilisation of natural enemies for controlling noxious organisms. This book series is intended to accelerate these developments through exploring the progress made within the various aspects of biological control, and via documenting these advances to the benefit of fellow scientists, students, public officials, policymakers, and the public at large. Each of the books in this series is expected to provide a comprehensive, authoritative synthesis of the topic, likely to stand the test of time.

Heikki M.T. Hokkanen, Series Editor

Contents

Contributors

Ramon Albajes
University of Lleida, Centre UdL-IRTA, Rovira Roure, 191, 25198 Lleida, Spain, ramon.albajes@irta.es

Barbara I.P. Barratt
AgResearch Ltd., Invermay Agricultural Centre, Private Bag 50034, Mosgiel 9035, New Zealand, barbara.barratt@agresearch.co.nz

J.S. Bentur
Directorate of Rice Research, Rajendranagar, Hyderabad 500 030, Andhra Pradesh, India, jbentur@yahoo.com

David Bergvinson
Program Officer, Global Development, Bill & Melinda Gates Foundation, PO Box 23350, Seattle, WA 98102, USA, david.bergvinson@gatesfoundation.org

Franz Bigler
Agroscope Reckenholz-Tänikon Research Station ART, Reckenholzstrasse 191, 8046 Zurich, Switzerland, franz.bigler@art.admin.ch

Mao Chen
Department of Entomology, Cornell University/NYSAES, 630 W. North Street, Geneva, NY 14456, USA, mc447@cornell.edu

Michael B. Cohen
11643 77 Ave., Edmonton, AB T6G 0M4, Canada, michael.cohen@shaw.ca

Galen P. Dively
University of Maryland, Entomology Department, College Park, MD 20742, USA, galen@umd.edu

David S. Douches
Department of Crop and Soil Sciences, Michigan State University, East Lansing, MI 48824, USA, douchesd@msu.edu

Juan Ferré
Department of Genetics, University of Valencia, Dr. Moliner 50,
46100 Burjassot (Valencia), Spain, juan.ferre@uv.es

Gary P. Fitt
CSIRO Entomology, 120 Meiers Road, Brisbane,
Queensland 4068, Australia, gary.fitt@csiro.au

Marc Fuchs
Department of Plant Pathology, Cornell University/NYSAES,
Geneva, NY 14456, USA, mf13@cornell.edu

Angharad M.R. Gatehouse
University of Newcastle, School of Biology, Institute for Research
on Environment and Sustainability, Devonshire Building,
Newcastle upon Tyne, NE1 7RU, UK, a.m.r.gatehouse@ncl.ac.uk

Edward J. Grafius
Department of Entomology, Michigan State University, East Lansing,
MI 48824, USA, grafius@msu.edu

Graham P. Head
Monsanto Company, 800 North Lindbergh Blvd., St. Louis,
MO 63167, USA, graham.p.head@monsanto.com

Richard L. Hellmich
USDA–ARS, Corn Insects and Crop Genetics Research Unit
and Department of Entomology, Iowa State University,
110 Genetics Laboratory c/o Insectary, Ames, IA 50011, USA
richard.hellmich@ars.usda.gov

K.L. Heong
International Rice Research Institute (IRRI), DAPO 7777,
Metro Manila, Philippines, k.heong@cgiar.org

Rod A. Herman
Dow AgroSciences LLC, 9330 Zionsville Road, Indianapolis, IN 46268, USA,
raherman@dow.com

George G. Kennedy
Department of Entomology, Box 7630, North Carolina State University,
Raleigh, NC 27695-7630, USA, George_Kennedy@ncsu.edu

Susan C. MacIntosh
MacIntosh & Associates, Inc., 1203 Hartford Ave., Saint Paul,
MN 55116-1622, USA, macintosh-associates@comcast.net

Louise A. Malone
The Horticulture and Food Research Institute of New Zealand Limited,
Private Bag 92169, Auckland Mail Centre, Auckland 1142, New Zealand,
lmalone@hortresearch.co.nz

Sharlene R. Matten
United States Environmental Protection Agency, Office of Science Coordination and Policy (7201M), 1200 Pennsylvania Ave., NW, Washington, DC 20460, USA, matten.sharlene@epa.gov

Steven E. Naranjo
USDA-ARS, Arid-Land Agricultural Research Center, 21881 North Cardon Lane Maricopa, AZ 85238, USA, Steve.Naranjo@ars.usda.gov

Jarrad R. Prasifka
USDA–ARS, Corn Insects and Crop Genetics Research Unit, 110 Genetics Laboratory c/o Insectary, Iowa State University, Ames, IA 50011, USA, jarrad.prasifka@ars.usda.gov

Carl E. Pray
Department of Agricultural, Food, and Resource Economics, Rutgers University, 55 Dudley Road, New Brunswick, NJ 08901, USA, pray@aesop.rutgers.edu

Matin Qaim
Department of Agricultural Economics and Rural Development, Georg-August-University of Göttingen, Platz der Göttinger Sieben 5, 37073 Göttingen, Germany, mqaim@uni-goettingen.de

Hector D. Quemada
Department of Biology, Calvin College, 1726 Knollcrest Circle, S.E., Grand Rapids, MI 49546-4403, USA, hdq2@calvin.edu

Jörg Romeis
Agroscope Reckenholz-Tänikon Research Station ART, Reckenholzstrasse 191, 8046 Zurich, Switzerland, joerg.romeis@art.admin.ch

John R. Ruberson
Department of Entomology, University of Georgia, 122 So. Entomology Dr., Tifton, GA 31794, USA, ruberson@uga.edu

Hari C. Sharma
International Crops Research Institute for the Semi-Arid Tropics (ICRISAT), Patancheru 502 324, Andhra Pradesh, India, H.Sharma@cgiar.org

Anthony M. Shelton
Department of Entomology, Cornell University/NYSAES, 630 W. North Street, Geneva, NY 14456, USA, ams5@cornell.edu

Frank A. Shotkoski
International Programs/ABSPII, Cornell University, 213 Rice Hall, Ithaca, NY 14853, USA, fas23@cornell.edu

Nicholas P. Storer
Dow AgroSciences LLC, 9330 Zionsville Road, Indianapolis, IN 46268, USA, nstorer@dow.com

Gary Toenniessen
The Rockefeller Foundation, 420 5th Ave., New York, NY 10018-2702, USA, GToenniessen@rockfound.org

Roy G. Van Driesche
Department of Plant, Soil & Insect Sciences, Agricultural Engineering Building 320, University of Massachusetts, Amherst, MA 01003, USA, vandries@nre.umass.edu

Jeroen Van Rie
Bayer BioScience N.V., Technologiepark 38, 9052 Ghent, Belgium, Jeroen.VanRie@bayercropscience.com

Zhen-Ying Wang
Institute of Plant Protection, Chinese Academy of Agricultural Sciences, West Yuanmingyuan Road, Beijing 100094, China, zywang@ippcaas.cn

Michael J. Weiss
Golden Harvest Seeds, W 4166 County Road H, Pine River, WI 54965, USA, michael.weiss@ghseeds.com

Lewis Wilson
CSIRO Plant Industry & Cotton Catchment Communities Co-operative Research Centre, Locked Bag 59, Narrabri, New South Wales, Australia, 2390, lewis.wilson@csiro.au

Kongming Wu
Institute of Plant Protection, Chinese Academy of Agricultural Sciences, West Yuanmingyuan Road, Beijing 100094, China, kmwu@ippcaas.cn

Gongyin Ye
Institute of Insect Sciences, College of Agriculture and Biotechnology, Zhejiang University, Hangzhou 310029, China, chu@zju.edu.cn

David Zilberman
Department of Agricultural and Resource Economics, University of California, 207 Giannini Hall, Berkeley, CA 94720, USA, zilber@are.berkeley.edu

Chapter 1
Integration of Insect-Resistant Genetically Modified Crops within IPM Programs

George G. Kennedy*

Abstract Although host plant resistance has long been an important insect management tactic, its wide-spread use has been constrained by the limited availability of elite cultivars possessing high levels of resistance to key pest species. The application of recombinant DNA technology to genetically engineer insect-resistant crop plants has provided a way to eliminate this constraint and make host plant resistance a prominent component of integrated pest management (IPM) in major cropping systems world-wide. It is within the framework of IPM, rather than as a stand-alone insect control measure, that insect-resistant GM crops have the greatest potential to contribute to the establishment of sustainable crop protection systems. This chapter reviews the defining elements of IPM and examines the attributes of insect-resistant GM crops as IPM tools. Insect-resistant GM crops available to date, like their counterparts developed through conventional plant breeding, are proving to be safe, effective and easy to use insect suppression tools that are compatible with other IPM tactics, including cultural and chemical controls and the conservation of natural enemies as important agents of biological control. Because of their high level of efficacy against the key pest species that they target, GM *Bt* cotton and *Bt* maize varieties expressing cry genes derived from *Bacillus thuringiensis* (*Bt*) have been widely adopted and have led to significant reductions in insecticide use. Experience in *Bt* cotton has revealed the potential for reductions in insecticide use to be accompanied by the emergence of secondary pests and the need to adjust the pest management systems to address these "new" pests. Emphasis on the importance of resistance management to mitigate selection for pest adaptation to *Bt* crops has elevated the role of resistance management to a position of fundamental importance in the implementation of IPM.

Department of Entomology, North Carolina State University, Raleigh, NC, USA

* To whom correspondence should be addressed. E-mail: george_kennedy@ncsu.edu

J. Romeis, A.M. Shelton, G.G. Kennedy (eds.), *Integration of Insect-Resistant Genetically Modified Crops within IPM Programs.*

1.1 Introduction

When highly effective, synthetic insecticides were introduced beginning in the late 1940s and 1950s, it became possible to achieve unprecedented levels of insect control easily, reliably and inexpensively. Lured by the power and promise of insecticides, agricultural entomologists focused heavily on the development and use of chemical controls (Newsom, 1980; Perkins, 1982; Kogan, 1998; Smith and Kennedy, 2002). Despite early concerns about the risks associated with near-exclusive reliance on insecticides for pest control, the prophylactic use of insecticides grew until an array of serious problems became apparent. Included among these were: outbreaks of secondary pests and resurgence of target pest populations following destruction of beneficial arthropods; dramatic control failures following the development of insecticide resistance; hazards to pesticide applicators, consumers, and wildlife; and a general simplification of the biotic component of the agroecosystem (Smith, 1970).

Integrated pest management (IPM), as a concept and set of principles for crop protection, developed in response to these problems (Huffaker and Smith, 1980; Kogan, 1998; Kennedy, 2004; Koul et al., 2004). Since its formalization as a concept over 40 years ago, IPM has profoundly influenced the development and implementation of crop protection throughout much of the world (e.g., Blommers, 1994; Luttrell et al., 1994; Abate et al., 2000; Matteson, 2000; Wu and Guo, 2005). Although host plant resistance has long been an important insect management tactic, the application of recombinant DNA technology to produce genetically modifed (GM), insect-resistant crop plants is altering how agricultural insect pests are managed on a scale unprecedented since the introduction of synthetic organic insecticides over 50 years ago. It is within the framework of IPM, rather than as stand-alone insect control measures, that insect-resistant GM crops have the greatest potential to contribute significantly to the establishment of sustainable crop protection systems.

Effectively integrating insect-resistant GM crops into IPM programs requires an understanding of the basic principles of IPM as well as the factors that influence the structure of agricultural production systems and the adoption of crop protection practices. This chapter presents a very brief overview of the defining elements of IPM, followed by a discussion of the general attributes of insect-resistant GM crops and the issues relating to their use as IPM tools. Because the only GM crops that have been widely grown commercially express one or more Cry toxins of *Bacillus thuringiensis* (*Bt*), much of the discussion draws on experiences with these crops.

1.2 Integrated Pest Management

IPM has as its defining elements the use of decision rules to identify the need for and selection of appropriate control actions, which may be used singly or in combination to provide economic benefits to growers and society, and benefits to the environment (Kogan, 1998). IPM focuses on populations, communities and ecosystems, and emphasizes that multiple methods should be used to control single pests as well as pest complexes (Rabb, 1970; Huffaker and Smith, 1980; Rabb et al., 1984; Kogan, 1986; Kogan and Jepson, 2007).

The life system concept (Clark et al., 1967) provides a valuable framework for understanding the array of factors and processes that influence insect populations and pest outbreaks, and which are important in defining viable pest management approaches. The life system of an organism represents that part of the ecosystem that determines the existence, abundance and evolution of a particular population. It includes the subject population and the totality of biotic factors (parasites, predators, pathogens, competitors, host abundance, host quality, etc.) and abiotic factors (weather, day length, light intensity, soil properties, chemicals, etc.) that influence the population. The spatial scale of a pest's life system is determined by the mobility of the pest and the other organisms that affect it. For some species, such as the African armyworm (*Spodoptera exempta*) and black cutworm (*Agrotis ipsulon*) (Lepidoptera: Noctuidae); the rice planthoppers *Sogatella furcifera* and *Nilaparvata lugens* (Hemiptera: Delphacidae), and the green peach aphid (*Myzus persicae*, Hemiptera: Aphididae) the distances may be vast (Taylor, 1977; Rose and Khasimuddin, 1979; Showers, 1997; Otuka et al., 2005). For others, such as Colorado potato beetle (*Leptinotarsa decemliniata*, Coleoptera: Chrysomelidae), distances are much smaller (French et al., 1993). The dimensions of a life system are defined by biological interactions that typically transcend farm units (Kennedy and Storer, 2000).

In contrast, the farm units in which IPM is implemented are economic enterprises, defined largely by factors unrelated to pest life systems. The selection and placement of crops grown during any given season and over years, as well as the production practices employed on a farm, represent business decisions by the farmer. These decisions are influenced by many factors including the financial status and managerial skills of the farmer, land ownership, land quality, tradition, government regulations and price support structures, and markets. The decisions that are made often influence pest life systems as well as the array of pest management options available to the farmer.

Farms are components of agroecosystems that are defined by the processes and interactions among the biotic and abiotic components that affect them. The structure of agroecosystems influences the particular pest problems that affect crops within the system. Changes in that structure influence pest problems and pest management in a manner that is determined by the intersection of the agroecosystem and pest life systems. Pest management measures that are widely implemented have the potential to significantly alter agroecosystem structure, as in the case of highly effective insecticides and herbicides that allow crop rotation intervals to be extended.

While the principles of IPM are general, the implementation of IPM is site-specific, reflecting spatial and temporal variation in the population dynamics of pest species as well as the crop and the context in which the crop and its pests must be managed. Pest management is rarely the highest priority and never the only priority in crop production. Consequently, pest management systems must be cost effective and logistically compatible with the farming operation, or they will not be implemented.

IPM programs address multiple pest species. The pest complex to be managed typically includes one or more species that are severe and regularly encountered (i.e., key pests). It also includes an array of occasional pests, which may periodically reach damaging levels due to factors such as the occurrence of unusually favorable weather conditions, and secondary pests, which may reach damaging levels if their natural enemies are destroyed by an insecticide application or other pest management measures directed against a key or occasional pest (e.g., Pedigo, 1996). The general approach is to reduce the mean level of pest abundance in the crop to sub-economic levels and to intervene only when necessary with remedial measures to suppress populations that approach damaging levels. Accomplishing this generally involves various combinations of cultural practices (e.g. site selection, crop rotation, tillage, water and nutrient management, planting and harvest date manipulation, cultivar selection, manipulation of plant and row spacing), biological control, manipulation of pest behavior, and host plant resistance, which act to prevent or minimize exposure of the crop to damaging pest populations. These are used in conjunction with monitoring of pest populations and crop condition through sampling to determine if and when pest populations reach threshold levels and suppressive measures, usually chemical controls (insecticides or acaricides), are needed to suppress populations that have reached threshold levels.

The specific combinations of pest management tools that are used depend on the production requirements (e.g., soil-type, nutrient, water, temperature, number of days to maturity, equipment and labor) and value of the crop and the pest species to be managed, as well as the cost, effectiveness, and complexity of the available management options. Also important are the infrastructure supporting agriculture and IPM, the political and regulatory environment in which agriculture and IPM operates, the availability of information regarding management technologies, and the resources and education level of the farmer (Bergvinson, 2004; Dhaliwal et al., 2004). Therefore, it is not surprising that the specific tools, tactics, and strategies widely used in IPM vary greatly among crops and between lesser developed, developing and developed countries (Bergvinson, 2004).

1.3 Insect-Resistant GM Crops and IPM

Among available pest management technologies, insect pest resistant cultivars developed through conventional plant breeding methods have been used with great effectiveness against important pests in numerous cropping systems including wheat, maize, rice, sorghum, alfalfa and *Phaseolus* beans (Dhaliwal et al., 2005; Smith, 2005). Smith (2005) estimated the economic value of genetic resistance to the major arthropod pests of wheat in the USA to be ca. US$192 million per year. Similarly, the value of arthropod resistant cultivars of pearl millet, sorghum and chickpea in Africa, Asia and Latin America has been estimated at over US$580 million per year (Heinrichs and Adensina, 1999), and the value of *Phaseolus* cultivars resistant to *Empoasca krameri* (Homopteras: Cicadellidae) in Latin America has been estimated at US$500 per acre per year (Cardona and Cortes, 1991 as cited in Smith, 2005, p. 6). While the most widely publicized examples of the successful use of host plant resistance have involved cultivars with exceptionally high levels of resistance that provide complete control of the pest population (e.g. Hessian fly [*Mayetiola destructor*, Diptera: Cecidomyiidae] resistant wheat [Painter, 1951; Panda and Khush, 1995]), cultivars having moderate levels of resistance to important pest species have made enormous contributions to crop production in both major and minor crops worldwide, despite the fact that the underlying chemical and/or physical mechanisms conferring resistance are often poorly understood (Koul et al., 2004; Dhaliwal et al., 2005; Smith, 2005).

Used within the context of IPM, insect-resistant cultivars offer a number of advantages. They are safe and easy to use, requiring only planting seeds of an adapted, resistant cultivar. In general, resistant cultivars have been compatible with other IPM tactics, including cultural, biological, and chemical controls (Smith, 2005). They have been most widely used in agronomic crops, which because of their low per hectare value do not support intensive or costly pest management inputs. Despite the many advantages of host plant resistance as an IPM tool, the widespread adoption of non-transgenic, insect-resistant cultivars has been constrained by the limited availability of elite cultivars possessing high levels of resistance to key pest species. The application of recombinant DNA technology to develop insect-resistant crop plants has provided a way to eliminate this constraint and make host plant resistance a prominent component of IPM programs in more crops.

1.3.1 Host Plant Resistance Through Genetic Engineering

Recombinant DNA technology greatly increases the potential array of available resistance traits that can be used to obtain insect-resistant crops (Malone et al., chapter 13). It also greatly reduces the time required to develop commercially acceptable resistant cultivars. The development of commercially viable, insect-resistant cultivars using conventional plant breeding procedures is a complex process that can take many years (Smith, 2005). Because the sources of resistance genes generally are limited to plants that can be cross-pollinated with the crop plant, potential sources of naturally occurring resistance are limited to other cultivars, land races, and wild plants of the same species or closely related species. In some cases, however, it is possible to use crosses involving bridge species, manipulate ploidy levels, and employ other sophisticated techniques such as embryo rescue to transfer resistance genes from more distantly related plant species. In addition, naturally occurring resistance is often polygenic involving multiple alleles on separate chromosomes and may involve complex genetic mechanisms (Kennedy and Barbour, 1992; Smith, 2005), thus necessitating the use of sophisticated and complex plant breeding procedures. Polygenic resistance and resistance derived from wild relatives of crops often involve genes having negative, pleiotropic effects or linkages with genes conferring undesirable traits. Breaking these linkages can be difficult and time consuming. In most cases, neither the specific genes coding for resistance nor the underlying chemical or physical mechanisms responsible for resistance are known. Consequently, progeny screening in each generation requires the use of insect bioassays or measurement of insect populations or damage (Smith, 2005). The variation inherent in such procedures interferes with efficient selection of resistant parents for the next generation of crosses and slows progress. The use of molecular genetic markers tightly linked to resistance genes is helping to improve selection efficiency, especially for polygenic resistance traits (Yencho et al., 2000; Smith, 2005).

With recombinant DNA technology, we are no longer limited to using resistance traits occurring naturally in plants that are genetically compatible with the crop. It is now possible to identify and use genes from virtually any organism that, when expressed in a plant, will confer pest resistance. Because the techniques of genetic engineering allow genes to be inserted directly into advanced crop breeding lines or cultivars, linkage drag is minimized and the time required to transfer the trait into commercial cultivars can be greatly reduced. Further, because the gene products that confer resistance can be well defined, it is possible to test them directly to address questions regarding health and environmental effects. Finally, because transgenic resistance traits can be patented, there is an economic incentive for unprecedented private sector investment in the development of pest resistant GM-crop cultivars.

The first insect-resistant transgenic plants were produced in 1987, when genes coding for a Cry toxin of *Bacillus thuringiensis* Berliner were expressed in tobacco and conferred resistance to *Manduca sexta* L. (Lepidoptera: Sphingidae) (Vaeck et al., 1987). Subsequently, synthetic genes modeled on *Bt* genes but designed to be more compatible with plant expression systems were found to boost levels of toxin expression resulting in plants having higher levels of resistance (Perlak et al., 1990, 1991; Koziel et al., 1993; Carozzi and Koziel, 1997). The first insect-resistant transgenic crop cultivars of maize, cotton and potato were approved for commercial release in the USA in 1995 and were first planted in 1996. Since then, the global area planted to *Bt* crops has grown dramatically. In 2007, 42.1 million hectares were planted to *Bt* maize and *Bt* cotton in 22 countries (James, 2007).

The first *Bt* crops to be commercialized expressed a single toxin, but more recently, cultivars expressing multiple *Bt* toxins have been commercialized to enhance efficacy, expand the spectrum of pest species controlled, and delay the development of pest resistance to *Bt* crops (see Ferré et al., chapter 3; Hellmich et al., chapter 5; Naranjo et al., chapter 6). To date, only crops expressing *Bt* toxins that target selected species of lepidopteran or coleopteran pests have been commercialized. Early and continued emphasis on the use of *Bt cry* genes to obtain insect-resistant plants results from the high but selective toxicity of *Bt* Cry toxins to key pest species and the fact that the molecular genetics of *B. thuringiensis* is well understood. Equally important is the long regulatory history with *Bt*, owing to its use as a microbial insecticide, which has provided a level of confidence regarding the limited potential for adverse human and environmental effects. Other toxins from other organisms, which are active against additional pest taxa, are under investigation (Malone et al., chapter 13).

From an IPM perspective, transgenic *Bt* crops have appeal because they are highly effective against the targeted pests, but their toxicity is specific to a very limited range of species. The toxins are biodegradable and do not accumulate in the environment. Because they are expressed throughout most or all of the season in plant tissues affected by the targeted pests, the pests are exposed to the toxin during their most vulnerable stages and even pests that feed in plant parts normally sheltered from insecticide sprays are exposed to the plant produced toxins. Unlike insecticide sprays, the toxin is contained in the plant, which reduces exposure of non-target organisms to the toxin (Gatehouse et al., 1991; Romeis et al., chapter 4).

Bt crops were among the first transgenic crops to be commercialized. As such they were the subject of ethical, socio-economic, and regulatory scrutiny before they were approved for commercial sale. This scrutiny was particularly intense not only because *Bt* crops were at the vanguard of the application of GM technology to agricultural crops, but also because they had the potential to be widely grown on a global scale due to their anticipated ability to effectively and efficiently manage some of the most important insect pests of major agricultural crops.

1.3.2 Ethical Concerns

The ethical issues surrounding GM crops centered generally on genetic engineering and gene transfer among species in the context of world agriculture and food security, human and environmental welfare, and "unease about the unnatural status of the technology."(Nuffield Council on Bioethics, 1999; Comstock, 2000; Thompson, 2000). More recently the debate has shifted to issues relating to the use of GM crops in developing countries and the need to examine possible costs, benefits and risks associated with particular GM crops on a case-by-case basis relative to other alternatives, including maintaining the status quo (Nuffield Council on Bioethics, 2003).

1.3.3 Socio-Economic Issues

The socio-economic issues surrounding insect-resistant GM crops reflect the differing perspectives of farmers who benefit directly from the technology because it is easy to use and increases their profit, and consumers who do not benefit directly. Whereas many farmers have embraced this technology, there has been considerable consumer resistance to GM crops based on concerns about the ethics and safety of the genetic engineering technology used to produce them and the safety of the GM crops themselves. Additional concerns reflect broader issues relating to the potential for agricultural biotechnology to accelerate the consolidation and corporate control of agriculture (Shelton et al., 2002). Regulatory systems for GM crops in general and for pest resistant, GM crops in particular have been developed in many countries to address human and environmental safety concerns. However, the absence of functioning regulatory systems for GM crops in some countries is a constraint to their adoption and affects their role in IPM (Matten et al., chapter 2; Qaim et al., chapter 12).

Regulatory issues and consumer resistance to *Bt* crops have profoundly affected the commercialization of *Bt* potato and *Bt* maize. *Bt* potato cultivars expressing the *Bt* Cry3A toxin conferring resistance to *L. decemlineata* were approved for sale in the USA in 1995. These cultivars were sold under the trade name NewLeaf® until potato processors, concerned about consumer resistance and loss of market share in Europe and Japan, suspended contracts for *Bt* potatoes with growers in 2000 (Grafius and Douches, chapter 7). Similarly, *Bt* maize expressing the Cry9C toxin active against several lepidopteran pests was approved under the trade name StarLink® for use as animal feed, but was not approved for human consumption. Although it represented less than 1 percent of the total maize harvested in the USA in 2000, it was detected in taco shells and other food products. In response, the registration of StarLink® maize was voluntarily withdrawn; the registrant, Aventis, paid millions of dollars in compensation to U.S. farmers; and the U.S. government bought several hundred thousand bags of maize seed containing traces of Cry9C to ensure a stable and predictable market. In response to the StarLink® episode, the U.S. Environmental Protection Agency (USEPA) ceased to issue registrations for only feed or food use (Shelton et al., 2002). With the increasing adoption of GM crops in developing countries, there is also concern that they will displace agricultural labor, which is an important source of income in rural economies (Nuffield Council on Bioethics, 2003).

1.3.4 Health and Environmental Concerns

Early in the development of GM crops it became apparent that concerns over their safety and potential environmental effects would have to be addressed through regulatory oversight. The regulatory framework and processes that have been implemented are described by NRC (2000), Conner et al. (2003) and Nap et al. (2003). In the USA, the regulatory process focuses on the GM product (i.e. the transgenic plant) not the process (i.e. genetic transformation) that was used to produce it (NRC, 1987, 2000). This focus allows transgenic resistance traits to be registered when produced by plants, provided that they meet the regulatory requirements for human and environmental safety. In the case of *Bt* crops, which were the first insect-resistant GM crops, the long regulatory history of *Bt* pesticides and their safe use as foliar sprays to control insect pests on numerous food crops and in forest and aquatic systems greatly expedited the human health and environmental risk assessments required for regulatory approval in the USA. Future insect resistance traits are likely to require much more in-depth, regulatory scrutiny to ensure that they meet human health and environmental risk standards. In the case of Cry toxins, concerns about potential allergenicity were an important issue for the Cry9C toxin found in StarLink® maize and were the reason that its registration did not include use as a human food, although further research indicated this concern was unfounded. Environmental concerns focused on issues of gene transfer, potential weediness of *Bt* plants, environmental persistence of Cry toxins, and effects on non-target natural enemies, herbivores and detritivores. These issues have been the subject of extensive research (NRC, 2000; Shelton et al., 2002; O'Callaghan et al., 2005; Romeis et al., 2006; Sanvido et al., 2007).

1.3.4.1 Gene Transfer

Issues of gene transfer center around spread of transgenes through outcrossing to related, non-crop species and non-GM cultivars of the same crop. The scientific concerns center on the potential for the acquisition of a transgene by a non-crop plant through outcrossing to provide a fitness advantage that improves the plant's ability to compete with other plants in its habitat and leads to altered community structure or an increased potential for weediness. These concerns are not limited to transgenic traits but apply to any genetic trait that has the potential to confer a fitness advantage on non-crop plants that acquire it through gene transfer from a crop. Thus, gene transfer is also an issue for non-transgenic, herbicide-tolerant crops such as Clearfield canola produced through mutation breeding (BASF, 2008).

Other concerns relate to genetic contamination of non-transgenic crop varieties, especially in crops or locations where farmers save seed from year-to-year, or where the crop is produced for the organic market. In addressing the issue of gene transfer, the USEPA ruled that, except for specific and limited situations, the potential for gene transfer through outcrossing from *Bt* maize, cotton and potato to wild relatives of these crops was negligible due to differences in temporal and spatial distributions of the crops and their wild relatives or in chromosome number (USEPA, 2000). Under this ruling, the use of *Bt* cotton was restricted or prohibited in areas of Florida and Hawaii where related species of cotton (*Gossypium*) occur. In the case of maize, outcrossing to wild relatives is a concern only in regions of Mexico, Central and South America where they occur naturally. However, cross-fertilization may also be of concern in areas where GM maize is grown in proximity to non-transgenic maize in which adventitious presence of transgenes above a certain level is unacceptable. Isolation distances that minimize potential cross-fertilization between GM and non-GM maize have been identified as one measure to address this problem (e.g., Brookes et al., 2004; Devos et al., 2005; Sanvido et al., 2008; Hellmich et al., chapter 5). In the case of *Bt* rice, which has not yet been commercialized, the potential is high that transgenes will outcross to closely related wild rice species, as well as to non-transgenic rice varieties and weedy rice (Lu and Snow, 2005). The possible consequences of the spread of *Bt* genes from rice through outcrossing have not yet been fully assessed (Cohen et al., chapter 8).

1.3.4.2 Non-Target Effects

Issues relating to non-target effects have focused on the potential for the novel traits expressed in GM crops to produce adverse effects on non-target, plant feeding insects and beneficial species (O'Callaghan et al., 2005; Romeis et al., 2006; chapter 4). Extensive research on non-target effects of *Bt* crops has generally not detected significant adverse, population-level effects on these groups of non-target species (Romeis et al., chapter 4; Hellmich et al., chapter 5; Naranjo et al., chapter 6; Cohen et al., chapter 8). That research has highlighted the critical importance of appropriately designed experiments (O'Callaghan et al., 2005; Romeis et al., 2006) and the complexities involved in extrapolating from effects observed on individual insects in laboratory experiments to population-level consequences in the field (Kennedy and Gould, 2007). It has highlighted the importance of applying risk assessment methodologies that include both hazard identification and evaluation of the likelihood of exposure to the hazard (e.g., Sears et al., 2001; Raybould, 2007), and has led to the development of testing methods to assess the potential effects on nontarget organisms. These methods involve selection of appropriate organisms for testing based on ease of handling, abundance, importance and endangered status; and a tiered-testing approach that evaluates responses to a range of concentrations of the transgenic trait as well as the organisms' potential exposure to the trait in the field (Garcia-Alonso et al., 2006; Rose, 2007; Raybould et al., 2007; Romeis et al., 2008).

1.3.4.3 Pest Adaptation to Insect-Resistant GM Crops

A final issue of concern for insect-resistant GM crops involves the potential for the targeted pests to become resistant to the toxins expressed by the plants. The potential for extant production of insect-resistant GM crops and their ability to impose intense selection for adaptation by affected insect populations led to concern that their benefits would soon be lost to the development resistant pest populations (Gould, 1988a, b). Extensive research stimulated by this concern ultimately led to consensus that implementation of a high dose/refuge insect resistance management (IRM) strategy was needed to delay or prevent the selection of resistance in targeted pest populations. Implementation of this strategy requires toxin expression at a level sufficiently high to negate any resistance mechanisms that confer low to moderate levels of resistance and to kill all individuals heterozygous for the resistance allele. It further requires that a there is a refuge from exposure to the *Bt* toxin adequate in size to produce a sufficient number of homozygous susceptible insects to ensure that all homozygous resistant individuals surviving in the *Bt* crop mate with a susceptible insect to produce heterozygous offspring, which will be killed by the *Bt* crop. IRM in *Bt* crops is discussed in detail by Ferré et al. (chapter 3).

The threat of insect resistance to *Bt* crops is considered to be sufficiently great that the USEPA has required the implementation of IRM as a condition of registration for *Bt* crops. The specific details of resistance management plans in the USA and elsewhere have changed over time as new information became available and *Bt* crops expressing multiple toxins have been commercialized (USEPA, 2001a, b, 2005, 2007; Matten et al., chapter 2; Ferré et al., chapter 3). Additional information on IRM in *Bt* maize, cotton, potato and rice can be found in chapters 5, 6, 7 and 8, respectively.

1.4 Economic and Human Health Impacts of *Bt* Crops

Since their commercial introduction in 1995, *Bt* crops have provided important economic and human health benefits, which are discussed in detail by Qaim et al. (chapter 12) and briefly summarized here. These include reductions in insecticide use and increases in yields and gross margins ($/ha) (see also Fitt, chapter 11). The benefits vary greatly with location and year, reflecting in part differences in the severity of pest pressure, patterns of insecticide use in non-*Bt* crops and the added cost of *Bt* seed. On average, these benefits are greater for *Bt* cotton than for *Bt* maize, due to the greater intensity of insecticide use in cotton. For example, insecticide use averaged 51 percent (range = 33 to 77) less and effective yields averaged 22 percent (range = 9 to 34) greater in *Bt* than in non-*Bt* cotton in Argentina, China, India, Mexico, South Africa and the USA, while gross margin gains averaged US$163/ha (range = 23 to 470) greater in *Bt* cotton. By comparison, insecticide use averaged 20 percent (range 0 to 63) less and effective yields averaged 8 percent (range = 5 to 11) greater in *Bt* than in non-*Bt* maize, while gross margin gain averaged US$47/ha (range = 10 to 116) greater in *Bt* maize in Argentina, South Africa, Spain and the USA (See Tables 12.2 and 12.3 in Qaim et al., chapter 12). The impact of *Bt* maize on insecticide use is likely to increase dramatically with the widespread adoption of corn rootworm (*Diabrotica* spp., Coleoptera: Chrysomelidae) resistant varieties expressing *Bt* Cry3 or binary toxins because insecticide use on ca. 9.2 million hectares for control of rootworms in the USA accounts for 25 to 30 percent of the total insecticides applied to maize worldwide (Gianessi et al., 2002; James, 2003).

The proportion of the economic benefits that accrue to the farmer, the consumer and the technology company also vary among countries, depending on the degree of protection provided for intellectual property rights and the degree of government control over commodity prices. Direct health benefits accrue from the reductions in insecticide use on *Bt* crops as a result of lower pesticide residues in food and water, and reduced exposure of farm workers during pesticide applications. These benefits are especially great in developing countries in which pesticide regulation is weak, the education level of farmers is generally low, and pesticides are applied manually. Because pesticide residues on food are of greatest concern in fruits and vegetables, and no insect-resistant GM fruit and vegetable crops are as yet commercially available, the full potential of GM technology to reduce exposure to pesticide residues in foods has not yet been realized (see Shelton et al., chapter 9 for discussion of pest protected GM fruit and vegetable crops).

1.5 Impacts of Insect-Resistant GM Crops on IPM

Insect-resistant GM crops represent a form of host plant resistance (HPR) that differs from traditional HPR in the specific resistance traits and their source, and the method by which the resistance genes were introduced into the crop germplasm. The expression of host plant resistance on the insect/plant interaction is generally classified as antibiosis, antixenosis (= non-preference) or tolerance (Painter, 1951; Panda and Khush, 1995; Dhaliwal et al., 2005; Smith, 2005). Antibiotic resistance typically involves plant traits that interfere with the insect's metabolic processes. By reducing pest reproduction and survival, and increasing generation time, antibiotic resistance reduces the rate at which the affected species' populations increase in the crop. In extreme cases, survival rates may be so low that populations fail to become established (Luginbill, 1969). Antibiotic resistance may result from plant produced toxins that have lethal or sub-lethal effects; it may also result from certain physical or chemical/physical attributes of the plant involving trichomes or a hypersensitive response (Arora and Dhaliwal, 2005; Ram et al., 2005; Smith, 2005). Insect resistance conferred by *Bt* Cry and Vip toxins, as well as most if not all of the other transgenic insect resistance traits under development, represent examples of antibiosis resistance (see Malone et al., chapter 13).

Antixenotic resistance involves plant traits that interfere with selection of the resistant plant by the insect for feeding and/or oviposition. Antixenotic resistance may reduce the rate at which a pest population increases by reducing both the number of initial colonizers of the crop and the proportion of each successive generation remaining in the crop. The actual mechanism responsible for antixenosis may be chemical or physical (Arora and Dhaliwal, 2005; Ram et al., 2005; Smith, 2005). Because antixenosis also involves a behavioral response of the insect to the plant, its expression is context dependent in that the insect's response to the antixenotic plant may be significantly affected by the presence of alternative hosts. In the case of antixenotic resistance that interferes with host selection by the adult prior to oviposition, the population reduction in the resistant cultivar may be accompanied by a corresponding increase in the population in other susceptible crops that are more attractive than the antixenotic resistant cultivar, but less attractive than the susceptible cultivar it replaced (Kennedy et al., 1987). Although transgenic traits conferring antixenotic resistance could almost certainly be identified and expressed in crops to confer insect resistance, the reliance of this type of resistance on a complex, context dependent, behavioral response by the insect may limit its use as a resistance modality in insect-resistant GM crops.

Tolerance refers to the ability of a plant to sustain higher levels of injury due to insect feeding than susceptible plants before economic yield is adversely affected. Thus, tolerance has the effect of raising the economic injury level. It is an important component of host plant resistance in many crops, including wheat, sorghum, and alfalfa, where tolerance to aphid feeding is important in reducing losses (Panda and Khush, 1995; Dhaliwal and Singh, 2005; Smith, 2005). It is likely that transgenic approaches will ultimately be used to enhance plant tolerance to insect-induced stress, although there are currently no examples of this.

Within an IPM context, host plant resistance offers numerous advantages. Because pest suppression comes pre-packaged in the seed, it is easy to use and becomes a substitute for more labor-intensive or more insecticide-intensive practices (Shelton, 2007). Additionally, plant resistance mechanisms are generally highly selective in their activity; consequently, the use of resistant cultivars is generally compatible with other pest management tactics and generally poses little risk of non-target effects, although there are examples of negative tri-trophic effects (Bottrell et al., 1998; Smith, 2005; Kennedy and Gould, 2007). The effects of antibiotic resistance on the target pest species are density independent and cumulative over pest generations within the same crop. One drawback of this is that the resistant crop continuously suppresses the pest population and exerts selection pressure for adaptation by the pest to the resistant crop even when populations are at sub-economic levels. Currently available insect-resistant GM crops, based on *Bt* Cry toxins share these attributes. Additionally, the level of resistance expressed by *Bt* crops is unusually high compared to that expressed by most insect-resistant varieties developed through conventional breeding. *Bt* crops, like other GM crops, differ from insect-resistant crops developed through conventional breeding in that the price of seed includes an added charge for the transgenic trait, at least in countries in which intellectual property rights are protected. Thus, there is an identifiable cost associated with purchasing the insect control provided by the *Bt* crop.

1.6 *Bt* Crops and IPM

Based on a decade of experience with commercial production of *Bt* cotton and *Bt* maize, it is apparent that their role in insect management is generally consistent with the that of conventionally bred, insect-resistant cultivars in a other crops (Dhaliwal et al., 2005; Smith, 2005). However, because *Bt* cotton and *Bt* maize have been very widely grown and exhibit very high levels of resistance against some of the targeted pest species but only moderate levels of resistance or no resistance against other species, it is possible to see in these crops a breadth and level of influence on insect management programs that rarely has been seen with conventional insect-resistant crop varieties.

1.6.1 Decision Rules

One of the fundamental principles of IPM is the use of decision rules based on cost/benefit analyses to determine the need for and appropriate set of pest management tactics to protect the crop in a manner that provides economic, societal and environmental benefits. The most fundamental decision rule focuses on economic benefits and is based on the economic injury level (EIL) and economic threshold (ET) concepts (Stern et al., 1959). The EIL defines the level of pest abundance above which the cost of implementing a management tactic is less than the value of crop yield that would be lost if the control measure were not implemented. The ET represents the pest population level, or an index thereof, at which the management tactic should be applied to prevent the pest population from exceeding the EIL. Although simple in concept, EILs and ETs are difficult to define and complex to implement (Pedigo et al., 1986; Higley and Pedigo, 1996). In practice, application of the threshold concept typically involves the use of a nominal threshold, which is based on experience, rather than a true ET based on an empirically defined, dynamic EIL that accounts for crop yield potential, plant stage-specific tolerance to pest injury, costs of the management tactic, and commodity price. Further, it is most easily applied to decisions regarding the application of population suppressive measures such as insecticides in response to existing pest infestations. The threshold concept is particularly difficult to apply in situations where the management tactics must be implemented before the pest is present, as in the selection of planting dates to avoid a pest, application of pre-plant or at-plant systemic insecticides, or planting a resistant crop variety, unless it is possible to foresee the risk that a damaging infestation of the pest in question will occur.

The decision to use an insect-resistant *Bt* crop must be made prior to planting. It involves weighing the cost of implementing the technology against the risk of experiencing a yield-suppressing infestation of the targeted pest species during the season. The costs of using a *Bt* crop for crop protection include both the fee premium charged for the *Bt* trait and the costs (if any) associated with any undesirable agronomic characteristics of the *Bt* cultivar compared to non-*Bt* cultivars. In the case of *Bt* cotton, the principal targets are bollworms, a complex of fruit-feeding lepidopteran species that are key pests in most cotton production areas of the world (Naranjo et al., chapter 6). Because they reach damaging levels in most years, the decision to plant *Bt* cotton can often be made on the basis of geographical location and past experience.

In the USA, *Bt* cotton is widely grown in production areas that regularly experience damaging populations of caterpillars, but is not grown in the San Jaoquin Valley of California where lepidopteran pests are rarely a problem (Naranjo et al., chapter 6). Because *Bt* cotton varieties expressing only Cry1Ac toxins, which do not completely control *Helicoverpa zea* and *H. armigera*, populations of these insects in *Bt* cotton must be monitored in areas where they are problems. However, thresholds based on egg abundance, which were used for both *Heliothis* spp. and *Helicoverpa* spp. populations to determine the need for insecticide applications in conventional cotton, are no longer appropriate in *Bt* cotton which kills some but not all of the larvae. New sampling procedures and thresholds have been developed for *Bt* cotton, which focus on populations of older larvae. These larval-based thresholds allow identification of populations that are not being controlled by the *Bt* crop at a time when they can still be controlled with insecticides (Naranjo et al., chapter 6).

In *Bt* maize, the situation is somewhat different. Stalk boring lepidopteran species (Crambidae or Noctuidae) are the primary target in most areas where *Bt* maize expressing Cry1Ab or Cry1F toxins is grown for grain. In conventional maize varieties, host plant resistance and tolerance, which keep losses to modest levels, are the primary means of managing stalk borers. Insecticides are used to control stalk borers by only a limited proportion of growers because properly timing applications to contact larvae before they bore into the stalk is difficult (Hellmich et al., chapter 5). In this situation, the decision to plant *Bt* maize to manage stalk borers must be based on an assessment of the risk that a damaging stalk borer population will develop during the coming year. In the USA there is a risk/benefit assessment model (*Bt* Evaluation Tool) available on the internet (http://www.*Bt*et.psu.edu/; accessed 4 January 2008) to assist maize producers in deciding whether planting *Bt* maize varieties is a favorable investment. This model estimates the net benefits likely to be derived from planting *Bt* maize based on historic or projected infestation levels of *Ostrinia nubilalis* (Lepidoptera: Crambidae), seeding rate, seed premium charge for the *Bt* trait (i.e., technology fee), projected yield, price, and expected level of population suppression.

In the case of *Bt* maize expressing the Cry 3 or binary toxins for resistance to corn rootworms (*Diabrotica* spp.; Coleoptera: Chrysomelidae), the situation is different in that in areas where rootworms are a problem, they are capable of causing significant yield losses and are the target of extensive insecticide use. The options for managing rootworms involve planting a rootworm-resistant, *Bt* maize hybrid; using insecticides; or crop rotation. The effectiveness of crop rotation has eroded in areas where *D. virgifera virgifera* populations have adapted to crop rotation by ovipositing in the principal rotation crop, soybean (Hellmich et al., chapter 5). The decision to apply a rootworm control measure must be based on past experience and the populations of adult rootworms in the preceding year's crop.

For both *Bt* cotton and *Bt* maize, there is evidence that area-wide populations of at least some targeted pests can be suppressed by widespread planting of *Bt* crops (Carrière et al., 2003; Wilson et al., 2004; Hutchison et al., 2007; Storer et al., chapter 10). Thus, it is possible that, in such cases, the area-wide populations of the target pest may be suppressed to the point that historical infestation levels will become a poor indicator of the potential for damaging populations to develop in non-*Bt* crops.

The ability to use threshold or risk-based decision criteria in assessing the appropriateness of a particular insect-resistant GM-trait requires the availability of cultivars that do not express that trait. Currently, both GM maize and cotton cultivars expressing herbicide tolerance and *Bt* toxins in combination (stacked events) are widely available (James, 2007). Because of supply constraints, growers desiring to purchase only herbicide-tolerant cultivars in some instances have had to purchase cultivars expressing both herbicide tolerance and a *Bt* toxin. As the number of value-added, GM traits increases, the number of potential combinations of traits that could be stacked within individual cultivars increases geometrically, as do the costs associated with maintaining inventories of geographically adapted cultivars expressing different combinations of traits. Consequently, we can expect that commercially available, GM cultivars of the future will express multiple, unrelated, transgenic traits, and farmers in many cases likely will not have the option of planting cultivars expressing only single traits. To the extent that this occurs, insect-resistant GM crops are likely to be widely used in situations where they are neither needed nor appropriate; making IRM more difficult.

1.6.2 Reduced Insecticide Use, Enhanced Natural Enemy Populations, and Pest Shifts in Bt Crops

As indicated previously, there have been dramatic reductions in insecticide use in *Bt* cotton and significant reductions in *Bt* maize. The future commercialization of *Bt* rice and *Bt* vegetable and fruit crops will almost certainly lead to significant reductions in insecticide use in those crops as well (Cohen et al., chapter 8; Shelton et al., chapter 9). This reduced insecticide use, in conjunction with the selective activity of the *Bt* toxins, results in a more favorable environment for beneficial insects, including natural enemies of pests. It also provides an opportunity for populations of secondary pest species previously controlled by applications of insecticides directed against key pests to reach damaging levels in *Bt* crops. Both of these consequences have important pest management implications.

Numerous field studies have documented the general compatibility of *Bt* crops, including maize, cotton, potato and rice, with the natural enemy complex present in those crops (see chapters 5, 6, 7, and 8). *Bt* maize has little or no effect on populations of most predators, parasitoids and pollinators present in maize fields, with the exception of parasitoids that specialize on pest species that are effectively controlled by the *Bt* maize. In the latter case, the parasitoid populations respond largely to declines in their hosts' population, although *Bt* crops can also cause reductions in an individual's fitness when feeding on *Bt*-intoxicated hosts. (Romeis et al., 2006; Marvier et al., 2007; Romeis et al., chapter 4). In cotton, natural enemies play an important role in suppressing pest populations. Several field studies have found that biological control capacity in *Bt* cotton fields was comparable to that in fields planted to non-*Bt* cotton, which were not treated with insecticides, and greater than in non-*Bt* cotton fields in which insect pests were managed using conventional insecticides (Obrycki et al., 2004; Sisterson et al., 2004; Naranjo, 2005; Head et al., 2005; Romeis et al., chapter 4; Naranjo et al., chapter 6). In rice, biological control by naturally occurring parasitoids and predators is particularly important to insect management. Although *Bt* rice has not yet been commercialized, numerous studies have been conducted to evaluate potential impacts on the natural enemy complex and other non-target species. Consistent with the experiences in *Bt* cotton and *Bt* maize, these studies have failed to detect significant adverse impacts of *Bt* rice (Cohen et al., chapter 8). The use of *Bt* crops along with the increased availability of highly effective, selective insecticides is enhancing the opportunity for biological control to play a greater role in IPM. This is especially the case in cotton, where *Bt* cultivars are leading to reduced problems with some secondary pests such as cotton aphids and whiteflies (Naranjo et al., chapter 6).

Despite the general compatibility of *Bt* crops with biological control, reductions in the use of broad-spectrum insecticides in *Bt* crops and selectivity of the *Bt* toxins expressed in the plants create an environment that is more favorable to some pest species than is the case in non-*Bt* crops in which key pests are managed primarily with broad-spectrum insecticides. This is dramatically illustrated by the increase in significance of true bugs as pests of cotton following the widespread adoption of *Bt* cultivars to control the bollworm complex. The elevation of true bugs to key pest status in *Bt* cotton almost certainly reflects the absence of effective biological control of these species in the cotton agroecosystem, rather than a significant effect of *Bt* cotton on the natural enemy complex (Naranjo et al., chapter 6). In Australia, China, USA, and elsewhere, plant bugs (Miridae) and stinkbugs (Pentatomidae) have become key pests in *Bt* cotton, where the reduction or elimination of insecticide applications targeting lepidopteran pests has allowed their populations to reach damaging levels regularly if not treated with insecticides. In Australia, up to three applications of broad-spectrum insecticides per season may be used to control the green mired, *Creontiades dilutus*, in *Bt* cotton, and outbreaks of spider mites, aphids and whitefly have been attributed to the disruptive effects of these insecticide treatments on the natural enemy complex (Wilson et al., 1998; Doyle et al., 2006; Farrell et al., 2006; Khan et al., 2006; Naranjo et al., chapter 6). The recent elevation of true bugs to key pest status in *Bt* cotton in some agroecosystems is leading to the development of new tools and combinations of new and existing tactics, including cultural controls, selective insecticides and habitat manipulation to manage the insect complex in *Bt* cotton (Ellsworth and Barkley, 2005; Sharma, 2005; Wu and Guo, 2005; Carrière et al., 2006; Naranjo and Luttrell, 2008; Naranjo et al., chapter 6).

1.6.3 Landscape-Level Effects

In major farming regions, much of the landscape can be occupied by a few crop species. In these settings, patterns of crop placement and crop and pest management practices can be a major determinant of the population dynamics of many important pest species at both a local and a landscape scale (Kennedy and Storer, 2000). *Bt* maize and *Bt* cotton are now extensively planted in several countries. In 2007, *Bt* maize represented 49, 64, and 54 percent of the total area under maize production in the USA, Argentina, and Canada, respectively; and *Bt* cotton represented 72, 66, and 99 percent of the total area under cotton production in the USA, India, and China, respectively (Qaim et al., chapter 12). As *Bt* crops become registered in additional countries and as current and novel, insect-resistant GM crop technologies (Malone et al., chapter 13) are extended to additional crops, the proportion of total crop area planted to insect-resistant crops globally can be expected to increase dramatically.

Landscape-level effects of pest management practices implemented on an area-wide basis have been shown to dramatically suppress targeted pest populations and form the basis for area-wide pest management programs (Ellsworth and Martinez-Carillo, 2001; Calkins and Faust, 2003; French et al., 2007; Koul and Cuperus, 2008). Insect-resistant GM crops have the potential to exert agroecosystem-level effects on populations of targeted pests as well as on sensitive, non-target species because the technology used to produce them enables pest resistance genes conferring very high levels of resistance to targeted pests to be widely deployed in multiple crops that have the potential to be planted over extant areas.

A variety of landscape-level effects of insect-resistant GM crops have been postulated (Kennedy and Gould, 2007; Storer et al., chapter 10). Potential landscape-level effects of greatest importance involve area-wide population suppression of pest, beneficial, or endangered species. Such effects have the greatest likelihood of occurring in situations where a significant portion of the landscape is occupied by the GM crop and the affected herbivores are highly sensitive to the toxin and highly mobile, and for which the crop is a principal food plant (Storer et al., chapter 10). Using a computer modeling approach, Kennedy et al. (1987) demonstrated how widespread planting of insect-resistant maize, which suppressed early season populations of *Helicoverpa zea* (Lepidoptera: Noctuidae), could influence the occurrence of damaging populations of *H. zea* populations in soybean later in the season. Other modeling studies, specifically focused on *Bt* maize and *Bt* cotton (Storer et al., 2003), indicated that in the agroecosystem of eastern North Carolina, *H. zea* populations could be reduced by 50 to 60 percent when the proportion of the total land area planted to either crop exceeded 50 percent (Storer et al., 2003). In practice, it is possible that area-wide suppression of affected populations may be mitigated by reductions in density-dependent mortality in response to declining population size and by the presence of alternate host plants in the landscape, including plantings of susceptible cultivars of the same crop intended to serve as refuges for resistance management (Ferré et al., chapter 3).

Experience with *Bt* crops in the USA provides some evidence for area-wide suppression of populations of pink bollworm (*Pectinophora gossypiella*; Lepidoptera: Gelichiidae) in Arizona and *H. virescens* in Mississippi associated with production of *Bt* cotton, and of *O. nubilalis* in the upper Midwest associated with *Bt* maize (Carrière et al., 2003; Chu et al., 2006; Adamczyk and Hubbard, 2006; Hutchinson et al., 2007). The significance of any such landscape level effects would obviously depend on their magnitude and the spatial and temporal scale over which they occur.

Large scale population-level effects on non-target species, especially those that are threatened or endangered would be particularly serious. Several studies have examined potential effects of *Bt* crops on sensitive, non-target lepidopteran species, including the Monarch butterfly (*Danaus plexippus*; Lepidoptera; Nymphalidae) and several endangered species. In-depth studies of effects of *Bt* maize on the Monarch butterfly concluded that despite susceptibility of larvae to the *Bt* toxin expressed in *Bt* maize, exposure to the toxin is very limited and the potential for significant effects on Monarch populations is negligible (Sears et al., 2001). Other studies similarly concluded that the potential for significant population-level effects on a number of other non-target or endangered lepidopteran species that may be exposed to maize or maize pollen is negligible (e.g., Wolt et al., 2005; Peterson et al., 2006).

The potential for landscape-level effects of GM crops on populations of natural enemies remains an important concern. This is especially the case for GM traits that adversely affect natural enemy populations within crops that serve as an important habitat for the increase of natural enemy populations, which subsequently disperse to other crops where they are important in suppressing pest populations (Kennedy and Gould, 2007). In the case of *Bt* crops there is no evidence for such population level effects, with the possible exception of parasitoid species that specialize on pest species (Romeis et al., 2006; chapter 4). Because risk to natural enemies is heavily scrutinized when evaluating candidate insect resistance traits for use in GM crops (Garcia-Alonso et al., 2006; Rose, 2007; Romeis et al., 2008) problems derived from significant landscape level effects of insect-resistant GM crops on natural enemies are not likely to become an issue, although continued awareness of the potential for such effects is essential.

Pest adaptation to insect-resistant GM crops involves a shift in the genetic composition of the pest population at the landscape scale (Storer et al., chapter 10). Implementation of the high dose-refuge strategy for managing resistance to *Bt* crops is based on manipulating the spatial arrangement of *Bt* and non-*Bt* crops within the landscape (Ferré et al., chapter 3; Storer et al., chapter 10). Ensuring that the area and distribution of refuges is adequate is virtually impossible in countries such as China and India, where *Bt* crops are produced on very small parcels of land by millions of farmers (Naranjo et al., chapter 6). More broadly, as the number of different crops expressing the same resistance traits or traits for which there is a high potential for cross resistance increases, the ability to manage the deployment of resistance traits to ensure the appropriate abundance and positioning of refuges across the landscape may become increasingly limited. Pyramiding multiple resistance traits into each resistant variety (Ferré et al., chapter 3) and expanding the number and types of resistance traits deployed in GM crops will be very important for successful resistance management for GM crops in the future.

1.7 Concluding Remarks

By increasing the potential array of traits that can be used to obtain insect-resistant crops and greatly reducing the time required to develop insect-resistant cultivars, genetic engineering is making it possible for host plant resistance to become the primary insect management tool in many cropping systems. Consequently, it is important that insect-resistant GM crops are deployed in a manner that improves the economic, environmental and social sustainability of agriculture. Because of the fundamentally novel nature of genetic engineering and the scale over which insect-resistant GM crops were expected to be deployed, the commercialization of *Bt* crops raised novel, socio-economic, environmental and health concerns, as well as regulatory challenges. Addressing these concerns for *Bt* crops has provided not only assurances of their safety and effectiveness, but also has documented their significant benefits and provided a framework for anticipating the challenges posed by the next generations of pest-protected, GM crops.

Bt cultivars have become a primary tool for managing key pests in cotton and maize. Experience in those crops has demonstrated significant reductions in insecticide use and changes in the way insecticides are used. It has also revealed the potential for these reductions to be accompanied by the emergence of secondary pests and the need to adjust the pest management systems to address these "new" pests. *Bt* crops have proven to be compatible with other pest management tactics, including cultural and chemical controls, and the conservation of natural enemies as important agents of biological control. Emphasis on the importance of IRM to mitigate selection for pest adaptation to *Bt* crops and the institutionalization of IRM requirements through regulation has highlighted the importance of resistance management and sustainability within the conceptual framework of IPM, and elevated the role of IRM to a position of fundamental importance in the implementation of IPM.

In addition to dramatic reductions in insecticide use, *Bt* crops have provided health and environmental benefits due to reduced pesticide residues on food and exposure by farmers and farm laborers, especially in developing countries. As new, transgenic, insect resistance traits are developed and deployed commercially and the array of crops in which they are deployed increases, the spectrum of pests controlled will increase and quantities of broad-spectrum insecticides used will decrease. Accompanying this, we can expect to see increased economic, health, and environmental benefits.

The resulting large-scale reductions in insecticide use on a global scale are likely to significantly affect research and development efforts on new insecticides. Given the tremendous cost of developing and registering new pesticides (Huckaba, 2004), and the loss of market share to insect-resistant GM crops, investment in insecticide research and development will almost certainly decline significantly, resulting in fewer new insecticides and new modes of action. Potential consequences of this may be increased reliance on transgenic, insect resistance traits as a primary insect management tool. However, because insecticides provide the only fast-acting, easy-to-use and highly effective tool for suppressing insect populations that have reached damaging levels, they are vitally important. Insect-resistant crop cultivars, regardless of whether they are GM or developed through conventional breeding, can only be used preventatively; they cannot be deployed mid-season to control an unanticipated insect problem. Unless insect-resistant GM crops of the future express broad-spectrum, insecticidal activity due either to expression of a broad-spectrum toxin or a broad array of selective toxins, they will be vulnerable to unanticipated outbreaks of non-affected pest species. Although it is theoretically possible to develop insect-resistant GM crops with very broad-spectrum, insecticidal activity, the plant would then be the delivery system for the widespread, preventative use of broad-spectrum insecticides, and arguably would not be compatible with the principles of IPM.

Agricultural biotechnology provides the ability to produce a broad array of insect-resistant, disease resistant, and herbicide-tolerant crop cultivars that also express a variety of other value-added traits. The stacking of multiple, transgenic traits in single cultivars may soon limit the ability of farmers to plant cultivars expressing a particular suite of insect-resistant and other pest management relevant traits based on need determined by using threshold and risk analysis criteria. This may make it difficult or impossible to meet IRM guidelines for refuge size and placement, and shift the primary IRM strategy to reliance on expression of multiple toxins within the same cultivar (Ferré et al., chapter 3). To the extent that these changes take place and are effective, they will represent a fundamental change in the implementation of IPM. It remains uncertain whether they will negate or otherwise compromise the fundamental goal of IPM, which is to use appropriate control actions, singly or in combination, to provide economic benefits to growers and society, and benefits to the environment.

References

Abate, T., van Huis, A., and Ampofo, J.K.O., 2000. Pest management strategies in traditional agriculture: An African perspective. Annual Review of Entomology 45: 631–659.

Adamczyk, J.J., and Hubbard, D., 2006. Changes in populations of *Heliothis virescens* (F.) (Lepidoptera: Noctuidae) and *Helicoverpa zea* (Boddie) (Lepidoptera: Noctuidae) in the Mississippi Delta from 1986 to 2005 as indicated by adult male pheromone traps. Journal of Cotton Science 10: 155–160.

Arora, R., and Dhaliwal, G.S., 2005. Biochemical bases of resistance in plants to insects. In: Host Plant Resistance to Insects: Concepts and Applications, G.S. Dhaliwal and R. Singh, eds., Panima Publishing, New Delhi, India, pp. 84–125.

BASF, 2008. CLEARFIELD production system. http://www.agsolutions.ca/basf/agprocan/agsolutions/WebASClearfield.nsf/defaultWest.htm (accessed 3 January 2008).

Bergvinson, D., 2004. Opportunities and challenges for IPM in developing countries. In: Integrated Pest Management: Potential, Constraints and Challenges, O. Koul, G.S. Dhaliwal and G.W. Cuperus, eds., CABI, Wallingford, UK, pp. 281–312.

Blommers, L.H.M., 1994. Integrated pest management in European apple orchards. Annual Review of Entomology 39: 213–242.

Bottrell, D.G., Barbosa, P., and Gould, F., 1998. Manipulating natural enemies by plant variety selection and modification: A realistic strategy? Annual Review of Entomology 43: 347–367.

Brookes, G., Barfoot, P., Melé, E., Messeguer, J., Bénétrix, F., Bloc, D., Foueillassar, X., Fabié, A., and Poeydomenge, C., 2004. Genetically Modified Maize: Pollen Movement and Crop Co-existence. PG Economics, Dorchester, UK. http://www.pgeconomics.co.uk/pdf/Maizepollennov2004final.pdf (accessed 2 November 2007).

Calkins, C.O., and Faust, R.J., 2003. Overview of areawide programs and the program for suppression of codling moth in the western USA directed by the United States Department of Agriculture – Agricultural Research Service. Pest Management Science 59: 601–604.

Carozzi, N., and Koziel, M., eds., 1997. Advances in Insect Control: The Role of Transgenic Plants. Taylor & Francis, London, UK, 301 p.

Carrière, Y., Ellers-Kirk, C., Sisterson, M., Antilla, L., Whitlow, M., Dennehy, T.J., and Tabashnik, B.E., 2003. Long-term regional suppression of pink bollworm by *Bacillus thuringiensis* cotton. Proceedings of the National Academy of Sciences of the USA 100: 1519–1523.

Carrière, Y., Ellsworth, P.C., Dutilleul, P., Ellers-Kirk, C., Barkley, V., and Antilla, L., 2006. A GIS-based approach for areawide pest management: The scales of *Lygus hesperus* movements to cotton from alfalfa, weeds, and cotton. Entomologia Experimentalis et Applicata 118: 203–210.

Chu, C.-C., Natwick, E.T., Lopez, R.L., Dessert, J.R., and Henneberry, T.J., 2006. Pink bollworm moth (Lepidoptera: Gelechiidae) catches in the Imperial Valley, California from 1989 to 2003. Insect Science 13: 469–475.

Clark, L.R., Geier, P.W., Hughes, R.D., and Morris, R.F., 1967. The Ecology of Insect Populations in Theory and Practice. Methuen, London, UK.

Comstock, G., 2000. Vexing Nature? On the Ethical Case Against Biotechnology. Kluwer, Boston, MA, USA.

Conner, A.J., Glare, T.R., and Nap, J.P., 2003. The release of genetically modified crops into the environment - Part II. Overview of ecological risk assessment. Plant Journal 33: 19–46.

Devos, Y., Reheul, D., and de Schrijver, A., 2005. The co-existence between transgenic and non-transgenic maize in the European Union: A focus on pollen flow and cross-fertilization. Environmental Biosafety Research 4: 71–87.

Dhaliwal, G.G., and Singh, R., eds., 2005. Host Plant Resistance to Insects: Concepts and Applications. Panima Publishing Corporation, New Delhi, India.

Dhaliwal, G.S., Koul, O., and Arora, R., 2004. Integrated pest management: Retrospect and prospect. In: Integrated peat management: Potential, constraints and challenges, O. Koul, G.S. Dhaliwal and G.W. Cuperus, eds., CABI, Wallingford, UK, pp. 1–20.

Dhaliwal, G.G., Singh, R., and Jindal, V., 2005. Host plant resistance and insect pest management: Progress and potential. In: Host Plant Resistance to Insects: Concepts and Applications, G.S. Dhaliwal and R. Singh, eds., Panima Publishing Corporation, New Delhi, India, pp. 517–558.

Doyle, B., Reeve, I., and Coleman, M., 2006. The CCA 2005 Bollgard Comparison Report: A Survey of Cotton Growers' and Consultants' Experience with Bollgard in the 2004–2005 season. The Cotton Research and Development Corporation and the Cotton Catchment Community Cooperative Research Centre, Institute for Rural Futures, Armidale, New South Wales, Australia.

Ellsworth, P.C., and Barkley, V., 2005. Transitioning Lygus chemical controls to more selective options for Arizona Cotton. In: Cotton, A College of Agriculture Report, Series P-142, University of Arizona, Tucson, AZ, USA, pp. 165–178.

Ellsworth, P.C., and Martinez-Carillo, J.L., 2001. IPM for *Bemisia tabaci*: A case study from North America. Crop Protection 20: 853–869.

Farrell, T., Mensah, R., Sequeira, R., Wilson, L., and Dillon, M., 2006. Key insect and mite pests of Australian Cotton. In: Cotton Pest Management Guide 2006–07, T. Farrell, ed., New South Wales Department of Primary Industries, Australia, pp. 1–17.

French, B.W., Chandler, L.D., and Riedel, W.E., 2007. Effectiveness of corn rootworm (Coleoptera: Chrysomelidae) areawide pest management in South Dakota. Journal of Economic Entomology 100: 1542–1554.

French, N.M., II, Follett, P., Nault, B.A., and Kennedy, G.G., 1993. Colonization of potato fields in eastern North Carolina by Colorado potato beetle. Entomologia Experimentalis et Applicata 68: 247–256.

Garcia-Alonso, M., Jacobs, E., Raybould, A., Nickson, T.E., Sowig, P., Willekens, H., van der Kouwe, P., Layton, R., Amijee, F., Fuentes, A.M., and Tencalla, F., 2006. A tiered system for assessing the risk of genetically modified plants to non-target organisms. Environmental Biosafety Research 5: 57–65.

Gatehouse, A.M.R., Boulter, D., and Hilder, V.A., 1991. Novel insect resistance using protease inhibitor genes. In: Molecular Approaches to Crop Improvement, E.S. Dennis and D.J. Llewellyn, eds., Springer, Dordrecht, The Netherlands, pp. 63–77.

Gianessi, L.P., Silvers, C.S., Sankula, S., and Carpenter, J.E., 2002. Plant Biotechnology: Current and Potential Impact for Improving Pest Management in U.S. Agriculture, an Analysis of 40 Case Studies. National Center for Food and Agricultural Policy, Washington, DC, USA.

Gould, F., 1988a. Evolutionary biology and genetically engineered crops. BioScience 38: 26–33.

Gould, F., 1988b. Genetic engineering, integrated pest management and the evolution of pests. Trends in Ecology & Evolution 6: S15–S19.

Head, G., Moar, M., Eubanks, M., Freeman, B., Ruberson, J., Hagerty, A., and Turnipseed, S., 2005. A multiyear, large-scale comparison of arthropod populations on commercially managed *Bt* and non-*Bt* cotton fields. Environmental Entomology 34: 1257–1266.

Heinrichs, E.A., and Adensina, A., 1999. Contribution of multiple-pest resistance to tropical crop production. In: Economic, Environmental, and Social Benefits of Resistance in Field Crops, B.R. Wiseman and J.A. Webster, eds., Thomas Say Publications in Entomology: Proceedings, Entomological Society of America, Lanham, MD, USA, pp. 15–21.
Higley, L.G., and Pedigo, L.P., eds., 1996. Economic Thresholds for Integrated Pest Management. University of Nebraska Press, Lincoln, NB, USA.
Huckaba, R., 2004. Pesticide Development from Discovery to Registration of a Pesticide in the United States. http://www.ento.vt.edu/~mullins/pestus2004/notes/lecture/Lec25.html (accessed 27 November 2007).
Huffaker, C.B., and Smith, R.F., 1980. Rationale, organization, and development of a national integrated pest management project. In: New Technology of Pest Control, C.B. Huffaker, ed., Wiley, New York, USA, pp. 1–24.
Hutchison, W.D., Burkness, E.C., Moon, R., Leslie, T., Fleischer, and Abrahamson, M., 2007. Evidence of regional suppression of European corn borer populations in transgenic maize in the Midwestern USA: Analysis of long-term time series data from three states. Proceedings of the XVI International Plant Protection Congress 2007, pp. 512–513.
James, C., 2003. Global Review of Commercialized Transgenic Crops: 2002 Feature: *Bt* Maize. ISAAA Brief No. 29, International Service for the Acquisition of Agri-Biotech Applications, Ithaca, NY, USA.
James, C., 2007. Global Status of Commercialized Biotech/GM Crops: 2007. ISAAA Brief No. 37, International Service for the Acquisition of Agri-biotech Applications, Ithaca, NY, USA.
Kennedy, G.G., 2004. Carl Barton Huffaker: Theoretician, experimentalist, and practitioner. American Entomologist 50: 76–81.
Kennedy, G.G., and Barbour, J.D., 1992. Resistance variation in natural and managed systems. In: Plant Resistance to Herbivores and Pathogens: Ecology, Evolution, and Genetics, R.S. Fritz and E.L. Simms, eds., The University of Chicago Press, Chicago, IL, USA, pp. 13–41.
Kennedy, G.G., and Gould, F., 2007. Ecology of natural enemies and genetically engineered host plants. In: Perspectives in Ecological Theory and Integrated Pest Management, M. Kogan and P. Jepson, eds., Cambridge University Press, Cambridge, UK, pp. 269–300.
Kennedy, G.G., and Storer, N.P., 2000. Life systems of polyphagous arthropod pests in temporally unstable cropping systems. Annual Review of Entomology 45: 467–495.
Kennedy, G.G., Gould, F., dePonti, O.M.B., and Stinner, R.E., 1987. Ecological, agricultural, genetic, and commercial considerations in the deployment of insect-resistant germplasm. Environmental Entomology 16: 327–338.
Khan, M., Quade, A., and Murray, D., 2006. Mirid management - effect of salt rate when mixed with reduced rates of chemical. In: Proceedings of the 13th Australian Cotton Conference, Australian Cotton Growers Research Association, Narrabri, New South Wales, Australia, pp. 537–542.
Kogan, M., ed., 1986. Ecological Theory and Integrated Pest Management Practice. Wiley, New York, USA.
Kogan, M., 1998. Integrated pest management: Historical perspectives and contemporary developments. Annual Review of Entomology 43: 243–270.
Kogan, M., and Jepson, P., 2007. Ecology, sustainable development and IPM: The human factor. In: Perspectives in Ecological Theory and Integrated Pest Management, M. Kogan and P. Jepson, eds., Cambridge University Press, Cambridge, UK, pp. 1–44.
Koul, O., and Cuperus, G., eds., 2008. Areawide Pest Management: Theory and Implementation. CABI, Wallingford, UK.
Koul, O., Dhaliwal, G.S., and Cuperus, G.W., eds., 2004. Integrated Pest Management: Potential, Constraints and Challenges. CABI, Wallingford, UK.
Koziel, M.G., Beland, G.L., Bowman, C., Carozzi, N., Crenshaw, R., Crossland, L., Dawson, J., Dasai, N., Hill, M., Kadwell, S., Launis, K., Lewis, K., Maddox, D., McPherson, K., Keghji, M.R., Merlin, E., Rhodes, R., Warren, G.W., Wright, M., and Evola, S.V., 1993. Field performance of elite transgenic maize plants expressing an insecticidal protein gene derived from *Bacillus thuringiensis*. BioTechnology 11: 195–200.

Lu, B.R., and Snow, A.A., 2005. Gene flow from genetically modified rice and its environmental consequences. BioScience 55: 669–679.

Luginbill, P., 1969. Developing resistant plants – the ideal method of controlling insects. USDA-ARS Production Research Report 111, United States Department of Agriculture, Agricultural Research Service, Washington, DC, USA.

Luttrell, R.G., Fitt, G.P., Ramalho, F.S., and Sugonyaev, E.S., 1994. Cotton pest management: Part 1, A worldwide perspective. Annual Review of Entomology 39: 517–526.

Marvier, M., McCreedy, C., Regetz, J., and Kareiva, P., 2007. A meta-analysis of effects of *Bt* cotton and maize on nontarget invertebrates. Science 316: 1475–1477.

Matteson, P.C., 2000. Insect pest management in tropical Asian irrigated rice. Annual Review of Entomology 45: 549–574.

Nap, J.P., Metz, P.L.J., Escaler, M., and Conner, A.J., 2003. The release of genetically modified crops into the environment - Part I. Overview of current status and regulations. Plant Journal 33: 1–18.

Naranjo, S.E., 2005. Long-term assessment of the effects of transgenic *Bt* cotton on the function of the natural enemy community. Environmental Entomology 34: 1211–1223.

Naranjo, S.E., and Luttrell, R.G., 2008. Cotton arthropod IPM. In: Integrated Pest Management, E.B. Radcliff and W.D. Hutchison, eds., Cambridge University Press, Cambridge, UK.

NRC (National Research Council), 1987. Introduction of Recombinant DNA-Engineered Organisms into the Environment: Key Issues. National Academy Press, Washington, DC, USA, 187 p.

NRC (National Research Council), 2000. Genetically Modified Pest-protected Plants: Science and Regulation. National Academy Press, Washington, DC, USA, 292 p.

Newson, L.D., 1980. The next rung up the integrated pest management ladder. Bulletin of the Entomological Society of America 26: 369–374.

Nuffield Council on Bioethics, 1999. Genetically Modified Crops: The Ethical and Social Issues. Nuffield Council on Bioethics, London, UK, 164 pp. http://www.nuffield.org.uk/bioethics/publications/modified crops/rep0000000075.html (accessed 27 December 2007).

Nuffield Council on Bioethics, 2003. The Use of Genetically Modified Crops in Developing Countries: A Follow-Up Discussion Paper. Nuffield Council on Bioethics, London, 122 pp. http://www.nuffieldbioethics.org/fileLibrary/pdf/GM_Crops_Discussion_Paper_2004.pdf (accessed 27 December 2007).

Obrycki, J.J., Ruberson, J.R., and Losey, J.E., 2004. Interactions between natural enemies and transgenic insecticidal crops. In: Genetics, Evolution and Biological Control, L.E. Ehler, R. Sforza and T. Mateille, eds., CABI, Wallingford, UK, pp. 183–206.

O'Callaghan, M., Glare, T.R., Burgess, E.P.J., and Malone, L.A., 2005. Effects of plants genetically modified for insect resistance on nontarget organisms. Annual Review of Entomology 50: 271–292.

Otuka, A., Watanabe, T., Suzuki, Y., Matsumura, M., Furuna, A., and Chino, M., 2005. Real-time prediction system for migration of rice planthoppers *Sogatella furcifera* (Horvath) and *Nilaparvata lugens* (Stal) (Homoptera: Delphacidae). Applied Entomology and Zoology 40: 221–229.

Painter, R.H., 1951. Insect Resistance in Crop Plants. University of Kansas Press, Lawrence, KS, USA.

Panda, N., and Khush, G.S., 1995. Host Plant Resistance to Insects. CABI, Wallingford, UK.

Pedigo, L.P., 1996. Entomology and Pest Management, 2nd edition. Prentice-Hall, Upper Saddle River, NJ, USA.

Pedigo, L.P., Hutchins, S.H., and Higley, L.G., 1986. Economic injury levels in theory and practice. Annual Review of Entomology 31: 341–368.

Perkins, J.H., 1982. Insects, Experts, and the Insecticide Crisis: The Quest for New Pest Management Strategies. Plenum, New York, USA.

Perlak, F.J., Deaton, R.W., Armstrong, T.A., Fuchs, R.L., Sims, S.R., Greenplate, J.T., and Fischhoff, D.A., 1990. Insect-resistant cotton plants. BioTechnology 8: 939–943.

Perlak, F.J., Fuchs, R.L., Dean, D.A., McPherson, S.L., and Fischhof, D.A., 1991. Modification of coding sequence enhances plant expression of insect control protein genes. Proceedings of the National Academy of Sciences of the USA 88: 3323–3328.

Peterson, R.K., Meyer, S.J., Wolf, A.T., Wolt, J.D., and Davis, P.M., 2006. Genetically engineered plants, endangered species, and risk: A temporal and spatial exposure assessment for Karner blue butterfly larvae and Bt maize pollen. Risk Analysis 26: 845–858.

Rabb, R.L., 1970. Introduction to the conference. In: Concepts of Pest Management, R.L. Rabb and F.E. Guthrie, eds., North Carolina State University, Raleigh, NC, USA, pp. 1–5.

Rabb, R.L., DeFoliare, G.R., and Kennedy, G.G., 1984. An ecological approach to managing insect populations. In: Ecological Entomology, C.B. Huffaker and R.L. Rabb, eds., Wiley, New York, USA, pp. 697–728.

Ram, P., Singh, R., and Dhaliwal, C.S., 2005. Biophysical bases of resistance in plants to insects. In: Host Plant Resistance to Insects: Concepts and Applications, G.S. Dhaliwal and R. Singh, eds., Panima Publishing, New Delhi, India, pp. 42–83.

Raybould, A., 2007. Ecological versus ecotoxicological methods for assessing the environmental risks of transgenic crops. Plant Science 173: 589–602.

Raybould, A., Stacey, D., Vlachos, D., Graser, G., Li, X., and Joseph, R., 2007. Non-target organisms risk assessment of MIR604 maize expressing mCry3A for control of corn rootworms. Journal of Applied Entomology 131: 391–399.

Romeis, J., Dutton, A., and Bigler, F., 2004. *Bacillus thuringiensis* toxin (Cry1Ab) has no direct effect on larvae of the green lacewing *Chrysoperla carnea* (Stephens) (Neuroptera: Chrysopidae). Journal of Insect Physiology 50: 175–183.

Romeis, J., Meissle, M., and Bigler, F., 2006. Transgenic crops expressing *Bacillus thuringiensis* toxins and biological control. Nature Biotechnology 24: 63–71.

Romeis, J., Bartsch, D., Bigler, F., Candolfi, M.P., Gielkens, M.M.C., Hartley, S.E., Hellmich, R.L., Huesing, J.E., Jepson, P.C., Layton, R., Quemada, H., Raybould, A., Rose, R.I., Schiemann, J., Sears, M.K., Shelton, A.M., Sweet, J., Vaituzis, Z., and Wolt, J.D., 2008. Assessment of risk of insect-resistant transgenic crops to nontarget arthropods. Nature Biotechnology 26 (2): 203–208.

Rose, R.I., 2007. White Paper on Tier-Based Testing for the Effects of Proteinaceous Insecticidal Plant-Incorporated Protectants on Non-target Arthropods for Regulatory Risk Assessments. USDA-APHIS and US Environmental Protection Agency, Washington, DC, USA. http://www.aphis.usda.gov/brs/pdf/NTO_White_Paper_1.pdf (accessed 1 December 2007).

Rose, D.J.W., and Khasimuddin, S., 1979. Wide-area monitoring of the African armyworm, *Spodoptera exempta* (Walker) (Lepidoptera: Noctuidae). In: Movement of Highly Mobile Insects: Concepts and Methodology in Research, R.L. Rabb and G.G. Kennedy, eds., North Carolina State University, Raleigh, NC, USA, pp. 212–219.

Sanvido, O., Romeis, J., and Bigler, F., 2007. Ecological impacts of genetically modified crops: Ten years of field research and commercial cultivation. Advances in Biochemical Engineering and Biotechnology 107: 235–278.

Sanvido, O., Widmer, F., Winzeler, M., Streit, B., Szerencsits, E., and Bigler, F., 2008. Definition and feasibility of isolation distances for transgenic maize cultivation. Transgenic Research 17: 317–335.

Sears, M.K., Hellmich, R.L., Siegfried, B.D., Pleasants, J.M., Stanley-Horn, D.E., Oberhauser, K.S., and Dively, G.P., 2001. Impact of *Bt* corn pollen on monarch butterfly populations: A risk assessment. Proceedings of the National Academy of Sciences of the USA 98: 11937–11942.

Sharma, H.C., ed., 2005. *Heliothis/Helicoverpa* Management: Emerging Trends and Strategies for Future Research. Oxford/IBH, New Delhi, India.

Showers, W.B., 1997. Migratory ecology of the black cutworm. Annual Review of Entomology 42: 393–425.

Shelton, A.M., 2007. Considerations on the use of transgenic crops for insect control. Journal of Developmental Studies 43: 890–900.

Shelton, A.M., Zhao, J.-Z., and Roush, R.T., 2002. Economic, ecological, food safety, and social consequences of the deployment of *Bt* transgenic plants. Annual Review of Entomology 47: 845–881.

Sisterson, M.S., Biggs, R.W., Olson, C., Carrière, Y., Dennehy, T.J., and Tabashnik, B.E., 2004. Arthropod abundance and diversity in *Bt* and non-*Bt* cotton fields. Environmental Entomology 33: 921–929.

Smith, C.M., 2004. Plant resistance against pests: Issues and strategies. In: Integrated Pest Management: Potential, Constraints and Challenges, O. Koul, G.S. Dhaliwal and G.W. Cuperus, eds., CABI, Wallingford, UK, pp. 147–168.

Smith, C.M., 2005. Plant Resistance to Arthropods: Molecular and Conventional Approaches. Springer Science & Business Media, Dordrecht, The Netherlands.

Smith, E.H., and Kennedy, G.G., 2002. History of pesticides. In: Encyclopedia of Pest Management, D. Pimentel, ed., CRC, Boca Raton, FL, USA, pp. 376–380.

Smith, R.F., 1970. Pesticides: Their use and limitations in pest management. In: Concepts of Pest Management, R.L. Rabb and F.E. Guthrie, eds., North Carolina State University, Raleigh, NC, USA, pp. 103–118.

Stern, V.M., Smith, R.F., van den Bosch, R., and Hagen, K.S., 1959. The integrated control concept. Hilgardia 29: 81–101.

Storer, N.P., Peck, S.L., Gould, F., Van Duyn, J.W., and Kennedy, G.G., 2003. Spatial processes in the evolution of resistance in *Helicoverpa zea* (Lepidoptera: Noctuidae) to Bt transgenic corn and cotton: a biology-rich stochastic simulation model. Journal of Economic Entomology 96: 156–172.

Taylor, L.R., 1977. Migration and spatial dynamics of an aphid, *Myzus* persicae. Journal of Animal Ecology 46: 411–423.

Thompson, P.B., 2000. Bioethics issues in a bio-based economy. In: The Biobased Economy of the Twenty-First Century: Agriculture Expanding into Health, Energy, Chemicals, and Materials. NABC Rep. 12. Ithaca, NY. National Agricultural Biotechnology Council. pp 113–123.

USEPA (United States Environmental Protection Agency), 2000. Biopesticides Registration Document: Preliminary Risks and Benefits Sections; *Bacillus thuringiensis* Plant-Pesticides. US Environmental Protection Agency, Office of Pesticide Programs, Biopesticides Pollution Prevention Division, Washington, DC, USA.

USEPA (United States Environmental Protection Agency), 2001a. *Bt* Biopesticides Registration Action Document D. Insect Resistance Management. http://www.epa.gov/oppbppd1/biopesticides/pips/bt_brad2/4-irm.pdf. (accessed 4 January 2008).

USEPA (United States Environmental Protection Agency), 2001b. *Bt* Cotton Refuge Requirements for the 2001 Growing Season. www.epa.gov/pesticides/biopesticides/pips/bt_cotton_refuge_2001.htm. (accessed 4 January 2008).

USEPA (United States Environmental Protection Agency), 2005. *Bt* Cry 1F/Cry1Ac Widestrike® Cotton Registration Action Document. www.epa.gov/pesticides/biopesticides/ingredients/tech_docs/brad_006512-006513.pdf. (accessed 4 January 2008).

USEPA (United States Environmental Protection Agency), 2007. Notice of Pesticide Registration, No. 68467-3, approved 17 July 2007.

Vaeck, M., Reynaerts, A., Hofte, H., Jansens, S., De Beuckeleer, M., Dean, C., Zabeau, M., Van Montagu M., and Leemans, J., 1987. Transgenic plants protected from insect attack. Nature 328: 33–37.

Wilson, L.J., Bauer, L.R., and Lally, D.A., 1998. Effect of early season insecticide use on predators and outbreaks of spider mites (Acari: Tetranychidae) in cotton. Bulletin of Entomological Research 88: 477–488.

Wilson, L.J., Mensah, R.K., and Fitt, G.P., 2004. Implementing integrated pest management in Australian cotton. In: Novel Approaches to Insect Pest Management in Field and Protected Crops, A.R. Horowitz and I. Ishaaya, eds., Springer, Berlin, Germany, pp. 97–118.

Wolt, J.D., Conlan, C.A., and Majima, K., 2005. An ecological risk assessment of Cry1F maize pollen impact to pale grass blue butterfly. Environmental Biosafety Research 4: 243–251.

Wu, K.M., and Guo, Y.Y., 2005. The evolution of cotton pest management practices in China. Annual Review of Entomology 50: 31–52.

Yencho, G.C., Cohen, M.B., and Byrne, P.F., 2000. Applications of tagging and mapping insect resistance loci in plants. Annual Review of Entomology 45: 393–422.

Chapter 2
How Governmental Regulation Can Help or Hinder the Integration of *Bt* Crops within IPM Programs

Sharlene R. Matten[1,*], Graham P. Head[2], and Hector D. Quemada[3]

Abstract Regulatory risk assessments are an important part of the introduction of insect-resistant genetically modified (GM) crops (e.g., *Bacillus thuringiensis* [*Bt*] crops) into the environment to ensure the safe use of such products. In doing so, the regulatory assessment process can be clearly beneficial to integrated pest management (IPM) programs. In general, the regulatory framework for insect-resistant GM crops includes an assessment of the following: effects of the insecticidal trait on non-target organisms, other potential adverse environmental impacts, evolution of resistance to target pests, and environmental and agronomic benefits of the insecticidal trait. Each country's regulatory system is dependent on the overall environmental risk management goals, relevant and available risk information, scientific capacity, and the available financial resources. A number of regulatory activities can help to ensure that new products such as *Bt* crops fit well within IPM programs: (1) evaluation of the environmental safety of new products, and their ability to enhance IPM; (2) encouragement of the adoption of new technologies with improved environmental safety profiles; (3) adoption of an expedited regulatory review system; and (4) encouragement and appropriate oversight of sustainable use of such products. Governmental regulation of insect-resistant GM crops can also hinder IPM programs by creating significant barriers to the adoption of such technologies. Such barriers include: (1) absence of functioning regulatory systems in many developing countries; (2) meeting the obligations and understanding the various interpretations of international treaties, e.g., Cartagena Protocol on Biosafety; (3) lack of public sector research to generate data supporting the safety of these crops; and (4) regulatory costs involved in the development and commercialization of novel products for small market sectors. Ways in which regulatory data requirements can be globally harmonized need to be considered to decrease the regulatory

[1] United States Environmental Protection Agency, Office of Science Coordination and Policy, Washington, DC, USA

[2] Monsanto Company, St. Louis, MO, USA

[3] Department of Biology, Calvin College, Grand Rapids, MI, USA

* To whom correspondence should be addressed. E-mail: Matten.Sharlene@epamail.epa.gov

J. Romeis, A.M. Shelton, G.G. Kennedy (eds.), *Integration of Insect-Resistant Genetically Modified Crops within IPM Programs.*

barriers for insect-resistant GM crops and comparable technologies. International organizations can play a key role in rationalizing regulatory systems; however, public sector research will also be needed to make sure that the risk assessment process is scientifically sound and transparent.

2.1 Regulatory Risk Assessment of *Bt* Crops

The regulatory framework for *Bt* crops in the United States of America and other countries has been developed and deployed within the broad context of risk assessment and integrated pest management (IPM). This framework typically includes consideration of environmental and agronomic benefits, such as reductions in applications of broad-spectrum insecticides, yield improvements and mycotoxin reduction in grain (USEPA, 2001a; Carpenter et al., 2002; Brookes and Barfoot, 2006; Cattaneo et al., 2006; Fernandez-Cornejo and Caswell, 2006; Fitt, chapter 11; Qaim et al., chapter 12). Furthermore, it includes assessments of the potential for the evolution of *Bt* resistance in target pests (Gould, 1998; USEPA, 2001a; Glaser and Matten, 2003; Tabashnik et al., 2003; Matten and Reynolds, 2003; Matten et al., 2004; Ferré et al., chapter 3), and effects on non-target organisms and other potential human and environmental impacts (USEPA, 2001a; Johnson et al., 2007). An assessment of environmental risk and management of insect resistance to *Bt* crops is critical to the sustainability of IPM programs.

2.1.1 The Nature of Regulatory Risk Assessment

An environmental risk assessment is conducted to facilitate regulatory decision-making with regard to identifying potential undesirable impacts and options for mitigating them. Assessing the potential for environmental risks of *Bt* crops, or any other insect-resistant crop, involves estimating the likelihood that the presence of the *Bt* gene(s) will have adverse effects on the environment. A risk exists if the exposure to the protein produced by the inserted gene has hazardous effects on non-target organisms. A risk assessment is a synthesis of sufficient information to determine whether the risks of a proposed course of action are acceptable. Estimates of potential hazard(s) and exposure allow regulatory decision-makers to determine the likelihood that a *Bt* protein, for example, may cause a problem and also to gauge the scale of that problem. There are several components of the environmental risk assessment that are particularly important: (1) Overall risk management goals and assessment endpoints; (2) Hazard identification; (3) Exposure identification; (4) Test endpoints; and (5) Iterative or tiered approach (Rose, 2007; Johnson et al., 2007; Raybould, 2007).

One of the more controversial areas of the regulation of *Bt* crops is post-market monitoring (PMM) after commercialization. In some parts of the world, for example Europe, it is required to monitor the environmental impacts of GM crops in commercial cultivation, as described in European Community Directive 2001/18/EC (EC,

2001). PMM includes both case-specific monitoring and general surveillance. Case-specific monitoring focuses on anticipated effects of a specific GM crop on the environment and aims to assess whether these effects do occur. A typical example for a case-specific monitoring area is insect resistance management of insect-resistant GM crops where major target pest populations are monitored to detect changes in the frequency of resistance alleles (USEPA, 2001a; Glaser and Matten, 2003; Tabashnik et al., 2003; Matten and Reynolds, 2003; Matten et al., 2004). General surveillance, in contrast, has the aim to detect adverse effects on the environment that were not anticipated during pre-market risk assessment. However, general surveillance lacks specific hypotheses concerning what one should monitor and why, as well as any baseline comparison to alternative practices such as the use of conventional pesticides. Therefore, the implications of finding an effect or change are unclear and causality to the cultivation of GM crops must be determined in separate risk assessment studies. Sanvido et al. (2005) describe a practical framework for the design of general surveillance of genetically-modified crops and propose to establish appropriate reporting systems to collect reports on adverse incidents that come from existing environmental monitoring programs. Ecological monitoring may also be conducted for a limited period of time to fill in data gaps (e.g., USA) or as a risk management option based on pre-market assessment of environmental risk.

2.1.2 Regulation and Insect Resistance Management

Insect resistance management (IRM) adds another dimension to IPM programs when *Bt* crops are deployed. IPM is one of the principal strategies for preventing resistance development because it uses diverse tactics to suppress pest populations and, conversely, IRM is a critical component of IPM programs because it ensures that important pest control tools remain viable for long-term use. In the USA, there has been substantial policy interest in maintaining the productivity of *Bt* as an important public resource to agricultural production systems, unlike any conventional pesticide (Berwald et al., 2006). One regulatory policy that attempts to maintain productivity of *Bt* is the institution of specific IRM requirements. Regulation of IRM for *Bt* crops is unprecedented in the pesticide world; there is no equivalent requirement for any conventional pesticide though the European Union (EU) recognizes the importance of pesticide resistance and requires registrants to address the risk of resistance development as part of dossiers submitted for EU registration (Anonymous, 2003; MacDonald et al., 2003). Voluntary IRM activities have been encouraged for conventional insecticides with some success. In the USA and Canada, for example, voluntary resistance management labeling guidelines for all agricultural pesticides based on the rotation of mode of action were developed as a joint activity under the auspices of the North American Free Trade Agreement (NAFTA). These guidelines were published as EPA Pesticide Registration Notice 2001–2005 (USEPA, 2001b) and Canada Pesticide Regulatory Directive DIR 99–06 (Health Canada, 1999). While the EU, USA, and Canada have adopted a combination of mandatory and voluntary approaches to regulating insect resistance to chemical insecticides, the role of mandatory approaches is one of great debate amongst the various stakeholders from industry, grower organizations, academia, and government (Thompson and Head, 2001).

IRM strategies for *Bt* crops differ among countries based primarily on the target pests, agricultural practices, cropping patterns, adoption of the technology, and cost. Both mandatory and voluntary regulatory systems are in place in countries where *Bt* crops are grown. Mandatory IRM programs are required in the USA, Canada, and Australia, for example. In contrast, voluntary IRM programs exist in China, for example. At a global level, some form of baseline susceptibility/monitoring studies are conducted in all countries prior to commercialization. Regardless of the country, the basic elements needed to develop and implement an IRM plan remain the same: (1) knowledge of pest biology and ecology; (2) toxin dose; (3) the genetics of potential resistance; (4) cropping patterns and potential use of the *Bt* crop; (5) baseline susceptibility; (6) a resistance monitoring plan; (7) grower education; and (8) remedial action plan should resistance develop. Simple field studies, work on model systems, and simulation models can allow for a qualitative comparison of possible IRM strategies. The area of IRM requirements for *Bt* crops has not been without controversy and has stimulated much interest among academic researchers, government, industry, and growers. IRM programs for *Bt* crops are discussed in several other chapters in this book in general (Ferré et al., chapter 3), for maize (Hellmich et al., chapter 5), for cotton (Naranjo et al., chapter 6), for potato (Grafius and Douches, chapter 7), and for rice (Cohen et al., chapter 8).

The overall environmental risk management goals, relevant and available risk information, and the technical tools available influence the way *Bt* crops are regulated in each country. This chapter focuses on how regulatory systems can either enhance or hinder the use of *Bt* crops within an IPM program, and provides insights into regulatory issues that will arise from non-*Bt* insect-resistant crops developed in the future (Malone et al., chapter 13).

2.2 Regulation as an Enhancement to IPM Programs

Government regulation helps to ensure that new agricultural pest-control products will fit well into IPM programs in a number of ways, either through explicit consideration of IPM needs or because of common objectives between environmental regulations and IPM. We discuss some of these direct and indirect positive impacts of regulation on IPM programs in the sections below. These impacts are not specific to *Bt* crops, or other products of biotechnology, but rather will apply to any technology being considered for commercial use in agriculture. Due to the rapid and widespread adoption of *Bt* crops on a global level, *Bt* crops represent the most important set of novel insect control technologies to be approved by regulatory agencies in the last two decades.

2.2.1 *Regulatory Assessments Focus on the Environmental Safety of New Products*

In assessing the environmental safety of any *Bt* crop, governmental regulatory agencies explicitly consider the potential risk posed to various groups of non-target organisms. For example, before allowing a *Bt* crop to be grown commercially, data on the risk posed to representatives of economically and ecologically important guilds such as organisms important for biological control (predators and parasitoids such as ladybird beetles, lacewings, and parasitic wasps), pollinators (such as honeybees), and organisms involved in soil processes (such as springtails and earthworms) are reviewed by agencies such as the United States Environmental Protection Agency (USEPA) and the United States Department of Agriculture (USDA) in the United States, and the European Food Safety Authority (EFSA) in the European Union, and comparable agencies in other countries (Rose, 2007). These risk assessments focus on non-target organisms that are locally important and require data generated in the relevant crop and country. By ensuring that new technologies will not have significant adverse impacts on these non-target groups, regulatory agencies indirectly ensure that these technologies also will fit well into IPM programs, complementing biological control functions and minimizing non-target pest flare-ups. By applying such standards, regulatory agencies also encourage the development of future pest control technologies with these characteristics.

2.2.2 *Regulatory Assessments Encourage the Advancement of Technologies with Improved Environmental Profiles*

In the environmental risk assessment process carried out by regulatory agencies, the potential risks associated with these new agricultural pest control technologies generally are compared with the observed impacts of alternative pest control technologies that farmers may currently use. For *Bt* crops, these alternative technologies usually will be conventional insecticides. These assessments of relative risk help to ensure that *Bt* crops and other new pest control technologies being introduced are superior, or at least equivalent, to existing technologies in their environmental profiles. In doing these assessments, the components of agro-ecosystems that are examined and the criteria that are applied are largely coincident with the needs of IPM programs, and thereby ensure that new technologies, such as *Bt* crops, will be useful additions to existing IPM programs. For instance, environmental impact quotients (EIQ) for *Bt* crop systems typically have been found to be significantly lower than the EIQs for alternative technologies such as conventional insecticides (Kleter et al., 2007). EIQ is a measure designed to summarize the impacts of a pesticide on various ecosystems components, as well as effects on human health. As a consequence, no significant adverse environmental effects have been associated with the global adoption of *Bt* crops (Sanvido et al., 2007), while the technologies that they have replaced (i.e., insecticides) often had adverse impacts on the environment and human health (for example, Naranjo et al., 2005; Wu and Guo, 2005; Cattaneo et al., 2006; Qaim et al., chapter 12).

2.2.3 New Technologies with Superior Environmental Profiles Can Be Fast-Tracked

Because environmental agencies are focused on the impact of pest control technologies on the environment, they often have developed specific mechanisms to expedite the approval of environmentally safer pest control alternatives to conventional chemical pesticides, such as *Bt* crops. For example, in the USA, the Pesticide Registration Improvement Act of 2003 promotes shorter decision review periods for applications for reduced-risk. At the same time, approvals for the use of products with unfavorable environmental profiles may be withdrawn. For example in the USA, the Federal Food, Drug, and Cosmetic Act was amended in 1996 to include the Food Quality Protection Act or FQPA. This Act required EPA to reassess by August 2006 all of the pesticide tolerances that were in place in early August 1996 to ensure that they met current safety standards and were supported by up-to-date scientific data. FQPA also mandated a registration review process. Every 15 years, EPA will reassess each pesticide to see whether it still meets the registration standards required under the Federal Insecticide, and Rodenticide Act (FIFRA). The net effect of these initiatives will be to provide safer tools for IPM programs.

2.2.4 Sustainable Product Use Can Be Encouraged

As regulatory agencies identify and approve the use of new environmentally friendly technologies, they also look for ways to ensure that these products are used in a sustainable way so that their continued availability is assured. In the case of *Bt* crops, regulatory agencies such as the US EPA have worked with the product developers to construct and implement IRM programs for each product that will serve to delay the evolution of target pest resistance and thereby protect the durability of these products (Gould, 1998; Glaser and Matten, 2003). These IRM programs have included the implementation of structured non-*Bt* refuges and resistance monitoring programs and, in the case of the United States, Australia and India, the replacement of single *Bt* gene cotton (Bollgard I®) with the more durable dual *Bt* cotton (Bollgard II®) when it became available. IRM programs have now been implemented for *Bt* crops on a world-wide basis in both developed (for example, USA, Canada and Australia) and developing countries (such as India and the Philippines). Many of these programs were initially implemented voluntarily by the product developers in these countries; however, they now are typically required by regulators as part of regulatory packages for *Bt* crops in most countries. IRM activities have aided IPM programs by ensuring that effective and reliable tools are available for sustainable control of certain key lepidopteran and coleopteran pests. IRM programs for *Bt* crops have contributed to mitigating field resistance to *Bt* crops in the world during the past decade (Tabashnik et al., 2003; Ferré et al., chapter 3). While IRM programs have been invaluable, programs that are effective for one or more target pests in one geographical region may not be as effective against other economically-important pests in other geographical regions. A case in point is the recent report of fall armyworm (*Spodoptera frugiperda*, J.E. Smith) resistance to the Cry1F protein expressed in TC1507 maize fields in 2006 in Puerto Rico. Fall armyworm is the most important pest of maize in Puerto Rico where the tropical climate allows year-round production of maize and multiple pest generations each year. The mountainous island also creates a more closed pest population than is the case for other pests and other geographies. In 2007, USEPA reviewed unpublished data submitted by Dow AgroSciences and Pioneer Hi-Bred International that detailed their investigation of unexpected fall armyworm damage found in TC1507 maize fields in 2006 in Puerto Rico and whether such damage was caused by resistant insects (Matten, 2007). Based on review of the screening level and concentration-dependent bioassays, the conclusions was that the unexpected performance failures of TC1507 maize observed in 2006 in Puerto Rico were due to Cry1F-resistant fall armyworm. Because of this finding, sales of this product have been suspended in Puerto Rico, consistent with the IRM program. Fall armyworm resistance to TC1507 maize is much less likely to occur in the continental USA because it can only overwinter in the extreme south of Texas and Florida, and therefore, selection in maize-growing regions exerts no long term selection pressure.

2.3 Regulation as a Hindrance to IPM Programs

While the goals of environmental regulation often have much in common with IPM goals, regulation can hinder IPM programs by creating significant barriers to the introduction of important new technologies under certain conditions. This is particularly true of products of agricultural biotechnology such as *Bt* crops because of the specific and complex regulatory systems that have been created to deal with these products. Unfortunately, the barriers created often are greatest where the technologies are potentially most needed, for example in developing countries in Africa and Asia (Gressel et al., 2004; Thomson, 2008). The circumstances under which regulation can adversely affects IPM programs are discussed below.

2.3.1 The Absence of Functioning Regulatory Systems in Many Developing Countries

Experience to date with *Bt* crops has shown that they can play a role in the implementation of IPM practices in developed and developing countries (Obando-Rodriguez et al., 1999; Bambawale et al., 2004; Sanvido et al., 2007; Kennedy, chapter 1). However, a critical step in the application of these crops is the regulatory approvals that must be obtained before they can be used, based on appropriate risk assessments by regulatory authorities. Therefore, a sound and functional regulatory system must be established before the full potential of these crops can be realized. This system must be capable of making the necessary scientific evaluations in order to arrive at a reasoned and scientifically supportable decision. However, a regulatory decision also ultimately involves non-scientific issues to a greater or lesser extent. Regulatory systems should be able to manage non-science issues, such as labor (Shelton, 2007), in such a way that appropriate and beneficial technologies are not prevented from reaching the market.

Functional biotechnology regulatory systems are largely absent in most developing countries. In Africa, for example, relatively few have established biosafety frameworks. According to the Biosafety Clearinghouse mechanism of the Cartagena Protocol on Biosafety, only 14 countries on the African continent have written laws, regulations, guidelines, or policies concerning genetically engineered crops (http://bch.cbd.int). With the exception of South Africa, which has approved *Bt* maize and *Bt* cotton for commercial release (http://www.agbios.com/dbase.php), none of those countries have had experience in the assessment of applications for commercialization of any genetically engineered plant variety. Consequently, their ability to conduct a risk assessment connected with an application for commercial release of a *Bt* crop has yet to be tested. By contrast, more developing countries in Asia have established functioning regulatory systems. China and India have commercialized *Bt* cotton, while the Philippines has commercialized *Bt* maize (James, 2007). The absence of a regulatory system, or even one that has demonstrated functionality, has prevented many developing countries from experiencing the benefits that have been experienced by those countries where *Bt* crops have been approved. Most of these countries have not even been able to conduct confined field trials to determine efficacy or conduct studies that are prerequisites for any regulatory decision concerning these crops.

There are many reasons for the absence of functional regulatory systems in developing countries, but a primary factor is the lack of scientific capacity in many regulatory agencies. Risk assessment to support regulatory decisions requires a multi-disciplinary approach, encompassing such fields as toxicology, eco-toxicology, genetics, molecular biology, chemistry, taxonomy and ecology. While most developing countries possess expertise in many of these fields, few of them have expertise in the complete range of scientific disciplines that may be required, particularly within the regulatory agencies themselves. Furthermore, many developing country regulatory systems are composed of part-time members rather than full-time professional staff, a situation made necessary by the lack of government resources to support such a staff, and because of the involvement of a broad range of government ministries. Broad representation requires capacity building in ministries staffed by decision makers, many of whom do not possess the basic understanding of the biological disciplines underlying the development of *Bt* crops. Even in those ministries and scientific bodies involved in the process that may have the necessary expertise, the focusing of this expertise into the discipline of risk assessment requires capacity building as well.

2.3.2 Meeting the Obligations of International Treaties

At the international level, the Cartagena Protocol on Biosafety (CPB) has the potential to further hinder the introduction of *Bt* crops. While the original intent of this international agreement was to facilitate the safe trans-boundary movement of genetically engineered crops and other commodities, in order to assure fair and equitable access to the benefits of biotechnology (Cartagena Protocol on Biosafety to the Convention on Biological Diversity, 2000), elements of the implementation of this protocol could, in some countries, severely affect the transfer of this technology. Primary among these elements is the implementation of Article 27 of the CPB, regarding the establishment of rules and procedures concerning liability and redress. The negotiations surrounding this provision of the CPB are at a critical stage, and fundamental questions regarding such issues as the scope of this provision (whether limited to damage to biodiversity or more broadly to traditional and socioeconomic damage), the definition of who is liable, the limits of liability, and the requirements for insurance, have implications for the introduction of genetically engineered crops. For example, if liability were to be extended to developers of genetically engineered crops for an indefinite period of time, and for an indefinite amount (a possible scenario under the current negotiations), the sharing of *Bt* crops developed in countries that are party to the CPB with other party countries could be severely restricted. Similar effects would be seen on crops developed in non-party countries as well. This is particularly problematic because most countries that are parties to the CPB are developing countries that have invested heavily in public sector research to develop genetically engineered local crops to address local needs. The sharing of the benefits of this research between developing countries would be severely affected by overly restrictive liability regimes.

2.3.3 Special Barriers to Products Coming from the Public Sector

For the private sector, significant experience has been gained over the years in the procedures to generate data supporting the safety of these crops. On the other hand, the public sector has had very little experience in the commercialization of transgenic crops. There are only two examples of transgenic crops developed by the public sector – papaya and plums – and these examples do not provide good guidance for the regulatory requirements governing *Bt* crops. Unlike *Bt* proteins, the viral coat protein expressed by the transgenic papaya have no known toxicity (Gonsalves et al., 1996), and therefore do not raise questions of hazard to non-target organisms. The plum transgenic lines do not produce detectable levels of protein (Scorza, 2004). Therefore, questions about the impact of a novel protein on the environment – a major consideration with *Bt* crops – were not even considered in these cases.

Public sector initiatives face additional hurdles in trying to introduce new technologies with IPM applications. These hurdles are exacerbated by the cost of the regulatory approval process. The regulatory requirements for *Bt* crops, because they have been based to a large degree on the requirements covering conventional pesticides, have imposed significant costs on the approval process, estimated to be between $7 million and $15 million for *Bt* maize for approval in ten major market countries (Kalaitzandonakes et al., 2006, 2007). Thus, approval in even one country could cost between $700,000 and $1.5 million. This cost is well beyond the reach of public sector projects, even in the developed world. Therefore, if current regulatory models continue to be applied to *Bt* crops, the ability to develop these crops on a worldwide basis, particularly those that address developing country needs, will be hindered. Developing country public research is focused on crops for the poor, and therefore is a government investment, with returns coming back to the public in less definable ways – food security, better health, greater subsistence farmer income – than for a product developed by the private sector. The challenge, therefore, especially for developing country regulatory agencies, is to examine where data requirements can be reduced or streamlined without compromising the level of safety achieved by current developed-world regulatory requirements, in order that the investments made by governments are fully realized.

2.3.4 Barriers to Developing Products for Small Markets

Because of the regulatory costs currently involved with *Bt* crops, it is difficult for either the public or private sector to develop novel products specifically for small markets, including specialty crops in the developed and developing world and almost any crop in countries with relatively small agricultural sectors. However, efforts involving private-public partnerships may prove fruitful in bringing some *Bt* crops like eggplant and vegetable crucifers to market in India and other developing countries (Shelton et al., chapter 9). Technologies developed primarily for use in other systems or countries may still make it into these smaller markets, but this will dramatically limit the problems that can be addressed through biotechnology in the short term. Many developing countries urgently need safe and reliable pest control alternatives, and *Bt* crops provide a good solution to these needs. Here, too, adaptations to existing regulatory systems and standards will need to be considered if the benefits of *Bt* crops and comparable technologies are to be more broadly realized.

2.4 Future Considerations

Regulatory risk assessments are an important part of the introduction of any new agricultural technology, and can help to ensure that new technologies meet certain standards with respect to environmental safety. In doing so, the regulatory assessment process can be clearly beneficial to IPM programs, but a balance between regulatory rigor and efficiency must be achieved. Functioning regulatory systems need to adequately assess the potential risks associated with new technologies but should not be so burdensome as to be a barrier to the introduction of valuable technologies. This balance is more difficult to reach when resources and scientific capacity are more limited, as is the case in many developing countries.

Ways to harmonize regulatory requirements across regions and to allow data generated in one country to be recognized in other countries will need to be investigated if pest management programs in developing countries are to fully realize the benefits of *Bt* crops (Romeis et al., 2008). For example, laboratory data showing that there is an absence of a toxic effect of a particular Cry protein on a certain non-target specific is generally valid and could be used for risk assessments in any country. International organizations like the Organization for Economic Cooperation and Development (OECD) can play an important role in rationalizing regulatory systems. Public sector scientists will also need to make sure that their voices are heard as part of this process.

Disclaimer

The views expressed in this article are those of the individual authors and do not necessarily reflect the views and policies of the US Environmental Protection Agency, Monsanto Company, or Crop Technology Consulting, Inc. The use of trade names does not imply endorsement by the US Government.

References

Anonymous, 2003. Efficacy evaluation of plant protection products. PP1/123(2) Resistance Risk Analysis. EPPO Bulletin 33: 37–63.

Bambawale, O., Singh, A., Sharma, O., Bhosle, B., Lavekar, R., Dhandapani, A., Kanwar, V., Tanwar, R., Rathod, K., Patange, N., and Pawar, V., 2004. Performance of Bt cotton (MECH-162) under Integrated Pest Management in farmers' participatory field trial in Nanded district, Central India. Current Science 86: 1628–1633.

Berwald, D., Matten, S., and Widawsky, D., 2006. Economic analysis and regulating pesticide biotechnology at the U.S. Environmental Protection Agency. In: Regulating Agricultural Biotechnology: Economics and Policy, R.E. Just, J.M. Alston and D. Zilberman, eds., Springer, New York, USA, pp. 1–15.

Brookes, G., and Barfoot, P., 2006. Global impact of biotech crops: Socio-economic and environmental effects in the first ten years of commercial use. AgBioForum 9: 139–151.

Carpenter, J., Felsot, A., Goode, T., Hammig, M., Onstad, D., and Sankula, S., 2002. Comparative environmental impacts of biotechnology-derived and traditional soybean, corn, and cotton crops. Council for Agricultural Science and Technology. Ames, IA, USA. http://www.cast-science.org.

Cartagena Protocol on Biosafety to the Convention on Biological Diversity, 2000. Secretariat of the Convention on Biological Diversity, Montreal, Canada.

Cattaneo, M.G., Yafuso, C., Schmidt, C., Huang, C., Rahman, M., Olson, C., Ellers-Kirk, C., Orr, B.J., Marsh, S.E., Antilla, L., Dutilleul, P., and Carrière, Y., 2006. Farm-scale evaluation of transgenic cotton impacts on biodiversity, pesticide use, and yield. Proceedings of the National Academy of Sciences of the USA 103: 7571–7576.

European Community (EC), 2001. Directive 2001/18/EC of the European Parliament and of the Council of 12 March 2001 on the deliberate release into the environment of genetically modified organisms and repealing Council Directive 90/220/EEC. Official Journal of the European Communities L 106: 1–39.

Fernandez-Cornejo, J., and Caswell, M., 2006. The First Decade of Genetically Engineered Crops in the United States. United States Department of Agriculture, Economic Research Service. Economic Information Bulletin Number 11, April 2006. http://www.ers.usda.gov/publishers/eib11 (accessed 26 December 2007).

Glaser, J.A., and Matten, S.R., 2003. Sustainability of insect resistance management strategies for transgenic Bt corn. Biotechnology Advances 22: 45–69.

Gonsalves, D., Manshardt, R., and Yepes, M., 1996. Petition for Determination of Regulatory Status. Transgenic Papaya Lines 55-1 and 63-1 and Their Derivatives. USDA Petition 96-051-01p. http://www.aphis.usda.gov/brs/aphisdocs/96_05101p.pdf (accessed 1 December 2007).

Gould, F., 1998. Sustainability of transgenic insecticidal cultivars: integrating pest genetics and ecology. Annual Review of Entomology 43: 701–726.

Gressel, J., Hanafi, A., Head, G., Marasas, W., Obilana, A.B., Ochanda, J., Souissi, T., and Tzotzos, G., 2004. Major heretofore intractable biotic constraints to African food security may be amenable to novel biotechnological solutions. Crop Protection 23: 661–689.

Health Canada, 1999. Voluntary Pesticide Resistance-Management Labeling Based on Target Site/Mode of Action, DIR99-06. http://www.hc-sc.gc.ca/pmra-arla/English/pdf/dir/dir9906-e.pdf (accessed 30 November 2007).

James, C., 2007. Global Status of Commercialized Biotech/GM Crops: 2007. ISAAA Brief No. 37, International Service for the Acquisition of Agri-Biotech Applications, Ithaca, NY, USA.

Johnson, K.L., Raybould, A.F., Hudson, M.D., and Poppy, G.M., 2007. How does scientific risk assessment of GM crops fit within the wider risk analysis? Trends in Plant Sciences 12: 1–5.

Kalaitzandonakes, N., Alston, J., and Bradford, K., 2006. Compliance costs for regulatory approval of new biotech crops. In: Regulating Agricultural Biotechnology: Economics and Policy, R.E. Just, J.M. Alston and D. Zilberman, eds., Springer, New York, USA, pp. 37–57.

Kalaitzandonakes, N., Alston, J., and Bradford, K., 2007. Compliance costs for regulatory approval of new biotech crops. Nature Biotechnology 25: 509–511.

Kleter, G.A., Bhula, R., Bodnaruk, K., Carazo, E., Felsot, A.S., Harris, C.A., Katayama, A., Kuiper, H.A., Racke, K.D., Rubin, B., Shevah, Y., Stephenson, G.R., Tanaka, K., Unsworth, J., Wauchope, R.D., and Wong, S.S., 2007. Altered pesticide use on transgenic crops and the associated general impact from an environmental perspective. Pest Management Science 63: 1107–1115.

MacDonald, O.C., Meakin, I., and Richardson, D.M., 2003. The role and impact of the regulator in resistance management. Proceedings of the 2003 BCPC International Congress – Crop Science and Technology 2: 703–708.

Matten, S.R., 2007. Review of Dow AgroScience's (and Pioneer HiBred's) Submission (dated 12 July 2007) Regarding Fall Armyworm Resistance to the Cry1F Protein Expressed in TC1507 Herculex® I Insect Protection Maize in Puerto Rico (EPA registrations 68467-2 and 29964-3). Memorandum from S.R. Matten, USEPA/OPP/BPPD to M. Mendelsohn, USEPA/OPP/BPPD, 24 August 2007.

Matten, S.R., and Reynolds, A.H., 2003. Current resistance management requirements for Bt cotton in the United States. Journal of New Seeds 5: 137–178.

Matten, S.R., Hellmich, R.L., and Reynolds, A.H., 2004. Current resistance management strategies for Bt corn in the United States. In: Transgenic Crop Production: Concepts and Strategies, O. Koul and G.S. Dhaliwal, eds., Oxford/IBH Publishing, New Delhi, India, pp. 261–288.

Naranjo, S., Head, G., and Dively, G., 2005. Field studies assessing arthropod nontarget effects in Bt transgenic crops: Introduction. Environmental Entomology 34: 1178–1180.

Obando-Rodriguez, A., Garcia, J., Garica, J., Magana, J., and Garcia, A., 1999. Bolligard gen. Cotton as an alternative for IPM in Delicias, Chihuahua, Mexico. In Proceedings of the

Beltwide Cotton Conference, Orlando, FL, 3–7 Jan. 1999, National Cotton Council of America, Memphis, TN, USA, pp. 967–968.
Raybould, A., 2007. Ecological versus ecotoxicological methods for assessing the environmental risks of transgenic crops. Plant Science 173: 589–602.
Romeis, J., Bartsch, D., Bigler, F., Candolfi, M.P., Gielkens, M.M.C., Hartley, S.E., Hellmich, R.L., Huesing, J.E., Jepson, P.C., Layton, R., Quemada, H., Raybould, A., Rose, R.I., Schiemann, J., Sears, M.K., Shelton, A.M., Sweet, J., Vaituzis, Z., and Wolt, J.D., 2008. Assessment of risk of insect-resistant transgenic crops to nontarget arthropods. Nature Biotechnology 26: 203–208.
Rose, R.I., 2007. White Paper on Tier-Based Testing for the Effects of Proteinaceous Insecticidal Plant-Incorporated Protectants on Non-Target Arthropods for Regulatory Risk Assessments. USDA-APHIS and US Environmental Protection Agency, Washington, DC, USA. http://www.aphis.usda.gov/brs/pdf/NTO_White_Paper_1.pdf (accessed 1 December 2007).
Sanvido, O., Widmer, F., Winzeler, M., and Bigler, F., 2005. A conceptual framework for the design of environmental post-market monitoring of genetically modified plants. Environmental Biosafety Research 4: 13–27.
Sanvido, O., Romeis, J., and Bigler, F., 2007. Ecological impacts of genetically modified crops: Ten years of field research and commercial cultivation. Advances in Biochemical Engineering and Biotechnology 107: 235–278.
Scorza, R., 2004. Application for Determination of Non-Regulatory Status for C5 ('Honeysweet') Plum (*Prunus domestica* L.) Resistant to *Plum Pox* Virus. USDA Petition 04-264-01p. http://www.aphis.usda.gov/brs/aphisdocs/04_26401p.pdf (accessed 1 December 2007).
Shelton, A.M., 2007. Considerations on the use of transgenic crops for insect control. Journal of Developmental Studies 43: 890–900.
Tabashnik, B.E., Carrière, Y., Dennehy, T.J., Morin, S., Sisterson, M., Roush, R.T., Shelton, A.M., and Zhao J.-Z., 2003. Insect resistance to transgenic Bt crops: lessons from the laboratory and field. Journal of Economic Entomology 96: 1031–1038.
Thompson, G.D., and Head, G. 2001. Implications of regulating insect resistance management. American Entomologist 47: 6–10.
Thomson, J.A., 2008. The role of biotechnology for agricultural sustainability in Africa. Philosophical Transactions of the Royal Society B 363: 905–913.
USEPA (United States Environmental Protection Agency), 2001a. Biopesticides Registration Action Document: *Bacillus thuringiensis* Plant-Incorporated Protectants (10/16/01), posted at http://www.epa.gov/oppbppd1/biopesticides/pips/bt_brad.htm (accessed 30 November 2007).
USEPA (United States Environmental Protection Agency), 2001b. Guidance for Pesticide Registrants on Pesticide Resistance Management Labeling (PR Notice 2001–5) posted at http://www.epa.gov/PR_Notices/pr2001-5.pdf (accessed 30 November 2007).
Wu, K.M., and Guo, Y.Y., 2005. The evolution of cotton pest management practices in China. Annual Review of Entomology 50: 31–52.

Chapter 3
Insecticidal Genetically Modified Crops and Insect Resistance Management (IRM)

Juan Ferré[1,*], Jeroen Van Rie[2], and Susan C. MacIntosh[3]

Abstract Economically important crops, such as maize and cotton, have been transformed with genes encoding insecticidal proteins from *Bacillus thuringiensis* (*Bt*) to confer them protection against the most important insect pests. Of the 114 million hectares globally planted with GM crops in 2007, over one third are insect-resistant *Bt* crops, and the area keeps increasing every year. The potential for insects to evolve resistance to GM insecticidal plants is considered to be one of the main threats to this technology, since resistance to *Bt* sprayable products has been demonstrated. Insect resistance management plans for this new class of pesticides are encouraged and became mandatory in the USA. Of the several strategies considered, a high dose of the insecticidal protein along with an adjacent refuge plot of non-*Bt* plants has been chosen as the most effective. Second generation *Bt* cotton combines two insecticidal proteins with unique target sites. Such "pyramided" *Bt* crops hold great promise and, in combination with the high dose/refuge strategy, will likely confer maximum protection to the *Bt* crop technology against insect resistance. So far, no case of resistance evolution to *Bt* crops has been reported.

3.1 Introduction

Crops genetically modified using genes that confer protection to insects have proven to be an effective insect control tool and are used on a significant scale in agriculture. Currently, all such commercially available GM crops express genes encoding Cry proteins from *Bacillus thuringiensis* (*Bt*). Other genes encoding insecticidal proteins such as Vip proteins from *Bt*, lectins and protease inhibitors have been evaluated in transgenic plants and crops expressing such genes may

[1] University of Valencia, Burjassot, Spain

[2] Bayer BioScience N.V., Ghent, Belgium

[3] MacIntosh & Associates, Inc., Saint Paul, MN, USA

* To whom correspondence should be addressed. E-mail: Juan.Ferre@uv.es

J. Romeis, A.M. Shelton, G.G. Kennedy (eds.), *Integration of Insect-Resistant Genetically Modified Crops within IPM Programs.*

become commercialized in the future (e.g., cotton expressing Vip3Aa [Naranjo et al., chapter 6; Malone et al., chapter 13]). Probably the largest threat to the continued success of insecticidal GM crops is the potential development of insect resistance. This paper provides information on *Bt* crops and their management in terms of preventing or delaying the development of resistant insect populations, but the general principles should be applicable to any insecticidal GM crop.

3.1.1 *B. thuringiensis* as a Source of Insecticidal Proteins

Bt is a motile, gram-positive, spore forming bacterium found widely in natural environments, such as soil, grain bins and other havens for insects. Ishiwata was the first to link *Bt* with insect disease when he observed sotto disease in *Bombyx mori* (Lepidoptera: Bombycidae, silkworm) in the early 1900s. A few years later, Berliner identified and named the insecticidal bacterium, *Bacillus thuringiensis* (Berliner, 1915). In the 1930s, the first *Bt* product was commercialized for a few years in France under the trade name of Sporeine (Lambert and Peferoen, 1992). It was not until the late 1950s that *Bt* was successfully commercialized, by Pacific Yeast Products, under the trademark of Thuricide® (Briggs, 1963). Since that time, *Bt* formulated products have become the most successful biopesticide used in agriculture, forestry and public health.

Upon sporulation, *Bt* produces proteinaceous crystalline inclusions that are the primary basis for its insecticidal activity. These crystal proteins (designated Cry proteins), contained within these inclusions, are δ-endotoxins. A multitude of *Bt* isolates have been discovered since the mid-1950s that produce more than 170 different types of Cry proteins (Crickmore et al., 1998, 2007). Despite this diversity, only a few *Bt* strains belonging to five serovars have found commercial success, namely: *Bt* var. *kurstaki*, *Bt* var. *thuringiensis* and *Bt* var. *aizawai* for Lepidoptera; *Bt* var. *israelensis* for mosquitoes and black flies (Diptera), and *Bt* var. *tenebrionis* for Coleoptera.

The narrow insect spectrum of activity has been both a blessing and a curse for *Bt* insecticides. *Bt* has a long history of safe use, demonstrating a benign environmental profile, causing no harm to non-target beneficial insects, animals or humans (Sjoblad et al., 1992). One major aspect of this safety profile is due to the short half-life of *Bt* formulated products when applied topically, primarily caused by inactivation of the Cry proteins by high temperatures, humidity, and ultraviolet light (Leong et al., 1980; Pozsgay et al., 1987). The lack of persistence has been a research focus for improved formulations, which have found marginal success (Burges and Jones, 1998). The unique mode of action of Cry proteins, which are only soluble in the conditions of the target insect midgut and bind specifically to insect midgut receptors, defines insect specificity. And since vertebrates or even non-target insects and other invertebrates, are unable to properly process the Cry proteins and do not possess midgut membrane receptors for them, Cry proteins have no impact on vertebrates or human safety.

However, the narrow spectrum and reduced persistence has limited formulated *Bt* product commercial success when compared to synthetic chemical insecticides, which typically have wider insect spectrums and display longer environmental persistence. Synthetic insecticides are generally easier to use and cheaper than biopesticides. Thus, the market for bioinsecticides is less than 5% of the total global insecticide market of roughly $8B. An improved agricultural delivery method, by introducing *Bt cry* genes into crops by genetic modification, has maintained the safety profile while at the same time eliminating the issues of short environmental persistence.

3.1.2 *Bt Crops*

One early application of the plant molecular techniques developed in the 1980s was to introduce, in plants, genes conferring new traits of agronomical importance. Pest tolerance has always been one of the challenges of plant breeders. For this reason, *Bt* genes coding for insecticidal proteins have been transferred to agronomically relevant crops to confer them protection to their most important insect pests (Shelton et al., 2002). These genetically modified (GM) crops are known as *Bt* crops and constitute the most extensively planted GM crops after those transformed for herbicide tolerance. In 2007, the global area of GM crops planted for commercial purposes was 114 million hectares, of which 20.3 million hectares were planted to *Bt* crops and 21.8 million hectares to crops combining herbicide tolerance and insect resistance (James, 2007). The benefits of planting *Bt* crops instead of conventional varieties are widely accepted, which is reflected by their rapid rate of adoption (Qaim et al., chapter 12).

Maize, cotton, potato and rice have been transformed with *cry* genes coding for proteins highly active against the most important pests. Thus, *Bt* maize has been transformed with either *cry1Ab*, *cry1Ac* or *cry9C* to protect it against *Ostrinia nubilalis* (Lepidoptera: Crambidae, European corn borer) and *Sesamia nonagriodes* (Lepidoptera: Noctuidae, Mediterranean corn borer), with *cry1F* to protect it against *Spodoptera frugiperda* (Lepidoptera: Noctuidae), and with *cry3Bb*, *cry34Ab* and *cry35Ab* (producing a binary toxin) to protect it against the rootworms of the genus *Diabrotica* (Coleoptera: Chrysomelidae). By far, most of the *Bt* maize currently planted is *cry1Ab* maize. Field trials with *Bt* crops expressing other genes are under way, including maize expressing a variant *cry3A* gene for the control of rootworms. Combinations of the above genes are at the field trial stage or very close to commercialization (USEPA, 2006).

Most *Bt* cotton planted commercially contains *cry1Ac* or a fusion gene of *cry1Ac* and *cry1Ab*. Cry1Ac is highly active against the lepidopterans that feed on the cotton bolls: *Heliothis virescens* (Lepidoptera: Noctuidae, tobacco budworm), *Pectinophora gossypiella* (Lepidoptera: Gelechiidae, pink bollworm), and reasonably effective against *Helicoverpa armigera* (Lepidoptera: Noctuidae, Old World bollworm) and *Helicoverpa zea* (Lepidoptera: Noctuidae, cotton bollworm). In China, the cowpea trypsin inhibitor gene (*CpTi*) was combined with a *cry1Ac* gene to produce a pyramided gene product commercialized in the early 2000s. The *vip3A* gene (a *Bt* gene producing a secretable insecticidal protein) has also been introduced in cotton and should be commercialized soon; it also confers protection against the above pests. The *cry1Ac* and *cry2Ab* genes have been combined in the same plant (Bollgard II), giving rise to second generation *Bt* cotton, which is extensively planted in Australia and also adopted in the USA. Another combination of *Bt* genes that has been recently commercialized is the one combining the *cry1Ac* and *cry1F* genes, conferring additional protection against *Spodoptera* spp. The reason to combine two Cry proteins is not only to broaden the spectrum of protection but also for resistant management purposes, as will be discussed in detail in section 3.3.2.

Bt potatoes protected against *Leptinotarsa decemlineata* (Coleoptera: Chrysomelidae, Colorado potato beetle) have also been planted commercially in North America and Europe, and contained the *cry3Aa* gene. However, their commercialization was stopped in 2001 due to many issues, including marketing (Grafius and Douches, chapter 7).

A *Bt* rice product is under development that contains a fusion gene of *cry1Ac* and *cry1Ab* to confer protection against the rice stem borers, mainly *Scirpophaga incertulas* (Lepidoptera: Pyralidae, yellow stem borer) and *Chilo suppressalis* (Lepidoptera: Pyralidae, striped stem borer), major stem borer pests widely distributed in Asia (Cohen et al., chapter 8). Two other *Bt* crops are under development: eggplant and cruciferous vegetables (Shelton et al., chapter 9). *Bt* eggplant is targeted for control of *Leucinodes orbonalis* (Lepidoptera: Pyralidae, eggplant fruit and shoot borer) and *Bt* crucifer vegetables are targeted against *Plutella xylostella* (Lepidoptera: Plutellidae, diamondback moth).

3.2 Evaluation of the Potential of Insects to Develop Resistance to *Bt* Insecticidal Proteins

3.2.1 Early Evidences of the Potential of Insects to Evolve Resistance to B. thuringiensis

Early attempts to develop *Bt*-resistant insect colonies by laboratory selection were generally unsuccessful, as summarized by both Briese (1981) and Georghiou (1988). One stored grain pest, *Ephestia cautella* (Lepidoptera: Pyralidae, almond moth), developed only a marginal level of resistance (7.5-fold) after 21 generations of selection pressure (McGaughey and Beeman, 1988) with *Bt* var. *kurstaki*. Another low level of resistance development (twofold) to *Bt* var. *kurstaki* was observed when *Homeosoma electellum* (Lepidoptera: Pyralidae, sunflower moth) was selected for 12 generations (Brewer, 1991).

In 1979, Kinsinger and McGaughey first documented wide variations of susceptibility to Dipel® in *Plodia interpunctella* (Lepidoptera: Pyralidae, Indian meal moth), collected from grain bins, although none of the grain bins had been previously treated with *Bt* formulations (Dipel is a formulated *Bt* var. *kurstaki* biopesticide containing Cry1Aa, Cry1Ab, Cry1Ac and Cry2A proteins). Susceptibility to

Dipel ranged up to 42-fold when comparing grain bin collected populations to a laboratory reared colony never exposed to *Bt*. A larger survey of grain bins (McGaughey, 1985), both treated with *Bt* and untreated, confirmed the high natural variability found in the earlier study (Kinsinger and McGaughey, 1979), and extended the finding by laboratory selection experiments. McGaughey (1985) was successful in developing a 100-fold level of resistance to Dipel when he subjected *P. interpunctella* populations to only 15 generations of selection in the laboratory. Further studies demonstrated varying levels and rates of resistance development (15-fold to 250-fold), with five different *P. interpunctella* colonies that were selected for 40 generations (McGaughey and Beeman, 1988).

The first incidence of laboratory resistance of a field crop pest, *H. virescens*, was reported by Stone et al. (1989). After only three generations of selection on Cry1Ab, susceptibility was reduced by threefold. Selection was continued using either a single *Bt* protein (Cry1Ab) or Dipel, which pushed the level of resistance to greater than 70-fold for the purified Cry1Ab protein and 57-fold for Dipel after 22 generations of selection pressure.

For years many concluded that resistance to *Bt*-based biopesticides was unlikely since no reports of field resistance were documented despite more than 20 years of use. Laboratory selection experiments, while showing the potential for resistance, represented a unique set of circumstances that rarely mimic field conditions, especially the small number of insects in the selected colonies with limited genetic diversity. The complex mode of action of *Bt* involving multiple toxins and multiple target sites was thought to be the basis for a lack of resistance development (Whalon and McGaughey, 1998). However, during the grain bin survey, McGaughey (1985) identified a small but significant reduction in *Bt* susceptibility in *P. interpunctella* populations collected from *Bt*-treated grain bins as compared to untreated populations. Perhaps the development of resistance should not have been surprising since grain bins create a unique closed environment that keeps insects in close proximity to the biopesticide.

The first evidence of resistance to *Bt* in the open field occurred in the Philippines with the diamondback moth, *P. xylostella* (Kirsch and Schmutterer, 1988) followed by cases in Hawaii (Tabashnik et al., 1990), Thailand (Zoebelein, 1990), the Philippines (Ferré et al., 1991), Korea (Song, 1991), and Japan (Tanaka and Kimura, 1991). Since then many other cases of field development of resistance in this insect species have been reported (Ferré and Van Rie, 2002). Most of these populations arose in tropical locations where *P. xylostella* can produce up to 25 generations per year and where the fields were heavily sprayed with *Bt*-based insecticides (Ferré et al., 1991; Tabashnik, 1994; Tang et al., 1996).

3.2.2 Evolution of Resistance in the Field vs. in the Laboratory

Selection of insect populations under laboratory conditions has shown that, provided enough initial variability, any insect species can evolve resistance to *Bt* formulated products and/or their Cry proteins (Ferré and Van Rie, 2002). Laboratory selection has shown that the main pests targeted by current *Bt* crops can develop resistance to Cry proteins. Laboratory resistant populations have been obtained for *O. nubilalis*, *H. virescens*, *H. armigera*, *P. gossypiella* (Liu et al., 1999, 2001; Tabashnik et al., 2002), and *L. decemlineata* (Whalon et al., 1993), among others (Table 3.1).

In contrast to the laboratory situation, field resistance to *Bt* commercial formulations has occurred in only one insect species, *P. xylostella*, although this has occurred several times independently, mainly in South East Asia, Hawaii, and Southeastern USA (Ferré and Van Rie, 2002). The prolificacy of this insect species, coupled with intensive use of a single pesticide over a period of several years, was the basis for development of resistance in all cases. For example, in the case of the initial observation of *Bt*-resistant *P. xylostella* in a watercress field in Hawaii, the watercress was treated with *Bt* sprays 50–100 times from 1978–1982 (Tabashnik et al., 1990). When considering that populations of *P. xylostella* are generally confined to discrete areas, sometimes even specific isolated fields, this development of resistant insects can in no way be construed as a widespread product failure for *Bt* pesticides. More recently, *Trichoplusia ni* (Lepidoptera: Noctuidae, cabbage looper) has evolved resistance to *Bt* insecticides in Canadian greenhouses where *Bt* sprays were routinely used (Janmaat and Myers, 2003).

Despite the cases of evolution of resistance to *Bt* formulations or its purified Cry proteins, no case of resistance to *Bt* crops has ever been reported. However, the cases of laboratory selection experiments and field resistance development indicate that the risk of resistance exists, especially with individual Cry proteins.

3.2.3 Factors That Influence the Evolution of Resistance to Bt Sprays vs. Transgenic Plants

There are two major differences between *Bt* sprays and *Bt* crops in relation to the selection pressure they pose on insect populations. One is that *Bt* sprays almost invariably contain a combination of insecticidal proteins because *Bt* strains normally carry several insecticidal protein genes (Iriarte et al., 1998). Furthermore, the presence of the spore has been shown to act synergistically against some pests (Moar et al., 1995; Dubois and Dean, 1995; Tang et al., 1996). In contrast, currently marketed *Bt* crops express one insecticidal protein gene, or two at the most. The second major difference is that persistence of *Bt* spores and crystals once sprayed is relatively short, depending on the environmental conditions (insecticidal crystals are washed off by rain and their proteins are gradually inactivated by UV radiation and/or degraded by phylloplane microorganisms). In contrast, the current *Bt* crops express the *cry* genes constitutively, and their protein products are stable within the plant cell environment.

The above two differences, between *Bt* sprays and *Bt* crops, can result in a differential speed of the development of resistant insect populations between these two insect control approaches. The combination of more than one insecticidal protein in a *Bt* product diminishes the chances of finding a resistant individual. In fact,

Table 3.1 Laboratory-selected Lepidoptera species and strains, of importance to currently deployed *Bt* crops, which have developed resistance to *Bacillus thuringiensis* insecticidal proteins

Species (family)	Origin	Selecting agent[a] (form[b])	Strain name	Resistance and cross-resistance levels			Genetics of resistance (dominance levels estimated as D_{LC})[e]	Reference
				ES[c]	Toxin/ Formulation[a]	RR[d]		
Heliothis virescens (Noctuidae)	North Carolina (USA)	Cry1Ab (CC)/ Dipel (FSC)	SEL	14	Cry1Ab protoxin	20	–	Stone et al., 1989
				22	Cry1Ab protoxin	69	Polygenic, PD (0.71)	Sims and Stone, 1991
	North Carolina (USA)	Cry1Ac (T)	CP73-3	17	Cry1Ab	13	–	Gould et al., 1992
				17	Cry1Ac	50	PR (0.31)	
	North Carolina (USA)	Cry1Ac (T/CC)	YHD2	19	Cry1Ac	>10,000	–	Gould et al., 1995
				–	Cry1Ab	>2,300	Monogenic, PR (0.24)	
				–	Cry2Aa	25	PD (0.80)	
	North Carolina (USA)	Cry1Ac (T)	KCB	–	Cry1Ac	400	–	Forcada et al., 1999
Helicoverpa armigera (Noctuidae)	Various locations (India)	Cry1Ac (IB)	–	10	Cry1Ac protoxin	76	–	Kranthi et al., 2000
	Gujarat (India)	Cry1Ac (*Bt* cotton/CC)	Res-Bt	15	Cry1Ac protoxin	93	Monogenic, SD (0.42)	Kranthi et al., 2006
	Maharastra (India)	Cry1Ac (CC)	Res-AC	14	Cry1Ac protoxin	205	Monogenic, SD (0.56)	Kranthi et al., 2006
	Various locations (Australia)	Cry1Ac (SC)	BX	21	Cry1Ac protoxin	321	PR	Akhurst et al., 2003

(continued)

Table 3.1 (continued)

Species (family)	Origin	Selecting agent[a] (form[b])	Strain name	Resistance and cross-resistance levels: ES[c]	Toxin/ Formulation[a]	RR[d]	Genetics of resistance (dominance levels estimated as D_{LC})[e]	Reference
	New South Wales (Australia)	Cry2Ab (SC)	SP15	–	Cry2Ab protoxin	6,830	Monogenic, CR	Mahon et al., 2007
	China	Cry1Ac (*Bt* cotton)	–	16	Cry1Ac	43*	Monogenic, PD (0.64)[f]	Liang et al., 2000
	Henan (China)	Cry1Ac (*Bt* cotton)	–	42	Cry1Ac protoxin	1,680	–	Meng et al., 2003
	Hebei (China)	Cry1Ac (T)	GYBT	28	Cry1Ac	564	PR (0.24)	Xu et al., 2005
	Henan (China)	Cry1Ac (CC)	–	52	Cry1Ac	425*	–	Luo et al., 2006
Helicoverpa zea (Noctuidae)	Mississippi and Texas (USA)	Cry1Ac (PT)	FZ	7	Cry1Ac	119	–	Luttrell et al., 1999
Pectinophora gossypiella (Gelechiidae)	Various locations (USA)	Cry1Ac (*Bt* cotton/CC)	APHIS-98R	–	Cry1Ac protoxin	>100	SD to CR depending on the toxin conc. CR on *Bt* cotton	Liu et al., 1999, 2001
	Arizona (USA)	Cry1Ac (CC)	AZP-R	28	Cry1Ac protoxin	3,100	Monogenic, PR (0.20)	Tabashnik et al., 2002
Ostrinia nubilalis (Crambidae)	Kansas (USA)	Dipel (FSC)	KS-SC-R	7	Dipel	65	Monogenic, PD (0.86)	Huang et al., 1999
	Minnesota (USA)	Cry1Ac (CC)	S-I	17	Cry1Ac protoxin	162 (FR)[g]	–	Bolin et al., 1999
	Minnesota (USA)	Cry1Ac (CC)	S-II	14	Cry1Ac protoxin	58 (FR)	–	Bolin et al., 1999
	Minnesota (USA)	Cry1Ac (CC)/ Cry1Ab (*Bt* maize)	S-IV	40	Cry1Ac protoxin	8.4	–	Bolin et al., 1999
	Nebraska (USA)	Cry1Ab (SC)	N	7	Cry1Ab protoxin	14* (FR)	–	Chaufaux et al., 2001

	France and Switzerland	Cry1Ab (C)	LAS	9	Cry1Ab protoxin	32 (FR)	–	Chaufaux et al., 2001
	Italy	Cry1Ab (SC)	Europe-R[h]	9	Cry1Ab protoxin	13* (FR)	–	Chaufaux et al., 2001
				95	Cry1Ab	2,000	Polygenic, SD (0.44)	Alves et al., 2006
	Nebraska (USA) and Italy	Cry1Ab (SC)	RSTT-R	41	Cry1Ab protoxin	9	–	Siqueira et al., 2004
				56	Cry1Ab	1,300	Polygenic, SD (0.60)	Alves et al., 2006
	Spain	Cry1Ab (SC)	R-strain	8	Cry1Ab protoxin	10	–	Farinós et al., 2004
Sesamia nonagrioides (Noctuidae)	Spain	Cry1Ab (SC)	R-strain	8	Cry1Ab protoxin	21	–	Farinós et al., 2004
Diatraea saccharalis (Crambidae)	Louisiana (USA)	Cry1Ab (PM)	Isoline 52	–	Cry1Ab	Completed larval devel. on *Bt* maize	Monogenic, CR	Huang et al., 2007

[a] Dipel is a tradename for a commercial formulation of *B. thuringiensis* var. *kurstaki*.

[b] Different forms of selecting agent have been used: formulated spore-crystal preparations (FSC), spore-crystal preparations (SC), microencapsulated recombinant *P. fluorescens* cells expressing a *cry* gene (CC), parasporal crystals (C), inclusion bodies from recombinant *E. coli* cells expressing a *cry* gene (IB), protoxin (PT), activated toxin (T), and transformed plant material expressing a *cry* gene (PM).

[c] When available, the number of episodes of selection after which the insects were tested, is given.

[d] RR = resistance ratio. For *Bt* formulations or (pro)toxins this is defined as the LC_{50} (or LD_{50}) of resistant strain divided by the LC_{50} (or LD_{50}) of susceptible control strain; when the value is followed by '*' it refers instead to EC_{50} values (the concentration responsible for 50% growth inhibition). All values for Cry proteins refer to activated toxins unless otherwise indicated.

[e] In all reported cases resistance was autosomal and its type of inheritance: completely recessive (CR), partially recessive (PR), semi-dominant (SD), or partially dominant (PD). Dominance levels have been calculated from LC_{50} values (D_{LC}, dominance of insecticide resistance) (Bourguet et al., 2000) as $D_{LC} = (\log LC_{RS} - \log LC_{SS})/(\log LC_{RR} - \log LC_{SS})$. The range for D_{LC} is 0 (complete recessivity) to 1 (complete dominance). When LC_{50} values were available from the two reciprocal crosses (RR females × SS males and SS females × RR males), the mean value was used to calculate D_{LC}.

[f] The authors reported partial recessive inheritance, but according to the LC_{50} values, the calculated D_{LC} is 0.64, which corresponds to partial dominant inheritance.

[g] FR indicates that resistance levels fluctuated considerably from generation to generation, and the value given is the maximum value obtained.

[h] *O. nubilalis* Europe-R strain was originally referred to as strain I.

examples exist of insect populations becoming resistant to a single Cry protein but being still fully susceptible to *Bt* formulations containing the same Cry protein along with other Cry proteins (Ferré et al., 1991; Moar et al., 1995). However, it has been shown that insects can become resistant simultaneously to several Cry proteins if an alteration of their interaction with a shared target site occurs (Lee et al., 1995; Tabashnik et al., 1997; Wang et al., 2007). In practice, it is not rare to find that, in *Bt* products, the Cry proteins most toxic to a specific insect pest share the same target site (Ballester et al., 1999; Estela et al., 2004; Ibargutxi et al., 2006). In this case, the *Bt* product will act almost as if a single insecticidal protein was produced and evolution of resistance would become much more likely to occur.

The second difference resides in the persistence of the insecticidal proteins and, thus, relates to the chance an insect has of ingesting a sublethal dose of the protein. Assuming that the dose of Cry protein sprayed on a plant or produced by a GM plant is enough to kill the insects feeding on them, the chance of an insect to escape from death is much higher in the sprayed field than in the GM field. One reason is that the *Bt* crop is supposed to express the insecticidal protein at a consistent (high) level throughout the growing season. However, not only persistence of the sprayed product is responsible for the escapes in the *Bt* sprayed field, but also the fact that, in practice, it is almost impossible to cover uniformly the whole plant surface. The effect of this on the evolution of resistance is double: first, the insects which have not ingested a lethal dose of the bioinsecticide will have the same effect as the "refuge" in *Bt* fields in terms of "diluting" the resistance alleles. Second, insects carrying resistance genes not conferring total resistance to a high dose of bioinsecticide may be exposed to sublethal doses and survive; this, in the long run, will give rise to the combination of different resistance genes leading to higher levels of resistance. While the first phenomenon may result in slower development of resistance to *Bt* sprays than to *Bt* crops, the second phenomenon will likely have the opposite effect.

Due to these concerns, *Bt* crops were introduced with strict resistance management plans, which are lacking for *Bt* formulated products. Should resistance develop to *Bt* crops, even with these management safeguards, there should only be limited impact to *Bt* formulated products, since the markets are almost completely segregated. The largest markets for *Bt* formulated products are in fruits, vegetables and forestry, none of which currently have crop species that express *Bt* proteins. Alternatively, maize, cotton, potatoes and rice, which have been transformed with *Bt* genes, are rarely sprayed with *Bt* formulations. Actually, *Bt* sweet corn is grown in the US, although it is <5% of the total sweet corn market and <1% of the total corn market.

3.2.4 Estimation of the Frequency of Resistance Alleles in Field Populations

One key element for estimating the rate of evolution of resistance in a population exposed to an insecticide is the initial frequency of resistance alleles. However, the problem of estimating this is not trivial, since most resistance alleles are recessive and the frequency of such alleles is very low before resistance becomes evident. A population genetics law tells us that, in an ideal scenario, the frequency of recessive resistance alleles (q) is the square root of the frequency of resistant individuals (q^2). The practical problem for the application of this law for the estimation of initial frequencies of resistant alleles is that, in populations not previously exposed to insecticides, the frequency of homozygous resistant insects may be so extremely low that in practice we may be unable to detect them. A study which applied this direct approach on field populations of *P. gossypiella* from *Bt* cotton fields in Arizona obtained an estimate of the average frequency of resistance alleles of 0.16 from samples collected in 1997, an unexpectedly high frequency for such type of alleles. While this high level cannot be readily explained, the same study reported frequencies of $<8 \times 10^{-4}$ in samples collected in 1998 and 1999 from the same cotton fields (Tabashnik et al., 2000b). Follow up studies of these populations have confirmed that the resistance frequencies of resistance alleles must have been originally very low since evolution of resistance has not built up yet despite the continuous planting of *Bt* cotton (Tabashnik et al., 2005b).

An indirect estimate of the frequency of such alleles can be obtained from the laboratory selection experiments that have succeeded in obtaining resistance. In these cases at least one copy of a resistance allele had to be present at the start of selection (unless it appeared by mutation during selection and this is considered a very unlikely event). Given that most successful selection experiments started with 100 to 700 insects from the field (McGaughey and Johnson, 1992; Gould et al., 1992, 1995), the frequencies of resistance alleles in the original populations must have been around 1 to 5×10^{-3}. However, we have to be cautious with the estimates obtained with this approach, since the values can be overestimated if we do not consider other selection attempts in these same populations leading to unsuccessful results (Gould et al., 1995) or if the populations had been previously exposed to *Bt*, either as a natural infestation or to inadvertent bioinsecticide treatments (McGaughey and Johnson, 1992; Estada and Ferré, 1994).

The frequency of a major *Bt*-resistance allele in field populations of *H. virescens* has been estimated making use of a resistant strain almost completely homozygous for a recessive resistance allele (Gould et al., 1997). Field collected males (over 2,000) were individually mated to females of the resistant strain and the F_1 and F_2 offspring from over 1,000 successful single pair matings was tested for resistance. The estimated frequency of resistance alleles in the field sample was of 1.5×10^{-3}, in close agreement with a preliminary estimate obtained from a selection experiment (Gould et al., 1995). This direct approach has the main disadvantage of being only applicable to recessive alleles of the locus for which the laboratory strain is homozygous for resistance, since recessive alleles at any other locus escape detection.

A different approach is based on testing for resistance the F_2 progeny of insects collected in the field. Since most recessive alleles are carried in heterozygosis, the F_2 progeny allows the detection of the recessive allele in homozygosis. The F_2 screening method is far more sensitive (more than ten times) than a discriminating-dose assay for detection of recessive traits and it does not require obtaining previously a resistant laboratory strain (Andow and Alstad, 1998). This method has been applied to several populations of different insect species. In field populations of *O. nubilalis*, no resistant homozygotes were found for major resistance genes, and the estimated frequency of Cry1Ab-resistance alleles was <0.013 for a Minnesota population (Andow et al., 1998), <0.0039 for an Iowa population (Andow et al., 2000), and a global frequency of <0.0044 for populations from Texas and Kansas (Stodola et al., 2006). In European populations (from Spain and Greece) of the Mediterranean corn borer, *S. nonagrioides*, no major Cry1Ab resistance allele was found either, and the frequency was estimated as <0.0097 (Andreadis et al., 2007). In a study with populations from Australia, this method detected a frequency of 4×10^{-3} for low level resistance alleles but $< 10^{-3}$ for high level resistance in *P. xylostella*, and $<7 \times 10^{-4}$ for any type of resistance alleles in *H. armigera* (Ahmad and Roush, 1999).

Recently, because of the increased knowledge on the genetic basis of resistance, DNA-based screening has been applied to estimate the resistance allele frequency in field populations. In contrast with previous estimates, lower frequencies have been obtained using the molecular approach, with values of <0.0003 in *P. gossypiella* (Tabashnik et al., 2006) and 7×10^{-5} in *H. virescens* (Gahan et al., 2007). The reason for this may be that the molecular approach detects solely the frequency of the gene tested, whereas the estimates by other methods may be influenced by other resistance genes.

In contrast to recessive alleles, non-recessive alleles (dominant or partially dominant) can be detected in both homozygous and heterozygous individuals, and this results in a lower, and more practically feasible, number of insects to be tested in order to have a realistic probability of detecting resistance. A toxin challenge test is sufficient to determine whether the progeny of field collected females carry resistance alleles or not. *H. zea* populations from North Carolina were estimated to carry non-recessive alleles for major resistance genes for Cry1Ac at a frequency of 0.00043 and for Cry2Aa at a frequency of 0.00039 (Burd et al., 2003).

3.2.5 Mode of Action of Cry Proteins

Cytological studies have demonstrated that Cry proteins act by destroying the midgut epithelium, leading to starvation, paralysis, septicemia and death. Mechanism of action studies have focused on Cry1A proteins and Lepidoptera and have revealed the major steps between the ingestion of crystal proteins and the disruption of the gut. In general terms, crystal proteins are dissolved in the highly alkaline lepidopteran midgut upon ingestion and are proteolytically activated to a trypsin resistant core fragment of about 60 kDa. This protein passes through the pores in the peritrophic membrane, binds to a membrane protein (complex) in the brush border of the midgut epithelial cells and inserts into the membrane, resulting in the formation of pores (Peyronnet et al., 2004; Vachon et al., 2006). Ultimately, the midgut cells swell and lyse.

Two aspects of the mode of action have been studied in considerable detail: interaction of the Cry proteins with binding sites on the midgut membrane, and the process of membrane insertion.

As a soluble protein, a Cry protein must clearly undergo a structural change in order to insert into the membrane. In the case of the interaction of Cry1A proteins with *Manduca sexta* (Lepidoptera: Sphingidae, tobacco hornworm) membranes, the following model for the pathway of toxic action has been proposed (Bravo et al., 2005). A first structural change in the Cry protein would be the cleavage of alpha-helix 1 following binding to a cadherin-like protein. This is thought to expose hydrophobic regions in the monomeric protein, resulting in the oligomerization into a – probably tetrameric (Vié et al., 2001; Gómez et al., 2002b; Puntheeranurak et al., 2005) – pre-pore complex (Gómez et al., 2002b). The highly alkaline pH in the midgut appears to result in increased toxin flexibility and thereby contribute to these structural changes (Rausell et al., 2004b). This pre-pore oligomeric complex gains affinity for glycosylphosphatidyl-inositol (GPI)-anchored aminopeptidase-N and the interaction with this receptor molecule drives the complex into membrane microdomains or lipid rafts where it undergoes further conformational changes and is converted into a membrane inserted pore (Zhuang et al., 2002; Bravo et al., 2004; Rausell et al., 2004a, b; Pardo-López et al., 2006). GPI-anchored proteins preferentially partition into these microdomains (Sangiorgio et al., 2004; Rajendran and Simons, 2005). This model has recently gained support from the engineering of Cry1A proteins able to form the pre-pore complex without the need to bind to the cadherin-like receptor. These toxins were active against insects in which the *cadherin*-like gene had been silenced or with mutations in this gene which prevented toxin binding (Soberón et al., 2007). Whether the model of sequential binding and its components proposed for *M. sexta* (i.e. first to cadherin and then to GPI-anchored aminopeptidase-N) apply as a general phenomenon in Cry proteins mode of action in insects remains to be seen. For example, it has been suggested that GPI-anchored alkaline phosphatase rather than aminopeptidase-N may play the role of the second interaction site in *H. virescens* (Jurat-Fuentes and Adang, 2006a).

The functional role of aminopeptidase-N in Cry protein toxicity has been indicated by RNAi experiments in *Spodoptera litura* (Lepidoptera: Noctuidae, cluster caterpillar) (Rajagopal et al., 2002) and *H. armigera* (Sivakumar et al., 2007), and by ectopic expression in *Drosophila melanogaster* (Diptera: Drosophilidae) larvae (Gill and Ellar, 2002). The significance of cadherin-like proteins as receptors has been demonstrated by ectopic expression in different cell lines (Nagamatsu et al., 1999; Tsuda et al., 2003; Hua et al., 2004b; Zhang et al., 2005; Flannagan et al., 2005; Jurat-Fuentes and Adang, 2006b) and is further corroborated by the association between a mutant form of a *cadherin* gene and resistance to *Bt* Cry proteins in resistant *H. virescens* (Gahan et al., 2001; Jurat-Fuentes et al., 2004), *H. armigera* (Xu et al., 2005; Yang et al., 2006) and *P. gossypiella* (Morin et al., 2003; Tabashnik et al., 2004; Tabashnik et al., 2005a). Using specific antibodies, it was recently demonstrated that both types of receptors (aminopeptidase-N and cadherin) are involved in *in vivo* toxicity to *M. sexta* (Gómez et al., 2006). In addition, glycolipids

(Griffitts et al., 2005) and glycosylated GPI-anchored alkaline phosphatase (McNall and Adang, 2003; Jurat-Fuentes and Adang, 2004; Fernández et al., 2006) have been implicated in Cry protein binding (Pigott and Ellar, 2007). Several studies have mapped binding epitopes of Cry1A proteins for their receptors (Gómez et al., 2001; Gómez et al., 2002a; Nakanishi et al., 2002; Gómez et al., 2003; Hua et al., 2004a; Xie et al., 2005; Atsumi et al., 2005).

The process of membrane insertion is frequently described by the umbrella model (Knowles, 1994; Gazit et al., 1998). According to this model, alpha-helices 4 and 5 insert into the membrane as the handle of an umbrella, while the other helices are rearranged on the membrane surface. An alternative model, the penknife model (Hodgman and Ellar, 1990), proposes that alpha-helices 5 and 6 flip out of domain I and insert into the membrane as a helical hairpin. Some recent studies however suggest that the whole protein partitions into the membrane, rather than individual hairpins as in the above models (Alzate et al., 2006; Tomimoto et al., 2006).

The pores are presumed to be the direct cause of cell death due to osmotic lysis. However, in insect cells expressing a Cry1Ab cadherin receptor, cytolytic activity due to protein monomers was observed. In this experimental system, the oligomeric form of Cry1Ab was not involved in the cytotoxic pathway (Zhang et al., 2005). Subsequent studies demonstrated that Cry1Ab induces an adenylyl cyclase/protein kinase A cell death pathway, resulting in cell swelling and lysis (Zhang et al., 2006). The relative importance of formation of lytic pores versus induction of a cell death pathway as a toxicity mechanism may differ for different insect species – Cry protein combinations or different experimental assay systems. Indeed, the two proposed mechanisms may not necessarily be mutually exclusive. In this context it could be hypothesized that lipid rafts, which are characterized by a high content of GPI-anchored proteins and have been suggested to serve as a platform for various signaling complexes (Sangiorgio et al., 2004; Rajendran and Simons, 2005) may be a common component of both toxicity mechanisms.

3.2.6 Mechanisms of Resistance to Cry Proteins

Insects could, in principle, become resistant to Cry proteins due to mutations in genes encoding proteins involved in any of the different steps in the mode of action (Heckel et al., 2007). Several mechanisms have been observed in laboratory selected insect strains (Ferré and Van Rie, 2002), such as altered binding to midgut receptors (see references in Ferré and Van Rie, 2002), altered protoxin activation (Forcada et al., 1996; Oppert et al., 1997; Li et al., 2004; Karumbaiah et al., 2007), toxin degradation (Forcada et al., 1996), more efficient repair (or replacement) of damaged midgut cells (Forcada et al., 1999; Martínez-Ramírez et al., 1999), esterase sequestration (Gunning et al., 2005) and elevated immune status (Ma et al., 2005). Altered regulation of phosphatases involved in intracellular signaling pathways could be yet another mechanism in insects to avoid the toxic effects of Cry proteins (Jurat-Fuentes and Adang, 2006a). Similarly, modifications in mitogen-activated protein kinase (MPAK) pathways involved in cell defense in *Caenorhabditis elegans* against Cry5B (Huffman et al., 2004) may result in reduced susceptibility to nematicidal Cry proteins. Genetic linkage between resistance and a biochemical modification has only been demonstrated in several cases of altered receptor binding and in one case of altered protoxin activation (Oppert et al., 1997). In contrast to the variety of mechanisms observed in laboratory selected insect strains, only one major mechanism, i.e. altered binding, has so far been detected in insect species that have developed resistance in the field (*P. xylostella*) (Ferré and Van Rie, 2002) or in greenhouses (*T. ni*) (Wang et al., 2007).

3.2.7 Binding Site Competition Studies and Their Use at Predicting Cross-Resistance

The first study on the molecular basis of Cry protein resistance in insects involved a *P. interpunctella* strain resistant to Cry1Ab, but not to Cry1Ca. Whereas a dramatic reduction in binding affinity of Cry1Ab to brush border membrane vesicles (BBMV) from resistant insects was observed, high affinity binding of Cry1Ca was demonstrated to BBMV from both the susceptible and resistant insects (Van Rie et al., 1990). Importantly, these findings showed that resistance to one Cry protein did not necessarily imply general cross-resistance to Cry proteins. Subsequent binding studies using BBMV from field-selected populations of *P. xylostella*, have shown that the pattern of cross resistance generally parallels the pattern of binding specificity of the Cry proteins tested. For example, a *P. xylostella* strain collected from fields in Hawaii which had been treated with a sprayable *Bt* product (Dipel) and further selected in the lab, had high levels of (cross-) resistance to Cry1Aa, Cry1Ab, Cry1Ac, Cry1Fa and Cry1Ja but no significant (cross-) resistance to Cry1Ba, Cry1Bb, Cry1Ca, Cry1Da, Cry1Ia or Cry2Aa (Tabashnik et al., 1996). Binding of Cry1Ab and Cry1Ac, but not Cry1Ca, was strongly reduced in this resistant strain (Tabashnik et al., 1997). A very similar pattern of resistance and binding characteristics was observed in *P. xylostella* strains from Pennsylvania (Tabashnik et al., 1997), the Philippines (Ferré et al., 1991) and Florida (Tang et al., 1996). Indeed, while there was a complete lack of Cry1Ab binding in the *P. xylostella* strain from Florida, binding of Cry1B and Cry1C was unaltered (Tang et al., 1996). These cross-resistance and binding data in *P. xylostella* can be understood in view of the model for the Cry binding sites in this species (see Fig. 3.1): according to this model, one site (site 1) is recognized only by Cry1Aa; another (site 2) is shared among Cry1Aa, Cry1Ab, Cry1Ac, Cry1F and Cry1J, and two additional sites bind Cry1Ba (site 3) and Cry1Ca (site 4) (Ferré and Van Rie, 2002). Site 1 appears to be a non-functional binding site.

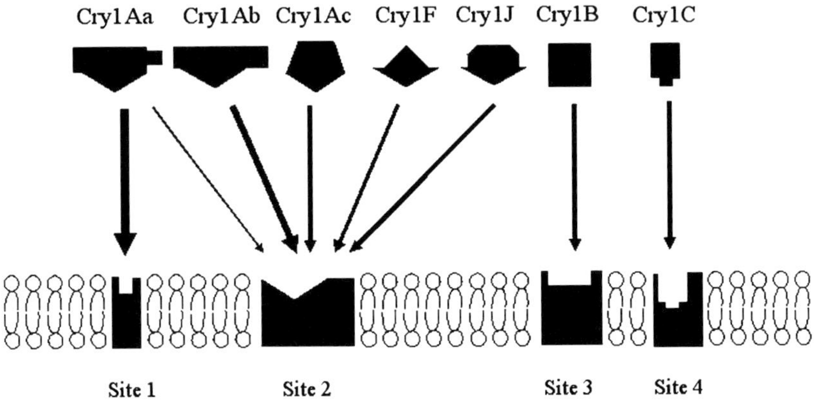

Fig. 3.1 Proposed model for the Cry protein binding sites in the midgut brush border membrane of the diamondback moth, *Plutella xylostella*

Resistant *P. xylostella* strains, selected using Cry1A-containing *Bt* products, appear to have an altered site 2, explaining their cross-resistance to Cry1F and Cry1J, while their site 3 and site 4 have remained unaltered, explaining their full susceptibility to these two Cry proteins.

Two *P. gossypiella* strains, resistant to Cry1Ac, displayed cross-resistance to Cry1Aa, Cry1Ab, but not to Cry1Ca, Cry1Da, Cry2Aa and Cry9Ca (Tabashnik et al., 2000a). Binding studies in one of these resistant strains demonstrated that Cry1Ab binding was greatly reduced. Cry1Aa and Cry1Ac bind to the same site as Cry1Ab, whereas Cry1Ca, Cry1Da, Cry2Aa or Cry9Ca do not compete for this site (González-Cabrera et al., 2003). Thus, also in the case of *P. gossypiella*, there is a correspondence between binding specificity and cross-resistance. Remarkably however, Cry1Ac binding was unaffected in the resistant strain, possibly indicating that post-binding events are disrupted, rather than binding itself as in the case of Cry1Ab.

In *H. virescens*, three populations of Cry binding sites have been detected: the A binding site, with the cadherin-like binding protein as a key component (Jurat-Fuentes et al., 2004), is recognized by Cry1Aa, Cry1Ab, Cry1Ac, Cry1F and Cry1J; the B binding site is recognized by Cry1Ab and Cry1Ac, and the C binding site is recognized only by Cry1Ac (Van Rie et al., 1989; Jurat-Fuentes and Adang, 2001). Cry2Aa does not compete with any of these sites (Jurat-Fuentes and Adang, 2001). In the YHD2 strain, selected for high levels of Cry1Ac resistance, cross-resistance to Cry1Aa, Cry1Ab and Cry1Fa but not to Cry2Aa was observed. Binding of Cry1Aa was almost completely abolished in this strain, whereas Cry1Ab and Cry1Ac binding remained essentially unchanged (Lee et al., 1995). These data suggested that toxicity of Cry1A toxins in *H. virescens* is mainly due to interaction with the A (cadherin) site, rather than with the B or C site. Again, these data indicated a correspondence between competition binding and cross-resistance. A parallel between binding specificity and cross-resistance patterns has also been observed in *H. armigera* (Akhurst et al., 2003; Estela et al., 2004) and *T. ni* (Estada and Ferré, 1994; Iracheta et al., 2000; Wang et al., 2007). Such correspondence between binding and resistance patterns in several insect species indicates that heterologous competition studies are a useful method to assess the value of combinations of toxins with independent binding sites to be used in combination as a resistance management strategy.

It should be noted that in some laboratory selected strains, deviations from such correspondence have been observed. For example, two *H. virescens* strains selected for Cry1Ac resistance, CP73-3 and KCB, showed cross-resistance to Cry2Aa. Following further selection with Cry2Aa, resulting in strains CXC and KCBhyb respectively, the mechanism(s) of resistance have been studied in both strains. Resistance to both toxins in these strains is due to the presence of two or more distinct resistance mechanisms rather than due to a single 'universal' resistance gene (Jurat-Fuentes et al., 2003; Gahan et al., 2005; Karumbaiah et al., 2007). In the CXC strain, the main Cry1Ac resistance gene is different from the one in the YHD2 strain, while Cry2Aa resistance appears to be determined by a combination of many genes with small effects (Gahan et al., 2005). In the KCBhyb strain, the main Cry1Ac resistance gene is not responsible for Cry2Aa resistance (Jurat-Fuentes and Adang, 2006a). However, as mentioned above, only one major mechanism, i.e. altered binding, has so far been detected in field-selected resistant insects and the observations regarding the molecular basis of resistance in such insect strains are likely to be more relevant to resistance management tactics than observations that have been made only for laboratory selected insect strains. For example, some of the mechanisms observed for laboratory selected insect strains may be associated with significant fitness costs that would prevent the corresponding genes to achieve a high frequency in field populations.

3.2.8 Approaches to the Characterisation of Major Resistance Genes

Unraveling the identity of Cry protein resistance genes in insects was first accomplished in *H. virescens* strain YDH2 resistant to Cry1Ac. This quest was initiated by application of the methods of quantitative trait locus (QTL) mapping. A linkage was established between a major Cry protein resistance locus, termed BtR-4, and a genetic marker on linkage group 9 (Heckel et al., 1997). The use of additional polymorphic markers on this linkage group allowed QTL mapping of BtR-4 to a region of about 19 cM. Further efforts to identify the resistance gene were based on a candidate gene mapping approach, leading to the identification of a fragment of the *H. virescens cadherin* gene which mapped to the QTL region. Further molecular analysis demonstrated that disruption of this *cadherin* gene by retrotransposon-mediated insertion which prematurely terminates transcription was linked to high levels of resistance to Cry1Ac (Gahan et al., 2001). Complementation tests using crosses between field-collected males and YDH2 females suggested that the resistance alleles occur in field populations (Gould et al., 1997; Gahan et al., 2001). From an analysis of *cadherin* alleles and Cry1A toxin binding in susceptible and resistant *H. virescens* strains it was concluded that a wild type allele was necessary for cadherin production and Cry1Aa binding, while most of the Cry1Ab and Cry1Ac binding was independent of the *cadherin* allele (Jurat-Fuentes et al., 2004). These observations corroborated the hypothesis put forward by Lee et al. (1995) that the binding site that is shared between Cry1Aa, Cry1Ab and Cry1Ac (site A) is mainly involved in toxic function, while sites B and C are less important. In *P. gossypiella*, three mutant alleles of a *cadherin* gene (r1, r2 and r3), that encode a truncated protein or a protein with internal deletions and are linked to Cry1Ac resistance, were first identified in the AZP-R resistant strain (Morin et al., 2003) and later also found in another resistant strain (Tabashnik et al., 2004). Different combinations of these alleles were subsequently detected in two additional resistant strains (Tabashnik et al., 2005a). Also in *H. armigera*, disruption of a *cadherin* gene was found to be associated with high levels of Cry1Ac resistance (Xu et al., 2005; Yang et al., 2006). In various insect species, regions of cadherin proteins involved in Cry1A binding have been mapped, resulting in the identification of three such sites (Nagamatsu et al., 1999; Gómez et al., 2001, 2002a, 2003; Dorsch et al., 2002; Hua et al., 2004a; Xie et al., 2005; Fabrick and Tabashnik, 2007). The lesions in the *cadherin* genes are upstream of either all three Cry1A binding sites (*H. armigera* r1, *H. virescens*, *P. gossypiella* r2) or only sites 2 and 3 (*P. gossypiella* r1, *P. gossypiella* r3). Although detailed information about gene lesions linked to resistance is available for some lepidopteran species, other, as yet undetected lesions may also lead to resistance. Thus, it will likely remain a significant challenge to define the subset of lesions for which to design PCR based resistance monitoring tools (Gahan et al., 2007).

Whereas the above studies relied on resistance genes present in the starting insect populations, studies on resistance genes in the nematode *C. elegans* were based on the creation of resistance alleles by mutagenesis followed by selection of resistant lines using the nematicidal Cry5B protein. This resulted in the characterization of mutations at five loci, designated *bre-1* through *bre-5*, associated with resistance (*bre* = *Bt* toxin resistant). Positional cloning methods identified these *bre* genes as genes encoding enzymes participating in a glyosylation pathway implicated in the production of glycolipid receptors to which Cry5B binds (Griffitts et al., 2001, 2003; Barrows et al., 2007). Thus, in contrast to the elucidation of the identity of resistance genes in insects, the identification of such genes in *C. elegans* was established without prior knowledge of candidate genes. Some recent studies suggest that genomic and proteomic studies may well contribute to the identification of genes involved in resistance mechanisms in insects or nematodes (Candas et al., 2003; McNall and Adang, 2003; Huffman et al., 2004).

3.3 Managing Insect Resistance to *Bt* Crops: From Theory to Practice

3.3.1 High Expression of the Bt Gene Combined with the Use of Refuges

The most favoured strategy for managing resistance to *Bt* crops is the combination of two independent concepts to delay development of resistance: high expression of the *cry* gene and the use of refuges. The high expression of the insecticidal protein will make ineffective any mechanism conferring to the insect low to moderate levels of resistance and it is assumed to kill all heterozygotes. The refuge will permit a certain fraction of the population to escape selection; these susceptible insects will mate with any resistant insects having survived from the exposure to *Bt* plants and will produce susceptible (heterozygote) offspring.

For this strategy to be effective resistance has to be recessive, the susceptible individuals have to outnumber the resistant survivors, the refuges have to be at an appropriately close distance from all *Bt* plants, sexual maturity of resistant and susceptible insects must be reached more or less synchronically and mating between them must be at random, the initial frequency of resistance alleles must be low, and the toxin concentration in plants has to be high enough to kill all resistance heterozygous insects (Andow, 2002). If these conditions are fulfilled, practically all resistant individuals will mate with susceptible ones, producing heterozygous offspring which will die upon exposure to the *Bt* plants (Figure 3.2).

The distribution of the refuge areas has to ensure that any resistant insect that would survive the exposure to the toxin has the chance to meet and mate with susceptible insects from the refuges. The structure of the refuges can be perimetral (surrounding the field), embedded (intercalating rows or blocks of non transformed plants), in a separate field (near the transgenic field), or a mixture of the preceding. Seed mixtures are not an alternative for those pests in which larvae can move from plant to plant, since planting non-transformed seeds among transgenic seeds could favour the selection of resistance. If caterpillars feeding on a transgenic plant could migrate to a non-transformed plant before ingesting a lethal dose of Cry protein, insects with low levels of resistance could escape selection, and mating among them could combine resistance genes and confer higher levels of resistance to the offspring (Cohen et al., chapter 8). Conversely, if totally susceptible larvae feeding on non-transformed plants would migrate to *Bt* plants then they would die, thus not contributing to the refuge effect. Additionally, larger larvae may be able to survive on a *Bt* crop after feeding first on a non-transformed plant. However, seed mixtures must be evaluated on a case-by-case basis taking into account both the insect pest and crop biology. There is no doubt that seed mixtures have a distinct advantage since it would ensure 100% grower compliance, eliminating the need for refuges, which creates a financial burden on the grower.

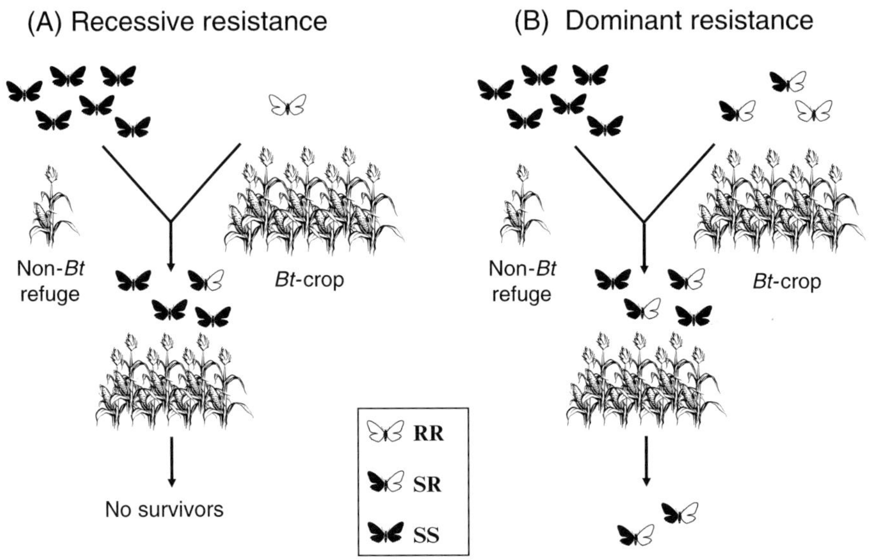

Fig. 3.2 Schematic representation of the high dose/refuge strategy under two assumptions: resistance being recessive (A) or dominant (B). Butterflies represent adult survivors of different genotypes: RR, homozygous resistant; RS, heterozygous; SS, homozygous susceptible

Most cases of resistance follow a mode of inheritance that is neither completely recessive nor completely dominant (Ferré and Van Rie, 2002). However, adjusting the expression of the *cry* gene to high enough levels can convert an incompletely dominant resistance to a functionally recessive resistance (Bourguet et al., 2000), readily killing all or most of the heterozygotes (Fig. 3.3). This is why the "high dose" plays a very important role in this strategy. Figure 3.3 shows ideal cases of different types of inheritance of resistance. In addition to complete recessivity (labelled as 1), other modes of inheritance which deviate from this condition are shown (those labelled as 2, 3, 4, and 5). The refuge strategy would fail in all cases except for case 1 (complete recessivity) if the concentration of Cry protein produced by the plant is low, such as the one labelled as A. However, in those cases where the concentration-mortality line is relatively close to that of susceptible homozygotes (cases 2 and 3) the refuge strategy could succeed if plants produced a higher concentration of the insecticidal protein, such as the one labelled as B, because the incompletely recessive inheritance would become functionally recessive.

Two other aspects to consider in this strategy are the initial frequency of the resistance allele and random mating. The frequency of the resistance allele must be low so that the frequency of resistant homozygotes is low enough to make it extremely unlikely that two resistant individuals could find each other and mate. The behaviour of the insects may also affect the possibilities of random mating, since this requires migration from the refuge to any part of the *Bt* crop and *vice versa*. If females mate near the place of emergence before migrating then the odds to mate with genetically related males are high and the resulting inbreeding would

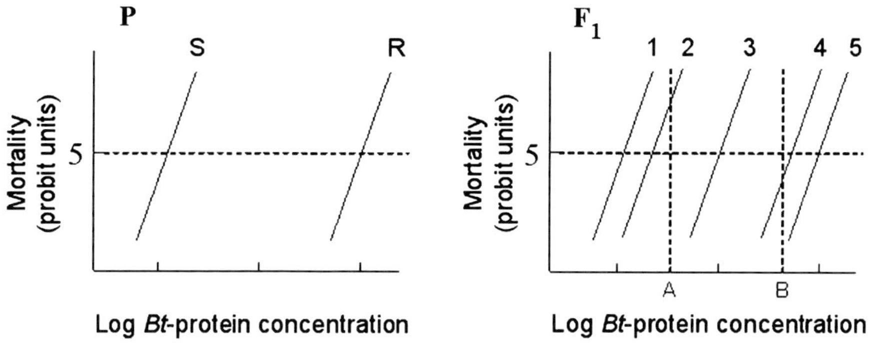

Fig. 3.3 Ideal concentration-mortality responses to *Bt* insecticidal proteins. P: Parental strains, susceptible (S) and resistant (R); F_1: Offspring from the S × R cross; the probit curves represent different situations in which the inheritance of resistance is either completely recessive (1), incompletely recessive (2), codominant (3), incompletely dominant (4), or completely dominant (5)

increase considerably the frequency of resistant homozygotes. This type of behaviour has been found in *O. nubilalis* (Dalecky et al., 2006). Differences in the development time of resistant and susceptible insects may also affect random mating (Liu et al., 1999).

Different high-dose *Bt* crop cultivars are commercially available. The maize events Mon810 and Bt11 both constitutively express Cry1Ab and are sold under the trade name 'Yieldgard' and provide excellent control of corn borers (Shelton et al., 2002). More recently in 2003, the maize event TC1507, containing a different Cry protein, Cry1F, was introduced in North America as 'Herculex I'. In addition to excellent corn borer control, it was reported to provide control of armyworms (*Spodoptera* spp.) and cutworms (*Agrotis* spp., Lepidoptera: Noctuidae). The *Bt* cotton event Mon531, containing the *cry1Ac* gene driven by a constitutive promoter and sold as 'Bollgard' in the US or as 'Ingard' in Australia, has been the most successful *Bt* cotton event. It has an excellent control against the tobacco budworm (*H. virescens*) (Jenkins et al., 1995) and pink bollworm (*P. gossypiella*). However against the cotton bollworm (*H. armigera* in the Old Word and *H. zea* in the New World) control was not always complete (Bacheler and Mott, 1997) and extra foliar insecticide applications may be needed when insect pressure is high (Burd et al., 1999). In China, besides the Mon531 event, cotton varieties GK using a modified *Bt* fusion gene *cry1Ab/cry1Ac* were developed by public research institutes lead by the Chinese Academy of Agricultural Sciences (CAAS) and have been grown since 1997 (Pray et al., 2001).

Are the assumptions of the high-dose/refuge strategy regarding the toxin dose, the dominance level and initial frequency of resistance genes, and random mating met in practice? Using modified *cry* genes, a high-dose (defined as 25-times the dose needed to kill all homozygous susceptible larvae) was achieved in cotton to control *H. virescens* and *P. gossypiella* and in maize for *O. nubilalis*. However, both the Cry1Ab in *Bt* maize and Cry1Ac expression in *Bt* cotton are not likely to represent a high dose for *H. zea* (Agi et al., 2001) or *H. armigera* (Wu and Guo, 2005), especially later in the growing season. Resistance to Cry proteins, as tested on high dose transgenic *Bt* plants is indeed functionally recessive (Metz et al., 1995; Liu et al., 1999; Zhao et al., 2000; Morin et al., 2003; Bird and Akhurst, 2005). As mentioned in section 3.2.4, relatively low frequency of resistance alleles for Cry1A proteins have been measured in *H. armigera*, *H. virescens*, *O. nubilalis* and *P. gossypiella* field populations. Refugia that are temporally and spatially contiguous with the transgenic crop should ensure random mating between homozygous resistant and susceptible adults. Asynchronous development of resistant and susceptible insects on *Bt* crops and non-*Bt* crops, respectively, has been observed in *P. gossypiella* (Liu et al., 1999, 2001) and *H. armigera* (Bird and Akhurst, 2005). The fact that there is partial overlap between consecutive insect generations in these species suggests that such asynchronous development may not necessarily result in assortive mating. It is recommended by the USEPA (US Environmental Protection Agency) that refugia should produce a least 500 susceptible moths for each homozygous resistant moth emerging from the *Bt* crop (USEPA, 2001a).

The usefulness of the high-dose/refuge strategy was originally based on projections from computer models simulating insect population growth under various conditions. More recently, some laboratory studies (Liu and Tabashnik, 1997), controlled greenhouse trials (Tang et al., 2001) and field trials (Shelton et al., 2000) have provided experimental support for this strategy. Unprecedented in the field of insect control, the EPA required a compulsory insect resistance management plan, based on the high-dose/refuge strategy, in 1995 with the introduction of *Bt* crops (see Matten et al., chapter 2). Based on the experience with *Bt* crops grown under different agronomic conditions the plan is further optimized on a regular basis. Currently, the insect resistance management plan for Lepidoptera controlling *Bt* maize requires a structured refuge of at least 20% non-*Bt* maize, but 50% in cotton growing areas due to the extra potential selection pressure on *H. zea* from Cry1A expressing cotton. The refuge maize can be treated with insecticides only when the level of pest pressure meets or exceeds the economic threshold and sprayable *Bt* insecticides must not be applied to the refuge. The refuge must be placed within 0.5 mile (0.25 mile preferred) near the *Bt* maize field and it can be a separate field, a block within the maize field, the field perimeters or an alternation of four or more consecutive rows of refuge maize with *Bt* maize (USEPA, 2001a).

Three options are possible for *Bt* cotton that expresses a single gene: 5% external structured unsprayed refuge, 20% external sprayed refuge and 5% embedded refuge (USEPA, 2001b, 2005). Additional data on pest dispersal ability and oviposition preference, as well as on the availability of alternate hosts serving as a refuge will be useful to further fine-tune size, placement and management of refuges. In countries with no formal refuge requirements, such as China and India, mixed planting systems of *Bt* and suitable non-*Bt* crops on small-scale farms may play an important role in delaying resistance development. Indeed, various non-*Bt* crops may act as important sources of nonstructured natural refuges (Sequeira and Playford, 2001; Wu and Guo, 2005; Ravi et al., 2005; Vijaykumar et al., 2007). The EPA recently approved the use of a natural refuge for Texas and regions east of Texas for pyramided *Bt* cotton varieties, i.e. 'Bollgard II' and 'Widestrike' cotton (USEPA, 2007b). Areas where *P. gossypiella* is considered a pest (e.g., East Texas, Arizona) will still require a structured refuge. Other requirements of the insect resistance management plan include annual resistance monitoring, grower education, compliance assurance, research, reporting and the availability of a remedial action plan should insect resistance develop in the field (Bourguet et al., 2005).

No major insect control failure has occurred since *Bt* crops were commercialized, 11 years ago (Bates et al., 2005b; Wu and Guo, 2005; Tabashnik et al., 2005b; Tabashnik et al., 2006). It is highly likely that the high-dose/refuge strategy has contributed to this, although some of the assumptions underlying this strategy are not completely fulfilled for some crop-pest insect combinations. Fitness costs associated with resistance and incomplete resistance (i.e., the disadvantage for resistant insects on *Bt* cotton versus non-*Bt* cotton) appear to be an important aspect in explaining this delay in resistance in some insects, but not in others (Bates et al., 2005b). Transgenic maize expressing Cry3Bb1 (Vaughn et al., 2005) does not provide a high dose for corn rootworm control (USEPA, 2007a). Efficacy data shows 17% to 62% larval survival on such maize plants. This is related to the relatively low susceptibility of (western) corn rootworm to Cry3Bb1 (Siegfried et al., 2005). The IRM plan for this maize product includes a 20% refuge, although some have recommended larger refuge sizes in view of the low dose (Powell, 2003). Based on results from a simulation model, Crowder and Onstad (2005) have argued that significant (i.e. higher than 10%) survival of susceptible beetles acts to slightly slow the rate of evolution of resistance to transgenic maize.

3.3.2 Expression, in the Same Plant, of More Than One Bt Gene ("Pyramided Plants")[1]

A second approach to delay resistance is the combination in the same plant of two insecticidal proteins effective against the same pest. This strategy is based on the concept that, considering that resistance to the two proteins is conferred independently by different genes, the odds of finding a "double" resistant individual is the product of the odds of finding "single" resistant individuals (Roush, 1997). In other words, if resistant insects to any single toxin are found to occur at, for example, a phenotypic frequency of 10^{-6}, then the expected phenotypic frequency of "double" resistant insects is 10^{-12}. Therefore, if resistance to any one protein is rare, the chance of finding individuals simultaneously resistant to the two proteins may

[1] We have used the term "pyramiding" for combinations of genes conferring the same trait (e.g., insect resistance to a given insect pest) and the term "stacking" for combinations of genes with either multiple traits (e.g., insect resistance and herbicide tolerance) or targeting different insect pests (e.g., corn borers and corn rootworms).

become negligible. Similar to the high-dose/refuge strategy, the strategy of using a combination of two insecticidal proteins also requires that the initial frequency of resistance alleles is low, otherwise "single" resistant insects would not be rare and, as a consequence, the chance of finding "double" resistant insects would no longer be negligible.

A key requirement of this strategy is the lack of cross-resistance. Cross-resistance occurs when resistance to one insecticide is acquired by selection with a different insecticide. In the case of *Bt* crops, cross-resistance occurs when insects become resistant to one Cry protein by exposing them to plants expressing a different Cry protein. This can occur when the two Cry proteins share a common step in their mode of action and this step has been altered in the resistant insects, such as a common membrane receptor. As indicated in section 3.2.7, cross-resistance is common among proteins of the Cry1A family, and also between them and Cry1F and/or Cry1J (Tabashnik et al., 1994, 2000a; Gould et al., 1995). However, outside this group, cross-resistance with the rest of *Bt* proteins is either absent or appears at very low levels (Gould et al., 1995; Tabashnik et al., 2000a, c; Wang et al., 2007).

The high-dose/refuge strategy can be combined with the strategy of pyramiding two or more *Bt* proteins, with a different mode of action, into one variety. The first commercial *Bt* crop with pyramided Cry proteins is 'Bollgard II', a cotton variety expressing Cry1Ac and Cry2Ab. The 'Bollgard II' event Mon15985 was created by inserting the *cry2Ab* gene into the 'Bollgard' event Mon531. 'Bollgard II' expresses more Cry protein and the levels of insect control are higher than in 'Bollgard' (Greenplate et al., 2000). Field studies showed that 'Bollgard II' provided excellent cotton bollworm control and an increased efficacy against armyworms and loopers (Adamczyk et al., 2001). Another cotton variety expressing two insecticidal proteins, 'Widestrike', was obtained by cross-breeding two insect-resistant cotton events: DAS-24236-5, producing constitutively Cry1F protein and DAS-21023-5, producing constitutively Cry1Ac protein. However, these two proteins may not act sufficiently independent to fulfill the basic assumption of this resistance management strategy (Hernández and Ferré, 2005). Introduced in the US market in 2005, it was reported to give excellent control of budworm and bollworm (Huckaba et al., 2003) and the presence of the Cry1F protein in the transgenic variety increases control of fall armyworm (*S. frugiperda*) and beet armyworm (*S. exigua*) (Adamczyk and Gore, 2004). 'VipCot' cotton is expected to be commercialized soon and combines Cry1Ab and Vip3Aa proteins.

Computer models (Roush, 1997) have shown that refuge size could potentially be reduced from 30–40% when using single *Bt* plants sequentially to 5–10% for pyramided or 'dual' *Bt* plants. EPA however did not reduce its refuge requirements when the 'dual' *Bt* cotton event 'Bollgard II' was introduced in the US (Matten and Reynolds, 2003), but amended their decision recently by allowing a natural refuge (e.g., no structured cotton refuge) for 'Bollgard II' and 'Widestrike' (USEPA, 2007b). Zhao et al. (2003) showed experimentally in the greenhouse that transgenic plants expressing two Cry proteins, binding to different sites in the target insect, delay development of resistance: a population of *P. xylostella* containing resistance genes for Cry1Ac and Cry1C developed slower resistance to the pyramided

Cry1Ac/Cry1C broccoli with 20% refuge than to the Cry1Ac broccoli with 20% refuge. Also compared to single *Bt* plants deployed in mosaics (with 20% refuge) the resistance development was delayed. In a subsequent study using one- and two-gene broccoli plants, resistance developed more slowly in selection treatments containing only two-gene plants in comparison to selection treatments containing both one- and two-gene plants (Zhao et al., 2005). This potentially has important implications for the deployment of single gene plants expressing a single *Bt* gene, once pyramided *Bt* plants also expressing such gene are commercialized. The authors of the study argue that it would be advantageous from a resistance management standpoint for regulatory agencies to consider canceling registrations for single-gene plants as soon as pyramided plants are available. In this context, it is interesting to note that Australia has replaced the 'Ingard' cotton (with just the *cry1Ac* gene) with 'Bollgard II' (combining the *cry1Ac* and *Cry2Ab* genes) in a very short period of time (only two seasons) (Naranjo et al., chapter 6).

Apart from pyramiding two *Bt* proteins, crops could be developed by pyramiding a Cry protein with a non-*Bt* insecticidal protein. Recently CAAS introduced the cotton variety SGK321 in China, expressing both the *Bt* fusion gene *cry1Ab/cry1Ac* and the cowpea trypsin inhibitor gene (CpTI) (Guo et al., 1999; Wu and Guo, 2005). It is not clear however if the second protein provides sufficient levels of toxicity to target insect species for this cotton variety to adequately function as a pyramided variety for resistance management purposes.

3.3.3 Temporal Rotation of Cultivars Expressing Different Bt Genes

This approach is similar to the long ago adopted practice, in conventional agriculture, of crop and/or pesticide rotation. Insects (and their resistant offspring) that escape the effect of the Cry protein of the first planted cultivar will be killed by the different Cry protein carried by the second cultivar. Similarly, insects resistant to the second cultivar will be killed when this is replaced by the first cultivar. One requirement for the success of this approach is that there must be no cross-resistance between the Cry proteins expressed by the two *Bt* cultivars. In addition, to avoid the gradual build up of resistance against the two insecticidal proteins, it is required that resistance is associated with a fitness cost. For example, every season planted with a *Bt* cultivar producing protein A will select for resistance to this toxin; if insects that eventually become resistant to protein A have a lower fitness than susceptible insects in the absence of exposure to this protein (either when feeding on plants expressing protein B or on plants that fail to express protein A), then the frequency of the allele conferring resistance to protein A will tend to decrease in the following season. This will permit reversion of resistance (to protein A) and will counteract the effect of the positive selection exerted by plants expressing protein A. The same situation applies when considering resistance to the second insecticidal protein (protein B in this example). While temporal rotation of cultivars expressing different *Bt* genes may seem promising, this strategy appears to be inferior to using pyramided *Bt* genes in the first place (Zhao et al., 2003).

3.3.4 Wound-Induced/Tissue-Specific Expression

Constitutive expression of transgenes poses an energy expenditure load on the plants and a strong selection pressure on the insect populations that feed on them. An alternative to constitutive expression is to express the insecticidal genes only in those critical parts of the plant that require protection (for example, the bolls in cotton, the kernel and stem in maize, or the tubers in potatoes) and only when protection is required (i.e. when the insect damage exceeds a determined threshold). Expressing *Bt* genes under tissue-specific promoters or under wound-induced promoters has been considered since the beginning of *Bt* plant transformation. Nevertheless, this technology lags still behind that of constitutive promoters due to concerns that sufficient timely expression may not occur. A good candidate for wound-induced expression is the promotor from the maize proteinase inhibitor gene (*mpi*), which proved to confer protection to rice (Breitler et al., 2001, 2004) and sorghum (Girijashankar et al., 2005) when driving *cry1A* genes. Chemically inducible promoters can also be used to express the *Bt* gene by the application of a benign inducer whenever the pest pressure becomes high enough (Cao et al., 2001; Bates et al., 2005a).

A singularity of the tissue-specific approach is that it requires low mobility of the pest at the larval stage, as discussed above in the case of seed mixtures. Larval movement from some parts of the plant to other parts could significantly decrease the benefits pursued unless the wound induction of insecticidal protein expression is swift.

From a resistance management standpoint, the advantages of the use of tissue-specific or inducible promoters include a lower selection pressure, since a decreasing number of generations are exposed to the insecticidal proteins. Also, because within-plant or within-field refuges are created, potential problems of growers' compliance with the requirement of refuges are avoided. However, before wide adoption as a resistance management tool, the effectiveness of this strategy has to be empirically tested.

3.4 The Impact of Government Regulations on the Management of Insect Resistance

Perhaps the most remarkable aspect surrounding the registration of *Bt* crops in the US has been the requirement by the EPA for the registrant to submit a detailed insect resistance management plan (IRM). Historically, the EPA had not mandated IRM plans for registration of synthetic chemicals, biochemicals or biopesticides, expecting that the producer should manage the stewardship of their own products. However, resistance development became more common necessitating the need for emergency registrations (known as Section 18s to reflect the code of regulation) of new active ingredients to try and overcome field failures, causing swift EPA reviews that greatly stretched EPA resources. As resistance development became more widespread, industry and scientific groups (e.g., Insect Resistance Action Committee; The Bt Working Group; ILSI, 1999) formed to find practical solutions to delay resistance development and combat resistant insect populations, such as funding research to develop and test IRM plans, new product labelling systems to clarify groupings of active ingredients, educational materials for growers and monitoring for insect susceptibility. Beginning in 1992, the EPA provided an updated draft guidance on reporting requirements on information concerning unreasonable adverse effects of their products (known as Section 6(a)(2) reporting), which primarily focuses on health and safety aspects of the product, by extending this legal reporting requirement to include insect resistance. Finally in 1996, this new guidance became final and registrants now must report cases of insect resistance within 30 days (PR Notice 98-3, http://www.epa.gov/PR_Notices/pr98-3.pdf; accessed 11 January 2008).

While the code of regulations was being reviewed and updated by the EPA, they were also reviewing the first Plant Incorporated Protectant (PIP) registration applications. And at the same time, resistance development was first identified for the *Bt* biopesticide in field populations (Tabashnik et al., 1990). Scientific researchers and non-government associations (e.g., Greenpeace, Union of Concerned Scientists) shared their concern that insects would develop resistance to these highly effective *Bt* crops in 5 short years. These events taken together caused the EPA to link a PIP registration with the first mandated IRM plan. Yet, despite the confirmation of resistance development to *Bt* formulated products, there are no IRM plans required for these products.

All PIP registrations were originally time limited (e.g., 10 years), since risk assessment information and processes are evolving. When the *Bt* PIP registrations were nearing expiration, EPA would reassess the IRM plans and examine whether ongoing monitoring and grower compliance efforts are robust enough to maximize the lifetime of the product. The elements of the IRM plan include the biology of the pest(s), potential strategies and their deployment options (e.g., rotation, dual gene constructs, destruction of crop residue), product fit with integrated pest management practices, field monitoring for pest susceptibility, communication/education for growers to implement the plan, and proof of compliance. For maize, an industry team, known as the Agricultural Biotechnology Stewardship Technical Committee (ABSTC), helped to establish a uniform refuge size and deployment options across *Bt* maize products and refine IRM plans (USEPA, 2001a).

The first *Bt* crop was registered in 1995 and quick adoption of *Bt* crops by growers has surpassed all expectations (James, 2006). Yet, insect resistance has not developed, and in some cases, overall suppression of certain insect pests, such as *P. gossypiella* has been recorded (Tabashnik et al., 2005b; Storer et al., chapter 10). Despite the early concerns of rapid resistance evolution, and the fact that the IRM plans were based on theoretical data, these plans evidently are working to delay or avoid the development of resistance to *Bt* crops (USEPA, 2001a).

Outside of the US, *Bt* crop cultivation brings similar concerns about insect resistance development. In Australia, *Bt* cotton was introduced in 1996 along with a strict Resistance Management Strategy (RMS) that is similar to that required by the EPA. Furthermore, the rate of adoption of *Bt* cotton has been closely controlled, with a maximum of 30% of all cotton hectares for the single gene (*cry1Ac*) variety (James, 2002). As described above, China and India do not have strict refuge requirements but recommend a 20% refuge for *Bt* cotton (Wu and Guo, 2005). While there is no monitoring for grower compliance, there is yearly monitoring for insect susceptibility in both India (http://cicr.nic.in/; accessed 4 January 2008) and China (Wu et al., 2002). While grower compliance is uncertain, rapid introduction of pyramided *Bt* cotton will reduce the overall risk of insect resistance evolution.

In the European Union, IRM plans are examined to ensure that they are scientifically adequate as part of the overall risk assessment under the European Food Safety Authority (EFSA). EFSA will decide if a case-specific monitoring plan is necessary. However, once the EU grants an authorization to cultivate a *Bt* crop, then the individual EU states, where cultivation will occur, are responsible for supervising the IRM plans and monitoring requirements. Therefore, the potential exists for a patchwork of supervision and requirements linked to 27 different EU competent authorities.

3.4.1 Methods to Assess High-Dose Expression

An essential aspect of the high dose/refuge strategy is the confirmation that the potential commercial event expresses a true high dose. As described above, the goal is to have a sufficient level of *Bt* protein expression to kill any heterozygote insects that might emerge through the mating of rare *Bt*-resistant insects to susceptible insects, reared in the non-*Bt* refuge.

The most obvious method to demonstrate that a *Bt* plant expresses a dose of insecticidal protein sufficient to kill heterozygous resistant larvae would be to assess the mortality of tissues of such plants to larvae that are the offspring of a cross between insects from a susceptible strain and from an insect strain that is resistant to the *Bt* protein. However, an insect population resistant to *Bt* is not always available. Furthermore, a laboratory resistance colony may not be a good predictor of field resistance. Therefore, EPA defined a high dose as a *Bt* protein level that is 25 times the dose needed to kill susceptible larvae. While it is possible that a level of resistance could exceed the 25-fold level, it was acknowledged by the EPA that such a level of resistance development was unlikely, at least over the timeframe of the *Bt* registrations (e.g., 10 years) (SAP, 1998).

According to the US Scientific Advisory Panel (SAP), convened in February 1998 by the EPA, there are five different imperfect ways to prove that a potential commercial cultivar can be considered to provide a high dose (SAP, 1998). Each registrant must demonstrate that their product expresses a high dose using at least two of the five methods described:

a. Make serial dilutions of lyophilized *Bt* plant materials (e.g., leaf tissue) into an artificial insect diet. The tissue from non-*Bt* plants of the same crop will serve as a control. Neonate larvae of the target species are bioassayed to determine the lowest dilution of *Bt* plant material that gives 100% mortality.
b. Bioassays using plant lines with expression levels approximately 25-fold lower than the commercial cultivar, determined by a reliable quantitative technique. A series of different *Bt* events should be compared for the level of *Bt* protein expression. Two or three events are selected that have a range of *Bt* protein expression of at least 25-fold less than that of the potential commercial *Bt* variety. Neonate larvae of the target species are bioassayed on these low expressing events.
c. Survey large numbers of commercial plants on sentinel plots in the field (e.g., sentinel sweet corn method) to make sure that the cultivar is at the $LD_{99.99}$ or higher level to assure that 95% of heterozygote insects would probably be killed. For example, this approach using *Bt* sweet corn hybrids can be used, since they attract high densities of *O. nubilalis* and *H. zea* moths. Sampling can be limited to sweet corn ears in the *Bt* plot and a frequency of resistant phenotypes can be estimated as the ratio of density of larvae/plant in *Bt* sweet corn to density of larvae/plant in an adjacent planting of non-*Bt* sweet corn (Andow et al., 1998; Hutchison, unpublished data). This technique is more difficult for crop species that do not attract high levels of target insect species and certainly more difficult that any of the other methods due to the imprecise nature of scouting efforts.
d. Similar to (c) above, but would use controlled infestation with a laboratory strain of the pest that had an LD_{50} value similar to field strains. This method overcomes the issue of finding sufficient numbers of insects, but still requires detailed scouting efforts.
e. Determine if an older instar larvae of the targeted pest could be found with an LD_{50} that was at least 25-fold higher than that of neonate larvae. If so, that stage could be tested on the crop plants to determine if 95% or more of the older stage were killed. This method is perhaps one of the most practical techniques, since there are typically wide ranges of LD_{50} values between different instars of lepidopteran larvae. For example, susceptibility of *O. nubilalis* larvae to the Cry9C protein ranges from 0.22 μg/ml for 1st instar larvae to 94.4 μg/ml for 4th instar larvae, representing a 429-fold difference in susceptibility. When 4th instar *O. nubilalis* larvae were exposed to Cry9C-containing maize, 100% mortality was measured (MacIntosh et al., 1998).

These techniques were again reviewed and confirmed in 2000 by another scientific advisory panel (SAP, 2001).

3.4.2 Monitoring Insect Susceptibility

Monitoring encompasses a broad range of topics that focus primarily on the observation of unexpected damage to a *Bt* crop. This damage could result from a number of different factors, with insect resistance development being but one of the possibilities. As with any new commercial product launch, good product stewardship by the registrant will be taken to ensure that the product performs as expected. Any unexpected damage reports must be carefully investigated. A useful tool for in-field monitoring is lateral flow strip technology that can identify *Bt* plants and confirm a high level of Cry protein expression.

Direct monitoring of insect population susceptibility, especially in high risk areas where *Bt* crop adoption has been the greatest, is an aggressive method to test the effectiveness of the resistant management programs (SAP, 2001). It is believed that yearly monitoring will be able to detect small shifts in resistance gene frequency prior to the onset of wide spread crop failures. Baseline data is collected on the target field insect populations, preferably before the first introduction of a *Bt* crop, but at least during the initial years of launch prior to high market penetration (Sims et al., 1996; Siegfried et al., 1999, 2000; Siegfried and Spencer, 2000; Song et al., 2000). This baseline data is critical to determine a discriminating dose, or a dose that ensures 100% mortality of a fully susceptible insect population (Sims et al., 1996; Marçon et al., 2000). The greatest value of a discriminating dose assay is that large number of insects can be tested in an efficient manner providing the best opportunity for finding that rare resistant insect.

Following the establishment of the baseline data, annual sampling of field insect populations should be linked to regions of the high product sales, which would present the highest risk of resistance development. One of the most difficult aspects is to find sufficient numbers of insects to establish a colony for testing. However, annual monitoring of maize and cotton pests has demonstrated no shift in insect susceptibility (USEPA, 2001a).

Other monitoring methods, such as the F_2 screen, may have some value as a research tool to provide estimations of the initial allele frequency of an insect population (Andow and Alstad, 1998). The utility and reliability of this proposed technique has not been validated by other academic labs, and has, in fact, been criticized as a very labor intensive effort and not suitable for routine screening purposes (Andow in USEPA/USDA, 1999; Hawthorne et al., 2001). The F_2 screen can also be dramatically influenced by the doses used and is also not appropriate for cases in which resistance is due to more than one gene (Zhao et al., 2002).

3.4.3 Gaining Grower Compliance: Education and Incentives

Mandated IRM plans were not only new to product registrants, but also for the growers that must implement such plans. Notably, EPA's authority is over the product registration and registrant, but not individual growers. Yet, grower compliance with IRM requirements is critical for insect resistance management, as non-compliance will likely increase the rate at which resistance develops. The development of robust IRM plans must balance the scientific basis and technical capabilities with the pragmatic realities of crop production. Thus the registrant is responsible for ensuring that the product performs as expected, but it is the grower that must plant and manage the crop according to IRM plans, which in many cases puts financial burdens primarily on the grower. And there continue to be disagreements on appropriate refuge size and what level of grower non-compliance would lead to a greater risk of resistance development. For all these reasons, assessing grower compliance to IRM plans is an important requirement for all registrants.

To gain grower compliance, a number of efforts are being utilized within the overall IRM strategy for a specific *Bt* crop. The basis of a successfully implemented IRM plan is grower education, which should include the following elements: how the *Bt* plant protects against insect damage, the components and requirements of the IRM strategy and the importance of compliance. In order to reach as many growers as possible, education is offered through grower group meetings, technical brochures, on-farm visits, and through Internet-based tools, such as the National Corn Growers Association education module for *Bt* maize (NCGA, http://209.98.199.114/ncga-irm/; accessed 4 January 2008).

A compliance plan should include practical mechanisms to maximize adoption of IRM, such as grower contracts, education, certification tests, audits, rewards for compliance, crop insurance for refuges, databases of non-compliance growers, sales restrictions and fines for non-compliance. The elements of a compliance plan should be tailored to the crop and as stated above, grower education and contracts provide a strong basis for compliance.

Measuring the effectiveness of the compliance may identify other useful methods, depending on the underlying sources of non-compliance. Therefore, the EPA requires an annual audit of grower compliance. Ideally third parties, other than the registrant, should carry out audits, but since grower lists are considered highly confidential, to date, all audits have been industry driven.

Under the auspices of the Agricultural Biotechnology Stewardship Technical Committee (ABSTC), *Bt* maize compliance surveys have been conducted in the US by marketing research firms since 1999, with phone surveys of more than 500 maize growers, farming at least 200 acres. In 2002, the monitoring effort was expanded with the development of a Compliance Assurance Program (CAP) to promote IRM awareness and determine on-farm compliance. Grower compliance has increased each year to more than 90% for the 2005 growing season (USEPA, 2001a; NCGA, 2006). For the first time in 2005, a small group of non-compliant growers were identified and enhanced educational efforts, including on-farm compliance assessments, were undertaken. A grower that continues to be non-compliant could lose assess to *Bt* crop technology under the uniform standards outlined in the CAP (NCGA, 2006) (see Bates et al., 2005b for more critical assessments of compliance).

Up until the 2005 season, Monsanto was the sole registrant for *Bt* cotton and they carried out compliance audits during on-farm visits of growers. Since the cotton grower pool is much smaller than the number of maize growers, one on one contact has been possible. The percentage of *Bt* cotton growers that were following refuge guidelines exceeded 90% from 1996–2000 (USEPA, 2001a). Since 2000 compliance has remained strong.

3.4.4 *Implications of Pyramiding Bt Traits with Other Value-Added Traits, Including but Not Limited to Herbicide Tolerance, for Resistance Management*

Plants expressing multiple genes are becoming an ever more important part of the *Bt* crop market, both the pyramiding of two or more insecticidal proteins as well as stacking insect control traits with other agronomical traits, such as herbicide tolerance. As already explained above, pyramiding insect control proteins, such as two *Bt* proteins that bind unique sites in the same insect, can greatly reduce the risk of resistance development. *Bt* protein combinations can also expand the insect spectrum of control as seen with stacks of Cry1 protein for *O. nubilalis* control and Cry3Bb or Cry34Ab/Cry35Ab proteins for *Diabrotica* spp. control.

Weed management is another major grower concern and thus the adoption of herbicide tolerant crops has also been swift. When herbicide tolerance traits are stacked with a *Bt* trait, the management of weeds, especially in the required refuge acreage, becomes a complex situation. The crop variety of the refuge must be closely matched to the *Bt* variety, in order to minimize agricultural management. Seed companies and distributors should ensure that properly matched non-*Bt* herbicide tolerant varieties are available to be used in refuge plots. The refuge and *Bt* varieties should be uniform so that the plots mature at similar rates, which allows for random mating of insect populations emerging from the two plot types. Likewise, any introduced agronomic trait that may impact the crop rate of maturation or other crop quality should be carefully evaluated and a matched cultivar utilized for the refuge.

3.5 Concluding Remarks

Bt genes encoding insecticidal proteins have been successfully expressed in economically important crops, such as maize and cotton, to confer them protection against the most important insect pests. Of the 114 million hectares globally planted with GM crops in 2007, over one third are insect-resistant *Bt* crops, and the area keeps increasing every year. Most planted *Bt* maize varieties express the Cry1Ab protein, highly toxic to corn borers, but other proteins have also been expressed in maize to confer protection against armyworms and rootworms. Most globally planted *Bt* cotton has been transformed with *cry1Ac* and also with a combination of *cry1Ac* and *cry2Ab*. The combination of these two genes was not only to broaden the spectrum of protection but to serve as a tool in resistance management.

The potential for insects to develop resistance to insecticides is well known. Early attempts to develop *Bt*-resistant insect colonies in the laboratory were generally unsuccessful, but by the early 1990s *Bt* resistance had been reported in *P. interpunctella*, *H. virescens*, and *L. decemlineata* among others. More importantly, resistance to *Bt* in an open field population of *P. xylostella* was also observed. Numerous cases of *P. xylostella* resistance to *Bt* have now been documented in tropical regions around the world. Thus the potential for insects to evolve resistance to *Bt* insecticides was clearly demonstrated, and it is considered to be one of the main threats posed to GM insecticidal plants since the initial introductions utilized single gene constructs.

Given the threat of resistance, the EPA made an unprecedented move to require insect resistance management plans for this new class of pesticides, know as Plant-Incorporated Protectants (PIP). While many techniques are available for delaying resistance development, the insect resistance management plans for these first *Bt* crops typically incorporated a high dose of the insecticidal protein, along with a refuge plot adjacent to the PIP crop. Pyramiding insecticidal proteins with unique sites of action holds great promise, but has only recently been commercialized in cotton. Such "pyramided" *Bt* crops, in combination with the high dose/refuge strategy, will likely confer maximum protection to the *Bt* crop technology against insect resistance. Susceptibility of the target insect species is monitored annually to try to detect small shifts in susceptibility. Growers are offered various educational opportunities to better understand why resistance management is so important, assist them to implement IRM plans on their farm, and are annually audited to ensure that they are carrying out the required refuge acreage and proper placement. It is important to note that managing the refuge acreage requires attention to ensure that the crop is of the same maturity and matched for herbicide tolerance, should that trait be included in the PIP. However, no case of resistance evolution to *Bt* crops has been reported despite the rapid adoption in some regions and the fact that some of the IRM strategies relied mostly on theoretical assumptions. Continued research on insect biology and IRM strategies combined with consistent and robust oversight to ensure IRM compliance will secure the long-term use of this valuable technology.

References

Adamczyk, J.J., Jr., and Gore, J., 2004. Laboratory and field performance of cotton containing Cry1Ac, Cry1F, and both Cry1Ac and Cry1F (Widestrike®) against beet armyworm and fall armyworm larvae (Lepidoptera: Noctuidae). Florida Entomologist 87: 424–432.

Adamczyk, J.J., Jr., Adams, L.C., and Hardee, D.D., 2001. Field efficacy and seasonal expression profiles for terminal leaves of single and double *Bacillus thuringiensis* toxin cotton genotypes. Journal of Economic Entomology 94: 1589–1593.

Agi, A.L., Mahaffey, J.S., Bradley J.R., Jr., and Van Duyn, J.W., 2001. Efficacy of seed mixes of transgenic Bt and nontransgenic cotton against bollworm, *Helicoverpa zea* Boddie. Journal of Cotton Science 5: 74–80.

Ahmad, M., and Roush, R., 1999. Estimation of allele frequencies for *Bacillus thuringiensis* resistance in diamondback moth, *Plutella xylostella* and cotton bollworm, *Helicoverpa armigera*: An isofemale line (F_2) approach. BCPC Symposium Proceedings 72: 281–286.

Akhurst, R.J., James, W., Bird, L.J., and Beard, C., 2003. Resistance to the Cry1Ac δ-endotoxin of *Bacillus thuringiensis* in the cotton bollworm, *Helicoverpa armigera* (Lepidoptera: Noctuidae). Journal of Economic Entomology 96: 1290–1299.

Alves, A.P., Spencer, T.A., Tabashnik, B.E., and Siegfried, B.D., 2006. Inheritance of resistance to the Cry1Ab *Bacillus thuringiensis* toxin in *Ostrinia nubilalis* (Lepidoptera: Crambidae). Journal of Economic Entomology 99: 494–501.

Alzate, O., You, T., Claybon, M., Osorio, C., Curtiss, A., and Dean, D.H., 2006. Effects of disulfide bridges in domain I of *Bacillus thuringiensis* Cry1Aa δ-endotoxin on ion-channel formation in biological membranes. Biochemistry 45: 13597–13605.

Andow, D.A., 2002. Resisting resistance to *Bt*-corn. In: Genetically Engineered Organisms: Assessing Environmental and Human Health Effects, D.K. Letourneau and B.E. Burrows, eds., CRC, Boca Raton, FL, USA, pp. 99–124.

Andow, D.A., and Alstad, D.N., 1998. F_2 screening for rare resistance alleles. Journal of Economic Entomology 91: 572–578.

Andow, D.A., Alstad, D.N., Pang, Y.H., Bolin, P.C., and Hutchison, W.D., 1998. Using an F_2 screen to search for resistance alleles to *Bacillus thuringiensis* toxin in European corn borer (Lepidoptera: Crambidae). Journal of Economic Entomology 91: 579–584.

Andow, D.A., Olson, D.M., Hellmich, R.L., Alstad, D.N., and Hutchison, W.D., 2000. Frequency of resistance to *Bacillus thuringiensis* toxin Cry1Ab in an Iowa population of European corn borer (Lepidoptera: Crambidae). Journal of Economic Entomology 93: 26–30.

Andreadis, S.S., Álvarez-Alfageme, F., Sánchez-Ramos, I., Stodola, T.J., Andow, D.A., Milonas, P.G., Savopoulou-Soultani, M., and Castánera, P., 2007. Frequency of resistance to *Bacillus thuringiensis* toxin Cry1Ab in Greek and Spanish populations of *Sesamia nonagrioides* (Lepidoptera: Noctuidae). Journal of Economic Entomology 100: 195–201.

Atsumi, S., Mizuno, E., Hara, H., Nakanishi, K., Kitami, M., Miura, N., Tabunoki, H., Watanabe, A., and Sato, R., 2005. Location of the *Bombyx mori* aminopeptidase N type 1 binding site on *Bacillus thuringiensis* Cry1Aa toxin. Applied and Environmental Microbiology 71: 3966–3977.

Bacheler, J.S., and Mott, D.W., 1997. Efficacy of grower-managed Bt cotton in North Carolina. In: Beltwide Cotton Conference Proceedings, P. Dugger and D. Richter, eds., National Cotton Council, Memphis, TN, USA, pp. 858–861.

Ballester, V., Granero, F., Tabashnik, B.E., Malvar, T., and Ferré, J., 1999. Integrative model for binding of *Bacillus thuringiensis* toxins in susceptible and resistant larvae of the diamondback moth (*Plutella xylostella*). Applied and Environmental Microbiology 65: 1413–1419.

Barrows, B.D., Haslam, S.M., Bischof, L.J., Morris, H.R., Dell, A., and Aroian, R.V., 2007. Resistance to *Bacillus thuringiensis* toxin in *Caenorhabditis elegans* from loss of fucose. Journal of Biological Chemistry 282: 3302–3311.

Bates, S.L., Cao, J., Zhao, J.Z., Earle, E.D., Roush, R.T., and Shelton, A.M., 2005a. Evaluation of a chemically inducible promoter for developing a within-plant refuge for resistance management. Journal of Economic Entomology 98: 2188–2194.

Bates, S.L., Zhao, J.Z., Roush, R.T., and Shelton, A.M., 2005b. Insect resistance management in GM crops: Past, present and future. Nature Biotechnology 23: 57–62.

Berliner, E., 1915. Über die Schlaffsucht der Mehlmottenraupe (*Ephestia kuhniella*, Zell.) und ihren Erreger *Bacillus thuringiensis*, n.sp. Zeitschrift für Angewandtes Entomologie 2: 29–56.

Bird, L.J., and Akhurst, R.J., 2005. Fitness of Cry1A-resistant and -susceptible *Helicoverpa armigera* (Lepidoptera: Noctuidae) on transgenic cotton with reduced levels of Cry1Ac. Journal of Economic Entomology 97: 1699–1709.

Bolin, P.C., Hutchison, W.D., and Andow, D.A., 1999. Long-term selection for resistance to *Bacillus thuringiensis* Cry1Ac endotoxin in a Minnesota population of European corn borer (Lepidoptera: Crambidae). Journal of Economic Entomology 92: 1021–1030.

Bourguet, D., Genissel, A., and Raymond, M., 2000. Insecticide resistance and dominance levels. Journal of Economic Entomology 93: 1588–1595.
Bourguet, D., Desquilbet, M., and Lemarié, S., 2005. Regulating insect resistance management: The case of non-*Bt* corn refuges in the US. Journal of Environmental Management 76: 210–220.
Bravo, A., Gómez, I., Conde, J., Muñoz-Garay, C., Sánchez, J., Miranda, R., Zhuang, M., Gill, S.S., and Soberón, M., 2004. Oligomerization triggers binding of a *Bacillus thuringiensis* Cry1Ab pore-forming toxin to aminopeptidase N receptor leading to insertion into membrane microdomains. Biochimica et Biophysica Acta 1667: 38–46.
Bravo, A., Gill, S.S., and Soberón, M., 2005. *Bacillus thuringiensis*: Mechanisms and use. In: Comprehensive Molecular Insect Science, L.I. Gilbert, I. Kostas and S.S. Gill, eds., Elsevier, Oxford, UK, pp. 175–205.
Breitler, J.C., Cordero, M.J., Royer, M., Meynard, D., San Segundo, B., and Guiderdoni, E., 2001. The −689/+197 region of the maize protease inhibitor gene directs high level, wound-inducible expression of the *cry1B* gene which protects transgenic rice plants from stemborer attack. Molecular Breeding 7: 259–274.
Breitler, J.C., Vassal, J.M., Catala, M.M., Meynard, D., Marfà, V., Melé, E., Royer, M., Murillo, I., San Segundo, B., Guiderdoni, E., and Messeguer, J., 2004. *Bt* rice harbouring *cry* genes controlled by a constitutive or wound-inducible promoter: Protection and transgene expression under Mediterranean field conditions. Plant Biotechnology Journal 2: 417–430.
Brewer, G.J., 1991. Resistance to *Bacillus thuringiensis* subsp. *kurstaki* in the sunflower moth (Lepidoptera: Pyralidea). Environmental Entomology 20: 316–322.
Briese, D.T., 1981. Resistance of insect species to microbial pathogen. In: Pathogenesis of Invertebrate Microbial Diseases, E. Davidson and O. Allenheld, eds., Totowa, NJ, USA, pp. 511–545.
Briggs, J., 1963. Commercial production of insect pathogens. In: Insect Pathology: An Advanced Treatise, Vol. 2., E.A. Steinhaus, ed., Academic, New York, USA, pp. 519–548.
Burd, T., Bradley, J.R., Jr., and Van Duyn, J.W., 1999. Performance of selected Bt cotton genotypes against bollworm in North Carolina. In: Beltwide Cotton Conference Proceedings, P. Dugger and D. Richter, eds., National Cotton Council, Memphis, TN, USA, pp. 931–934.
Burd, A.D., Gould, F., Bradley, J.R., Van Duyn, J.W., and Moar, W.J., 2003. Estimated frequency of nonrecessive *Bt* resistance genes in bollworm, *Helicoverpa zea* (Boddie) (Lepidoptera: Noctuidae) in eastern North Carolina. Journal of Economic Entomology 96: 137–142.
Burges, H.D., and Jones, K.A., (1998). Formulation of bacteria, viruses and protozoa to control insects. In: Formulation of Microbial Biopesticides; Beneficial Microorganisms, Nematodes and Seed Treatments, H.D. Burges, ed., Kluwer, Dordrecht, The Netherlands, pp. 33–127.
Candas, M., Loseva, O., Oppert, B., Kosaraju, P., and Bulla, L.A., Jr., 2003. Insect resistance to *Bacillus thuringiensis*: Alterations in the Indianmeal moth larval gut proteome. Molecular and Cellular Proteomics 2: 19–28.
Cao, J., Shelton, A.M., and Earle, E.D., 2001. Gene expression and insect resistance in transgenic broccoli containing a *Bacillus thuringiensis cry1Ab* gene with the chemically inducible PR-1a promoter. Molecular Breeding 8: 207–216.
Chaufaux, J., Seguin, M., Swanson, J.J., Bourguet, D., and Siegfried, B.D., 2001. Chronic exposure of the European corn borer (Lepidoptera: Crambidae) to Cry1Ab *Bacillus thuringiensis* toxin. Journal of Economic Entomology 94: 1564–1570.
Crickmore, N., Zeigler, D.R., Feitelson, J., Schnepf, E., Van Rie, J., Lereclus, D., Baum, J., and Dean, D.H., 1998. Revision of the nomenclature for the *Bacillus thuringiensis* pesticidal crystal proteins. Microbiology and Molecular Biology Reviews 62: 807–813.
Crickmore, N., Zeigler, D.R., Schnepf, E., Van Rie, J., Lereclus, D., Baum, J., Bravo, A., and Dean, D.H., 2007. Bacillus thuringiensis Toxin Nomenclature. http://www.lifesci.sussex.ac.uk/Home/Neil_Crickmore/Bt/ (accessed 4 January 2008).
Crowder, D.W., and Onstad, D.W., 2005. Using a generational time-step model to simulate dynamics of adaptation to transgenic corn and crop rotation by western corn rootworm (Coleoptera: Chrysomelidae). Journal of Economic Entomology 98: 518–533.

Dalecky, A., Ponsard, S., Bailey, R., Pélissier, C., and Bourguet, D., 2006. Resistance evolution to *Bt* crops: Predispersal mating of European corn borers. PLoS Biology 4: 1–10.

Dorsch, J.A., Candas, M., Griko, N.B., Maaty, W.S., Midboe, E.G., Vadlamudi, R.K., and Bulla, L.A., Jr., 2002. Cry1A toxins of *Bacillus thuringiensis* bind specifically to a region adjacent to the membrane-proximal extracellular domain of BT-R(1) in *Manduca sexta*: Involvement of a cadherin in the entomopathogenicity of Bacillus thuringiensis. Insect Biochemistry and Molecular Biology 32: 1025–1036.

Dubois, N.R., and Dean, D.H., 1995. Synergism between Cry1A insecticidal crystal proteins and spores of *Bacillus thuringiensis*, and vegetative cells against *Lymantria dispar* (Lepidoptera: Lymantriidae) larvae. Environmental Entomology 24: 1741–1747.

Estada, U., and Ferré, J., 1994. Binding of insecticidal crystal proteins of *Bacillus thuringiensis* to the midgut brush border of the cabbage looper, *Trichoplusia ni* (Hübner) (Lepidoptera: Noctuidae), and selection for resistance to one of the crystal proteins. Applied and Environmental Microbiology 60: 3840–3846.

Estela, A., Escriche, B., and Ferré, J., 2004. Interaction of *Bacillus thuringiensis* toxins with larval midgut binding sites of *Helicoverpa armigera* (Lepidoptera: Noctuidae). Applied and Environmental Microbiology 70: 1378–1384.

Fabrick, J.A., and Tabashnik, B.E., 2007. Binding of *Bacillus thruringiensis* toxin Cry1Ac to multiple sites of cadherin in pink bollworm. Insect Biochemistry and Molecular Biology 37: 97–106.

Farinós, G.P., de la Poza, M., Hernández-Crespo, P., Ortego, F., and Castañera, P., 2004. Resistance monitoring of field populations of the corn borers *Sesamia nonagrioides* and *Ostrinia nubilalis* after 5 years of Bt maize cultivation in Spain. Entomologia Experimentalis et Applicata 110: 23–30.

Fernández, L.E., Aimanova, K.G., Gill, S.S., Bravo, A., and Soberón, M., 2006. A GPI-anchored alkaline phosphatase is a functional midgut receptor of Cry11Aa toxin in *Aedes aegypti* larvae. Biochemical Journal 394: 77–84.

Ferré, J., and Van Rie, J., 2002. Biochemistry and genetics of insect resistance to *Bacillus thuringiensis*. Annual Review of Entomology 47: 501–533.

Ferré, J., Real, D.M., Van Rie, J., Jansens, S., and Peferoen, M., 1991. Resistance to the *Bacillus thuringiensis* bioinsecticide in a field population of *Plutella xylostella* is due to a change in a midgut membrane receptor. Proceedings of the National Academy of Sciences of the USA 88: 5119–5123.

Flannagan, R.D., Yu, C.G., Mathis, J.P., Meyer, T.E., Shi, X., Siqueira, H.A., and Siegfried, B.D., 2005. Identification, cloning and expression of a Cry1Ab cadherin receptor from European corn borer, *Ostrinia nubilalis* (Hubner) (Lepidoptera: Crambidae). Insect Biochemistry and Molecular Biology 35: 33–40.

Forcada, C., Alcácer, E., Garcerá, M.D., and Martínez, R., 1996. Differences in the midgut proteolytic activity of two *Heliothis virescens* strains, one susceptible and one resistant to *Bacillus thuringiensis* toxins. Archives of Insect Biochemistry and Physiology 31: 257–272.

Forcada, C., Alcácer, E., Garcerá, M.D., Tato, A., and Martínez, R., 1999. Resistance to *Bacillus thuringiensis* Cry1Ac toxin in three strains of *Heliothis virescens*: Proteolytic and SEM study of the larval midgut. Archives of Insect Biochemistry and Physiology 42: 51–63.

Gahan, L.J., Gould, F., and Heckel, D.G., 2001. Identification of a gene associated with Bt resistance in *Heliothis virescens*. Science 293: 857–860.

Gahan, L.J., Ma, Y.T., Coble, M.L., Gould, F., Moar, W.J., and Heckel, D.G., 2005. Genetic basis of resistance to Cry1Ac and Cry2Aa in *Heliothis virescens* (Lepidoptera: Noctuidae). Journal of Economic Entomology 98: 1357–1368.

Gahan, L.J., Gould, F., López, J.D., Jr., Micinski, S., and Heckel, D.G., 2007. A polymerase chain reaction screen of field populations of *Heliothis virescens* for a retrotransposon insertion conferring resistance to *Bacillus thuringiensis* toxin. Journal of Economic Entomology 100: 187–194.

Gazit, E., La Rocca, P., Sansom, M.S., and Shai, Y., 1998. The structure and organization within the membrane of the helices composing the pore-forming domain of *Bacillus thuringiensis* delta-endotoxin are consistent with an "umbrella-like" structure of the pore. Proceedings of the National Academy of Sciences of the USA 95: 12289–12294.

Gill, M., and Ellar, D., 2002. Transgenic Drosophila reveals a functional in vivo receptor for the *Bacillus thuringiensis* toxin Cry1Ac1. Insect Molecular Biology 11: 619–625.

Georghiou, G.P., 1988. Implications of potential resistance to biopesticides. In: Biotechnology, Biological Pesticides, and Novel Plant Pest Resistance for Insect Pest Management, D.W. Roberts and R.R. Granados, eds., Boyce Thompson Institute, Ithaca, NY, USA, pp.137–146.

Girijashankar, V., Sharma, H.C., Sharma, K.K., Swathisree, V., Prasad, L.S., Bhat, B.V., Royer, M., San Segundo, B., Narasu, M.L., Altosaar, I., and Seetharama, N., 2005. Development of transgenic sorghum for insect resistance against the spotted stem borer (*Chilo partellus*). Plant Cell Reports 24: 513–522.

Gómez, I., Oltean, D.I., Gill, S.S., Bravo, A., and Soberón, M., 2001. Mapping the epitope in cadherin-like receptors involved in *Bacillus thuringiensis* Cry1A toxin interaction using phage display. Journal of Biological Chemistry 276: 28906–28912.

Gómez, I., Miranda-Rios, J., Rudino-Pinera, E., Oltean, D.I., Gill, S.S., Bravo, A., and Soberón, M., 2002a. Hydropathic complementarity determines interaction of epitope (869)HITDTNNK(876) in *Manduca sexta* Bt-R(1) receptor with loop 2 of domain II of *Bacillus thuringiensis* Cry1A toxins. Journal of Biological Chemistry 277: 30137–30143.

Gómez, I., Sánchez, J., Miranda, R., Bravo, A., and Soberón, M., 2002b. Cadherin-like receptor binding facilitates proteolytic cleavage of helix alpha-1 in domain I and oligomer pre-pore formation of *Bacillus thuringiensis* Cry1Ab toxin. FEBS Letters 513: 242–246.

Gómez, I., Dean, D.H., Bravo, A., and Soberón, M., 2003. Molecular basis for *Bacillus thuringiensis* Cry1Ab toxin specificity: Two structural determinants in the *Manduca sexta* Bt-R1 receptor interact with loops alpha-8 and 2 in domain II of Cy1Ab toxin. Biochemistry 42: 10482–10489.

Gómez, I., Arenas, I., Benítez, I., Miranda-Rios, J., Becerril, B., Grande, R., Almagro, J.C., Bravo, A., and Soberón, M., 2006. Specific epitopes of domains II and III of *Bacillus thuringiensis* Cry1Ab toxin involved in the sequential interaction with cadherin and aminopeptidase-N receptors in *Manduca sexta*. Journal of Biological Chemistry 281: 34032–34039.

González-Cabrera, J., Escriche, B., Tabashnik, B.E., and Ferré, J., 2003. Binding of *Bacillus thuringiensis* toxins in resistant and susceptible strains of pink bollworm (*Pectinophora gossypiella*). Insect Biochemistry and Molecular Biology 33: 929–935.

Gould, F., Martínez-Ramírez, A., Anderson, A., Ferré, J., Silva, F.J., and Moar, W.J., 1992. Broad-spectrum resistance to *Bacillus thuringiensis* toxins in *Heliothis virescens*. Proceedings of the National Academy of Sciences of the USA 89: 7986–7990.

Gould, F., Anderson, A., Reynolds, A., Bumgarner, L., and Moar, W., 1995. Selection and genetic analysis of a *Heliothis virescens* (Lepidoptera: Noctuidae) strain with high levels of resistance to *Bacillus thuringiensis* toxins. Journal of Economic Entomology 88: 1545–1559.

Gould, F., Anderson, A., Jones, A., Sumerford, D., Heckel, D.G, Lopez, J., Micinski, S., Leonard, R., and Laster, M., 1997. Initial frequency of alleles for resistance to *Bacillus thuringiensis* toxins in field populations of *Heliothis virescens*. Proceedings of the National Academy of Sciences of the USA 94: 3519–3523.

Greenplate, J.T., Penn, S.R., Shappley, Z., Oppenhuizen, M., Mann, J., Reich, B., and Osborn, J., 2000. BollgardII efficacy: Quantification of total lepidopteran activity in a 2-gene product. In: Beltwide Cotton Conference Proceedings, P. Dugger and D. Richter, eds., National Cotton Council, Memphis, TN, USA, pp. 1041–1043.

Griffitts, J.S., Whitacre, J.L., Stevens, D.E., and Aroian, R.V., 2001. Bt toxin resistance from loss of a putative carbohydrate-modifying enzyme. Science 293: 860–864.

Griffitts, J.S., Huffman, D.L., Whitacre, J.L., Barrows, B.D., Marroquin, L.D., Muller, R., Brown, J.R., Hennet, T., Esko, J.D., and Aroian, R.V., 2003. Resistance to a bacterial toxin is mediated by removal of a conserved glycosylation pathway required for toxin-host interactions. Journal of Biological Chemistry 278: 45594–45602.

Griffitts, J.S., Haslam, S.M., Yang, T., Garczynski, S.F., Mulloy, B., Morris, H., Cremer, P.S., Dell, A., Adang, M.J., and Aroian, R.V., 2005. Glycolipids as receptors for *Bacillus thuringiensis* crystal toxin. Science 307: 922–925.

Gunning, R.V., Dang, H.T., Kemp, F.C., Nicholson, I.C., and Moores, G.D., 2005. New resistance mechanism in *Helicoverpa armigera* threatens transgenic crops expressing *Bacillus thuringiensis* Cry1Ac toxin. Applied and Environmental Microbiology 71: 2558–2563.

Guo, S., Cui, H., Xia, L., Wu, D., Ni, W., Zhang, Z., Zhang, B., and Xu, Y., 1999. Development of bivalent insect-resistant transgenic cotton plants. Scientia Agricultura Sinica 32: 1–6.

Hawthorne, D., Siegfried, B., Shelton, T., and Hellmich, R., 2001. Monitoring for resistance alleles: A report from an advisory panel on insect resistance monitoring methods for Bt corn (October 18–20, 2000). Unpublished study submitted to EPA (part of Monsanto's 2000 IRM monitoring report). MRID # 453205-02.

Heckel, D.G., Gahan, L.C., Gould, F., and Anderson, A., 1997. Identification of a linkage group with a major effect on resistance to *Bacillus thuringiensis* Cry1Ac endotoxin in the tobacco budworm (Lepidoptera: Noctuidae). Journal of Economic Entomology 90: 75–86.

Heckel, D.G., Gahan, L.J., Baxter, S.W., Zhao, J.-Z., Shelton, A.M., Gould, F., and Tabashnik, B.E., 2007. The diversity of Bt resistance genes in species of Lepidoptera. Journal of Invertebrate Pathology 95: 192–197.

Hernández, C.S., and Ferré, J., 2005. Common receptor for *Bacillus thuringiensis* toxins Cry1Ac, Cry1Fa, and Cry1Ja in *Helicoverpa armigera*, *Helicoverpa zea* and *Spodoptera exigua*. Applied and Environmental Microbiology 71: 5627–5629.

Hodgman, T.C., and Ellar, D.J., 1990. Models for the structure and function of the *Bacillus thuringiensis* delta-endotoxins determined by compilational analysis. DNA Sequence 1: 97–106.

Hua, G., Jurat-Fuentes, J.L., and Adang, M.J., 2004a. Bt-R1a extracellular cadherin repeat 12 mediates *Bacillus thuringiensis* Cry1Ab binding and cytotoxicity. Journal of Biological Chemistry 279: 28051–28056.

Hua, G., Jurat-Fuentes, J.L., and Adang, M.J., 2004b. Fluorescent-based assays establish *Manduca sexta* Bt-R(1a) cadherin as a receptor for multiple *Bacillus thuringiensis* Cry1A toxins in Drosophila S2 cells. Insect Biochemistry and Molecular Biology 34: 193–202.

Huang, F., Buschman, L.L., Higgins, R.A., and McGaughey, W.H., 1999. Inheritance of resistance to *Bacillus thuringiensis* toxin (Dipel ES) in the European corn borer. Science 284: 965–967.

Huang, F., Leonard, B.R., and Andow, D.A., 2007. Sugarcane borer (Lepidoptera: Crambidae) resistance to transgenic *Bacillus thuringiensis* maize. Journal of Economic Entomology 100: 164–171.

Huckaba, R.M., Lassiter, R.B., Huang, X., Blanco, C.A., Langston, V.B., Braxton, L.B., Haile, F.J., Richardson, J.M., and Pellow, J., 2003. Field efficacy of Dow-Agrosciences MXB-13 transgenic cotton for control of cotton bollworm and tobacco budworm. In: Beltwide Cotton Conference Proceedings, P. Dugger and D. Richter, eds., National Cotton Council, Memphis, TN, USA, pp. 1293–1298.

Huffman, D.L., Abrami, L., Sasik, R., Corbeil, J., van der Goot, F.G., and Aroian, R.V., 2004. Mitogen-activated protein kinase pathways defend against bacterial pore-forming toxins. Proceedings of the National Academy of Sciences of the USA 101: 10995–11000.

Ibargutxi, M.A., Estela, A., Ferré, J., and Caballero, P., 2006. Use of *Bacillus thuringiensis* toxins for control of the cotton pest *Earias insulana* (Boisd.) (Lepidoptera: Noctuidae). Applied and Environmental Microbiology 72: 437–442.

International Life Science Institute (ILSI), 1999. An Evaluation of Insect Resistance Management in Bt Field Corn: A Science-Based Framework for Risk Assessment and Risk Management. Report of an Expert Panel. November 23, 1998.

Iracheta, M.M., Pereyra-Alferez, B., Galan-Wong, L., and Ferré, J., 2000. Screening for *Bacillus thuringiensis* crystal proteins active against the cabbage looper, *Trichoplusia ni*. Journal of Invertebrate Pathology 76: 70–75.

Iriarte, J., Bel, Y., Ferrandis, M., Andrew, R., Murillo, J., Ferré, J., and Caballero, P., 1998. Environmental distribution and diversity of *Bacillus thuringiensis* in Spain. Systematic and Applied Microbiology 21: 97–106.

James, C., 2002. Global review of commercialized transgenic crops: 2001 feature: Bt cotton. ISAAA Brief No. 26, International Service for the Acquisition of Agri-Biotech Applications, Ithaca, NY, USA.

James, C., 2007. Global status of commercialized biotech/GM crops: 2007. ISAAA Brief No. 37, International Service for the Acquisition of Agri-Biotech Applications, Ithaca, NY, USA.

Janmaat, A.F., and Myers, J., 2003. Rapid evolution and the cost of resistance to *Bacillus thuringiensis* in greenhouse populations of cabbage loopers, *Trichoplusia ni*. Proceedings of the Royal Society of London, Series B 270: 2263–2270.

Jenkins, J.N., McCarty, J.C., Jr., and Wofford, T., 1995. Bt cotton a new era in cotton production. In: Beltwide Cotton Conference Proceedings, P. Dugger and D. Richter, eds., National Cotton Council, Memphis, TN, USA, pp. 171–173.

Jurat-Fuentes, J.L., and Adang, M.J., 2001. Importance of Cry1 delta-endotoxin domain II loops for binding specificity in *Heliothis virescens* (L.). Applied and Environmental Microbiology 67: 323–329.

Jurat-Fuentes, J.L., and Adang, M.J., 2004. Characterization of a Cry1Ac-receptor alkaline phosphatase in susceptible and resistant *Heliothis virescens* larvae. European Journal of Biochemistry 271: 3127–3135.

Jurat-Fuentes, J.L., and Adang, M.J., 2006a. Cry toxin mode of action in susceptible and resistant *Heliothis virescens* larvae. Journal of Invertebrate Pathology 92: 166–171.

Jurat-Fuentes, J.L., and Adang, M.J., 2006b. The *Heliothis virescens* cadherin protein expressed in *Drosophila* S2 cells functions as a receptor for *Bacillus thuringiensis* Cry1A but not Cry1Fa toxins. Biochemistry 45: 9688–9695.

Jurat-Fuentes, J.L., Gould, F.L., and Adang, M.J., 2003. Dual resistance to *Bacillus thuringiensis* Cry1Ac and Cry2Aa toxins in *Heliothis virescens* suggests multiple mechanisms of resistance. Applied and Environmental Microbiology 69: 5898–5906.

Jurat-Fuentes, J.L., Gahan, L.J., Gould, F.L., Heckel, D.G., and Adang, M.J., 2004. The HevCaLP protein mediates binding specificity of the Cry1A class of *Bacillus thuringiensis* toxins in *Heliothis virescens*. Biochemistry 43: 14299–14305.

Karumbaiah, L., Oppert, B., Jurat-Fuentes, J.L., and Adang, M.J., 2007. Analysis of midgut proteinases from *Bacillus thuringiensis*-susceptible and -resistant *Heliothis virescens* (Lepidoptera: Noctuidae). Comparative Biochemistry and Physiology B 146: 139–146.

Kinsinger, R.A., and McGaughey, W.H., 1979. Susceptibility of populations of Indian meal moth and almond moth to *Bacillus thuringiensis*. Journal of Economic Entomology 72: 346–349.

Kirsch, K., and Schmutterer, H., 1988. Low efficacy of a *Bacillus thuringiensis* (Berl.) formulation in controlling the diamondback moth, *Plutella xylostella* (L.) in the Philippines. Journal of Applied Entomology 105: 249–255.

Knowles, B.H., 1994. Mechanism of action of *Bacillus thuringiensis* insecticidal δ-endotoxins. Advances in Insect Physiology 24: 275–308.

Kranthi, K.R., Kranthi, S., Ali, S., and Banerjee, S.K., 2000. Resistance to "Cry1Ac δ-endotoxin of *Bacillus thuringiensis*" in a laboratory selected strain of *Helicoverpa armigera* (Hübner). Current Science 78: 1001–1004.

Kranthi, K.R., Dhawad, C.S., Naidu, S.R., Mate, K., Behere, G.T., Wadaskar, R.M., and Kranthi, S., 2006. Inheritance of resistance in Indian *Helicoverpa armigera* (Hübner) to Cry1Ac toxin of *Bacillus thuringiensis*. Crop Protection 25: 119–124.

Lambert, B., and Peferoen, M., 1992. Insecticidal promise of *Bacillus thuringiensis*. BioScience 42: 112–122.

Lee, M.K., Rajamohan, F., Gould, F., and Dean, D.H., 1995. Resistance to *Bacillus thuringiensis* CryIA δ-endotoxins in a laboratory-selected *Heliothis virescens* strain is related to receptor alteration. Applied and Environmental Microbiology 61: 3836–3842.

Leong, K.L.H., Cano, R.J., and Kubinski, A.M., 1980. Factors affecting *Bacillus thuringiensis* total field persistence. Environmental Entomology 9: 593–599.

Li, H., Oppert, B., Higgins, R.A., Huang, F., Zhu, K.Y., and Buschman, L.L., 2004. Comparative analysis of proteinase activities of *Bacillus thuringiensis*-resistant and -susceptible *Ostrinia nubilalis* (Lepidoptera: Crambidae). Insect Biochemistry and Molecular Biology 34: 753–762.

Liang, G., Tan, W., and Guo, Y., 2000. Study on screening and inheritance mode of resistance to Bt transgenic cotton in cotton bollworm. Acta Entomologica Sinica 43 (Suppl.): 57–62.

Liu, Y.B., and Tabashnik, B.E., 1997. Experimental evidence that refuges delay insect adaptation to *Bacillus thuringiensis*. Proceedings of the Royal Society of London, Series B 264: 605–610.

Liu, Y.B., Tabashnik, B.E., Dennehy, T.J., Patin, A.L., and Bartlett, A.C., 1999. Development time and resistance to *Bt* crops. Nature 400: 519.

Liu, Y.B., Tabashnik, B.E., Meyer, S.K., Carrière, Y., and Bartlett, A.C., 2001. Genetics of pink bollworm resistance to *Bacillus thuringiensis* toxin Cry1Ac. Journal of Economic Entomology 94: 248–252.

Luo, S., Wang, G., Liang, G., Wu, K.M., Bai, L., Ren, X., and Guo, Y., 2006. Binding of three Cry1A toxins in resistant and susceptible strains of cotton bollworm (*Helicoverpa armigera*). Pesticide Biochemistry and Physiology 85: 104–109.

Luttrell, R.G., Wan, L., and Knighten, K., 1999. Variation in susceptibility of Noctuid (Lepidoptera) larvae attacking cotton and soybean to purified endotoxin proteins and commercial formulations of *Bacillus thuringiensis*. Journal of Economic Entomology 92, 21–32.

Ma, G., Roberts, H., Sarjan, M., Featherstone, N., Lahnstein, J., Akhurst, R., and Schmidt, O., 2005. Is the mature endotoxin Cry1Ac from *Bacillus thuringiensis* inactivated by a coagulation reaction in the gut lumen of resistant *Helicoverpa armigera* larvae? Insect Biochemistry and Molecular Biology 35: 729–739.

MacIntosh, S., Jansens, S., Reed, J., and Newhouse, K., 1998. StarLink™ Corn: Insect Resistance Management Plan for a *Bacillus thuringiensis* Cry9C Corn Product, submitted to EPA October 1998. MRID#44679601.

Mahon, R.J., Olsen, K.M., Garsia, K.A., and Young, S.R., 2007. Resistance to the Bt toxin Cry2Ab in a strain of *Helicoverpa armigera* (Hübner) (Lepidoptera: Noctuidea) in Australia. Journal of Economic Entomology 100: 894–902.

Marçon, P., Siegfried, B., Spencer, T., and Hutchinson, W., 2000. Development of diagnostic concentrations for monitoring *Bacillus thuringiensis* resistance in European corn borer (Lepidoptera: Crambidae). Journal of Economic Entomology 93: 925–930.

Martínez-Ramírez, A.C., Gould, F., and Ferré, J., 1999. Histopathological effects and growth reduction in a susceptible and a resistant strain of *Heliothis virescens* (Lepidoptera: Noctuidae) caused by sublethal doses of pure Cry1A crystal proteins from *Bacillus thuringiensis*. Biocontrol Science and Technology 9: 239–246.

Matten, S.R., and Reynolds, A.H., 2003. EPA IRM requirements for Bollgard II cotton. In: Beltwide Cotton Conference Proceedings, P. Dugger and D. Richter, eds., National Cotton Council, Memphis, TN, USA, pp. 1111–1121.

McGaughey, W.H., 1985. Insect resistance to the biological insecticide *Bacillus thuringiensis*. Science 229: 193–195.

McGaughey, W.H., and Beeman, R.W., 1988. Resistance to *Bacillus thuringiensis* in colonies of Indianmeal moth and almond moth (Lepidoptera: Pyralidae). Journal of Economic Entomology 81: 28–33.

McGaughey, W.H., and Johnson, D.E., 1992. Indianmeal moth (Lepidoptera: Pyralidae) resistance to different strains and mixtures of *Bacillus thuringiensis*. Journal of Economic Entomology 85: 1594–600.

McNall, R.J., and Adang, M.J., 2003. Identification of novel *Bacillus thuringiensis* Cry1Ac binding proteins in *Manduca sexta* midgut through proteomic analysis. Insect Biochemistry and Molecular Biology 33: 999–1010.

Meng, F., Shen, J., Zhou, W., and Cen, H., 2003. Long-term selection for resistance to transgenic cotton expressing *Bacillus thuringiensis* toxin in *Helicoverpa armigera* (Hübner) (Lepidoptera: Noctuidae). Pest Management Science 60: 167–172.

Metz, T.D., Roush, R.T., Tang, J.D., Shelton, A.M., and Earle, E.D., 1995. Transgenic broccoli expressing a *Bacillus thuringiensis* insecticidal crystal protein: Implications for pest resistance management strategies. Molecular Breeding 1: 309–317.

Moar, W.J., Pusztai-Carey, M., van Faassen, H., Frutos, R., Rang, C., Luo, K., and Adang, M.J. 1995. Development of *Bacillus thuringiensis* CryIC resistance by *Spodoptera exigua* (Lepidoptera: Noctuidae). Applied and Environmental Microbiology 61: 2086–2092.

Morin, S., Biggs, R.W., Sisterson, M.S., Shriver, L., Ellers-Kirk, C., Higginson, D., Holley, D., Gahan, J.J., Heckel, D.G., Carrière, Y., Dennehy, T.J., Brown, J.K., and Tabashnik, B.E., 2003. Three cadherin alleles associated with resistance to *Bacillus thuringiensis* in pink bollworm. Proceedings of the National Academy of Sciences of the USA 100: 5004–5009.
Nagamatsu, Y., Koike, T., Sasaki, K., Yoshimoto, A., and Furukawa, Y., 1999. The cadherin-like protein is essential to specificity determination and cytotoxic action of the *Bacillus thuringiensis* insecticidal CryIAa toxin. FEBS Letters 460: 385–390.
Nakanishi, K., Yaoi, K., Nagino, Y., Hara, H., Kitami, M., Atsumi, S., Miura, N., and Sato, R., 2002. Aminopeptidase N isoforms from the midgut of *Bombyx mori* and *Plutella xylostella* – their classification and the factors that determine their binding specificity to *Bacillus thuringiensis* Cry1A toxin. FEBS Letters 519: 215–220.
NCGA (National Corn Growers Association), 2006. Corn Growers Value Insect Protection Technology, Implement IRM Plans in 2005. http://www.ncga.com/news/releases/2006/news042706.asp (accessed 4 January 2008).
Oppert, B., Kramer, K.J., Beeman, R.W., Johnson, D., and McGaughey, W.H., 1997. Proteinase-mediated insect resistance to *Bacillus thuringiensis* toxins. Journal of Biological Chemistry 272: 23473–23476.
Pardo-López, L., Gómez, I., Muñoz-Garay, C., Jiménez-Juarez, N., Soberón, M., and Bravo, A., 2006. Structural and functional analysis of the pre-pore and membrane-inserted pore of Cry1Ab toxin. Journal of Invertebrate Pathology 92: 172–177.
Peyronnet, O., Noulin, J.F., Laprade, R., and Schwartz, J.L., 2004. Patch-clamp study of the apical membrane of the midgut of *Manduca sexta* larvae: Direct demonstration of endogenous channels and effect of a *Bacillus thuringiensis* toxin. Journal of Insect Physiology 50: 791–803.
Pigott, C.R., and Ellar, D.J., 2007. Role of receptors in *Bacillus thuringiensis* crystal toxin activity. Microbiology and Molecular Biology Reviews 71: 255–281.
Powell, K., 2003. Concerns over refuge size for US EPA-approved Bt corn. Nature Biotechology 21: 467–468.
Pozsgay, M., Fast, P., Kaplan, H., and Carey, P.R., 1987. The effect of sunlight on the protein crystals from *Bacillus thuringiensis* var. *kurstaki* HD1 and NRD12: A Raman spectroscopic study. Journal of Invertebrate Pathology 50: 246–253.
Pray, C.E., Huang, J., Ma, D., and Qiao, F., 2001. Impact of Bt cotton in China. World Development 29: 813–825.
Puntheeranurak, T., Stroh, C., Zhu, R., Angsuthanasombat, C., and Hinterdorfer, P., 2005. Structure and distribution of the *Bacillus thuringiensis* Cry4Ba toxin in lipid membranes. Ultramicrosopy 105: 15–24.
Rajagopal, R., Sivakumar, S., Agrawal, N., Malhotra, P., and Bhatnagar, R.K., 2002. Silencing of midgut aminopeptidase N of *Spodoptera litura* by double-stranded RNA establishes its role as *Bacillus thuringiensis* toxin receptor. Journal of Biological Chemistry 370: 971–978.
Rajendran, L., and Simons, K., 2005. Lipid rafts and membrane dynamics. Journal of Cell Science 118: 1099–1102.
Rausell, C., Muñoz-Garay, C., Miranda-CassoLuengo, R., Gómez, I., Rudino-Pinera, E., Soberón, M., and Bravo, A., 2004a. Tryptophan spectroscopy studies and black lipid bilayer analysis indicate that the oligomeric structure of Cry1Ab toxin from *Bacillus thuringiensis* is the membrane-insertion intermediate. Biochemistry 43: 166–174.
Rausell, C., Pardo-López, L., Sánchez, J., Muñoz-Garay, C., Morera, C., Soberón, M., and Bravo, A., 2004b. Unfolding events in the water-soluble monomeric Cry1Ab toxin during transition to oligomeric pre-pore and membrane-inserted pore channel. Journal of Biological Chemistry 279: 55168–55175.
Ravi, K.C., Mohan, K.S., Manjunath, T.M., Head, G., Patil, B.V., Greba, A., Premalatha, K., Peter, J., and Rao, N.G.V., 2005. Relative abundance of *Helicoverpa amrigera* (Lepidoptera: Noctuidae) on different host crops in India and the role of these crops as natural refuge for *Bacillus thuringiensis* cotton. Environmental Entomology 34: 59–69.
Roush, R.T., 1997. Bt-transgenic crops: Just another pretty insecticide or a chance for a new start in resistance management? Pesticide Science 51: 328–334.

Sangiorgio, V., Pitto, M., Palestini, P., and Masserini, M., 2004. GPI-anchored proteins and lipid rafts. Italian Journal of Biochemistry 53: 98–111.

SAP (Scientific Advisory Panel), 1998. Subpanel on *Bacillus thuringiensis* (Bt) Plant-Pesticides: Transmittal of the Final Report of the FIFRA Scientific Advisory Panel Subpanel on *Bacillus thuringiensis* Plant-Pesticides and Resistance Management, Meeting held on February 9–10, 1998. Report dated, April 28, 1998. Docket Number: OPPTS- 00231. http://www.epa.gov/scipoly/sap/meetings/1998/index.htm (accessed 4 January 2008).

SAP (Scientific Advisory Panel), 2001. Subpanel on Insect Resistance Management Report: Sets of Scientific Issues Being Considered by the Environmental Protection Agency Regarding: Bt Plant-Pesticides Risk and Benefit Assessments. Report dated, March 12, 2001. http://www.epa.gov/oscpmont/sap/meetings/2000/index.htm (accessed 4 January 2008).

Sequeira, R.V., and Playford, C.L., 2001. Abundance of *Helicoverpa* (Lepidoptera: Noctuidae) pupae under cotton and other crops in central Queensland: Implications for resistance management. Australian Journal of Entomology 40: 264–269.

Shelton, A.M., Tang, J.D., Roush, R.T., Metz, T.D., and Earle, E.D., 2000. Field tests on managing resistance to Bt-engineered plants. Nature Biotechnology 18: 339–342.

Shelton, A.M., Zhao, J.-Z., and Roush, R.T., 2002. Economic, ecological, food safety, and social consequences of the deployment of Bt transgenic plants. Annual Review of Entomology 47: 845–881.

Siegfried, B., and Spencer, T., 2000. Monitoring Bt susceptibility of European corn borer to Cry1Ab: 2000 data summary. Unpublished study submitted to EPA (part of Monsanto's 2000 IRM report). MRID # 453205-02.

Siegfried, B., Spencer, T., and Nearman, J., 1999. Monitoring *Bt* susceptibility of European corn borer and corn earworm to Cry1Ab. Unpublished study submitted to EPA (part of Monsanto's 1999 IRM research report). MRID # 450369-02.

Siegfried, B.D., Spencer, T., and Nearman, J., 2000. Baseline susceptibility of the corn earworm (Lepidoptera: Noctuidae) to the Cry1Ab toxin from *Bacillus thuringiensis*. Journal of Economic Entomology 93: 1265–1268.

Siegfried, B.D., Vaughn, T.T., and Spencer, T., 2005. Baseline susceptibility of western corn rootworm (Coleoptera: Crysomelidae) to Cry3Bb1 *Bacillus thuringiensis* toxin. Journal of Economic Entomology 98: 1320–1324.

Sims, S.R., and Stone, T.B., 1991. Genetic basis of tobacco budworm resistance to an engineered *Pseudomonas fluorescens* expressing the δ-endotoxin of *Bacillus thuringiensis kurstaki*. Journal of Invertebrate Pathology 57: 206–210.

Sims, S.B., Greenplate, J.T. Stone, T.B. Caprio, M.A., and Gould, F.L., 1996. Monitoring strategies for early detection of Lepidoptera resistance to *Bacillus thuringiensis* insecticidal proteins. In: Molecular Genetics and Evolution of Pesticide Resistance, T.M. Brown, ed., American Chemical Society Symposium Series No. 645, American Chemical Society, Washington, DC, USA, pp. 229–242.

Siqueira, H.A.A., Moellenbeck, D., Spencer, T., and Siegfried, B.D., 2004. Cross-resistance of Cry1Ab-selected *Ostrinia nubilalis* (Lepidoptera: Crambidae) to *Bacillus thuringiensis* δ-endotoxins. Journal of Economic Entomology 97: 1049–1057.

Sivakumar, S., Rajagopal, R., Venkatesh, G.R., Srivastava, A., and Bhatnagar, R.K., 2007. Knockdown of aminopeptidase-N from *Helicoverpa armigera* larvae and in transfected SF21 cells by RNA interference reveals its functional interaction with *Bacillus thuringiensis* insecticidal protein Cry1Ac. Journal of Biological Chemistry 282: 7312–7319.

Sjoblad, R.D., McClintock, J.T., and Engler, R., 1992. Toxicological considerations for protein components of biological pesticide products. Regulatory Toxicology and Pharmacology 15: 3–9.

Soberón, M., Pardo-López, L., López, I., Gómez, I., Tabashnik, B.E., and Bravo, A., 2007. Engineering modified Bt toxins to counter insect resistance. Science 318: 1640–1642.

Song, S.S., 1991. Resistance of diamondback moth (*Plutella xylostella* L.: Yponomeutidae: Lepidoptera) against *Bacillus thuringiensis* Berliner. Korean Journal of Applied Entomology 30: 291–293.

Song, Q., Luppens, C., and Gan, X., 2000. Monitoring the susceptibility of the southwestern cornborer, *Diatraea grandiosella*, to *Bacillus thuringiensis* toxin Cry1Ab. Unpublished study submitted to EPA (part of Monsanto's 2000 IRM report). MRID # 453205-02.

Stodola, T.J., Andow, D.A., Hyden, A.R., Hinton, J.L., Roark, J.J., Buschman, L.L., Porter, P., and Cronholm, G.B., 2006. Frequency of resistance to *Bacillus thuringiensis* toxin Cry1Ab in southern United States Corn Belt population of European corn borer (Lepidoptera: Crambidae). Journal of Economic Entomology 99: 502–507.

Stone, T.B., Sims, S.R., and Marrone, P.G., 1989. Selection of tobacco budworm for resistance to a genetically engineered *Pseudomonas fluorescens* containing the δ-endotoxin of *Bacillus thuringiensis* subsp. *kurstaki*. Journal of Invertebrate Pathology 53: 228–234.

Tabashnik, B.E., 1994. Evolution of resistance to *Bacillus thuringiensis*. Annual Review of Entomology 39: 47–79.

Tabashnik, B.E., Cushing, N.L., Finson, N., and Johnson, M., 1990. Field Development of resistance to *Bacillus thuringiensis* in diamondback moth (Lepidoptera: Plutellidae). Journal of Economic Entomology 83: 1671–1676.

Tabashnik, B.E., Finson, N., Johnson, M.W., and Heckel, D.G., 1994. Cross-resistance to *Bacillus thuringiensis* toxin CryIF in the diamondback moth (Lepidoptera: Plutellidae). Applied and Environmental Microbiology 60: 4627–4629.

Tabashnik, B.E., Johnson, K.W., Engleman, J.T., and Baum, J.A., 1996. Cross-resistance of the diamondback moth indicates altered interactions with domain II of *Bacillus thuringiensis* toxins. Applied and Environmental Microbiology 62: 2839–2844.

Tabashnik, B.E., Liu, Y.B., Malvar, T., Heckel, D.G., Masson, L., Ballester, V., Granero, F., Ménsua, J.L., and Ferré, J., 1997. Global variation in the genetic and biochemical basis of diamondback moth resistance to *Bacillus thuringiensis*. Proceedings of the National Academy of Sciences of the USA 94: 12780–12785.

Tabashnik, B.E., Liu, Y.B., de Maagd, R.A., and Dennehy, T.J., 2000a. Cross-resistance of pink bollworm (*Pectinophora gossypiella*) to *Bacillus thuringiensis* toxins. Applied and Environmental Microbiology 66: 4582–4584.

Tabashnik, B.E., Patin, A.L., Dennehy, T.J., Liu, Y.B., Carrière, Y., Sims, M.A., and Antilla L., 2000b. Frequency of resistance to *Bacillus thuringiensis* in field populations of pink bollworm. Proceedings of the National Academy of Sciences of the USA 97: 12980–12984.

Tabashnik, B.E., Liu, Y.B., de Maagd, R., and Dennehy, T.J., 2000c. Cross-resistance of pink bollworm (*Pectinophora gossypiella*) to *Bacillus thuringiensis* toxins. Applied and Environmental Microbiology 66: 4582–4584.

Tabashnik, B.E., Liu, Y.B., Dennehy, T.J., Sims, M.A., Sisterson, M.S., Biggs, R.W., and Carrière, Y., 2002. Inheritance of resistance to Bt toxin Cry1Ac in a field-derived strain of pink bollworm (Lepidoptera: Gelechiidae). Journal of Economic Entomology 95: 1018–1026.

Tabashnik, B.E., Liu, Y.B., Unnithan, D.C., Carrière, Y., Dennehy, T.J., and Morin, S., 2004. Shared genetic basis of resistance to Bt toxin Cry1Ac in independent strains of pink bollworm. Journal of Economic Entomology 97: 721–726.

Tabashnik, B.E., Biggs, R.W., Higginson, D.M., Henderson, S., Unnithan, D.C., Unnithan, G.C., Ellers-Kirk, C., Sisterson, M.S., Dennehy, T.J., Carriere, Y., and Morin, S., 2005a. Association between resistance to Bt cotton and cadherin genotype in pink bollworm. Journal of Economic Entomology 98: 635–644.

Tabashnik, B.E., Dennehy, T.J., and Carrière, Y., 2005b. Delayed resistance to transgenic cotton in pink bollworm. Proceedings of the National Academy of Sciences of the USA 102: 15389–15393.

Tabashnik, B.E., Fabrick, J.A., Henderson, S., Biggs, R.W., Yafuso, C.M., Nyboer, M.E., Manhardt, N.M., Coughlin, L.A., Sollome, J., Carriere, Y., Dennehy, T.J., and Morin, S., 2006. DNA screening reveals pink bollworm resistance to Bt cotton remains rare after a decade of exposure. Journal of Economic Entomology 99: 1525–1530.

Tanaka, H., and Kimura, Y., 1991. Resistance to Bt formulation in diamondback moth, *Plutella xylostella* L., on watercress. Japanese Journal of Economic Entomology and Zoology 35: 253–255.

Tang, J.D., Shelton, A.M., Van Rie, J., De Roeck, S., Moar, W.J., Roush, R.T., and Peferoen, M., 1996. Toxicity of *Bacillus thuringiensis* spore and crystal protein to resistant diamondback moth (*Plutella xylostella*). Applied and Environmental Microbiology 62: 564–569.

Tang, J.D., Collins, H.L., Metz, T.D., Earle, E.D., Zhao, J.Z., Roush, R.T., and Shelton, A.M., 2001. Greenhouse tests on resistance management of Bt transgenic plants using refuge strategies. Journal of Economic Entomology 94: 240–247.

Tomimoto, K., Hayakawa, T., and Hori, H., 2006. Pronase digestion of brush border membrane-bound Cry1Aa shows that almost the whole activated Cry1Aa molecule penetrates into the membrane. Comparative Biochemistry and Physiology B 144: 413–422.

Tsuda, Y., Nakatani, F., Hashimoto, K., Ikawa, S., Matsuura, C., Fukada, T., Sugimoto, K., and Himeno, M., 2003. Cytotoxic activity of *Bacillus thuringiensis* Cry proteins on mammalian cells transfected with cadherin-like Cry receptor gene of *Bombyx mori* (silkworm). Biochemical Journal 369: 697–703.

USEPA (United States Environmental Protection Agency), 2001a. Bt Biopesticides Registration Action Document D. Insect Resistance Management. http://www.epa.gov/oppbppd1/biopesticides/pips/bt_brad2/4-irm.pdf (accessed 4 January 2008).

USEPA (United States Environmental Protection Agency), 2001b. *Bt* Cotton Refuge Requirements for the 2001 Growing Season. http://www.epa.gov/pesticides/biopesticides/pips/bt_cotton_refuge_2001.htm (accessed 4 January 2008).

USEPA (United States Environmental Protection Agency), 2005. Biopesticides Registration Action Document. *Bacillus thuringiensis* var. *aizawai* Cry1F and the Genetic Material (from the Insert of Plasmid PGMA281) Necessary for Its Production in Cotton and *Bacillus thuringiensis* var. *kurstaki* Cry1Ac and the Genetic Material (from the Insert of Plasmid PMYC3006) Necessary for Its Production in Cotton. http://www.epa.gov/pesticides/biopesticides/ingredients/tech_docs/brad_006512-006513.pdf (accessed 4 January 2008).

USEPA (United States Environmental Protection Agency), 2006. New Biopesticide Active Ingredients – 2006. http://www.epa.gov/pesticides/biopesticides/product_lists/new_ai_2006.htm (accessed 4 January 2008).

USEPA (United States Environmental Protection Agency), 2007a. Bacillus thuringiensis Cry3Bb1 Protein and the Genetic Material Necessary for Its Production (Vector ZMIR13L) in Event MON863 Corn (006484) Biopesticide Registration Action Document (BRAD). http://www.epa.gov/pesticides/biopesticides/ingredients/tech_docs/brad_006484.htm (accessed 4 January 2008).

USEPA (United States Environmental Protection Agency), 2007b. Notice of pesticide registration, No. 68467-3, approved July 17, 2007.

USEPA/USDA (United States Environmental Protection Agency/ United States States Department of Agriculture), 1999. EPA/USDA Position Paper on Insect Resistance Management for Bt Crops. Docket Number: D-16122A (posted 8/26/1999).

Vachon, V., Schwartz, J.L., and Laprade, R., 2006. Influence of the biophysical and biochemical environment on the kinetics of pore formation by Cry toxins. Journal of Invertebrate Pathology 92: 160–165.

Van Rie, J., Jansens, S., Höfte, H., Degheele, D., and Van Mellaert, H., 1989. Specificity of *Bacillus thuringiensis* δ-endotoxins: Importance of specific receptors on the brush border membranes of the mid-gut of target insects. European Journal of Biochemistry 186: 239–247.

Van Rie, J., McGaughey, W.H., Johnson, D.E., Barnett, B.D., and Van Mellaert, H., 1990. Mechanism of insect resistance to the microbial insecticide *Bacillus thuringiensis*. Science 247: 72–74.

Vaughn, T., Cavato, T., Brar, G., Coombe, T., DeGooyer, T., Ford, S., Groth, M., Howe, A., Johnson, S., Kolacz, K., Pilcher, C., Purcell, J., Romano, C., English, L., and Pershing, J., 2005. A method of controlling corn rootworm using a *Bacillus thuringiensis* protein expressed in transgenic maize. Crop Science 45: 931–938.

Vié, V., Van Mau, N., Pomarède, P., Dance, C., Schwartz, J.L., Laprade, R., Frutos, R., Rang, C., Masson, L., Heitz, F., and Le Grimellec, C., 2001. Lipid-induced pore formation of the *Bacillus thuringiensis* Cry1Aa insecticidal toxin. Journal of Membrane Biology 180: 195–203.

Vijaykumar, Bashasab, F., Krishnareddy, K.B., Kuruvinashetti, M.S., and Patil, B.V., 2007. Mating compatibility among *Helicoverpa armigera* (Lepidoptera: Noctuidae) occurring on selected host plants and Bt cotton survivors. Journal of Economic Entomology 100: 903–908.

Wang, P., Zhao, J.Z., Rodrigo-Simón, A., Kain, W. Janmaat, A.F., Shelton, A.M., Ferré, J., and Myers, J., 2007. Mechanism of resistance to *Bacillus thuringiensis* toxin Cry1Ac in a greenhouse population of cabbage looper, *Trichoplusia ni*. Applied and Environmental Microbiology 73: 1199–1207.

Whalon, M.E., and McGaughey, W.H., 1998 *Bacillus thuringiensis*: Use and resistance management. In: Insecticides with Novel Modes of Action, I. Ishaaya and D. Degheele, eds., Springer, New York, USA, pp. 106–137.

Whalon, M.E., Miller, D.L., Hollingworth, R.M., Grafius, E.J., and Miller, J.R., 1993. Selection of a Colorado potato beetle (Coleoptera: Chrysomelidae) strain resistant to *Bacillus thuringiensis*. Journal of Economic Entomology 86: 226–233.

Wu, K.M., and Guo, Y.Y., 2005. The evolution of cotton pest management practices in China. Annual Review of Entomology 50: 31–52.

Wu, K., Yuyuan, G., Lv, N., Greenplate, J.T., and Deaton, R., 2002. Resistance monitoring of *Heliocoverpa armigera* (Lepidoptera: Noctuidae) to *Bacillus thuringiensis* insecticidal protein in China. Journal of Economic Entomology 95: 826–831.

Xie, R., Zhuang, M., Ross, L.S., Gómez, I., Oltean, D.I., Bravo, A., Soberón, M., and Gill, S.S., 2005. Single amino acid mutations in the cadherin receptor from *Heliothis virescens* affect its toxin binding ability to Cry1A toxins. Journal of Biological Chemistry 280: 8416–8425.

Xu, X., Yu, L., and Wu, Y., 2005. Disruption of a cadherin gene associated with resistance to Cry1Ac δ-endotoxin of *Bacillus thuringiensis* in *Helicoverpa armigera*. Applied and Environmental Microbiology 71: 948–954.

Yang, Y., Chen, H., Wu, S., Yang, Y., Xu, X., and Wu, Y., 2006. Identification and molecular detection of a deletion mutation responsible for a truncated cadherin of *Helicoverpa armigera*. Insect Biochemistry and Molecular Biology 36: 735–740.

Zhang, X., Candas, M., Griko, N.B., Rose-Young, L., and Bulla, L.A., Jr., 2005. Cytotoxicity of *Bacillus thuringiensis* Cry1Ab toxin depends on specific binding of the toxin to the cadherin receptor BT-R1 expressed in insect cells. Cell Death and Differentiation 12: 1407–1416.

Zhang, X., Candas, M., Griko, N.B., Taussig, R., and Bulla, L.A., Jr., 2006. A mechanism of cell death involving an adenylyl cyclase/PKA signaling pathway is induced by the Cry1Ab toxin of *Bacillus thuringiensis*. Proceedings of the National Academy of Sciences of the USA 103: 9897–9902.

Zhao, J.Z., Collins, H.L., Tang, J.D., Cao, J., Earle, E.D, Roush, R.T., Herrero, S., Escriche, B., Ferré, J., and Shelton, A.M., 2000. Development and characterization of diamondback moth resistance to transgenic broccoli expressing high levels of Cry1C. Applied and Environmental Microbiology 66: 3784–3789.

Zhao, J.Z., Li, Y., Collins, H.L., and Shelton, A.M., 2002. Examination of the F2 screen for rare resistance alleles to *Bacillus thuringiensis* toxins in the diamondback moth. Journal of Economic Entomology 95: 14–21.

Zhao, J.Z., Cao, J., Li, Y., Collins, H.L., Roush, R.T., Earle, E.D., and Shelton, A.M., 2003. Transgenic plants expressing two *Bacillus thuringiensis* toxins delay insect resistance evolution. Nature Biotechnology 21: 1493–1497.

Zhao, J.Z., Cao, J., Collins, H.L., Bates, S.L., Roush, R.T., Earle, E.D., and Shelton, A.M., 2005. Concurrent use of transgenic plants expressing a single and two *Bacillus thuringiensis* genes speeds insect adaptation to pyramided plants. Proceedings of the National Academy of Sciences of the USA 102: 8426–8430.

Zhuang, M., Oltean, D.I., Gómez, I., Pullikuth, A.K., Soberón, M., Bravo, A., and Gill, S.S., 2002. *Heliothis virescens* and *Manduca sexta* lipid rafts are involved in Cry1A toxin binding to the midgut epithelium and subsequent pore formation. Journal of Biological Chemistry 277: 13863–13872.

Zoebelein, G., 1990. Twenty-three year surveillance of development of insecticide resistance in diamondback moth from Thailand (*Plutella xylostella* L., Lepidoptera, Plutellidae). Mededelingen Faculteit Landbouwwetenschappen Rijksuniversiteit, Gent 55: 313–322.

Chapter 4
Insect-Resistant Transgenic Crops and Biological Control

Jörg Romeis[1,*], Roy G. Van Driesche[2], Barbara I.P. Barratt[3], and Franz Bigler[1]

Abstract Natural enemies such as predators and parasitoids fulfil an important ecological and economic function by helping to keep herbivore populations below the economic injury level. Thus, they contribute to sustainable integrated pest management (IPM) systems. It is well established that plant resistance factors that affect herbivores also interact with natural enemies and consequently with the biological control function they provide. Similarly, host plant resistance derived from genetic engineering will have an impact on biological control. There is evidence today that the currently available transgenic crops that express Cry proteins derived from *Bacillus thuringiensis* (*Bt*) have no direct effects on natural enemies due to their narrow spectrum of activity. However, the fact that the target pests are efficiently controlled by the deployed *Bt* crops has inevitable consequences for natural enemies that specialize on these species as hosts or prey. On the other hand, it has become clear that in crop systems where the deployment of *Bt* varieties has lead to a decline in insecticide use, biological control organisms have benefited significantly. Consequently, this technology can contribute to natural enemy conservation and thus be a useful tool in IPM.

4.1 Introduction

On a global scale, the area planted to genetically modified (GM) crops is steadily increasing, passing 114 million hectares in 2007 (James, 2007). While the majority of the crops have been modified for herbicide tolerance, more than 42 million hectares express an insecticidal trait, i.e., Cry proteins (δ-endotoxins) derived from

[1] Agroscope Reckenholz-Tänikon Research Station ART, 8046 Zurich, Switzerland

[2] Department of Plant, Soil & Insect Sciences, University of Massachusetts, Amherst, MA, USA

[3] AgResearch Invermay, PB 50034, Mosgiel, New Zealand

* To whom correspondence should be addressed. E-mail: joerg.romeis@art.admin.ch

J. Romeis, A.M. Shelton, G.G. Kennedy (eds.), *Integration of Insect-Resistant Genetically Modified Crops within IPM Programs.*

the soil bacterium *Bacillus thuringiensis* (*Bt*). Since the commercial introduction of the first GM variety in 1996, a vast body of research on the potential environmental impacts of such crops has been conducted and has not revealed any harm beyond that encountered with traditional pest-resistant crops and far less harm than caused by conventional pesticides (Sanvido et al., 2007).

Host plant resistance is one of the major tactics used to protect crops against pests and diseases and is an important part of integrated pest management (IPM) that helps keep herbivore densities below the economic injury level (see Fig. 6.1 in Naranjo et al., chapter 6). Insect-resistant GM crops may be considered as having a specific form of host plant resistance and there is no reason to hypothesize that GM host plant resistance will affect biological control organisms in any other way than conventional resistance (Kennedy and Gould, 2007). While biological control services are provided by a range of organisms including arthropod predators and parasitoids, nematodes and pathogens, we focus on arthropod natural enemies in this chapter since they have received the most research effort with regard to insect-resistant GM crops (O'Callaghan et al., 2005; Romeis et al., 2006, 2008a).

Most IPM systems aim to enhance biological control through conservation of existing natural enemies, or to introduce new ones through inoculation or inundation (Bale et al., 2008). Thus, it is important to minimize the non-target effects of other components of IPM, such as pesticides, GM plants or habitat manipulation. Expectations that biological control can act effectively as a sole method of pest management in field crops are generally unrealistic. Biological control, however, is an important component of IPM systems. Conserving beneficial organisms alongside other crop management practices requires that the pest manager understands the role of biological control agents in regulating pests, their biology, environmental requirements and the ways in which they can be adversely affected by other practices. Only then can biological control be optimally combined with the other elements of an IPM system.

Insect-resistant plants, whether produced by conventional breeding or through genetic engineering, can have impacts on natural enemies (Hare, 2002; Kennedy and Gould, 2007). Such effects can stem from changes in the plant structure or primary or secondary plant metabolites. Adverse effects can occur, for example, if the natural enemy is exposed to the plant-born insecticidal factor and is susceptible to it. These factors can cause population level effects which might lead to changes in the level of biological control that natural enemies provide, something that is of relevance in the context of IPM (Kennedy and Gould, 2007; Kennedy, chapter 1). Due to their importance, natural enemies of crop pests and the biological control they provide are considered during the environmental risk assessment of GM crops that express insecticidal proteins (Mendelsohn et al., 2003; Garcia-Alonso et al., 2006; Raybould et al., 2007; Romeis et al., 2008a, b).

In this chapter we review the direct and indirect impacts of conventional and transgenic host plant resistance on natural enemies and biological control and indirect impacts that are due to changes in the agricultural system as, for example, reductions in the amount of insecticides used due to the adoption of GM crops.

4.2 Interactions of Insect-Resistant Plants and Herbivores

4.2.1 *Plant Resistance to Herbivores*

Plant resistance to insects has played a major role in the development of agricultural and horticultural crops. It is self-sustaining, requires little management, and is generally compatible with other pest management tactics. Economically, plant resistance can often yield higher returns on investment than insecticide development (Smith, 2005). Secondary plant metabolites such as allelochemicals provide natural plant defense to insect pests and the opportunity for enhancement of these products has been an important focus for the development of cultivars with increased pest resistance.

Plant resistance to herbivores can be intrinsic (governed by the plant itself) or extrinsic (induced by herbivores) (Price, 1986). Insect-resistance can be categorised as follows (Painter, 1951; Smith, 2005; Kennedy, chapter 1): (1) Antixenosis in which the plant adversely affects insect behaviour through morphological traits (such as leaf surface waxes or hairs which limit herbivore movement or feeding) or chemical traits (such as allelochemicals that repel insects or inhibit feeding). Such chemical traits can be induced by insect damage to the plant. (2) Antibiosis in which the plant adversely affects insect survival or development through allelochemicals. For example, specialist herbivores are often able to sequester such secondary plant compounds and use them for their own defense against natural enemies, whereas generalist herbivores are typically unable to do this. Antibiosis may also be achieved through growth inhibition as a result of reduced nutrient content. (3) Tolerance which occurs when plants can withstand and recover from a high level of insect attack, due either to morphological characteristics that reduce the impact of damage, or the plant's ability to maintain its photosynthetic capacity despite loss of considerable leaf area, or differential regulation of plant hormones such as auxins and abscissic acid.

The insect-resistant GM varieties that are commercialized today provide resistance of the antibiosis type, i.e., they kill the pest insects or significantly affect their development, but other resistance modalities may also be developed in the future for incorporating into IPM programs. For example, plant tolerance is often considered more advantageous in IPM systems than insect-induced resistance and antibiosis because there is no selection pressure and consequently counter-adaptation by the pests is unlikely (Rausher, 2001). Tolerance might involve sustained yield despite pest infestation by means of compensatory growth, rapid wound healing, or morphological characteristics. Plants with tolerance to insect feeding have sometimes developed in response to environmental stress such as low moisture. For example, tall fescue (*Festuca arundinacea*) is a useful forage plant in New Zealand because it has a very fibrous root system, which confers a degree of drought tolerance. This also provides tolerance to root-pruning scarab larvae because the root system can support a higher density of pests than other pasture grasses (East et al., 1980). Insect-tolerant crops are also less likely to adversely affect natural enemies; indeed, the higher pest densities sustained by insect tolerant crops may benefit natural enemy populations. A downside of insect-tolerant crops is the fact that elevated pest populations may spread to non-tolerant/resistant crops and cause damage there. However, to date no genes conferring tolerance to insect damage have been identified for any crop plant, and tolerance is thought to be comparatively complex genetically (Rausher, 2001). The potential to develop GM plants with an antixenosis type of resistance is discussed later in this chapter (section 4.7.2).

4.2.2 Plant Resistance and Tritrophic Interactions

Natural enemies have an important role to play in the co-evolution of plants and insects. "The third trophic level must be considered as part of a plant's battery of defenses against herbivores" (Price et al., 1980). Herbivore populations are consequently subject to regulation by both top-down (natural enemies) and bottom-up (host plant resistance) effects (*sensu* Power, 1992). As with plant-herbivore interactions, the tritrophic relationships are dynamic because of the shifting balance of selection pressures.

There have been several reviews of plant-herbivore-natural enemy interactions (e.g., Price, 1986; Price et al., 1980; Hare, 1992, 2002; Bottrell et al., 1998; Olff et al., 1999; Tollrian and Harvell, 1999; Groot and Dicke, 2002; Kennedy and Gould, 2007). These reviews give a number of examples of conventionally bred insect-resistant plants that have detrimental effects on different important life-table parameters of natural enemies. Conversely, there are studies that have provided examples of positive effects or enhancement of natural enemy activity on insect-resistant plants, and some plants with pest resistance that appear to have no impact on biological control agents.

Plants employ a range of physical defense systems based on morphology such as trichomes and epicuticular waxes that directly affect the efficacy of natural enemies (Obrycki, 1986; Eigenbrode and Espelie, 1995; Mohite and Uthamasamy, 1998; Simmons and Gurr, 2004; Romeis et al., 2005). Trichomes sometimes secrete sticky substances that limit mobility of natural enemies (Romeis et al., 1999; Lovinger et al., 2000) and sometimes also produce compounds that are repellent or toxic to natural enemies (Kennedy, 2003). While these effects are most often negative for the natural enemy, exceptions have also been reported. For example, it was observed that *Encarsia formosa* (Hymenoptera: Aphelinidae) encountered more whitefly hosts on hairy rather than glaborous-leaved cucumbers as a result of reduced movement (van Lenteren and de Ponti, 1990). Similarly, epicuticular waxes can affect access of predators or parasitoids to herbivores (Eigenbrode, 2004). For example, predation by *Hippodamia convergens* (Coleoptera: Coccinellidae) was increased on *Pisum sativum* or *Brassica oleracea* plants with reduced-wax leaves when compared to normal-wax plants due to the fact that the predators were better able to grip the plant's surface.

Multitrophic effects occur when plant traits indirectly affect natural enemies, for example via impacts on an herbivore. Such interactions can be beneficial, detrimental or neutral to natural enemies (Hare, 1992, 2002). They might be beneficial, for example, when the growth rate of the herbivore is prolonged so that it is potentially available to the natural enemies for longer or when the herbivore's defense behaviour or immune system is weakened (Benrey and Denno, 1997; Turlings and Benrey, 1998). However, if the quality of the herbivore as host/prey is reduced as a result of feeding on an insect-resistant plant, this can result in natural enemies with reduced fitness traits (Smith, 2005; see section 4.3.3). For example *Helicoverpa zea* (Lepidoptera: Noctuidae) feeding on resistant tomato plants containing 2-undecanone are unable to survive through the pupal stage, and hence the tachinid parasitoid *Archytus marmoratus* fails to complete its development (Farrar et al., 1992).

Toxic compounds sequestered by herbivores can also be toxic to natural enemies. Campbell and Duffey (1979) found that an alkaloid in tomatoes, α-tomatine, which was enhanced during plant breeding to help inhibit pests, is toxic to *Hyposoter exiguae* (Hymenoptera: Ichneumonidae), an endoparasitoid of a major tomato pest, *H. zea*. The parasitoid acquires the alkaloid from its less suceptible host after the host has ingested the compound. Furthermore, is it thought that α-tomatine might deter infection of caterpillars by entomopathogenic fungi and reduce the rate of predation by a predatory stinkbug (Gallardo et al., 1990; Traugott and Stamp, 1996). Another example of a negative effect on parasitoid fitness induced by the herbivore's host plant is the toxic effect of nicotine. The fitness of *Hypogaster annulipes* (Hymenoptera: Ichneumonidae), a generalist parasitoid of Noctuidae (Lepidoptera) caterpillars was reduced in hosts fed a diet with high nicotine levels (El-Heneidy et al., 1988). In contrast, *Cotesia congregata* (Hymenoptera: Braconidae), a specialist parasitoid of the tobacco hornworm, *Manduca sexta* (Lepidoptera: Sphingidae), was less severely affected (Barbosa et al., 1986)

Resistant plants can also alter herbivore behaviour and activity patterns, which can change the level of exposure to natural enemies. For example, increased movement of the rice brown planthopper, *Nilaparvata lugens* (Hemiptera: Delphacidae), on resistant rice plants increases its susceptibility to predatory spiders (Kartohardjono and Heinrichs, 1984).

4.3 Impact of Insect-Resistant GM Plants on Natural Enemies: Individual Level Effect

For the insecticidal proteins of insect-resistant GM plants to directly affect an individual natural enemy, the organism has to be exposed to the toxin and be susceptible to it. Consequently, an organism is not affected by the GM plant when either exposure or sensitivity (hazard) does not occur. For an effect to be of ecological relevance it must result in changes in population or community processes. Similarly, direct or indirect effects of the GM plant on individual natural enemies, natural enemy species or groups/guilds of natural enemies might not lead to a decreased biological control function. Those principles are the same as for insect-resistant plants that are bred by conventional techniques.

4.3.1 Exposure to Insecticidal Proteins

In contrast to applications of chemical insecticides with contact toxicity, insecticidal proteins expressed by GM plants have to be ingested to affect arthropods. This reduces the number of non-target species exposed to the toxin. Consequently, it becomes important to assess which organisms are exposed and at what level. It is evident that the level at which an organism is exposed to a plant-expressed insecticidal protein can vary depending on the concentration of the toxin in the plant or environment, the plant tissue in which the protein is expressed, and the feeding behaviour of the non-target organism. Therefore, exposure pathways and levels can be predicted only if the relevant information for the GM plant, the environment and the natural enemy is available (Dutton et al., 2003; Romeis et al., 2008a).

A number of routes have been identified through which natural enemies of herbivores can be exposed to insecticidal proteins expressed by GM plants (Romeis et al., 2008a). The most direct route is through plant feeding. Many predators are facultative feeders of pollen and plant sap, while both predators and parasitoids utilize extra-floral nectar (Wäckers, 2005; Wäckers and Van Rijn, 2005). Much research has been devoted to *Bt* Cry protein expression in the pollen of *Bt* maize plants following the alarm (later found not to be justified) raised for safety of monarch caterpillars (*Danaus plexippus*; Lepidoptera: Danaidae) exposed to pollen of *Bt* maize (Shelton and Sears, 2001). This exposure route through GM pollen is potentially important because maize pollen is actively consumed by a large number of predators (e.g., *Coleomegilla maculata* [Coleoptera: Coccinellidae], Lundgren et al., 2004, 2005), *Apis melifera* (Hymenoptera: Apidae) (Keller et al., 2005), web-building spiders (when recycling their web) (Ludy and Lang, 2006), or caterpillars feeding on foliage covered with pollen (Stanley-Horn et al., 2001; Zangerl et al., 2001). However, the amount of toxin expressed in today's Cry1Ab-expressing *Bt* maize varieties (Bt11 and MON810) is extremely low, i.e., a factor of more than one hundred times lower than in green leaf tissue (Dutton et al., 2003). In contrast to the Lepidoptera-resistant maize varieties, today's corn-rootworm (*Diabrotica* spp.)-resistant *Bt* (Cry3Bb1) maize varieties express toxin concentrations in the pollen that are close to that expressed in leave tissue (Monsanto Company, 2003, 2004).

Another important food source for natural enemies in agricultural fields is honeydew produced by sap-feeding Sternorrhyncha (Hemiptera) such as aphids (Aphididae), planthoppers (Delphacidae), and leafhoppers (Cicadellidae) (Wäckers, 2005). If insecticidal proteins appeared in honeydew, this could expose many natural enemies to the toxin. However, there is evidence that the Cry proteins expressed by today's *Bt* crops do not enter the phloem sap and thus do not appear in honeydew (Romeis et al., 2006, 2008a). The situation is different for plants that express toxins targeting sap-feeding pests since the toxin must be present in the phloem sap to reach the target pests. Insecticidal proteins that are ingested by sap-feeders are excreted in the insects' honeydew since they typically possess low proteolytic activity in the gut (Srivastava and Auclair, 1963; Rahbé et al., 1995). This has, for example, been shown for transgenic plants expressing lectins (Shi et al., 1994; Kanrar et al., 2002) or protease inhibitors (Rahbé et al., 2003). Similar findings were reported earlier for secondary plant compounds (Wink and Römer, 1986; Malcolm, 1990).

The second major route through which natural enemies are potentially exposed to plant-expressed insecticidal proteins is through their prey or host organisms. The amount of toxin that can be detected in herbivores feeding on *Bt* plants varies greatly among species (Harwood et al., 2005; Obrist et al., 2006a; Torres et al., 2006) and can even differ among life stages of herbivores (Fig. 4.1) (Howald et al., 2003; Obrist et al., 2005). This variation is due to a number of factors, including the plant parts or tissues that the organisms feed on and the fate of the toxin after ingestion. In general, herbivores contain lower amounts of insecticidal protein than the plant material they feed on. Comparable or even higher toxin levels have so far only been reported for spider mites (Dutton et al., 2002; Obrist et al. 2006a, b;

Fig. 4.1 Larvae of the green lacewing, *Chrysoperla carnea* (Neuroptera: Chrysopidae), preying on thrips. While this predator mainly attacks aphids it will also feed on other soft bodied prey when it is encountered. Studies by Obrist et al. (2005) have revealed that, given a choice, the non-feeding pre-pupal and pupal stages of the thrips, *Frankliniella tenuicornis* (Thysanoptera: Thripidae), are preferably attacked when compared to the moving larvae and adults. The *Bt* concentration in *Bt* (Cry1Ab) maize-fed thrips was reported to be much lower in the non-feeding pre-pupal and pupal stages (0.09 and 0.03 µg Cry1Ab g^{-1} fresh weight, respectively) when compared to the feeding adult and larval stages (1 and 21 µg Cry1Ab g^{-1} fresh weight, respectively) (Photo by Gabriela Brändle, Agroscope ART)

Torres and Ruberson, 2008). Two studies have shown that the Cry1Ab protein remains biologically active after ingestion by both a susceptible and a non-susceptible herbivore (Head et al., 2001; Obrist et al., 2006b).

Similar to plant-dwelling natural enemies, species living on the ground or in the soil are potentially exposed to the insecticidal proteins through various routes. Toxins from GM plants can be released into the soil via senescent plant material including pollen, via root exudates and via faeces of plant feeding organisms such as snails (Icoz and Stotsky, 2008). Thus, soil-dwelling natural enemies such as mites, predatory beetles or spiders will be exposed to the plant-produced insecticidal compounds. Despite the fact that *Bt* Cry proteins may persist in soil as a consequence of their binding to surface-active clay and humic acid compounds, none of the laboratory or field studies that have been conducted suggest accumulation of *Bt* toxins in soil even after several years of cultivation (Sanvido et al., 2007).

4.3.2 Direct Toxic Effects

Bt Cry proteins are known for their specificity, being active only against a narrow range of organisms. This host range limitation is due to the mode of action of these toxins. *Bt* Cry proteins need to be activated by gut proteases, bind to specific receptors in the brush border membrane of midgut epithelial cells, insert in the membrane and form a pore, which results in cell leakage (Schnepf et al., 1998). The Cry proteins expressed in today's *Bt*-transgenic maize and cotton varieties are known to be specific to Lepidoptera (e.g., Cry1A or Cry2A proteins) or Coleoptera (Cry3 proteins). The non-target toxicity studies conducted by biotechnology companies (as part of the regulatory risk assessment) or public research scientists have revealed no direct toxic effects on natural enemies (USEPA, 2001; Romeis et al., 2006).

It is most likely that insecticidal GM plants that will enter the market in the near future will express either single Cry proteins or combinations of such proteins. Besides new *Bt* maize and cotton varieties, other plants likely to be released in the foreseeable future include *Bt* rice (Cohen et al., chapter 8) and vegetables (Shelton et al., chapter 9). Other plants that are close to commercialization are maize and cotton lines expressing vegetative insecticidal proteins (Vip's), which are non-δ-endotoxins that are expressed by *Bt* in the vegetative growth stages before sporulation. The currently deployed Vip3A protein has been reported to be very specific to Lepidoptera (Estruch et al., 1996; Warren, 1997). Consequently, Vip3A-expressing transgenic plants are likely to cause minimal non-target effects, similar to today's transgenic crops that target Lepidoptera (Malone et al., chapter 13). In addition to the GM plants that express genes derived from *Bt*, experimental plants have been produced that express different proteins with insecticidal activity such as avidin, protease or α-amylase inhibitors, or lectins (Malone et al., chapter 13). In general these compounds have a much broader activity spectrum and consequently a higher potential to cause direct effects to non-target organisms (O'Callaghan et al., 2005; Malone et al., chapter 13).

Recently, two studies have demonstrated the potential of using RNA interference (RNAi) as a new method to protect plants from the attack by insect pests. Baum et al. (2007) showed that GM maize expressing a dsRNA targeting a subunit of the midgut enzyme vacuolar ATPase revealed strong protection against larvae of *Diabrotica virgifera virgifera* (Coleoptera: Chrysomelidae). Even though the studies on other coleopteran species indicate that gene silencing was potentially very selective, the potential of this technology to directly affect non-target herbivores or natural enemies still has to be critically evaluated. Another approach was followed by Mao et al. (2007). They used RNAi to silence a *Helicoverpa armigera* (Lepidoptera: Noctuidae) gene, which is induced by gossypol and appears to neutralize it. This silencing caused the insect's tolerance towards this powerful plant defense compound to be significantly reduced. Again it remains to be seen whether this approach would also render other organisms more sensitive to the plant chemical.

4.3.3 Prey/Host-Quality Mediated Effects

Natural enemies can also be affected indirectly by the GM plant when they feed on sublethally impaired herbivores ("sick prey"). Such effects appear to be caused by declines in the nutritional quality of the host/prey organisms. These so-called prey/host-quality mediated effects appear to account for most (if not all) of the *Bt* plant-effects on natural enemies that have been reported from laboratory and glasshouse studies (Romeis et al., 2006). It is well established that parasitoids are especially vulnerable to changes in their host's quality, since they usually complete their development in a single host (Godfray, 1994). Despite the fact that Cry1A-expressing plants are deployed to control pest Lepidoptera, a number of studies have focused on parasitoids that attack Lepidoptera larvae and it is not surprising that parasitoid life-table parameters are also significantly affected when the host suffers (Romeis et al., 2006). In extreme cases, parasitoids attack sublethally affected hosts that die before the parasitoid offspring completes development (Schuler et al., 2004; Jiang et al., 2004 cited in Chen et al., 2006; Davison et al., 2006). When tested under confined conditions, predators have also been found to be affected by altered prey quality when feeding on *Bt*-fed susceptible prey items. The most widely cited example is that of larvae of the green lacewing, *Chrysoperla carnea* (Neuroptera: Chrysopidae). When *C. carnea* larvae were fed lepidopteran larvae reared on Cry1Ab-expressing maize, a significantly prolonged larval development and an increased mortality were observed (Hilbeck et al., 1998; Dutton et al., 2002). Interestingly, predator larvae were not affected when feeding on non-Cry1Ab-susceptible *Tetranychus urticae* (two-spotted spider mite) (Acari: Tetranychidae) known to contain large amounts of biologically active *Bt* toxin (Dutton et al., 2002; Obrist et al., 2006b). These results suggested that the *C. carnea* larvae were affected by the reduced nutritional quality of the *Bt*-fed Lepidoptera larvae rather than directly by the *Bt* toxin. This was confirmed in direct feeding studies with high doses of purified Cry1A proteins and by the finding that the proteins do not bind to lacewing gut membranes, a prerequisite of toxicity, in both histopathological and *in vitro*-binding studies (Romeis et al., 2004; Rodrigo-Simón et al., 2006; Lawo and Romeis, 2008). Similar prey-quality mediated effects have, for example, also been reported for carabid beetles (Meissle et al., 2005; Ferry et al., 2006).

On the other hand, predators and parasitoids may also benefit from sublethally affected prey/host herbivores. For example, host defense behaviours against attacking parasitoids may be altered or the immune systems of potential hosts may be weakened, resulting in a lower rate of egg encapsulation. In addition, sublethally affected herbivorous arthropods may have longer development times, which could expose them to prolonged predation and parasitism that could translate into increased mortality of pests. Such positive effects were observed for partially resistant GM plants and parasitoids. Using low-expressing *Bt* tobacco plants as a model system, Johnson and Gould (1992) and Johnson (1997) showed that parasitism of *Heliothis virescens* (Lepidoptera: Noctuidae) larvae by *Campoletis sonorensis* (Hymenoptera: Ichneumonidae) was higher on the *Bt* plants compared to the non-transformed control plants. The authors suggested that this was due to prolonged larval development, which provided a larger 'window of opportunity' for the parasitoids to attack, and to increased larval movement on the plants. Bell et al. (1999) reported that the parasitoid *Eulophus pennicornis* (Hymenoptera: Eulophiae) benefited when its host, the tomato moth *Lacanobia oleracea* (Lepidoptera: Noctuidae), was fed a GNA-containing diet. These effects were later confirmed in an extensive glasshouse study where GNA-expressing potato plants and parasitoids acted synergistically to control *L. oleracea* (Bell et al., 2001).

4.4 Impacts of Insect-Resistant GM Crops on Natural Enemies: Population Level Effects

A large number of experimental field studies have been conducted with Cry protein-expressing maize, cotton, potato and rice varieties to address population level effects on target and non-target organisms. In addition, data have been collected in farmer fields to study the impact of GM crop deployment on insecticide use and consequences of insecticide reductions for arthropod biodiversity. With the exception of some recently published studies on plants expressing Vip3A or protease inhibitors, no non-target field investigations have been reported with plants expressing non-*Bt* Cry proteins (Malone et al., chapter 13). This section is therefore largely focusing on the experience with Cry toxin-expressing plants.

While most field studies have generally assessed arthropod abundance and species richness in *Bt* compared to non-*Bt* fields, some have focused particularly on natural enemies of the target pests. A number of review articles have compiled the available knowledge. Romeis et al. (2006) provide a comprehensive review of the information on the impact of *Bt*-transgenic crops on biological control published through 2005. Chen et al. (2006) provide an excellent overview on the Chinese research on non-target effects of *Bt* rice plants. Marvier et al. (2007) conducted a meta-analysis of the published field studies on non-target effects of *Bt* crops. Unfortunately, the authors did not differentiate between natural enemies (i.e., non-target species that should be protected) and pest herbivores, making it difficult to use their results to assess impacts on biological control agents.

4.4.1 Effect of Key-Pest Densities on Natural Enemies

Both conventionally bred and GM insect-resistant plants have been developed with the objective of reducing pest densities below damage thresholds. If successful, reduced pest densities will inevitably lead to a reduction in the abundance of some natural enemies, particularly the parasitoids and predators that are host/prey specific to the target pest(s). This is an obvious and unavoidable consequence of virtually any pest management system, irrespective of the mechanism and should not be of particular concern related to the use of GM plants (OECD, 1993; EFSA, 2006; Kennedy and Gould, 2007; Romeis et al., 2006, 2008a, b).

Overall, the available field results from *Bt* crops confirm the findings of the studies conducted under confined conditions: *Bt* plants provide good protection against the target pests but have no direct impact on natural enemies with the exception of density-dependent effects on specialists. For example, reduced parasitism levels of the target pest *O. nubilalis* have been reported from *Bt* maize for *Lydella thompsoni* and *Pseudoperichaeta nigrolineata* (Diptera: Tachinidae), which was attributed to lower populations of their host (Bourguet et al., 2002). Similarly, Pilcher et al. (2005) and Bruck et al. (2006) reported a significantly reduced abundance of *Macrocentrus cingulum* (Hymenoptera: Braconidae) in Cry1Ab-expressing maize fields. In Cry3A-expressing potatoes, Riddick et al. (1998) observed a significant reduction in the relative abundance of a specialist carabid predator of the target pest, the Colorado potato beetle (*Leptinotarsa decemlineata*, Coleoptera: Chrysomelidae), but not for a generalist adult coccinellid. For comparison, in *Bt* rice, where the target pests (stemborers) are not a dominant arthropod group, control did not lead to detectable changes in the parasitoid communities (Li et al., 2007). At the field level, the consequences of these density-dependent effects for the GM crop are negligible since the pest is controlled anyway. However, they might become important when a GM crop with a specific trait such as *O. nubilalis* resistance is planted on large areas and results in large scale suppression of specific natural enemies that are important in suppressing the pest population in other crops (see section 4.6; Kennedy and Gould, 2007).

Natural enemies with a broad host/prey range are expected to be much less affected if densities of one or a few key pests are substantially reduced, since they may compensate for the absence or reduced nutritional quality of one prey by eating more or switching to other prey species, if available. For example, even in Lepidoptera-resistant *Bt* crops where the density of caterpillars is significantly reduced, eggs of the same species are still available as prey since the adult moths do not appear to show an oviposition response to the insecticidal trait (Hellmich et al., 1999; Liu et al., 2002; Torres and Ruberson, 2006). Nevertheless, Naranjo (2005a) has reported a significant decline in some plant-dwelling generalist predators in *Bt* cotton fields, which he attributed to the decline in pest caterpillars. However, this decline was found to have no effect on the biological control of other cotton pests (see section 4.4.3).

Similar to Cry protein-expressing GM crops, field studies in maize (Dively, 2005; Fernandes et al., 2007) and cotton (Whitehouse et al., 2007) crops expressing Vip3A either alone or in combination with Cry1Ab revealed no major differences in either species richness or diversity of the beneficial non-target communities.

Besides the expected effects that result from the reduced density of the target herbivores, non-target arthropods have been found to respond to the fact that the *Bt* crop is less damaged and shows a different growth and health pattern when compared to the unprotected non-transformed crop. A good example is a three-year field study with *Bt* maize conducted by Toschki et al. (2007). One of the three years was characterized by a high *O. nubilalis* infestation that coincided with a period of high temperatures that caused browning and drying of the unprotected non-*Bt* plots while the *Bt* plants remained green and grew well. In this year, significant differences in the abundance of epigeic arthropods (i.e., carabids and spiders) between *Bt* and non-*Bt* plots were detected. While abundance of hygrophilic species increased in the *Bt* plots, that of xerophilic species decreased relative to non-*Bt* plots. The fact that no differences were detected in the other two years of the study with little *O. nubilalis* infestation suggests that the earlier observed differences in arthropod abundance were caused by differences in crop structure and microclimatic conditions. A similar effect, although not on natural enemies, has been reported by Candolfi et al. (2004) and Dively (2005). Due to the reduction of borer infestation in *Bt* maize, the abundance of saprophagous beetles and flies was significantly reduced.

4.4.2 Reduction of Insecticides in Insect-Resistant GM Crops and Impact on Natural Enemies

The data that are currently available clearly show that the adoption of *Bt*-transgenic varieties has led to substantial reductions in the use of chemical insecticides (Fitt, chapter 11; Qaim et al., chapter 12). Large per acre reductions in conventional insecticide use and large areas planted to *Bt* crops means that these varieties are reducing agricultural insecticide use on a scale that outstrips all other IPM efforts.

For the period from 1996 to 2005, use of *Bt* (Cry1Ac) cotton caused a 19.4% reduction in the total volume of insecticide active ingredient (a.i.) in global cotton production (Brookes and Barfoot, 2006a, b). Data from many countries that grow *Bt* (Cry1Ac) cotton show that the average insecticide use in *Bt* cotton is reduced by 25% to 80% when compared to non-*Bt* cotton (Fitt, chapter 11). In particular, significant reductions in insecticide use have been recorded in developing countries where use of pesticides is often accompanied by serious health effects on farm workers (Raney, 2006; Brookes and Barfoot, 2006b; Fitt, chapter 11; Qaim et al., chapter 12). The novel double gene varieties like Bollgard II require even less insecticide. Data from four seasons in Australia showed an average reduction in insecticide a.i. of 65–75% (with a 80–90% reduction in number of sprays) (Fitt, chapter 11). The potential for insecticide reduction depends on a number of factors including the targeted pest complex, the intensity of infestation and the general level of insecticide application before the introduction of *Bt* cotton. For example, a comparative study in farmer fields in different states of the USA revealed that the impact of *Bt* cotton on insecticide use is low in areas where insecticides are mainly applied to control pest Hemiptera (Head et al., 2005).

Similarly, *Bt* sweet corn can potentially reduce the number of insecticide applications for the control of *O. nubilalis*, *H. zea* and *Spodoptera frugiperda* (Lepidoptera: Noctuidae) by 70–90% (Musser and Shelton, 2003; Rose and Dively, 2007; Fitt, chapter 11). In contrast, the use of *Bt* maize (field corn) has caused a decline of only 4.1% in insecticide a.i., estimated for the period 1996–2005 for maize on a global scale (Brookes and Barfoot, 2006a, b). Most of this reduction is from Lepidoptera-active *Bt* maize to control *O. nubilalis* and other stemborers. However, Coleoptera-active *Bt* maize shows a much greater potential reduction in the near future, as insecticides used against *Diabrotica* spp. comprise 25–30% of global total insecticide a.i. used in maize (Rice, 2004). The low reduction in insecticide use by Lepidoptera-active *Bt* maize is due to the fact that stemborers with tunneling larvae are difficult to control by insecticide sprays (and so they are relatively little used) and the fact that the current *Bt* maize varieties do not provide effective control of other lepidopteran pests such as *H. zea* and *S. frugiperda* (Hellmich et al., chapter 5).

Similar to cotton and sweet corn, the deployment of insect-resistant *Bt* rice or vegetables such as eggplant or crucifers will likely lead to significant reductions in insecticide use (Cohen et al., chapter 8; Shelton et al., chapter 9). An experimental field study with *Bt* rice in China for control of stemborers has already shown a great potential for insecticide reductions (Huang et al., 2005, 2008).

Various experimental field studies of *Bt* crops have shown that natural enemies, with the exception of specialist species that depend on the targeted pest(s) (see section 4.4.1), either increase in abundance or remain the same in unsprayed *Bt* plots compared to plots of the same crop managed with chemical insecticides (Romeis et al., 2006). Of particular interest are studies that have been conducted in farmer fields. Natural enemy abundance was higher in *Bt* cotton fields compared to conventionally managed non-*Bt* cotton fields when the deployment of the transgenic varieties reduced the number of insecticide sprays (Head et al., 2005; Torres and Ruberson, 2005). In a recent study in commercial cotton fields in Arizona, Sisterson et al. (2007) recorded that the number of insecticide sprays was reduced by more than 50% in *Bt* cotton when compared to non-*Bt* cotton fields. Abundances of two generalist predators, *C. carnea* and *Orius tristicolor* (Heteroptera: Anthocoridae), were negatively associated with the number of insecticide sprays. Similar results were reported by Cattaneo et al. (2006) in which the reduction in insecticide use in *Bt* cotton was associated with significantly higher abundances of ants and beetles. With the effects of insecticides statistically removed, neither study detected any significant effect of *Bt* crops on natural enemies.

Musser and Shelton (2003) assessed the impact of *Bt* sweet corn and various commonly used insecticides on the target pest *O. nubilalis* and common natural enemies such as the ladybeetles *Coleomegilla maculata* and *Harmonia axyridis* (Coleoptera: Coccinellidae), and *Orius insidiosus* (Heteroptera: Anthocoridae). Overall, *Bt* sweet corn was much better at preserving these predators while controlling *O. nubilalis* than were the commonly used insecticides lambda cyhalothrin, indoxacarb and spinosad. Recent studies by Leslie et al. (2007), Hoheisel and Fleischer (2007) and Rose and Dively (2007) have confirmed the potential of *Bt* sweet corn to conserve natural enemies.

4.4.3 Impacts on Biological Control Function

Despite the fact that natural enemies of crop pests are valued because of the natural pest control they provide, few studies have directly measured this biological control function. While parasitism of target pests has been reported to be lower in *Bt* crops due to a reduced density of the host (see section 4.4.1), there is no indication that insect-resistance provided by the expression of *Bt* Cry proteins has an antagonistic effect on the biological control of non-target pest species. A five-year study in *Bt* cotton fields revealed a significant drop of about 20% in the abundance of five generalist predators, which was likely to have been caused by the decline in prey (i.e., Lepidoptera larvae) (Naranjo, 2005a). Despite this drop, predator:prey ratios for *Lygus hesperus* (Hemiptera: Miridae) and *Bemisia tabaci* (Hemiptera: Aleyrodidae) did not differ between unsprayed *Bt* and non-*Bt* cotton indicating that biological control of non-target pests was not impaired (Naranjo, 2005b). Furthermore, predation on sentinel eggs of *Pectinophora gossypiella* (Lepidoptera: Gelechiidae) and natural enemy impact on *B. tabaci* remained unchanged. Similar results have been reported from a two-year study in Chinese *Bt* rice fields, which revealed no negative effect on the populations of three different planthopper species and their predator *Cyrtorhinus lividipennis* (Hemiptera: Miridae) (Chen et al., 2007).

In addition, natural enemy activity was enhanced in *Bt* crops when they received fewer insecticide applications compared to a corresponding non-transgenic crop, with positive effects for the control of non-target herbivore species (see also section 4.5.3). For example, Head et al. (2005) reported lower populations of armyworms (*Spodoptera* spp.) in *Bt* cotton fields in South Carolina, which were attributed to enhanced abundance and activity of natural enemies due to reduced application of insecticides. Similarly, higher predator abundance in *Bt* sweet corn compared to insecticide-sprayed non-*Bt* crops resulted in an enhanced predation of sentinel *O. nubilalis* egg batches (Musser and Shelton, 2003). A number of studies have reported lower populations of aphids in *Bt* potato (Reed et al., 2001), *Bt* cotton (Wu and Guo, 2003), and *Bt* maize (Bhatti et al., 2005), which were probably caused by increased biological control activity in the *Bt* crops.

4.5 The Role of Natural Enemies in Controlling Pests in Insect-Resistant GM Crops

Pest outbreaks of the target pest due to a disruption of its natural enemies are referred to as pest resurgence, while outbreaks of other herbivores because of activities directed at other (key) pests are called secondary pest outbreaks. Here we discuss the potential of insect-resistant GM crops to provoke such outbreaks and the role that natural enemies can play in GM crops.

4.5.1 Pest Resurgence

A quick return of pests to damaging levels sometimes follows the routine use of broad-spectrum insecticides. This phenomenon of pest resurgence occurs because natural enemies are often more sensitive to insecticides than are the pests themselves (Croft, 1990). Increased sensitivity to insecticides may be due to lower levels of detoxification enzymes or a higher body surface to mass ratio (leading to greater relative absorption of residues of contact insecticides per unit of body weight). If the parasitoids and predators that normally attack a pest are destroyed, those pests that are still alive after insecticide residues dissipate will live in an environment with fewer natural enemies, leading to longer pest lifetimes and higher reproduction. This allows multivoltine pest populations to quickly return to damaging levels. Pest resurgence caused by insecticides has been observed in diverse crops, for many kinds of pests (Heinrichs et al., 1982; Gerson and Cohen, 1989; Buschman and DePew, 1990; Talhouk, 1991; Holt et al., 1992).

Could a GM crop attribute for insect suppression, such as the ability to express *Bt* toxins, cause pest resurgence? For insect resurgence to happen, several conditions must be met. First the pest-suppressing toxic residue or other suppressive factors must be temporary. With insecticides, toxic residues are present immediately after application, but later dissipate. This is not the case with *Bt* plants, which continue to produce the toxin throughout the crop cycle (even though expression levels in the plant may vary and may decline at later growth stages as has been reported for *Bt* cotton [Greenplate, 1999; Adamczyk et al., 2001; Olsen et al., 2005]). Second, the suppressing force must reduce populations of the pest's natural enemies more than the pest. With insecticides, this often happens because most conventional insecticides are broad-spectrum contact poisons that readily kill parasitoids and predators foraging on crop foliage at rates equal to or greater than the pests. In contrast, for *Bt* crops the suppressing force, the *Bt* toxins in the plant, is not a contact poison but rather a highly selective stomach poison (Schnepf et al., 1998). Since natural enemies are in general both less exposed and less susceptible to the *Bt* toxins than their herbivorous hosts/prey, i.e., the target pests (see section 4.3), *Bt* plants should either be harmless to the pest's natural enemies or kill them at a lower rate than the pest, thus preserving a favorable pest:natural enemy ratio. Consequently, *Bt* crops are unlikely to induce resurgence of target pests and there is no indication to date that this has happened.

Examples have been reported from non-GM crop systems that show stronger effects of plant resistance factors on a natural enemy than the pest, leading to so-called disruptive interactions (*sensu* Hare, 1992, 2002). Thus, one can not rule out the possibility that future GM crops expressing insecticidal proteins with broader activity (Malone et al., chapter 13) might cause pest resurgence. However, laboratory and glasshouse studies with plants expressing lectins (Bell et al., 1999) or protease inhibitors (Ferry et al., 2005; Álvarez-Alfageme et al., 2007) have revealed less impact on natural enemies than on target pests. Nonetheless, these plant-insect interactions would need to be assessed on a case-by-case basis, since they will differ among the expressed insecticidal proteins and the respective insects involved.

4.5.2 Secondary (Pesticide Induced) Pest Outbreak

Broad-spectrum insecticides and miticides are also well known to induce outbreaks of herbivores that are not normally pests. Secondary outbreaks occur because pesticides applied for key pests destroy the natural enemies of other herbivores and release them from regulation. Spider mites, scales, and leafminers are examples of such secondary or pesticide-created pests (Luck and Dahlsten, 1975; Van Driesche and Taub, 1983; DeBach and Rosen, 1991). Other prominent examples are outbreaks of brown planthopper (*N. lugens*) in rice (Gallagher et al., 1994) and outbreaks of sap-sucking pests in cotton (Naranjo et al., chapter 6). As new herbivores reach pest status, the crop's IPM system has to be altered to include control for these "new pests".

In the case of insect-resistant GM plants, there would be little chance of induced outbreaks of minor/secondary herbivores unless their primary natural enemies were also able to consume plant tissues/sap/exudates and were sensitive to the ingested insecticidal protein. As reported above (section 4.3.1) some groups such as predatory bugs feed on plant tissues to sustain themselves when prey are scarce and many predator groups feed on pollen, which may contain the insecticidal protein. Thus, direct exposure to plant-expressed toxins is possible. However, even if exposure and toxicity occur, enough predators would have to be killed or debilitated to lower their population density in order to cause secondary pest outbreaks. Similarly, the same outcome might be reached if the parasitoids of a secondary herbivore were more strongly affected by an insecticidal factor than its host. The critical condition required for such outcomes is not the width of the effect of the insecticidal factor, but whether its effect is differential, being more harmful to key natural enemies of a secondary herbivore than to the herbivore itself. For the currently available *Bt* crops such an effect has, however, not been observed.

4.5.3 Natural Enemies in a Multi-Pest Complex

GM crops with insecticidal traits specific for the crop's key pests, such as *Bt* crops that control larvae of key Lepidoptera and Coleoptera species, may experience pest populations of other herbivores. While this may appear to be secondary pest outbreaks, typically they are not. Rather, as GM crops are left less treated or untreated with conventional insecticides (see section 4.4.2), other herbivores that are not susceptible to the GM trait will no longer be chemically controlled. Some such herbivores will continue to remain rare because they are under natural biological control by local natural enemies. However, some herbivores among those not affected by the insecticidal trait of the GM crop may lack local effective natural enemies. Such species can become pests in GM crops. This phenomenon may also occur when more specific conventional insecticides replace broad-spectrum ones in crops with multi-pest complexes.

Most reports of secondary pest problems come from *Bt* cotton (see Naranjo et al., chapter 6 for a global overview). In different parts of the world, sucking pests such as mirid bugs, stinkbugs, leafhoppers and planthoppers have increased in abundance in *Bt* cotton fields (Greene et al., 2001; Wu et al., 2002; Lei et al., 2003; Men et al., 2005; Wilson et al., 2006) and these pests sometimes require insecticide applications with consequences for the natural control of other herbivores such as spider mites or aphids (Naranjo et al., chapter 6). Similarly, reduction of broad-spectrum insecticide use in *Bt* sweet corn has caused problems with the corn silk fly, *Euxesta stigmatias* (Diptera: Ulidiidae), in Florida and sporadic problems with dusky sap beetles, *Carpophilus lugubris* (Coleoptera: Nitidulidae) in some states (Shelton et al., chapter 9). Such problems are more likely to arise if the insecticidal traits are specific, such as *Bt* toxins, and would be less common if the traits confer more general forms of insect-resistance, such as for plants expressing lectins or protease inhibitors.

While the reduction in broad-spectrum insecticide use is likely the most important factor explaining the increase in secondary pests, other factors also appear to play a role. These factors include a reduction in competition with the target pest(s) and overall improved health of the plant. A recent field study of Vip cotton, for example, suggested that the higher abundance of mirids was likely due to the higher numbers of bolls and flowers in the Vip crop (Whitehouse et al., 2007). Such indirect interactions among herbivorous insects and their host plants are quite common and have been reported from a number of non-GM systems (Denno et al., 1995).

There is plenty of evidence to suggest that control of one pest can also affect the impact of natural enemies on a second pest. For example, a study from maize has revealed that *O. nubilalis* on maize plants damaged by *Diabrotica* spp. suffered less parasitism by *Macrocentrus grandii* (Hymenoptera: Braconidae) (White and Andow, 2006). Suggested reasons for this were changes in the habitat and low success of the parasitoid at the low host densities that were present on plants damaged by *Diabrotica* spp.

Options for gaining control of secondary pests of GM crops vary. In some cases, limited use of a conventional insecticide may provide control without provoking unwanted effects (Harris et al., 1998). In other cases, the secondary pest may be a suitable target for classical biological control, especially if it is not a direct pest of the marketed portion of the crop plant and is of concern over a large area. If the secondary pest is closely related to the target pest and the insect-resistance attribute is rather specific, it might be possible to expand the attribute's efficacy range. For example, with *Bt* plants this may be done by creating varieties with additional *cry* or *vip* genes, or by enhancing the toxin expression level. A recent study by Mehlo et al. (2005) has shown that the level of activity and the activity spectrum of Cry1Ac could be enhanced by fusion with the galactose-binding domain of ricin B toxin.

In addition, other methods such as mass releases of natural enemies, mating disruption with pheromones, habitat modification for natural enemy conservation, or use of microbial pesticides might be used. The potential to develop and establish such biologically-based, targeted alternatives in multi-pest crops where key pests are controlled by specific GM plants is higher given that less broad-spectrum insecticide has to be applied.

4.5.4 Applicability of Conservation Biocontrol Approaches in GM Crops

Concepts of conservation biological control, which have been developed to conserve natural enemies in crops treated with pesticides (Ehler, 1998), may not be adequate for understanding the effects of GM crop varieties on biological control, and may need to be reexamined in light of principles describing how varietal traits for pest resistance affect natural enemy populations at both the field and landscape scale. Conservation of natural enemies of pests in GM crops through reduced insecticides has a major impact on biological control and is discussed in section 4.4.2.

As discussed above, the GM crop may experience outbreaks of secondary pests if natural enemies are adversely affected. To suppress such outbreaks, in principle, any of the methods used in IPM programs might be employed, including the full array of conservation biological control measures (e.g., use of physiologically selective pesticides, ecologically selective ways of using pesticides, reduced dosages, selective formulations, limited application in time and space, intercropping, retaining desirable non-crop vegetation within crops, optimizing crop patterns in time and space, enhancement of sources of crop-derived natural enemy foods such as pollen, nectar, crop sap, manipulation of adjacent vegetation as sources of alternative hosts or prey, or provision of shelter or refuges for natural enemies [see Barbosa, 1998; Pickett and Bugg, 1998; Gurr et al., 2004; Van Driesche et al., 2008]). However, the attractiveness of pest-resistant crops (GM or not) is that pest control requires no further attention from growers. Growers in general are reluctant to adopt and implement management complicated plans that require additional financial purchases, use of labor, land, water or other inputs, unless the need is pressing and the approach is superior to all other options. For this reason, such manipulations, while possible, have not yet often been the growers' response to problems with secondary pests in GM crops.

4.6 Impact of Insect-Resistant GM Crops on Natural Enemies: Landscape Level Effects

Some insect-resistant GM crops, i.e., varieties of cotton and maize expressing *Bt* Cry proteins, have reached high adoption rates in parts of the world. This together with their high efficacy of controlling target pest insects makes landscape level effects on pest and natural enemy populations possible.

Bt crops provide a high level of resistance and area-wide reductions of populations of some target pests have been reported (Storer et al., chapter 10). The best documented example is that of the pink bollworm (*P. gossypiella*) in parts of the USA (Carrière et al., 2003; Chu et al., 2006). Other studies suggest that populations of *O. nubilalis* have been suppressed at the landscape level in some regions and such reductions will have implications for control of this pest in other crops (Storer et al., chapter 10).

It is likely that the large scale adoption of GM crops with a similar trait such as Lepidoptera-resistance will also affect natural enemies. Specialists might suffer from an area-wide reduction in their hosts or prey. This is especially likely for parasitoids of pests that do not occur on wild host plants in the region, such as *P. gossypiella* in Arizona. However, a landscape planted with insect-resistant GM crops will still contain some hosts, for a number of reasons: (1) the *Bt* crops may not provide total control of the target pest(s), (2) hosts may occur in non-GM refuges of the same crop, and (3) hosts or alternative hosts may occur on other crops or wild plants in the landscape. Therefore, the impact on a given parasitoid will also depend on its response to low host densities. For example, studies by White and Andow (2005) documented continued parasitism, albeit at a lower rate, of *O. nubilalis* larvae by *M. grandii* at low host densities.

On the other hand there is growing evidence that biological control *per se* benefits drastically from the large reductions in insecticide applications sometimes associated with adoption of insect-resistant GM crops (see section 4.4.2; Naranjo et al., chapter 6). Thus, it is likely that biological control at the landscape level will be enhanced by planting of GM crops, with potential benefits for non-GM crops.

It is a regulatory requirement in many countries that, at the landscape level, stands of GM crops are mixed with a certain area of the crop planted with non-GM varieties (Bates et al., 2005a; Matten et al., chapter 2; Ferré et al., chapter 3). The area planted to non-*Bt* crops varies with crop and region. Furthermore the size of the refuge can depend on whether it is sprayed with insecticides or not. These plots are intended to permit pests with susceptible genotypes (relative to the toxin expressed in the GM plant) to survive as part of a management plan to forestall development of resistance. While studies (e.g., White and Andow, 2005) have examined whether natural enemy attack of pests in such non-GM refuges affects the development of resistance, how such refuges might affect biological control of secondary pests in the GM-crop itself has not been investigated. If the refuge is treated with insecticides, its impact on biological control in the GM and non-GM crop would likely be negligible. In addition to the refuge area that is legally required in some countries, natural refuges for the pest insects exist in form of weeds and alternate crops. In China for example, a refuge for *Bt* cotton is not required since other crops provide a refuge for susceptible *H. armigera* over the entire cotton-growing season (Naranjo et al., chapter 6). Refuges provide habitat for target and secondary pests, as well as for their natural enemies. Refuges could thus potentially benefit two groups of natural enemies: (1) those parasitoids that are specialists of the pest that is effectively controlled by the GM crop and (2) generalist predators that are facultative plant feeders and also susceptible to the toxin expressed by the GM crop. On a landscape level, refuges can thus help to prevent the local extinction of specialist parasitoids of the crop's key pest. Refuges may also enhance the numbers of those generalist predators that are unable to live in the GM crop itself because of susceptibility to the insecticidal attribute. Both effects may contribute to pest control at the landscape level.

4.7 Novel Insect-Resistant GM Crops with Implications for Biological Control

Future insect-resistant GM crops are likely to express novel Cry proteins either alone or in combination. In addition the market might see GM plants with other classes of insecticidal proteins that target other herbivore groups and are less specific (Malone et al., chapter 13). These products have been addressed in the previous sections. There are, however, other developments that might lead to products that could affect biological control. Examples are given below.

4.7.1 Plants with More Targeted Expression of the Insecticidal Gene(s)

While in today's GM crops the transgenes are usually controlled by constitutive promoters such as CaMV35S, research efforts are underway to develop plants that more specifically express the insecticidal compound when and where it is needed to provide better control of the target pest species. This approach would also be of benefit for insect-resistant management (Bates et al., 2005a). A more targeted expression of the toxin will further reduce the number of non-target species that are exposed to the insecticidal proteins. For example, experimental plants have been produced that express lectins, such as GNA, under a phloem-specific promoter such as the rice sucrose synthase-1 promoter (Gatehouse, 1994; Rao et al., 1998; Dutta et al., 2005). The goal is to enhance toxin levels in phloem sap to provide better control of pests such as aphids. Such targeted expression would also ensure that herbivores that feed on other plant tissues would not ingest the toxin and consequently not pass it on to their natural enemies. Meiyalaghan et al. (2006) produced potato plants that express *cry* genes for control of potato tuber moth (*Phthorimaea operculella*, Lepidoptera: Gelechiidae) under control of a light inducible promoter which largely restricted the gene expression to the foliage with no or minimal expression in tubers. Another prominent example for a targeted toxin expression is GM peas that have been transformed to express the *Phaseolus vulgaris* (bean) alpha-amylase inhibitor-1 (α-AI1) for the control of bruchid beetles. The α*ai* gene construct used for transformation was regulated by flanking sequences from the seed-specific bean PHA (dlec2) gene to restrict the expression of the gene to the cotyledon and embryonic axis of the developing seed (Schroeder et al., 1995). This will restrict the potential exposure to natural enemies that attack herbivores feeding on the developing seeds. Of particular importance will be future plants where the gene conferring resistance is controlled by a chemically or wound inducible promoter to direct toxin production spatially and temporally. Such plants have been developed for Cry protein expressing rice (Breitler et al., 2001, 2004) and broccoli (Cao et al., 2001; Bates et al., 2005b) and for rice expressing the potato proteinase inhibitor II (Duan et al., 1996).

4.7.2 GM Plants with Characteristics Designed to Enhance Biological Control

There is substantial information on the ways in which plants influence biological control organisms and this knowledge could be used to develop plants for improved biological control (Bottrell et al., 1998; Cortesero et al., 2000). For example, plants interact positively with natural enemies by emitting compounds such as volatiles that attract the natural enemies to herbivore damaged plants (Turlings and Wäckers, 2004). Potentially, metabolic pathways of plants could be manipulated to produce plants that are more attractive to natural enemies. Studies with *Arabidopsis* have revealed that genetic engineering can allow the synthesis of additional signaling compounds that attract predators (Kappers et al., 2005) or parasitoids (Schnee et al., 2006). Ideally, the approach would be targeted to natural enemies that attack the young developmental stages of the pest before the damage to the plant is done. There are, however, a number of problems associated with this approach, including: (1) the natural enemies might come to associate these volatiles with the absence of their hosts (Agrawal, 2000), (2) there must be enough natural enemies at the right time in close vicinity to the crop to be attracted, and (3) plant volatiles are only one of many factors that affect the efficacy of a natural enemy.

As mentioned earlier, the efficacy of predators and parasitoids can be affected by morphological features such as plant surface structure and pubescence (Bottrell et al., 1998). Studies using plants derived through mutations that lack trichomes or are covered by reduced epicuticular waxes have shown that the searching behaviour or predatory efficacy of natural enemies can be enhanced with benefits for biological control (e.g., Romeis et al., 1999; Rutledge et al., 2003). Thus, once the genetic basis of trichome development and other morphological structures is identified, GM plants that lack certain structures that interfere with biological control could be developed. This benefit is, however, balanced by the costs associated with reduced trichomes or epicuticular waxes such as increased susceptibility to biotic (herbivores, diseases) or abiotic stresses (UV-light, drought) and may impact yield and other features of such plants. It has to be seen whether plants with enhanced properties for biological control organisms can be produced and play a vital role in IPM systems.

4.8 Conclusions

During the last two decades, resistance breeding for certain crops and traits has benefited enormously from recombinant DNA technology. Gene technology has allowed foreign genes to be incorporated into crop plants allowing plants to express completely new pest resistance properties, which has widened the available gene pool for breeders. Successful use of this approach has resulted in insect-resistant GM crops expressing Cry toxins derived from *Bt* being planted on a wide scale. To date, *Bt* crops obtain resistance by virtue of single or double genes whereas traditional plant breeding involves crossing related plant species or cultivars. This may involve a complex of several genes, often poorly defined. Consequently, the basis of such conventional plant resistance is often not well understood. In contrast, with GM plants, the gene and its products are well researched, and thus impacts on target and non-target species are easier to predict compared to the effects of natural pest resistance or that from conventional plant breeding programs. This has important ramifications for developing risk assessments for non-target arthropods (Romeis et al., 2008b).

Laboratory and field studies conducted thus far have shown that the currently used *Bt* crops do not cause any unexpected detrimental effects on predators or parasitoids or on the biological control function they provide. In addition, in crops such as cotton or sweet corn where the introduction of *Bt*-transgenic varieties results in significant reductions of insecticide applications, clear benefits on arthropod abundance in general, and biological control in particular, have been reported. Consequently, *Bt* crops can contribute to natural enemy conservation while at the same time protecting the crop from the targeted pests and are thus a useful component in IPM systems.

References

Adamczyk, J.J., Jr., Adams, L.C., and Hardee, D.D., 2001. Field efficacy and seasonal expression profiles for terminal leaves of single and double *Bacillus thuringiensis* toxin cotton genotypes. Journal of Economic Entomology 94: 1589–1593.

Agrawal, A.A., 2000. Mechanisms, ecological consequences and agricultural implications of tritrophic interactions. Current Opinion in Plant Biology 3: 329–335.

Álvarez-Alfageme, F., Martínez, M., Pascual-Ruiz, S., Castañera, P., Diaz, I., and Ortego, F., 2007. Effects of potato plants expressing a barley cystatin on the predatory bug *Podisus maculiventris* via herbivorous prey feeding on the plant. Transgenic Research 16: 1–13.

Bale, J.S., van Lenteren, J.C., and Bigler, F., 2008. Biological control and sustainable food production. Philosophical Transactions of the Royal Society B 363: 761–776.

Barbosa, P., ed., 1998. Conservation Biological Control. Academic, San Diego, CA, USA.

Barbosa, P., Saunders, J.A., Kemper, J., Trumbule, R., Olechno, J., and Martinat, P., 1986. Plant allelochemicals and insect parasitoids: effects of nicotine on *Cotesia congregata* Say (Hymenoptera, Braconidae) and *Hyposoter annulipes* (Cresson) (Hymenoptera, Ichneumonidae). Journal of Chemical Ecology 12: 1319–1328.

Bates, S.L., Zhao, J.-Z., Roush, R.T., and Shelton, A.M., 2005a. Insect resistance management in GM crops: past, present and future. Nature Biotechnology 23: 57–62.

Bates, S.L., Cao, J., Zhao, J.Z., Earle, E.D., Roush, R.T., and Shelton, A.M., 2005b. Evaluation of a chemically inducible promoter for developing a within-plant refuge for resistance management. Journal of Economic Entomology 98: 2188–2194.

Baum, J.A., Bogaert, T., Clinton, W., Heck, G.R., Feldmann, P., Ilagan, O., Johnson, S., Plaetinck, G., Munyikwa, T., Pleau, M., Vaughn, T., and Roberts, J., 2007. Control of coleopteran insect pests through RNA interference. Nature Biotechnology 25: 1322–1326.

Bell, H.A., Fitches, E.C., Down, R.E., Marris, G.C., Edwards, J.P., Gatehouse, J.A., and Gatehouse, A.M.R., 1999. The effect of snowdrop lectin (GNA) delivered via artificial diet and transgenic plants on *Eulophus pennicornis* (Hymenoptera: Eulophidae), a parasitoid of the tomato moth *Lacanobia oleracea* (Lepidoptera: Noctuidae). Journal of Insect Physiology 45: 983–991.

Bell, H.A., Fitches, E.C., Marris, G.C., Bell, J., Edwards, J.P., Gatehouse, J.A., and Gatehouse, A.M.R., 2001. Transgenic GNA expressing potato plants augment the beneficial biocontrol of *Lacanobia oleracea* (Lepidoptera; Noctuidae) by the parasitoid *Eulophus pennicornis* (Hymenoptera; Eulophidae). Transgenic Research 10: 35–42.

Benrey, B., and Denno, R.F., 1997. The slow-growth-high-mortality hypothesis: a test using the cabbage butterfly. Ecology 78: 987–999.

Bhatti, M.A., Duan, J., Head, G.P., Jiang, C.J., McKee, M.J., Nickson, T.E., Pilcher, C.L., and Pilcher, C.D., 2005. Field evaluation of the impact of corn rootworm (Coleoptera: Chrysomelidae)-protected Bt corn on foliage-dwelling arthropods. Environmental Entomology 34: 1336–1345.

Bottrell, D.G., Barbosa, P., and Gould, F., 1998. Manipulating natural enemies by plant variety selection and modification: a realistic strategy? Annual Review of Entomology 43: 347–367.

Bourguet, D., Chaufaux, J., Micoud, A., Delos, M., Naibo, B., Bombarde, F., Marque, G., Eychenne, N., and Pagliari, C., 2002. *Ostrinia nubilalis* parasitism and the field abundance of non-target insects in transgenic *Bacillus thuringiensis* corn (*Zea mays*). Environmental Biosafety Research 1: 49–60.

Breitler, J.C., Cordero, M.J., Royer, M., Meynard, D., San Segundo, B., and Guiderdoni, E., 2001. The −689/+197 region of the maize protease inhibitor gene directs high level, wound-inducible expression of the cry1B gene which protects transgenic rice plants from stemborer attack. Molecular Breeding 7: 259–274.

Breitler, J.C., Vassal, J.M., Catala, M.M., Meynard, D., Marfà, V., Melé, E., Royer, M., Murillo, I., San Segundo, B., Guiderdoni, E., and Messeguer, J., 2004. *Bt* rice harbouring *cry* genes controlled by a constitutive or wound-inducible promoter: protection and transgene expression under Mediterranean field conditions. Plant Biotechnology Journal 2: 417–430.

Brookes, G., and Barfoot, P., 2006a. Global impact of biotech crops: socio-economic and environmental effects in the first ten years of commercial use. AgBioForum 9: 139–151.

Brookes, G., and Barfoot, P., 2006b. GM Crops: The First Ten Years – Global Socio-Economic and Environmental Impacts. ISAAA Brief No. 36, International Service for the Acquisition of Agri-Biotech Applications, Ithaca, NY, USA.

Bruck, D.J., Lopez, M.D., Lewis, L.C., Prasifka, J.R., and Gunnarson, R.D., 2006. Effects of transgenic *Bacillus thuringiensis* corn and permethrin on nontarget arthropods. Journal of Agricultural and Urban Entomology 23: 111–124.

Buschman, L.L., and DePew, L.J., 1990. Outbreaks of banks grass mite (Acari: Tetranychidae) in grain sorghum following insecticide application. Journal of Economic Entomology 83: 1570–1574.

Campbell, B.C., and Duffey, S.S., 1979. Tomatine and parasitic wasps: potential incompatibility of plant antibiosis with biological control. Science 205: 700–702.

Candolfi, M.P., Brown, K., Grimm, C., Reber, B., and Schmidli, H., 2004. A faunistic approach to assess potential side-effects of genetically modified *Bt*-corn on non-target arthropods under field conditions. Biocontrol Science and Technology 14: 129–170.

Cao, J., Shelton, A.M., and Earle, E.D., 2001. Gene expression and insect resistance in transgenic broccoli containing a *Bacillus thuringiensis cry1Ab* gene with the chemically inducible PR-1a promoter. Molecular Breeding 8: 207–216.

Carrière, Y., Ellers-Kirk, C., Sisterson, M., Antilla, L., Whitlow, M., Dennehy, T.J., and Tabashnik, B.E., 2003. Long-term regional suppression of pink bollworm by *Bacillus thuringiensis* cotton. Proceedings of the National Academy of Sciences of the USA 100: 1519–1523.

Cattaneo, M.G., Yafuso, C., Schmidt, C., Huang, C.Y., Rahman, M., Olson, C., Ellers-Kirk, C., Orr, B.J., Marsh, S.E., Antilla, L., Dutilleu, P., and Carrière, Y., 2006. Farm-scale evaluation of the impacts of transgenic cotton on biodiversity, pesticide use, and yield. Proceedings of the National Academy of Sciences of the USA 103: 7571–7576.

Chen, M., Zhao, J.-Z., Ye, G.-Y., Fu, Q., and Shelton, A.M., 2006. Impact of insect-resistant transgenic rice on target insect pests and non-target arthropods in China. Insect Science 13: 409–420.

Chen, M., Liu, Z.C., Ye, G.Y., Shen, Z.C., Hu, C. Peng, Y. F., Altosaar, I., and Shelton, A.M., 2007. Impacts of transgenic *cry1Ab* rice on non-target planthoppers and their main predator *Cyrtorhinus lividipennis* (Hemiptera: Miridae) - A case study of the compatibility of Bt rice with biological control. Biological Control 42: 242–250.

Chu, C.-C., Natwick, E.T., Leon-Lopez, R., Dessert, J.R., and Henneberry, T.J., 2006. Pink bollworm moth (Lepidoptera: Gelechiidae) catches in the Imperial Valley, California from 1989 to 2003. Insect Science 13: 469–475.

Cortesero, A.M., Stapel, J.O., and Lewis, W.J., 2000. Understanding and manipulating plant attributes to enhance biological control. Biological Control 17: 35–49.

Croft, B.A., 1990. Arthropod Biological Control Agents and Pesticides. Wiley, New York, USA.

Davison, M.M., Butler, R.C., Wratten, S.D., and Conner, A.J., 2006. Impacts of insect-resistant transgenic potatoes on the survival and fecundity of a parasitoid and an insect predator. Biological Control 37: 224–230.

DeBach, P., and Rosen, D., 1991. Biological Control by Natural Enemies, 2nd ed. Cambridge University Press, Cambridge, UK, pp. 140–148.

Denno, R.F., McClure, M.S., and Ott, J.R., 1995. Interspecific interactions in phytophagous insects – competition re-examined and resurrected. Annual Review of Entomology 40: 297–331.

Dively, G.P., 2005. Impact of transgenic VIP3A x Cry1Ab lepidopteran-resistant field corn on the non-target arthropod community. Environmental Entomology 34: 1267–1291.

Duan, X., Li, X., Xue, Q., Abo-El-Saad, M., Xu, D., and Wu, R., 1996. Transgenic rice plants harboring an introduced potato proteinase inhibitor II gene are insect resistant. Nature Biotechnology 14: 494–498.

Dutta, I., Majumder, P., Saha, P., Ray, K., and Das, S., 2005. Constitutive and phloem specific expression of *Allium sativum* leaf agglutinin (ASAL) to engineer aphid (*Lipaphis erysimi*) resistance in transgenic Indian mustard (*Brassica juncea*). Plant Science 169: 996–1007.

Dutton, A., Klein, H., Romeis, J., and Bigler, F., 2002. Uptake of Bt-toxin by herbivores feeding on transgenic maize and consequences for the predator *Chrysoperla carnea*. Ecological Entomology 27: 441–447.

Dutton, A., Romeis, J., and Bigler, F., 2003. Assessing the risks of insect resistant transgenic plants on entomophagous arthropods: Bt-maize expressing Cry1Ab as a case study. BioControl 48: 611–636.

East, R., Kain, W.M., and Douglas, J.A., 1980. The effect of grass grub on the herbage production of different pasture species in the pumice country. Proceedings of the New Zealand Grassland Association 41: 105–115.

EFSA (European Food Safety Authority), 2006. Guidance document of the Scientific Panel on Genetically Modified Organisms for the risk assessment of genetically modified plants and

derived food and feed. EFSA Journal 99: 1–100. http://www.efsa.europa.eu/EFSA/Scientific_Document/gmo_guidance_gm_plants_en,0.pdf (accessed 21 January 2008).
Ehler, L.E., 1998. Conservation biological control: past, present and future. In: Conservation Biological Control, P. Barbosa, ed., Academic, San Diego, CA, USA, pp. 1–8.
Eigenbrode, S.D., 2004. The effects of plant epicuticular waxy blooms on attachement and effectiveness of predatory insects. Arthropod Structure & Development 33: 91–102.
Eigenbrode, S.D., and Espelie, K.E., 1995. Effects of plant epicuticular lipids on insect herbivores. Annual Review of Entomology 40: 141–194.
El-Heneidy, A.H., Barbosa, P., and Gross, P., 1988. Influence of dietary nicotine on the fall armyworm, *Spodoptera frugiperda* and its parasitoid, the Ichneumonid wasp *Hyposoter annulipes*. Entomologia Experimentalis et Applicata 46: 227–232.
Estruch, J.J., Warren, G.W., Mullins, M.A., Nye, G.J., Craig, J.A., and Koziel, A.G., 1996. Vip3A, a novel *Bacillus thuringiensis* vegetative insecticidal protein with a wide spectrum of activity against lepidopteran insects. Proceedings of the National Academy of Sciences of the USA 93: 5389–5394.
Farrar, R.R.J., Kennedy, G.G., and Kashyap, R.K., 1992. Influence of life history differences of two tachinid parasitopids of *Helicoverpa zea* (Boddie) (Lepidoptera: Noctuidae) on their interactions with glandular trichome/methyl ketone-based insect resistance in tomato. Journal of Chemical Ecology 18: 499–515.
Fernandes, O.A., Faria, M., Martinelli, S., Schmidt, F., Carvalho, V.F., and Moro, G., 2007. Short-term assessment of Bt maize on non-target arthropods in Brazil. Scientia Agricola 64: 249–255.
Ferry, N., Jouanin, L., Ceci, L.R., Mulligan, A., Emami, K., Gatehouse, J.A., and Gatehouse, A. M.R., 2005. Impact of oilseed rape expressing the insecticidal serine protease inhibitor, mustard trypsin inhibitor-2 on the beneficial predator *Pterostichus madidus*. Molecular Ecology 14: 337–349.
Ferry, N., Mulligan, E.A., Stewart, C.N., Tabashnik, B.E., Port, G.R., and Gatehouse, A.M.R., 2006. Prey-mediated effects of transgenic canola on a beneficial, non-target, carabid beetle. Transgenic Research 15: 501–514.
Gallagher, K.D., Kenmore, P.E., and Sogawa, K., 1994. Judicious use of insecticides deter planthopper outbreaks and extend the life of resistant varieties in Southeast Asian rice. In: Planthoppers: Their Ecology and Management, R.F. Denno and J.T. Perfect, eds., Chapman & Hall, New York, USA, pp. 599–614.
Gallardo, F., Boethel, D.J., Fuxa, J.R., and Richter, A., 1990. Susceptibility of *Heliothis zea* (Boddie) larvae to Nomuraea riley (Farlow) Samson: effects of alpha-tomatine at the third trophic level. Journal of Chemical Ecology 16: 1751–1759.
Garcia-Alonso, M., Jacobs, E., Raybould, A., Nickson, T.E., Sowig, P., Willekens, H., van der Kouwe, P., Layton, R., Amijee, F., Fuentes, A.M., and Tencalla, F., 2006. A tiered system for assessing the risk of genetically modified plants to non-target organisms. Environmental Biosafety Research 5: 57–65.
Gatehouse, J.A., 1994. Use of the rice sucrose synthase-1 promotor to direct phloem-specific expression of β-glucuronidase and snowdrop lectin genes in transgenic tobacco plants. Journal of Experimental Botany 45: 623–631.
Gerson, U., and Cohen, E., 1989. Resurgences of spider mites (Acari: Tetranychidae) induced by synthetic pyrethroids. Experimental and Applied Acarology 6: 29–46.
Godfray, H.C.J., 1994. Parasitoids. Princeton University Press, Princeton, NJ, USA.
Greene, J.K., Turnipseed, S.G., Sullivan, M.J., and May, O.L., 2001. Treatment thresholds for stink bugs (Hemiptera: Pentatomidae) in cotton. Journal of Economic Entomology 94: 403–409.
Greenplate, J.T., 1999. Quantification of *Bacillus thuringiensis* insect control protein Cry1Ac over time in Bollgard cotton fruit and terminaly. Journal of Economic Entomology 92: 1377–1383.
Groot, A.T., and Dicke, M., 2002. Insect-resistant transgenic plants in a multi-trophic context. The Plant Journal 31: 387–406.
Gurr, G.M., Wratten, S.D., and Altieri, M.A., eds., 2004. Ecological Engineering for Pest Management: Advances in Habitat Manipulation for Arthropods. CABI, Wallingford, UK.

Hare, J.D., 1992. Effects of plant variation on herbivore-natural enemy interactions. In: Plant Resistance to Herbivores and Pathogens: Ecology, Evolution and Genetics, R.S Fritz and E.L. Simms, eds., University of Chicago Press, Chicago, IL, USA, pp. 278–298.

Hare, J.D., 2002. Plant genetic variation in tritrophic interactions. In: Multitrophic Level Interactions, T. Tscharntke and B.A. Hawkins, eds., Cambridge University Press, Cambridge, UK, pp. 8–43.

Harris, J.G., Hershey, C.N., and Watkins, M.J., 1998. The usage of Karate (lambda-cyhalothrin) oversprays in combination with refugia, as a viable and sustainable resistance management strategy for Bt cotton. In: Proceedings of the Beltwide Cotton Conferences, San Diego, CA, USA, 5–9 January 1998, Vol. 2., P. Dugger and D. Richter, eds., National Cotton Council, Memphis, TN, USA, pp. 1217–1220.

Harwood, J.D., Wallin, W.G., and Obrycki, J.J., 2005. Uptake of Bt endotoxins by nontarget herbivores and higher order arthropod predators: molecular evidence from a transgenic corn agroecosystem. Molecular Ecology 14: 2815–2823.

Head, G., Brown, C.R., Groth, M.E., and Duan, J.J., 2001. Cry1Ab protein levels in phytophagous insects feeding on transgenic corn: implications for secondary exposure risk assessment. Entomologia Experimentalis et Applicata 99: 37–45.

Head, G., Moar, M., Eubanks, M., Freeman, B., Ruberson, J., Hagerty, A., and Turnipseed, S., 2005. A multi-year, large-scale comparison of arthropod populations on commercially managed *Bt* and non-*Bt* cotton fields. Environmental Entomology 34: 1257–1266.

Heinrichs, E.A., Aquino, G.B., Chelliah, S., Valencia, S.L., and Reissig, W.H., 1982. Resurgence of *Nilaparvata lugens* (Stål) populations as influenced by method and timing of insecticide applications in lowland rice. Environmental Entomology 11: 78–84.

Hellmich, R.L., Higgins, L.S., Witkowski, J.F., Campbell, J.E., and Lewis, L.C., 1999. Oviposition by European corn borer (Lepidoptera: Crambidae) in response to various transgenic corn events. Journal of Economic Entomology 92: 1014–1020.

Hilbeck, A., Baumgartner, M., Fried, P.M., and Bigler, F., 1998. Effects of transgenic *Bacillus thuringiensis* corn-fed prey on mortality and development time of immature *Chrysoperla carnea* (Neuroptera: Chrysopidae). Environmental Entomology 27: 480–487.

Hoheisel, G.A., and Fleischer, S.J., 2007. Coccinellids, aphids, and pollen in diversified vegetable fields with transgenic and isoline cultivars. Journal of Insect Science 7: 64. http://www.insectscience.org/7.61/.

Holt, J., Wareing, D.R., and Norton, G.A., 1992. Strategies of insecticide use to avoid resurgence of *Nilaparvata lugens* (Homoptera; Delphacidae) in tropical rice: a simulation analysis. Journal of Economic Entomology 85: 1979–1989.

Howald, R., Zwahlen, C., and Nentwig, W., 2003. Evaluation of Bt oilseed rape on the non-target herbivore *Athalia rosae*. Entomologia Experimentalis et Applicata 106: 87–93.

Huang, J.K., Hu, R.F., Rozelle, S., and Pray, C., 2005. Insect-resistant GM rice in farmers' fields: assessing productivity and health effects in China. Science 308: 688–690.

Huang, J., Hu, R., Rozelle, S. and Pray, C., 2008. Genetically modified rice, yields, and pesticides: Assessing farm-level productivity effects in China. Economic Development and Cultural Change 56: 241–264.

Icoz, I., and Stotsky, G., 2008. Fate and effects of insect-resistant *Bt* crops in soil ecosystems. Soil Biology and Biochemistry 40: 559–586.

James, C., 2007. Global Status of Commercialized Biotech/GM Crops: 2007. ISAAA Brief No. 37, International Service for the Acquisition of Agri-Biotech Applications, Ithaca, NY, USA.

Johnson, M.T., 1997. Interaction of resistant plants and wasp parasitoids of tobacco budworm (Lepidoptera: Noctuidae). Environmental Entomology 26: 207–214.

Johnson, M.T., and Gould, F., 1992. Interaction of genetically engineered host plant-resistance and natural enemies of *Heliothis virescens* (Lepidoptera, Noctuidae) in tobacco. Environmental Entomology 21: 586–597.

Kanrar, S., Venkateswari, J., Kirti, P.B., and Chopra, V.L., 2002. Transgenic Indian mustard (*Brassica juncea*) with resistance to the mustard aphid (*Lipaphis erysimi* Kalt.). Plant Cell Reports 20: 976–981.

Kappers, I.F., Aharoni, A., van Herpen, T.W.J.M., Luckerhoff, L.L.P., Dicke, M., and Bouwmeester, H.J., 2005. Genetic engineering of terpenoid metabolism attracts bodyguards to *Arabidopsis*. Science 309 (5743): 2070–2072.

Kartohardjono, A., and Heinrichs, E.A., 1984. Populations of the brown planthopper, *Nilaparvanta lugens* (Stal) (Homoptera: Delphacidae), and its predators on rice varieties with different levels of resistance. Environmental Entomology 13: 359–365.

Keller, I., Fluri, P., and Imdorf, A., 2005. Pollen nutrition and colony development in honey bees: part I. Bee World 86: 3–10.

Kennedy, G.G., 2003. Tomato, pests, parasitoids and predators: tritrophic interactions involving the genus *Lycopersicon*. Annual Review of Entomology 48: 51–72.

Kennedy, G.G., and Gould, F., 2007. Ecology of natural enemies and genetically engineered host plants. In: Perspectives in Ecological Theory and Integrated Pest Management, M. Kogan and P. Jepson, eds., Cambridge University Press, Cambridge, UK, pp. 269–300.

Lawo, N.C., and Romeis, J., 2008. Assessing the utilization of a carbohydrate food source and the impact of insecticidal proteins on larvae of the green lacewing, *Chrysoperla carnea*. Biological Control 44: 389–398.

Lei, T., Khan, M., and Wilson, L., 2003. Boll damage by sucking pests: an emerging threat, but what do we know about it? In: World Cotton Research Conference III: Cotton for the New Millennium, A. Swanepoel, ed., Agricultural Research Council - Institute for Industrial Crops, Cape Town, South Africa, pp. 1337–1344.

Leslie, T.W., Hoheisel, G.A., Biddinger, D.J., Rohr, J.R., and Fleisher, S.J., 2007. Transgenes sustain epigeal insect biodiversity in diversified vegetable farm systems. Environmental Entomology 36: 234–244.

Li, F.-F., Ye, G.-Y., Wu, Q., Peng, Y.-F., and Chen, X.-X., 2007. Arthropod abundance and diversity in *Bt* and non-*Bt* rice fields. Environmental Entomology 36: 646–654.

Liu, Y.-B., Tabashnik, B.E., Dennehy, T.J., Carrière, Y., Sims, M.A., and Meyer, S.K., 2002. Oviposition and mining in bolls of Bt- and non-Bt cotton by resistant and susceptible pink bollworm (Lepidoptera: Gelechiidae). Journal of Economic Entomology 95: 143–148.

Lovinger, A., Liewehr, D., and Lamp, W.O., 2000. Glandular trichomes on alfalfa impede searching behavior of the potato leafhopper parasitoid. Biological Control 18: 187–192.

Luck, R.F., and Dahlsten, D.L., 1975. Natural decline of a pine needle scale (*Chionaspis pinifoliae* [Fitch]), outbreak at South Lake Tahoe, California, following cessation of adult mosquito control with malathion. Ecology 56: 893–904.

Ludy, C., and Lang, A., 2006. Bt maize pollen exposure and impact on the garden spider, *Araneus diadematus*. Entomologia Experimentalis et Applicata 118: 145–156.

Lundgren, J.G., Razzak, A.A., and Wiedenmann, R.N., 2004. Population responses and food consumption by predators *Coleomegilla maculata* and *Harmonia axyridis* (Coleoptera: Coccinellidae) during anthesis in an Illinois cornfield. Environmental Entomology 33: 958–963.

Lundgren, J.G., Huber, A., and Wiedenmann, R.N., 2005. Quantification of consumption of corn pollen by the predator *Coleomegilla maculata* (Coleoptera: Coccinellidae) during anthesis in an Illinois cornfield. Agricultural and Forest Entomology 7: 53–60.

Malcolm, S.B., 1990. Chemical defence in chewing and sucking insect herbivores: plant-derived cardenolides in the monarch butterfly and orleander aphid. Chemoecology 1: 12–21.

Mao, Y.-B., Cai, W.-J., Wang, J.-W., Hong, G.-J., Tao, X.-Y., Wang, L.-J., Huang, Y.-P., and Chen, X.-Y., 2007. Silencing a cotton bollworm P450 monooxygenase gene by plant-mediated RNAi impairs larval tolerance of gossypol. Nature Biotechnology 25: 1307–1313.

Marvier, M., McCreedy, C., Regetz, J., and Kareiva, P., 2007. A meta-analysis of effects of Bt cotton and maize on nontarget invertebrates. Science 316: 1475–1477.

Mehlo, L., Gahakwa, D., Nghia, P.T., Loc, N.T., Capell, T., Gatehouse, J.A., Gatehouse, A.M.R., and Christou, P., 2005. An alternative strategy for sustainable pest resistance in genetically enhanced crops. Proceedings of the National Academy of Sciences of the USA 102: 7812–7816.

Meissle, M., Vojtech, E., and Poppy, G.M., 2005. Effects of Bt maize-fed prey on the generalist predator *Poecilus cupreus* L. (Coleoptera: Carabidae). Transgenic Research 14: 123–132.

Meiyalaghan, S., Jacobs, J.M.E., Butler, R.C., Wratten, S.D., and Conner, A.J., 2006. Expression of *cry*1Ac9 and *cry*9Aa2 genes under a potato light-inducible *Lhca*3 promoter in transgenic potatoes for tuber moth resistance. Euphytica 147: 297–309.

Men, X., Ge, F., Edwards, C.A., and Yardim, E.N., 2005. The influence of pesticide applications on *Helicoverpa armigera* Hubner and sucking pests in transgenic Bt cotton and non-transgenic cotton in China. Crop Protection 24: 319–324.

Mendelsohn, M., Kough, J., Vaituzis, Z., and Matthews, K., 2003. Are Bt crops safe? Nature Biotechnology 21: 1003–1009.

Mohite, P.B., and Uthamasamy, S., 1998. Host-plant resistance and natural enemies interaction in the management of *Helicoverpa armigera* (Hubner) on cotton. Indian Journal of Agricultural Research 32: 28–30.

Monsanto Company, 2003. Safety Assessment ox YieldGard Rootworm™ Corn. http://www.monsanto.com/pdf/products/yieldgard_rw_pss.pdf (accessed 8 January 2008).

Monsanto Company, 2004. Petition for the Determination of Nonregulated Status for MON 88017 Corn. Monsanto Petition # 04-CR-108U; OECD Unique Identifier: MON-88017-3. http://www.aphis.usda.gov/brs/aphisdocs/04_12501p.pdf (accessed 16 January 2008).

Musser, F.R., and Shelton, A.M., 2003. *Bt* sweet corn and selective insecticides: impacts on pests and predators. Journal of Economic Entomology 96: 71–80.

Naranjo, S.E., 2005a. Long-term assessment of the effects of transgenic Bt cotton on the abundance of non-target arthropod natural enemies. Environmental Entomology 34: 1193–1210.

Naranjo, S.E., 2005b. Long-term assessment of the effects of transgenic Bt cotton on the function of the natural enemy community. Environmental Entomology 34: 1211–1223.

Obrist, L.B., Klein, H., Dutton, A., and Bigler, F., 2005. Effects of Bt maize on *Frankliniella tenuicornis* and exposure of thrips predators to prey-mediated Bt toxin. Entomologia Experimentalis et Applicata 115: 409–416.

Obrist, L.B., Dutton, A., Albajes, R., and Bigler, F., 2006a. Exposure of arthropod predators to Cry1Ab toxin in Bt maize fields. Ecological Entomology 31: 143–154.

Obrist, L., Dutton, A., Romeis, J., and Bigler, F., 2006b. Fate of Cry1Ab toxin expressed by Bt maize upon ingestion by herbivorous arthropods and consequences for *Chrysoperla carnea*. BioControl 51: 31–48.

Obrycki, J.J., 1986. The influence of foliar pubescence on entomophagous insects. In: Interaction of Plant Resistance and Parasitoids and Predators of Insects, D.J. Boethel and R.D. Eikenbarry, eds., Ellis Horwood, Chichester, UK, pp. 61–83.

O'Callaghan, M., Glare, T.R., Burgess, E.P.J., and Malone, L.A., 2005. Effects of plants genetically modified for insect resistance on nontarget organisms. Annual Review of Entomology 50: 271–292.

OECD (Organisation for Economic Cooperation and Development), 1993. Safety Considerations for Biotechnology: Scale-Up of Crop Plants. Organisation for Economic Cooperation and Development, Paris, France.

Olff, H., Brown, V.K., and Drent, R.H., 1999. Herbivores: Between Plants and Predators. Blackwell Science, Oxford, UK.

Olsen, K.M., Daly, J.C., Holt, H.E., and Finnegan, E.J., 2005. Season-long variation in expression of the *cry1Ac* gene and efficacy of *Bacillus thuringiensis* toxin in transgenic cotton against *Helicoverpa armigera* (Lepidoptera: Noctuidae). Journal of Economic Entomology 98: 1007–1017.

Painter, R.H., 1951. Insect Resistance in Crop Plants. University of Kansas Press, Lawrence, KS, USA.

Pickett, C.H., and Bugg, R.L., eds., 1998. Enhancing Biological Control. University of California Press, Berkeley, CA, USA.

Pilcher, C.D., Rice, M.E., and Obrycki, J.J., 2005. Impact of transgenic *Bacillus thuringiensis* corn and crop phenology on five non-target arthropods. Environmental Entomology 34: 1302–1316.

Power, M.E., 1992. Top-down and bottom-up forces in food webs: do plants have primacy? Ecology 73: 733–746.

Price, P.W., 1986. Ecological aspects of host plant resistance and biological control: interactions among three trophic levels. In: Interactions of Plant Resistance and Parasitoids and Predators of Insects, D.J. Boethel and R.D. Eikenbarry, eds., Ellis Horwood, Chichester, UK, pp. 11–30.

Price, P.W., Bouton, C.E., Gross, P., McPheron, B.A., Thompson, J.N., and Weis, A.E., 1980. Interactions among three trophic levels: influence of plants on interactions between insect herbivores and natural enemies. Annual Review of Ecology and Systematics 11: 41–65.

Rahbé, Y., Sauvion, N., Febvay, G., Peumans, W.J., and Gatehouse, A.M.R., 1995. Toxicity of lectins and processing of ingested proteins in the pea aphid *Acyrthosiphon pisum*. Entomologia Experimentalis et Applicata 76: 143–155.

Rahbé, Y., Deraison, C., Bonade-Bottino, M., Girard, C., Nardon, C., and Jouanin, L., 2003. Effects of the cysteine protease inhibitor oryzacystatin (OC-I) on different aphids and reduced performance of *Myzus persicae* on OC-I expressing transgenic oilseed rape. Plant Science 164: 441–450.

Raney, T., 2006. Economic impact of transgenic crops in developing countries. Current Opinion in Biotechnology 17: 174–178.

Rao, K.V., Rathore, K.S., Hodges, T.K., Fu, X., Stoger, E., Sudhakar, D., Williams, S., Christou, P., Bharathi, M., Bown, D.P., Powell, K.S., Spence, J., Gatehouse, A.M.R., and Gatehouse, J.A., 1998. Expression of snowdrop lectin (GNA) in transgenic rice plants confers resistance to rice brown planthopper. Plant Journal 15: 469–477.

Rausher, M.D., 2001. Co-evolution and plant resistance to natural enemies. Nature 411: 857–864.

Raybould, A., Stacey, D., Vlachos, D., Graser, G., Li, X., and Joseph, R., 2007. Non-target organisms risk assessment of MIR604 maize expressing mCry3A for control of corn rootworms. Journal of Applied Entomology 131: 391–399.

Reed, G.L., Jensen, A.S., Riebe, J., Head, G., and Duan, J.J., 2001. Transgenic *Bt* potato and conventional insecticides for Colorado potato beetle management: comparative efficacy and non-target impacts. Entomologia Experimentalis et Applicata 100: 89–100.

Rice, M.E., 2004. Transgenic rootworm corn: assessing potential agronomic, economic, and environmental benefits. Online. Plant Health Progress. doi:10.1094/PHP-2004-0301-01-RV.

Riddick, E.W., Dively, G., and Barbosa, P., 1998. Effect of a seed-mix deployment of Cry3A-transgenic and nontransgenic potato on the abundance of *Lebia grandis* (Coleoptera: Carabidae) and *Coleomegilla maculata* (Coleoptera: Coccinellidae). Annals of the Entomological Society of America 91: 647–653.

Rodrigo-Simón, A., de Maagd, R.A., Avilla, C., Bakker, P.L., Molthoff, J., Gonzalez-Zamora, J.E., and Ferré, J., 2006. Lack of detrimental effects of *Bacillus thuringiensis* Cry toxins on the insect predator *Chrysoperla carnea*: a toxicological, histopathological, and biochemical approach. Applied and Environmental Microbiology 72: 1595–1603.

Romeis, J., Shanower, T.G., and Zebitz, C.P.W., 1999. Why *Trichogramma* (Hymenoptera: Trichogrammatidae) egg parasitoids of *Helicoverpa armigera* (Lepidoptera: Noctuidae) fail on chickpea. Bulletin of Entomological Research 89: 89–95.

Romeis, J., Dutton, A., and Bigler, F., 2004. *Bacillus thuringiensis* toxin (Cry1Ab) has no direct effect on larvae of the green lacewing *Chrysoperla carnea* (Stephens) (Neuroptera: Chrysopidae). Journal of Insect Physiology 50: 175–183.

Romeis, J., Babendreier, D., Wäckers, F.L., and Shanower, T.G., 2005. Habitat and plant specificity of *Trichogramma* egg parasitoids – underlying mechanisms and implications. Basic and Applied Ecology 6: 215–236.

Romeis, J., Meissle, M., and Bigler, F., 2006. Transgenic crops expressing *Bacillus thuringiensis* toxins and biological control. Nature Biotechnology 24: 63–71.

Romeis, J., Meissle, M., Raybould, R., and Hellmich, R.L., 2008a. Impact of insect-resistant genetically modified crops on non-target arthropods. In: Environmental Impact of Genetically Modified/ Novel Crops, N. Ferry and A.M.R. Gatehouse, eds., CABI, Wallingford, UK (in press).

Romeis, J., Bartsch, D., Bigler, F., Candolfi, M.P., Gielkens, M.M.C., Hartley, S.E., Hellmich, R.L., Huesing, J.E., Jepson, P.C., Layton, R., Quemada, H., Raybould, A., Rose, R.I., Schiemann, J., Sears, M.K., Shelton, A.M., Sweet, J., Vaituzis, Z., and Wolt, J.D., 2008b. Assessment of risk of insect-resistant transgenic crops to nontarget arthropods. Nature Biotechnology 26: 203–208.

Rose, R., and Dively, G.P., 2007. Effects of insecticide-treated and lepidopteran-active *Bt* transgenic sweet corn on the abundance and diversity of arthropods. Environmental Entomology 36: 1254–1268.

Rutledge, C.E., Robinson, A.P., and Eigenbrode, S.D., 2003. Effects of a simple plant morphological mutation on the arthropod community and the impact of predators on a principal insect herbivore. Oecologia 135: 39–50.

Sanvido, O., Romeis, J., and Bigler, F., 2007. Ecological impacts of genetically modified crops: ten years of field research and commercial cultivation. Advances in Biochemical Engineering and Biotechnology 107: 235–278.

Schnee, C., Köllner, T.G., Held, M., Turlings, T.C.J., Gershenzon, J., and Degenhardt, J., 2006. The products of a single maize sesquiterpene synthase form a volatile defense signal that attracts natural enemies of maize herbivores. Proceedings of the National Academy of Sciences of the USA 103: 1129–1134.

Schnepf, E., Crickmore, N., van Rie, J., Lereclus, D., Baum, J., Feitelson, J., Zeigler, D.R., and Dean, D.H., 1998. *Bacillus thuringiensis* and its pesticidal crystal proteins. Microbiology and Molecular Biology Reviews 62: 775–806.

Schroeder, H.E., Gollasch, S., Moore, A., Tabe, L.M., Craig, S., Hardie, D.C., Chrispeels, M.J., Spencer, D., and Higgins, T.J.V., 1995. Bean alpha-amylase inhibitor confers resistance to the pea weevil (*Bruchus pisorum*) in transgenic peas (*Pisum sativum* L). Plant Physiology 107: 1233–1239.

Schuler, T.H., Denholm, I., Clark, S.J., Stewart, C.N., and Poppy, G.M., 2004. Effects of Bt plants on the development and survival of the parasitoid *Cotesia plutellae* (Hymenoptera: Braconidae) in susceptible and Bt-resistant larvae of the diamondback moth, *Plutella xylostella* (Lepidoptera: Plutellidae). Journal of Insect Physiology 50: 435–443.

Shelton, A.M., and Sears, M.K., 2001. The monarch butterfly controversy: scientific interpretations of a phenomenon. The Plant Journal 27: 483–488.

Shi, Y., Wang, M.B., Powell, K.S., Van Damme, E., Hilder, V.A., Gatehouse, A.M.R., Boulter, D., and Gatehouse, J.A., 1994. Use of the rice sucrose synthase-1 promotor to direct phloem-specific expression of β-glucuronidase and snowdrop lectin genes in transgenic tobacco plants. Journal of Experimental Botany 45: 623–631.

Simmons, A.T., and Gurr, G.M., 2004. Trichome-based host plant resistance of *Lycopersicon* species and the biocontrol agent *Mallada signata*: are they compatible? Entomologia Experimentalis et Applicata 113: 95–101.

Sisterson, M.S., Biggs, R.W., Manhardt, N.M., Carrière, Y., Dennehy, T.J., and Tabashnik, B.E., 2007. Effects of transgenic Bt cotton on insecticide use and abundance of two generalist predators. Entomologia Experimentalis et Applicata 124: 305–311.

Smith, C.M., 2005. Plant Resistance to Arthropods: Molecular and Conventional Approaches. Springer Science & Business Media, Dordrecht, The Netherlands.

Srivastava, P.N., and Auclair, J.L., 1963. Characteristics and nature of proteases from the alimentary canal of the pea aphid, *Acyrthosiphon pisum* (Harr) (Homoptera, Aphididae). Journal of Insect Physiology 9: 469–474.

Stanley-Horn, D.E., Dively, G.P., Hellmich, R.L., Sears, M.K., Rose, R., Jesse, L.C.H., Losey, J.E., Obrycki, J.J., and Lewis, L.C., 2001. Assessing the impact of Cry1Ab-expressing corn pollen on monarch butterfly larvae in field studies. Proceedings of the National Academy of Sciences of the USA 98: 11931–11936.

Talhouk, A.S., 1991. On the management of the date palm and its arthropod enemies in the Arabian Penisula. Journal of Applied Entomology 111: 514–520.

Tollrian, R., and Harvell, C.D., 1999. The Ecology and Evolution of Inducible Defences. Princeton University Press, Princeton, NJ, USA.

Torres, J.B., and Ruberson, J.R., 2005. Canopy- and ground-dwelling predatory arthropods in commercial *Bt* and non-*Bt* cotton fields: patterns and mechanisms. Environmental Entomology 34: 1242–1256.

Torres, J.B., and Ruberson, J.R., 2006. Spatial and temporal dynamics of oviposition behavior of bollworm and three of its predators in Bt and non-Bt cotton fields. Entomologia Experimentalis et Applicata 120: 11–22.

Torres, J.B., and Ruberson, J.R., 2008. Interactions of *Bacillus thuringiensis* Cry1Ac toxin in genetically engineered cotton with predatory heteropterans. Transgenic Research 17: 345–354.

Torres, J.B., Ruberson, J.R., and Adang, M.J., 2006. Expression of *Bacillus thuringiensis* Cry1Ac protein in cotton plants, acquisition by pests and predators: a tritrophic analysis. Agricultural and Forest Entomology 8: 191–202.

Toschki, A., Hothorn, L.A., and Ross-Nickoll, M., 2007. Effects of cultivation of genetically modified Bt maize on epigeic arthropods (Araneae; Carabidae). Environmental Entomology 36: 967–981.

Traugott, M.S., and Stamp, N.E., 1996. Effects of chlorogenic acid and tomatine-fed caterpillars on the behavior of an insect predator. Journal of Insect Behavior 9: 461–476.

Turlings, T.C.J., and Benrey, B., 1998. Effects of plant metabolites on the behavior and development of parasitic wasps. Ecoscience 5: 321–333.

Turlings, T.C.J., and Wäckers, F.L., 2004. Recruitment of predators and parasitoids by herbivore-injured plants. In: Advances in Insect Chemical Ecology, R.T. Cardé and J.G. Millar, Cambridge University Press, Cambridge, UK, pp. 21–75.

USEPA (United States Environmental Protection Agency), 2001. Biopesticide Registration Action Document. *Bacillus thuringiensis (Bt)* Plant-Incorporated Protectants. 15 October 2001. http://www.epa.gov/oppbppd1/biopesticides/pips/bt_brad.htm (accessed 15 January 2008).

Van Driesche, R.G., and Taub, G., 1983. Impact of parasitoids on *Phyllonorycter* leafminers infesting apple in Massachusetts, USA. Protection Ecology 5: 303–317.

Van Driesche, R.G., Hoddle, M.S., and Center, T.D., 2008. Control of Pests and Weeds by Natural Enemies. Blackwell, Oxford, UK.

Van Lenteren, J.C., and de Ponti, O.M.B., 1990. Plant leaf morphology, host-plant resistance and biological control. Symposia Biologica Hungarica, Budapest 39: 365–386.

Wäckers, F.L., 2005. Suitability of (extra-) floral nectar, pollen, and honeydew as insect food sources. In: Plant-Provided Food for Carnivorous Insects: A Protective Mutualism and Its Applications, F.L. Wäckers, P.C.J. van Rijn and J. Bruin, Cambridge University Press, Cambridge, UK, pp. 17–74.

Wäckers, F.L., and Van Rijn, P.C.J., 2005. Food for protection: an introduction. In: Plant-Provided Food for Carnivorous Insects: A Protective Mutualism and Its Applications, F.L. Wäckers, P.C.J. van Rijn and J. Bruin, eds., Cambridge University Press, Cambridge, UK, pp. 11–14.

Warren, G.W., 1997. Vegetative insecticidal proteins: novel proteins for control of corn pests. In: Advances in Insect Control: The Role of Transgenic Plants, N. Carozzi and M. Koziel, eds., Taylor & Francis, London, UK, pp. 109–121.

White, J.A., and Andow, D.A., 2005. Host-parasitoid interactions in a transgenic landscape: spatial proximity effects of host density. Environmental Entomology 34: 1493–1500.

White, J.A., and Andow, D.A., 2006. Habitat modification contributes to associational resistance between herbivores. Oecologia 148: 482–490.

Whitehouse, M.E.A., Wilson, L.J., and Constable, G.A., 2007. Target and non-target effects on the invertebrate community of Vip cotton, a new insecticidal transgenic. Australian Journal of Agricultural Research 58: 273–285.

Wilson, L., Hickman, M., and Deutscher, S., 2006. Research update on IPM and secondary pests. In: Proceedings of the 13th Australian Cotton Research Conference, Broadbeach, Queensland, Australia, pp. 249–258.

Wink, M., and Römer, P., 1986. Acquired toxicity – the advantages of spezializing on alkaloid-rich lupins to *Macrosiphon albifrons* (Aphidae). Naturwissenschaften 73: 210–212.

Wu, K.M., and Guo, Y.Y., 2003. Influences of *Bacillus thuringiensis* Berliner cotton planting on population dynamics of the cotton aphid, *Aphis gossypii* Glover, in northern China. Environmental Entomology 32: 312–318.

Wu, K., Li, W., Feng, H., and Guo, Y., 2002. Seasonal abundance of the mirids, *Lygus lucorum* and *Adelphocoris* spp. (Hemiptera: Miridae) on Bt cotton in northern China. Crop Protection 21: 997–1002.

Zangerl, A.R., McKenna, D., Wraight, C.L., Carroll, M., Ficarello, P., Warner, R., and Berenbaum, M.R., 2001. Effects of exposure to event 176 *Bacillus thuringiensis* corn pollen on monarch and black swallowtail caterpillars under field conditions. Proceedings of the National Academy of Sciences of the USA 98: 11908–11912.

Chapter 5
The Present and Future Role of Insect-Resistant Genetically Modified Maize in IPM

Richard L. Hellmich[1,*], Ramon Albajes[2], David Bergvinson[3], Jarrad R. Prasifka[1], Zhen-Ying Wang[4], and Michael J. Weiss[5]

Abstract Commercial, genetically-modified (GM) maize was first planted in the United States (USA, 1996) and Canada (1997) but now is grown in 13 countries on a total of over 35 million hectares (>24% of area worldwide). The first GM maize plants produced a Cry protein derived from the soil bacterium *Bacillus thuringiensis* (*Bt*), which made them resistant to European corn borer and other lepidopteran maize pests. New GM maize hybrids not only have resistance to lepidopteran pests but some have resistance to coleopteran pests and tolerance to specific herbicides. Growers are attracted to the *Bt* maize hybrids for their convenience and because of yield protection, reduced need for chemical insecticides, and improved grain quality. Yet, most growers worldwide still rely on traditional integrated pest management (IPM) methods to control maize pests. They must weigh the appeal of buying insect protection "in the bag" against questions regarding economics, environmental safety, and insect resistance management (IRM). Traditional management of maize insects and the opportunities and challenges presented by GM maize are considered as they relate to current and future insect-resistant products. Four countries, two that currently have commercialize *Bt* maize (USA and Spain) and two that do not (China and Kenya), are highlighted. As with other insect management tactics (e.g., insecticide use or tillage), GM maize should not be considered inherently compatible or incompatible with IPM. Rather, the effect of GM insect-resistance on maize IPM likely depends on how the technology is developed and used.

[1] USDA–ARS, Corn Insects and Crop Genetics Research Unit, Ames, IA, USA

[2] University of Lleida, Centre Udl-IRTA, Lleida, Spain

[3] Bill & Melinda Gates Foundation, Seattle, WA, USA

[4] Institute of Plant Protection, Chinese Academy of Agricultural Sciences, Beijing, China

[5] Golden Harvest Seeds, Pine River, WI, USA

*To whom correspondence should be addressed. E-mail: Richard.Hellmich@ars.usda.gov

J. Romeis, A.M. Shelton, G.G. Kennedy (eds.), *Integration of Insect-Resistant Genetically Modified Crops within IPM Programs.*

5.1 Introduction

Maize, *Zea mays* (corn), is second only to rice, *Oryza* spp., as a world crop with over 140 million hectares planted. Annually, nearly 700 million metric tons (MMT) of grain are produced, primarily by the United States (USA; 39.5%), China (19.3%), Brazil (6.0%), Mexico (3.0%), Argentina (2.4%) and India (2.0%). As many as five other countries produce 10 MMT or more annually (FAOSTAT, 2007). Because many important pests of maize are lepidopteran stem borers, the first genetically-modified (GM) maize targeted a stem-boring pest, the European corn borer, *Ostrinia nubilalis* (Lepidoptera: Crambidae). Genetically-modified maize resistant to *O. nubilalis* was first commercially grown in the USA (1996) and Canada (1997); a decade later GM maize is grown in 13 countries on a total of over 35 million hectares, (>24% of maize area worldwide). Available varieties of GM maize now include combinations of traits to suppress lepidopteran and coleopteran pests and to provide herbicide tolerance. Additional traits are being tested to improve the efficacy and spectrum of GM insect-resistant maize.

Most maize growers, however, rely on traditional crop protection practices to manage insects, including cultural, biological or chemical (insecticidal) methods. As use of GM maize continues to spread, growers must weigh the appeal of buying insect protection "in the bag" against questions regarding economics, environmental safety, and insect resistance management (IRM). To agricultural scientists, GM maize provides another valuable option to manage pests, but as GM maize expands to include combined aspects of protection from multiple pests, herbicide tolerance, drought tolerance and nutrient enrichment, costs and benefits become more difficult for growers to evaluate. To place the role of insect-resistant GM maize into a broader context, this chapter will discuss traditional management of maize insects and the opportunities and challenges presented by GM maize as they relate to current and future (potential) insect-resistant products. Information specific to different maize-producing countries, including the United States, Spain, China, and Kenya, will be discussed.

5.2 Maize Integrated Pest Management

As in other crops, management of insect (and weed or pathogen) pests has changed greatly over the last several decades. While growers once relied primarily on cultural methods and (natural) biological control, the efficacy of new synthetic insecticides from the 1940s–1970s increased reliance on chemical pest suppression (Casida and Quistad, 1998). Along with grower dependence on insecticides, insect resistance and concerns that insecticides were harming the environment (and human health) led entomologists to develop integrated pest management (IPM) strategies (Stern et al., 1959; Kogan, 1998; Kennedy, chapter 1). The basic goal of IPM is to achieve effective crop protection through the integration of appropriate control actions in a manner that provides economic benefits to growers and society, and benefits to the environment.

Maize IPM includes both preventative and responsive pest management tactics. Preventative tactics are used prior to the occurrence of the injurious stage of the pest and include host plant resistance (HPR), cultural controls (e.g., modified planting dates, crop rotation, tillage), and natural biological control. Responsive management is used when levels of pests occur that are likely to produce crop losses that exceed the costs of suppression. This requires accurate measures of insect populations and an understanding of the relationship between pest injury and the crop plant damage response (Pedigo et al., 1986). The related IPM concepts of economic injury level (EIL) and economic threshold (ET) are discussed by Stern et al. (1959) and in Kennedy (chapter 1). Primary elements of IPM to integrate with GM maize include host plant resistance, cultural control, biological control and limited use of insecticides.

5.2.1 Host Plant Resistance

Host plant resistance refers to the heritable plant qualities that reduce pest losses, in this case from maize-feeding insects. The HPR in modern maize hybrids is the product of efforts by entomologists and plant breeders to enhance resistance. Resistance traits are generally separated into those that lower plant attractiveness to insects (nonpreference or antixenosis), impair development (antibiosis) or allow a plant to compensate for injury by an insect (tolerance) (Painter, 1968). Insect-resistant GM plants and plants bred for HPR may be considered relatively similar because resistance traits are delivered by the plant and are preventative forms of pest management.

Breeding for HPR in maize has focused on lepidopteran and coleopteran pests. In the USA, such efforts have emphasized resistance to *O. nubilalis*, corn earworm, *Helicoverpa zea* (Lepidoptera: Noctuidae), and western corn rootworm, *Diabrotica virgifera virgifera* (Coleoptera: Chrysomelidae). Antibiosis from hydroxamic acids and flavonoid glycosides in maize has been key for managing pests. The hydroxamic acid DIMBOA (2,4-dihydroxy-7-methoxy-1,4-benzoxazin-3-one) deters *O. nubilalis* leaf-feeding in vegetative-stage maize (Klun et al., 1967). DIMBOA also contributes to maize resistance to *D. v. virgifera*, leading to adults with low emergence, weight, and head-capsule width (Xie et al., 1990). Maysin, C-glycosyl flavone, in maize silks inhibits larval growth of *H. zea* and fall armyworm, *Spodoptera frugiperda* (Lepidoptera: Noctuidae) (Waiss et al., 1979; Wiseman et al., 1992). Because of their efficacy, increased levels of DIMBOA and maysin are common in commercially-available maize germplasm (Barry and Darrah, 1991; Widstrom and Snook, 2001). Some resistance to *O. nubilalis* feeding and tunneling is related to elevated levels of cell-wall fiber and lignin (Coors, 1987; Beeghly et al., 1997) or fortification of the epidermal cell wall (Bergvinson et al., 1995). Epidermal leaf toughness also can be used to identify resistant varieties effective across a wide range of lepidopterans, including tropical pests of maize (Bergvinson et al., 1994). Most research related to maize resistance to *D. v. virgifera*, has focused on tolerance, as plants with large root systems or high compensatory root growth are more tolerant to *D. v. virgifera* feeding (Prischmann et al., 2007).

Because many traits related to maize resistance to insects are multigenic (Frey et al., 1997; Meyer et al., 2007), incorporating them into breeding populations has been difficult. However, the use of marker-assisted selection can facilitate breeding once genes for these traits are identified (McMullen et al., 1998). One option to enhance maize HPR and transgenic insect-resistance includes efforts to combine natural traits with transgenic traits for maximum effectiveness (Warnock et al., 2001).

5.2.2 Cultural Control

Farming practices are used to help manage insect pests. Effectively, insect injury is reduced by physically destroying pests (cultivation), or limiting access to crops over time (planting or harvest dates) and space (crop rotation). For example, prior to synthetic insecticides and maize HPR, *O. nubilalis* management was essentially cultural, with stalk destruction in the fall or moldboard plowing of maize stubble in the spring prior to planting (Caffrey and Worthley, 1927; Umeozor et al., 1985). However, such methods only are effective if conducted over large areas. Late or early maize planting also can be used to reduce *O. nubilalis* injury for the first and second generations, respectively (Mason et al., 1996; Pilcher and Rice, 2001).

Crop rotation of maize with non-host crops, especially soybean, is common practice in the US Corn Belt because (in addition to its agronomic benefits) it has largely controlled *Diabrotica* spp. (Chiang, 1973). However, *Diabrotica* spp. have adapted to crop rotation; in areas of Minnesota, Iowa and South Dakota, northern corn rootworms, *Diabrotica barberi*, have extended their diapause for two or more years (Krysan et al., 1986). In Illinois, Indiana, Ohio, Michigan and Wisconsin, *D. v. virgifera* defeat rotation by ovipositing in non-maize crops such as soybeans (Levine and Oloumi-Sadeghi, 1996). For areas of Europe where western corn rootworm has invaded (from the USA), rotations are mandatory when *D. v. virgifera* are detected as a step towards local eradication (Byrne, 2003). Interestingly, crop rotation can be a responsive tactic if densities of soil insects are known before the crops are planted (as suggested for *Diabrotica* spp. by Gillette, 1912).

5.2.3 Biological Control

Populations of many maize pests are naturally suppressed by beneficial predators, parasitoids and pathogens. Natural enemies may be used in importation (classical), conservation and augmentative biological control to control crop pests. Importation of parasitoids has been used in the USA in response to the accidental introduction of *O. nubilalis* in the early 1900s; the tachinid fly, *Lydella thompsoni* (Diptera: Tachinidae), and the wasps, *Macrocentrus cingulum* (Hymenoptera: Braconidae) and

Eriborus terebrans (Hymenoptera: Ichneumonidae), have become established but do not consistently maintain *O. nubilalis* populations below economic levels (Anonymous, 1990; Mason et al., 1994). Similarly, the parasitoid *Cotesia flavipes* (Hymenoptera: Braconidae) was introduced to Kenya from Pakistan (Omwega et al., 1995) to control the spotted stem borer, *Chilo partellus* (Lepidoptera: Crambidae). This pest was accidentally introduced into Africa before the 1930s and has become one of the most damaging pests of maize (Tams, 1932; Overholt et al., 1997). *Cotesia flavipes* has spread in Kenya and Tanzania, becoming the dominant larval parasitoid of stem borers in southeastern Kenya (Zhou and Overholt, 2001; see section 5.4.4).

Conserving natural enemies of maize pests involves limiting negative factors, such as insecticides, and implementing habitat management to improve factors that support natural enemies such as the provision of alternative food sources (Landis et al., 2000). For example, the parasitoid, *E. terebrans* appears to be influenced by the local landscape, causing greater parasitism of *O. nubilalis* near wooded edges compared to field interiors or non-wooded edges (Landis and Haas, 1992); the wooded edges may provide food resources and a favorable microhabitat for adult wasps (Dyer and Landis, 1997). Many growers use modified field edges such as riparian buffers, filter strips, shelterbelts and living snow fences, which increase landscape diversity and may provide habitat for the natural enemies of maize pests.

In recent decades, augmentative biological control has become more feasible through development of efficient rearing protocols, allowing responsive pest management through inundative or inoculative releases of parasitoids or predators. This strategy may be more useful in high-value maize (grown for seed or fresh consumption) than in maize grown for grain. The egg parasitoids, *Tricogramma* spp. (Hymenoptera. Trichogrammatidae), are used for the inundative control of *O. nubilalis* in Switzerland (Bigler, 1986) and *Ostrinia furnacalis* (Lepidoptera: Crambidae) in China (see section 5.4.3), but this strategy has not been cost-effective in the USA (Andow et al., 1995; Gardner et al., 2007). To control *D. v. virgifera*, entomopathogenic nematodes have been tested with mixed results (Munson and Helms, 1970; Wright et al., 1993; Ellsbury et al., 1996; Jackson, 1996). However, it may be possible to breed maize that is more attractive to entomopathogenic nematodes to help manage *D. v. virgifera* populations (Rasmann et al., 2005).

5.2.4 Insecticides

The basic concept of IPM suggests that insecticide use may be appropriate when other methods cannot adequately suppress pest populations. Further, the decision to apply insecticides should be based on the use of sampling information and economic decision levels (e.g., EIL and ET; Kennedy, chapter 1). In major maize-producing areas like the US Corn Belt, sampling information and decision levels for certain pests are well established (for an overview see Steffey et al., 1999). However, similar guidelines are deficient or unavailable for many key maize pests throughout the world, effectively prohibiting judicious insecticide use.

5.3 Insect-Resistant GM Maize: Opportunities and Challenges

As of 2007, all available insect-resistant GM maize express one or more *cry* genes derived from the soil bacterium *Bacillus thuringiensis* (*Bt*). Consequently, the discussion below pertains specifically to maize varieties including *Bt* traits (*Bt* maize), though once other types and combinations of toxins become available, the opportunities and challenges for IPM are likely to be similar. *Bacillus thuringiensis* crystal (Cry) proteins differ from most conventional insecticides because they are toxic to only a small range of related insects. This is because specific pH levels, enzymes, and gut receptors are required to solubilize, activate and bind a given Cry toxin (Federici, 2002; Ferré et al., chapter 3). This specificity and its label as a "natural insecticide" have contributed to the use of *Bt* as a biologically-based insecticide by many organic growers. Certainly the history of safe grower use of *Bt* treatments has contributed to its commercial success in *Bt* plants.

5.3.1 Current Varieties of Bt Maize

Cry proteins are categorized by their spectrum of activity. For maize pests, primary Cry proteins are Cry1 and Cry2 for Lepidoptera and Cry3 proteins for Coleoptera (Schnepf et al., 1998). Registered types of *Bt* maize, called events, are shown in Tables 5.1 and 5.2 and their relative efficacies against key maize pests are shown in Table 5.3. Prior to 2002, lepidopteran resistance and herbicide tolerance often were combined (stacked); now triple stacks with lepidopteran resistance, coleopteran resistance and herbicide tolerance are available. Although not the focus of this chapter, herbicide tolerance traits increasingly will be stacked with *Bt* maize. This technology allows growers to control weeds by spraying with an herbicide without harming the crop. Growers are attracted to this technology because the companion herbicides replace more persistent herbicides, they are convenient to apply, and they can be used in no-till and minimum tillage systems (USDA-ERS, 2002). In the near future, collaborations between biotechnology companies potentially will produce GM maize with as many as eight different traits, including herbicide tolerance and insect resistance (Dow AgroSciences, 2007). While this may present maize growers with new options, it may also complicate the decision-making process on what to plant, especially if growers cannot pick-and-choose any desirable combination of traits.

5.3.2 Opportunities

Insect-resistant GM maize offers both economic and environmental advantages over using conventional insecticides to manage certain maize pests. Responses of US maize growers indicate an awareness of both types of benefits, as growers cite

Table 5.1 Current and previously registered *Bt* maize products for lepidopteran-resistance (LR) and coleopteran-resistance (CR) commercialized in the USA, field maize unless indicated (http://www.epa.gov/pesticides/biopesticides/pips/pip_list.htm, accessed 2 January 2008)

Events	Insecticidal proteins	Traits	Companies	First registered	Trade names
176[a]	Cry1Ab	LR, PAT[b]	Ciba Seeds	Aug 1995	KnockOut®
			Mycogen Seeds		NatureGuard®
Bt11	Cry1Ab	LR, PAT	Northrup King	Aug 1996	Agrisure™ CB
MON810	Cry1Ab	LR	Monsanto	Dec 1996	YieldGard®
DBT418[a]	Cry1Ac	LR, PAT	DeKalb Genetics	Mar 1997	Bt-Extra™
Bt11 (sweet corn)	Cry1Ab	LR, PAT	Novartis Seeds	Feb 1998	Attribute®
CBH351[a,c]	Cry9C	LR, PAT	Plant Genetic Systems	May 1998	StarLink™
TC1507	Cry1F	LR, PAT	Dow AgroSciences Pioneer Hi-Bred	May 2001	Herculex® I
MON863	Cry3Bb1	CR	Monsanto	Feb 2003	YieldGard® RW
MON863 X MON810[d]	Cry3Bb1, Cry1Ab	CR, LR	Monsanto	Oct 2003	YieldGard® Plus
DAS-59122-7	Cry34/35Ab1	CR, PAT	Dow AgroSciences Pioneer Hi-Bred	Aug 2005	Herculex® RW
TC1507 X DAS-59122-7[d]	Cry1F, Cry34/35Ab1	LR, CR, PAT	Dow AgroSciences Pioneer Hi-Bred	Oct 2005	Herculex® XTRA
MON88017	Cry3Bb1	CR, EPSPS[e]	Monsanto	Dec 2005	YieldGard® VT
MON810 X MON88017[d]	Cry1Ab, Cry3Bb1	LR, CR, EPSPS	Monsanto	Dec 2005	YieldGard® VT Triple
MIR604	modified Cry3A	CR	Syngenta Seeds	Oct 2006	Agrisure™ RW
Bt11 X MIR604[d]	Cry1Ab, modified Cry3A	LR, CR, PAT	Syngenta Seeds	Jan 2007	Agrisure™ CB/RW

[a] No longer registered
[b] PAT, phosphinothricin-N-acetyltransferase, which allows use of herbicide glufosinate ammonium (e.g., Liberty®)
[c] Registered for animal feed and non-food use only
[d] Stacks formed from conventional crosses
[e] EPSPS, 5-enolpyruvylshikimate-3-phosphate synthase, which allows use of herbicide glyphosate (e.g., Roundup®)

Table 5.2 *Bt* maize events for all countries that have commercially planted *Bt* maize with total annual grain maize production (MMT), total hectares (million), percentage *Bt* maize and year first produced for each country[a]

Country	MMT	Total ha	% Bt	1st Prod	Current commercial *Bt* maize events
USA	266.8	29.1	49[b]	1996	MON810, Bt11, TC1507, MON863, DAS-59122-7, MON88017, MIR604
Canada	9.2	1.2	49[c]	1997	MON810, Bt11, TC1507, MON863, DAS-59122-7, MON88017, MIR604
South Africa	9.6	3.1	44[d]	1997	MON810, Bt11
Argentina	15.9	2.5	63[d]	1998	MON810, Bt11, TC1507
Spain	4.2	0.4	21[e]	1998	MON810
France	14.3	1.7	1[e]	1998[f]	MON810
Portugal	0.7	0.1	<1[e]	1999[g]	MON810
Germany	3.7	0.4	<1[e]	2000	MON810
Honduras	0.5	0.3	<1[d]	2001	MON810
Philippines	5.1	2.5	5[d]	2003	MON810, Bt11
Uruguay	0.2	<0.1	<1[d]	2003	MON810, Bt11
Czech Republic	0.6	<0.1	<1[e]	2005	MON810
Slovakia	0.8	0.1	<1[e]	2006	MON810
Brazil	40.8	12.3	0	2008	MON810, Bt11

[a] MMT production, Total ha (million), average 2002–2006 (FAOSTAT, 2007)
[b] USDA-NASS, 2007
[c] Stratus Agri-Marketing Inc., 2006 figure
[d] James, 2007
[e] http://www.gmo-compass.org/eng/agri_biotechnology/gmo_planting/191.eu_growing_area.html (accessed 16 January 2008), 2007 figures.
[f] no planting 2001–2004
[g] no planting 2000–2004

unique opportunities to protect yield and reduce handling (and use) of insecticides to explain their rapid adoption of *Bt* maize (Pilcher et al., 2002). Economic benefits in the USA from *Bt* maize depend on maize prices and levels of pest populations; in some years, planting *Bt* maize can be an economic disadvantage (Carpenter and Gianessi, 2001), but under typical conditions should provide increased profits to *Bt* maize growers (Sankula, 2006). Research with *Bt* maize in Spain and the Philippines (Demont and Tollens, 2004; Yorobe and Quicoy, 2006; Gómez-Barbero et al., 2008) also suggests growers gain financially from using transgenic insect control. Brookes and Barfoot (2006) estimated that in the USA from 1996 to 2005 the cumulative decrease in insecticide active ingredient (a.i.) use on *Bt* maize was 4% (6,400 MT). Most of the reduction in insecticide a.i. was from lepidopteran-active *Bt* maize. However, coleopteran-active *Bt* maize shows a much greater potential benefit in the near future, as insecticides used against *Diabrotica* spp. comprise 25–30% of the global total in maize (James, 2003; Rice, 2004).

Table 5.3 Important lepidopteran and coleopteran (*Diabrotica* spp.) maize pests in USA, China, Spain, Kenya, Argentina (Arg) and Philippines (Phil). Cry proteins are rated 4 (excellent control), 3 (good control), 2 (some control/suppression) and 1 (no effect).[a] The "x" indicates the pest occurs in that country

Maize pests	Common names	Family	Cry Proteins			USA	China	Spain	Kenya	Arg.	Phil.
Lepidopteran borers			Cry1Ab	Cry1F							
Busseola fusca	African maize stem borer	Noctuidae	2.5	2					x		
Chilo partellus	Spotted stem borer	Crambidae	3	3					x		x
Diatraea grandiosella	Southwestern corn borer	Crambidae	4	4		x					
Diatraea saccharalis	Sugarcane borer	Crambidae	4	4		x				x	
Ostrinia nubilalis	European corn borer	Crambidae	4	4		x		x			
Ostrinia furnacalis	Asian corn borer	Crambidae	4	4			x				x
Sesamia nonagrioides	Mediterranean corn borer	Noctuidae	3.5	3				x			
Other Lepidoptera											
Agrotis spp.	Cutworm	Noctuidae	1[b]	2.5[b]		x	x	x	x	x	x
Helicoverpa armigera	Corn earworm	Noctuidae	2.5	2			x	x	x		x
Helicoverpa zea	Corn earworm	Noctuidae	2.5	2		x				x	
Mythimna separata	Oriental armyworm	Noctuidae	3	i.d.[c]			x				
Pseudaletia unipuncta	Armyworm	Noctuidae	2	2		x	x	x		x	
Spodoptera exigua	Beet armyworm	Noctuidae	2	3.5			x		x		
Spodoptera frugiperda	Fall armyworm	Noctuidae	2	3.5		x				x	
Striacosta albicosta	Western bean cutworm	Noctuidae	2	3		x					
Coleoptera			Cry3Bbl	34/35[d]	mCry3A[e]						
Diabrotica spp.	Corn rootworms	Chrysomelidae	3	3	3	x				x	

[a] Approximate ratings pooled from the authors and industry experts; control might vary depending on event, hybrid and environment

[b] Tests conducted with black cutworm, *A. ipsilon*

[c] insufficient data

[d] binary Cry34Abl+Cry35Abl

[e] modified Cry3A

Another benefit of insect-resistant GM maize is reduced occurrence of ear molds. Because insect damage provides a site for infection by molds, *Bt*-protected maize can have lower levels of the *Fusarium* mycotoxins, fumonisin and deoxynivalenol (Munkvold and Hellmich, 1999; Dowd, 2000). Consequences of contamination with mold may be serious, as fumonisins can cause fatal leucoencephalomalacia in horses, pulmonary edema in swine, and cancer in laboratory rats. Economic analysis suggests that US farmers save $23 million annually through reduced mycotoxins (Wu et al., 2004), though mycotoxin reduction could be a significant health benefit in other parts of the world where maize is a diet staple (Wu, 2006a, b).

One more potential benefit of *Bt* maize is area-wide suppression of pest populations. There is increasing evidence that *O. nubilalis* populations in the US Corn Belt have been suppressed by *Bt* maize (Hellmich, 2006; Storer et al., chapter 10). This phenomenon could have implications for refuge and IRM (see section 5.3.3.3).

5.3.3 Challenges

Detractors of *Bt* maize suggest several challenges, including the potential for effects on non-target organisms and gene flow between *Bt* maize and non-*Bt* maize, to outweigh any benefits. Other issues to consider include whether insect resistance to *Bt* can be managed, and whether the use of insect-resistant GM maize conflicts with the basic principles of IPM.

5.3.3.1 Effects on Non-target Organisms

With regard to non-target organisms, no surprising effects have been observed with *Bt* maize, which confirms the specificity of the *Bt* proteins. Most studies in the USA, Europe and China suggest *Bt* maize has little if any impact on predators and parasitoids and, when compared with maize treated with chemical insecticides, *Bt* maize often results in increased biodiversity (Bourguet et al., 2002; Candolfi et al., 2003; Dutton et al., 2003; Bhatti et al., 2005a, b; Daly and Buntin, 2005; de la Poza et al., 2005; Dively, 2005; Pilcher et al., 2005; Romeis et al., 2006; Fernandes et al., 2007; Marvier et al., 2007; Wang et al., 2007; for general reviews see O'Callaghan et al., 2005; Romeis et al., 2008; chapter 4). Although maize is not a major source of pollen for honey bees, *Apis mellifera* (Hymenoptera: Apidae), the US Environmental Protection Agency (USEPA) requires information on possible effects of *Bt* maize on honey bees. Feeding studies suggest pollen from *Bt* maize has no effect on honey bee larvae or adults (Hanley et al., 2003; Babendreier et al., 2005; Rose et al., 2007; Duan et al., 2008). Specialist insects that depend on target pests are the exception to the generalization that *Bt* maize does not impact non-target organisms. This is particularly true for some parasitoids, which may become less abundant along with their herbivorous hosts (Pilcher et al., 2005; Romeis et al., chapter 4; Storer et al., chapter 10). Also, fewer saprophagous dipterans have been observed in *Bt* maize fields, which has been attributed to the indirect effect of reduced lepidopteran plant injury (Candolfi et al., 2003; Dively, 2005). Studies on possible effects of *Bt* maize on soil microorganisms also suggest little if any impact (Blackwood and Buyer, 2004; Devare et al., 2004; Thies and Devare, 2007).

Two groups of studies raised questions about the possible effects of Cry toxins expressed in *Bt* maize on non-target organisms. First, research on the predatory lacewing, *Chrysoperla carnea* (Neuroptera: Chrysopidae), indicated lacewing larvae were negatively affected when they fed on lepidopteran larvae that consumed *Bt* maize expressing the toxin Cry1Ab (Hilbeck et al., 1998). However, subsequent research showed *C. carnea* was not directly affected by the toxin, but indirectly by feeding on intoxicated, moribund prey (Dutton et al., 2002; Romeis et al., 2004; Rodrigo-Simón et al., 2006; Lawo and Romeis, 2008). Later, studies with larvae of the monarch butterfly, *Danaus plexippus* (Lepidoptera: Danaidae), suggested monarch populations would be reduced from feeding on milkweed leaves coated with *Bt* maize pollen (Losey et al., 1999; Jesse and Obrycki, 2000). Again, more thorough research indicated the likely impact *Bt* maize on monarch was negligible because of limited exposure and low toxicity of *Bt* maize pollen to monarch larvae (Hellmich et al., 2001; Oberhauser et al., 2001; Pleasants et al., 2001; Sears et al., 2001; Stanley-Horn et al., 2001; Zangerl et al., 2001; Wolt et al., 2003; Dively et al., 2004). *Bt* maize event 176, which produces a high level of *Bt* protein in the pollen, had acute effects on monarch larvae fed milkweed foliage containing levels of pollen commonly encountered in maize fields during pollen shed (Hellmich et al., 2001; Stanley-Horn et al., 2001; Zangerl et al., 2001), but this early type of *Bt* maize was an exception, had limited planting and is no longer commercially available. Most recently, a preliminary study suggests that *Bt* maize pollen or detritus might have negative effects on caddisfly larvae (Trichoptera) in streams located in or near *Bt* maize fields (Rosi-Marshall et al., 2007), but the risk was not well established.

5.3.3.2 Gene Flow

The transfer of genetic material between populations (i.e., gene flow) is often considered to be a potential problem between GM crops and their wild relatives. In most areas of the world producing GM maize, however, production is isolated from related species that could hybridize with *Z. mays*. Gene flow as an environmental concern is thus restricted to those areas where wild relatives of maize occur (e.g., Mexico). In addition, in some areas such as the European Union (EU), gene flow issues with GM maize usually involve cross pollination or seed contamination of non-GM maize. Some growers, particularly of organic maize, demand little or no contamination from GM pollen or seed and generally object to production of any GM maize. This has been a particularly controversial issue in Europe. In 2003, the EU stipulated that labeling of food or feed as genetically modified was not required unless GM material exceeded a 0.9% threshold (European Union, 2003a, b). This legislation set the stage for the coexistence of GM and non-GM crops, but isolation distances and other measures to limit mixing or GM and non-GM products needed to be defined. Most research indicates separation of a minimum of 50 m between GM and non-GM maize is adequate to restrict outcrossing to less than the 0.9% threshold (Brookes et al., 2004; Devos et al., 2005; Sanvido et al., 2008), but others suggest as little as 20 m may be adequate, especially if several rows of non-GM maize are used as a buffer around GM maize (Messeguer et al., 2006; Weber et al., 2007).

5.3.3.3 Insect Resistance Management

Insect pests, including maize insects, commonly have developed resistance to conventional insecticides when they are overused (Georghiou, 1986). Larvae of *D. v. virgifera* evolved resistance to soil-applied cyclodiene insecticides by the 1960s (Ball and Weekman, 1962) and adults evolved resistance to methyl parathion in the 1990s (Meinke et al., 1998). Consequently, scientists and growers are concerned that overuse of *Bt* maize could produce pests resistant to *Bt* toxins (Tabashnik, 1994; Gould, 1998; Frutos et al., 1999). Though maize stem borers have not evolved resistance to insecticides (perhaps because insecticide exposure is limited once larvae bore into the plant), several important lepidopteran pests have been selected for resistance to *Bt* toxins in the laboratory (Tabashnik, 1994; Ferré and Van Rie, 2002; Ferré et al., chapter 3), including *O. nubilalis* (Huang et al., 1999; Alves et al., 2006) and *O. furnacalis* (Xu et al., 2006). In the field, only the diamondback moth, *Plutella xylostella* (Lepidoptera: Plutellidae), has evolved resistance to *Bt* sprays (Tabashnik et al., 1990) and with the possible exception of *S. frugiperda* resistance to Cry1F maize in Puerto Rico (Matten et al., chapter 2), no insects have evolved resistance to *Bt* crops (Tabashnik et al., 2003; Ferré et al., chapter 3).

Various strategies have been proposed for managing insect resistance to *Bt* maize, but currently the high-dose/refuge (HDR) strategy appears to be the most commonly recommend (Bates et al., 2005; Matten et al., chapter 2; Ferré et al., chapter 3). With this strategy, insects that feed on the *Bt* maize are exposed to an extremely high dose of toxin, which makes insect resistance functionally recessive (Gould, 1994). Refuges complement the high dose because they provide a population of susceptible insects that are not exposed to *Bt* toxin. Consequently, rare resistant moths that develop on *Bt* maize, instead of mating with each other, mate with the overwhelming number of susceptible moths from the refuge (Tabashnik and Croft, 1982; Gould, 1998). This process essentiality dilutes resistance genes and maintains a population of susceptible insects. This strategy should be effective as long as plants express a high dose of the toxin, genes conferring resistance are rare, and there are many insects from the refuge available to mate randomly with resistant insects (Gould, 1998). In addition to the biological factors, economic and social aspects of IRM cannot be ignored (Mitchell and Onstad, 2007; Hurley and Mitchell, 2007).

Studies have been conducted to establish baseline *Bt* susceptibility of maize pests in the USA (*O. nubilalis*, Marçon et al., 1999; *H. zea*, Siegfried et al., 2000; *D. v. virgifera*, Siegfried et al., 2005), European Union (*O. nubilalis*, González-Núñez et al., 2000; Saeglitz et al., 2006; *S. nonagrioides*, González-Núñez et al., 2000; Andreadis et al., 2007), and China (*O. furnacalis*, He et al., 2005). In general, these studies have found insect susceptibly to *Bt* varies little among populations.

If extensive planting of *Bt* maize results in an area-wide reduction of the pest population, then the size of *Bt* maize refuges could be adjusted to keep pest populations below EILs. Such an approach, however, would involve addressing IRM refuge requirements first so that insects do not evolve resistance. For example, the hypothetical insect pest in Fig. 5.1 (pre-*Bt* maize, A) is regularly above the EIL and requires annual IPM control measures. The use of *Bt* maize, however, could reduce the annual pest populations below the EIL. The question then becomes what percentages of *Bt* maize and corresponding non-*Bt* maize (or refuge) would keep the pest populations consistently below the EIL? In this example, insect populations resulting from a Refuge 1 strategy (B) are low, but from time to time they exceed the EIL, where the refuge maize might require treatment. On the other hand, insect populations from a Refuge 2 strategy (C) are consistently below the EIL. Obviously, growers would prefer a strategy that reduces or completely eliminates intervention. Such an approach, however, must be coordinated with IRM requirements because too little refuge could lead to pest resistance to *Bt* maize. But, hypothetically, for some pests there could be a balance between IRM requirements and reducing the need for control in refuge maize.

5.3.3.4 Conflicts with IPM Principles

A final challenge to consider for insect-resistant GM maize is the perception that current hybrids are used in ways that directly conflict with the underlying principles of IPM. For example, *Bt* maize varieties generally produce high levels of toxins

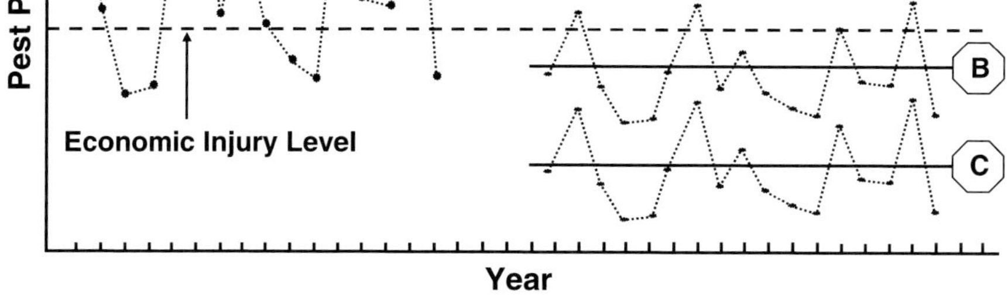

Fig. 5.1 Hypothetical equilibria for pest populations that are influenced by *Bt* maize. Populations vary annually but fluctuate either above the economic injury level (EIL) as in the pre-*Bt* maize pest equilibrium (A), near the EIL as in *Bt* maize with Refuge 1 (B), or well below the EIL as in Refuge 2 (C)

throughout the season whether pests are present or not. Similarly, the appeal of growing *Bt* maize as insurance against pest problems means that some growers will plant insect-resistant GM maize even if pests are not expected to reach damaging levels. Both of these points relate to the IPM concept that insecticides should only be used at times and locations where pests are expected to reach damaging levels. However, sustained high levels of plant resistance or unnecessary planting of resistant varieties have not been considered undesirable for HPR produced through conventional breeding. This suggests that perspectives on this issue are influenced by whether GM insect resistance is thought of as more similar to insecticides or to conventional host plant resistance. If insect-resistant GM maize is perceived as an insecticide, then such objections point out inappropriate uses of the technology. But if GM insect resistance is more accurately a form of host plant resistance, then objections that GM maize conflicts with IPM principles appear to be reflect a double standard for GM-derived plant resistance compared to conventional HPR.

Economic and practical constraints in production of hybrid maize seed also may result in GM maize traits being used when pests are not expected to reach damaging levels. This is a result of the fact that as the number of GM products increases, the number of GM-trait combinations increases geometrically. The high production costs for each hybrid (back-crossing into appropriate germplasm, etc.) and limited inventory space might compel seed providers to sell some stacked traits in areas where their use in pest management is not justified. For example, a grower may want to plant herbicide tolerant maize with *Diabrotica* spp. control, but does not need *O. nubilalis* control. If a stack containing all three is the only option, then the grower may be forced to accept the lepidopteran-active trait in the hybrid, but may or may not have to pay the associated technology fee. Does such a scenario promote the best IPM practice? If stacked maize products potentially lead to high use and compromise refuge requirements then, at least from an IRM perspective, this is not the best practice. On the other hand, is the use of an unneeded trait acceptable if IRM requirements are not compromised? As more products are developed, these situations will become more complex.

5.3.4 Future Types of Insect-Resistant GM Maize

Though commercially available insect-resistant maize varieties use single or multiple *Bt* (Cry) toxins to suppress lepidopteran and coleopteran pests, it seems likely that new technologies will appear continuously for several years. Two main areas of interest for future types of GM maize include developing insect-resistance products for additional pests and improving lepidopteran- and coleopteran-active products to delay the evolution of resistant insects.

To broaden the spectrum of insecticidal activity, maize varieties may include additional *cry* genes, vegetative insecticidal proteins (VIP), lectins, protease inhibitors, chitinases, RNA interference, and others (e.g., Baum et al., 2007; Malone et al., chapter 13). For example, the VIP proteins produced by *B. thuringiensus* show a different mode of action than Cry proteins (Estruch et al., 1996; Lee et al., 2003), and should allow management of black cutworm (*Agrotis ipsilon*; Lepidoptera: Noctuidae) and fall armyworm (*S. frugiperda*) (VIP3A; Lee et al., 2003). Other types of toxins likely will be useful in conferring insecticidal properties to more diverse maize pests; sucking insects like aphids and leafhoppers that are important virus vectors are likely targets.

Improving resistance to more effectively delay evolution of insects resistant to GM maize is a related research area. As in other crops, this may be accomplished by combining or "pyramiding" two or more toxins with different modes of action, including some of the relatively novel toxins noted above. Current commercial cotton varieties include multiple Cry toxins that target the same pest species (Bollgard II or Widestrike; Ferré et al., chapter 3; Naranjo et al., chapter 6). Experimental maize pyramids that target Lepidoptera include hybrids that produce VIP3A and Cry1Ab toxins (Dively, 2005) and hybrids that produce Cry2Ab2 toxin and a chimeric protein Cry1A.105 (USEPA, 2007). Of course, because the use of multiple, complementary toxins may delay resistance, it also may allow changes in the type or size of allowable refuges.

In general, it seems that combinations of multiple, complementary toxins will allow GM maize to protect against several arthropod pests and improve resistance management (Roush, 1998). Other strategies like the use of inducible promoters (which cause expression of traits in response to specific triggers) may be used to transform insect management in GM maize from a preventative strategy to a responsive one (Bates et al., 2005; Christou et al., 2006), perhaps eliminating some of the concerns noted above (see section 5.3.3.4).

5.4 Case Studies from GM and Non-GM Maize Producing Countries

The following case studies on the USA, Spain, China and Kenya provide a cross section of countries that use or are considering the use of *Bt* maize. Other early adopters of the GM technology include Canada, South Africa and Argentina, all of which produce >9 MMT of maize per year (Baute et al., 2002; James, 2003; Gouse et al., 2005, 2006; Trigo and Cap, 2006; FAOSTAT, 2007). Information also is available on the experiences of *Bt* maize growers in the Philippines (Yorobe and Quicoy, 2006).

5.4.1 United States of America Case Study

In 2007, USA growers harvested 34.8 million hectares of maize and produced 338 MMT of grain (USDA-NASS, 2007). Maize production is concentrated in the Corn Belt, especially Iowa, Illinois, Nebraska, Minnesota, and Indiana. In 1996, the first insect-resistant GM maize hybrids were sold, using *Bt* genes to suppress *O. nubilalis* and the southwestern corn borer, *Diatraea grandiosella* (Lepidoptera: Crambidae).

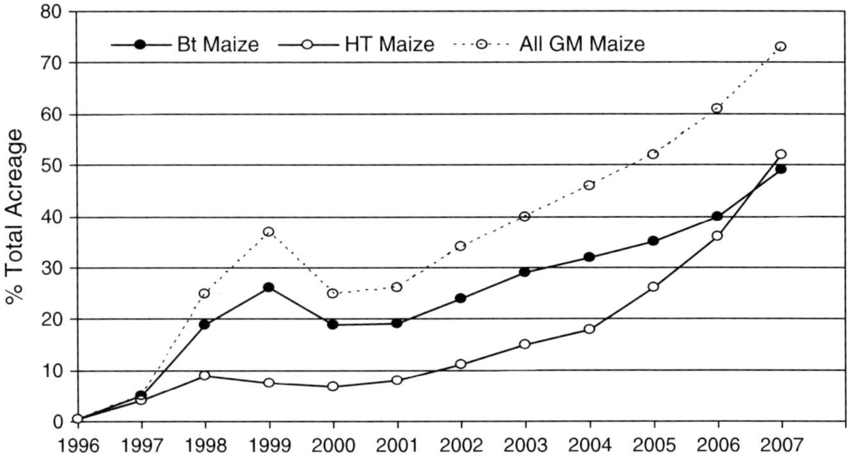

Fig. 5.2 Adoption of GM maize in the United States, 1995–2006. Data for insect-resistance (*Bt*) and herbicide-tolerance (HT) include stacked varieties that combine both types of GM traits (http://www.ers.usda.gov/Data/BiotechCrops, accessed 28 November 2007)

Adoption of transgenic maize in the USA has been rapid, especially after herbicide-resistant maize and *Bt* maize for the control of *Diabrotica* spp. were commercialized in 1997 and 2003, respectively (Fig. 5.2). Prior to *Bt* maize, host plant resistance, crop rotation and insecticides formed the foundation of IPM for key insect pests of maize in the Corn Belt. This case study though focuses on the most important Corn Belt pests, *O. nubilalis* and *Diabrotica* spp.

5.4.1.1 Major Insect Pests and Their Control

The European corn borer, accidentally introduced into the United States in the early 1900s, is the most important maize stem borer in the USA. The current range of *O. nubilalis* in North America covers the Corn Belt as well as southern states from Florida to east Texas, as far west as the Rocky Mountains, and into southern Canada. In the Corn Belt states, *O. nubilalis* is usually bivoltine, but there may be from one to four generations annually depending on latitude.

Prior to the introduction of *Bt* maize, cultural practices (i.e., changes to planting or harvest time, post-harvest stalk destruction) and HPR were major tools to reduce the devastating effects of *O. nubilalis* on maize yields. A combination of in-field monitoring of *O. nubilalis* and insecticide applications based on treatment thresholds could prevent losses of ~10–30% (Linker et al., 1990; Tollefson and Calvin, 1994; Mason et al., 1996). Many growers, however, elect not to use insecticides against *O. nubilalis* because applications must be timed after most eggs hatch but before larvae tunnel into the stalk (where they are protected from insecticides). Furthermore, most modern maize hybrids have some tolerance to *O. nubilalis* injury, so it is likely that without insecticide use, *O. nubilalis* usually represented a modest but chronic problem. In higher value crops, such as seed maize, popcorn and sweet corn (Shelton et al., chapter 9), management is more aggressive and in some cases may use biologically-based insecticides or biological control (e.g., Kuhar et al., 2003; Musser et al., 2006). Another stem borer, *D. grandiosella* can be a more destructive pest of maize than *O. nubilalis*. Southwestern corn borers were first reported as a pest of maize in 1913 and now are found from Arizona to Georgia and north to Missouri, and is an important maize pest in parts of Kansas, Missouri and Kentucky, where it can cause yield losses up to 50%, if not controlled (Chippendale and Sorenson, 1997).

It is debatable whether *O. nubilalis* or *Diabrotica* spp. have caused greater losses for US maize growers, but relative to insecticide use, the complex of western (*D. v. virgifera*), northern (*D. barberi*) and southern (*D. undecimpunctata howardi*) corn rootworms is unchallenged. In the USA estimates of insecticide a.i. applied annually to control this pest complex range from 2,400 to 3,500 MT (Gianessi et al., 2002; James, 2003; Rice, 2004). This represents approximately 60% of the total insecticides used on maize pests in the USA and, as mentioned previously, 25–30% of the insecticides used against maize pests worldwide.

Annual rotation of maize with other crops has been an effective management tool for *Diabrotica* spp. Yet the high efficacy of chlorinated hydrocarbons against soil-dwelling insects, especially *D. v. virgifera*, has led many growers to plant maize continuously. This was especially the practice in areas of Nebraska and Kansas, where irrigated maize has high-yield potential. However, the development and spread of insecticide-resistant *D. v. virgifera* during the late 1950s made the need for new *Diabrotica* spp. management strategies clear. Subsequently, an understanding of the relationship between adult populations in one year and larval damage the following year allowed producers to assign a risk level to larval injury and use responsive, rather then preventative, tactics (Pruess et al., 1974; Stamm et al., 1985). Either crop rotation or application of an insecticide was recommended if adult populations of more than one per plant were detected the previous year. However, discovery of *Diabrotica* spp. resistance to crop rotation has undermined the crop rotation tactic in many parts of the Corn Belt (see section 5.2.2).

5.4.1.2 Current Use of GM Maize in the USA

Since 1996 the use of GM maize has increased rapidly in the USA. There was a dip in grower use in 2000, but this has been followed by a steady increase to almost 75% adoption in 2007 (Fig. 5.2). Use of lepidopteran-active *Bt* maize approaches or exceeds 50% of the total area of production through much of the Corn Belt, with highest concentrations in northwest Iowa, and southwest Minnesota. High use percentages also occur in parts of Kansas, Oklahoma, Texas, Pennsylvania and Maryland (Fig. 5.3). Commercial, coleopteran-active *Bt* maize, which has demonstrated high consistency in suppressing corn rootworms (Moellenbeck et al., 2001; Vaughn et al., 2005), was first planted in 2003. Since then the adoption of *Bt* maize in eastern Corn Belt states, such as Illinois and Indiana, has greatly increased (USDA-ERS, 2007), at least partially in response to rotation-resistance in *D. v. virgifera*.

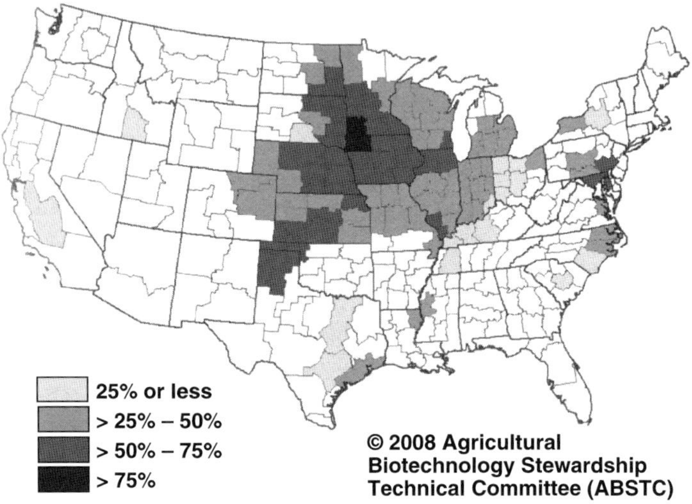

Fig. 5.3 Percentage of total maize hectares planted in 2006 to lepidopteran-active *Bt* maize hybrids in USA crop reporting districts in which > 40,468 hectares (100,000 acres) of maize were planted (Agricultural Biotechnology Stewardship Technical Committee)

The USEPA promotes IRM in *Bt* maize by mandating the use of structured refuges (also see Matten et al., chapter 2; Ferré et al., chapter 3). As of early 2008, in lepidopteran-active maize there is a mandate for a 20% refuge in the Corn Belt and 50% refuge in cotton-growing areas, with the refuges planted within one-half mile (~800 m) of the *Bt* maize (Matten et al., 2004). There is a higher refuge mandate in cotton-growing areas because maize serves as an important refuge source for *H. zea*, which often is a devastating pest of cotton (Naranjo et al., chapter 6). Coleopteran *Bt* maize in the USA also has a 20% refuge mandate, but the refuge maize must be planted adjacent to the *Bt* maize. To promote good IRM stewardship, registrants are required to monitor for resistance, educate growers about the importance of IRM, monitor for grower compliance, and develop remedial action plans in case resistance develops (Matten et al., 2004, chapter 2). Surveys suggest, at least in the USA, most growers understand the importance of planting refuges and most of them follow refuge recommendations (Goldberger et al., 2005; Alexander, 2007); although grower compliance could be lower in the future if Bt maize use percentages continue to increase. Thus far, ten years of resistance monitoring in the US Corn Belt has found no detectable changes in Cry1Ab susceptibility among *O. nubilalis* populations (Siegfried et al., 2007).

5.4.1.3 Effects on Integrated Pest Management

Growing GM insect-resistant maize is likely to impact several aspects of IPM, including the amount of insecticides used, potential problems with secondary pests, and basic decision-making processes used by maize growers with regard to crop production. The effectiveness of biological control (whether natural or through intervention) will almost certainly be increased as insecticide use decreases. The value of lepidopteran-active *Bt* in reducing insecticide use has been modest (Hunt et al., 2007), but novel products with multiple lepidopteran toxins may lead to greater reductions in insecticide use, especially in southern states, if better control of additional maize pests (e.g., *H. zea*, *S. frugiperda*) can be developed. As mentioned previously, the potential for coleopteran-active *Bt* maize to limit insecticide use is considerable; and if resistance to crop rotation for *D. v. virgifera* and *D. barberi* continues to spread, the value of *Bt* maize will become even greater.

Concerns that the use of insect-resistant *Bt* maize could lead to increased problems with secondary pests may stem from experiences with *Bt* cotton, where declining insecticide use against target lepidopteran pests allowed increases of some previously minor pest species (Naranjo et al., chapter 6). However, this seems less problematic for US maize growers with some minor exceptions. For coleopteran-active *Bt* maize, there are anecdotal reports of more problems with minor soil insect pests (e.g., grubs, wireworms). If such problems become widespread or persistent, the most likely result will be increased use of seed treated with systemic insecticide. Additionally, the recent eastward spread of western bean cutworm, *Striacosta albicosta* (Lepidoptera: Noctuidae), through the Corn Belt, which could be related to *Bt* maize or increased use of minimum tillage, poses a potential problem for growers of maize that rely on the *Bt* toxin Cry1Ab (Catangui and Berg, 2006; Storer et al., chapter 10). Although the use of current *Bt* maize with Cry1F or future hybrids with multiple lepidopteran-active toxins should allow this pest to be managed without insecticides.

Lastly, like reliance on insecticides, the use of insect-resistant GM maize could have undesirable impacts on how growers make decisions regarding pest management and crop production. In particular, over-use of *Bt* maize or complete reliance on genetic modifications for insect management could reduce the use of IPM practices that help control secondary pests. For example, since many minor pests are suppressed by crop rotation, recent trends towards more continuous maize production may contribute to new or worsening pest problems. To reduce the likelihood that *Bt* maize is relied upon exclusively and unnecessarily, a model available over the internet, the *Bt* maize Economic Tool (BET; www.btet.psu.edu) provides growers useful information by estimating the likelihood of net benefits from planting *Bt* maize. Growers are allowed to input specific information regarding their production plans and see predicted outcomes based on long-term averages for weather and *O. nubilalis* abundance. The combined information on pest and maize phenology, site-specific weather data and economics generate color-coded maps to help growers determine where *Bt* maize, on average, is economical.

5.4.2 Spain Case Study

After France and Italy, Spain is the third largest producer of grain maize in Western Europe. In recent years the area of maize production has varied from 400,000–500,000 ha.

Because most maize is irrigated with up to 700mm per year, the production area depends on availability of fresh water. Though maize is planted throughout most of Spain, cultivars, agronomic practices, and yield vary substantially among regions. Most of the production is devoted to livestock feed, with minor amounts for starch, sweet corn and popcorn.

5.4.2.1 Major Insect Pests and Their Control

Three groups of insects are targets of pest management by maize growers in Spain. In addition to two species of stem borers, wireworms (Coleoptera: Elateridae) and cutworms (Lepidoptera: Noctuidae) are primary soil pests. Finally, a group of sucking insects, aphids (Homoptera: Aphidae) and leafhoppers (Homoptera: Cicadellidae), are important because of their role as vectors of maize viruses.

Stem borers, including the Mediterranean corn borer, *Sesamia nonagrioides* (Lepidoptera: Noctuidae) and *O. nubilalis*, are the most damaging maize pests in all parts of the country. *Sesamia nonagrioides* is considered the more damaging species because it is more abundant and produces longer tunnels than *O. nubilalis*. For the stem borers, particularly for *S. nonagrioides*, insecticides are not generally used because larval tunneling limits the efficacy of insecticide-based management. When insecticides are used against stem borers, foliar applications are made (first generation) or insecticides are incorporated into irrigation water (second generation). As a cultural control, modification of planting dates is rarely used because the timing of flights for adults of the two stem borers are distinct, typically separated by four weeks. Most maize cultivars grown in Spain offer a low degree of resistance to stem borers, which is the main tactic used to limit losses due to these insects, and recent efforts have been devoted to looking for new sources of HPR (Butrón et al., 2006). Other control measures include tillage to prevent emergence of adult moths, but this is only effective when it is practiced over large areas. Ideally, tillage takes place after adults of the parasitoid *L. thompsoni* have emerged and exerted their suppressive effect on the stem borer population. In high-value seed maize, inundative biological control with *Trichogramma brassicae* (Hymenoptera: Trichogrammatidae) is sometimes used against *O. nubilalis*.

Injury from soil insects, such as wireworms or cutworms, is typically managed using insecticidal seed treatments. Though damage by these pests to maize seed and roots may be overestimated by growers, the absence of reliable economic thresholds and the low visibility of injuries caused by these insects present major obstacles to reducing the use of insecticide treated seed (Piqué et al., 1998). Also, because seeds treated with systemic insecticides appear to delay the development of aphid and leafhopper populations, there is an additional incentive for growers to buy treated seed (Pons and Albajes, 2002). Though the western corn rootworm *D. v. virgifera* has been found in Europe as far west as France, the pest is not yet present in Spain.

5.4.2.2 Current Use of GM Maize in Spain

For many years Spain has been the only country in the EU to grow a significant amount of *Bt* maize. Though *Bt* maize accounted for only about 21% of the total area planted to maize in Spain in 2007, it comprises approximately 50% of the maize grown in some areas. Spanish legislation concerning GM crops follows the general EU framework. Though EU legislation on genetically-modified organisms has been in place since the early 1990s, additional regulations have been developed since then. In the last decade, many experimental authorizations have been granted for GM crops, but few have been approved for cultivation, import and processing for feed and food. The EU Commission first allowed growers to cultivate Cry1Ab maize (Event 176) in 1997, but authorization for 176 was cancelled in 2005. Currently only event MON810 is authorized for cultivation, but other events have been allowed for import and use in processing or for grain (Bt11, 1998; NK603, 2004; MON863 and DAS1507, 2005; and MON863 × MON810, 2006). Periodic updates to the list of authorized GM crops in Europe can be found at http://www.gmo-compass.org/eng/gmo/db/ (accessed 3 January 2008).

The current approval of only MON810 for cultivation in Spain means only lepidopteran-active *Bt* maize is grown. Besides resistance to the two stem borers, the *Bt* varieties also reduce the occasional ear injury produced by *Helicoverpa armigera* (Lepidoptera: Noctuidae). There are no obligatory measures for IRM in Spain but recommendations are based on the high-dose/refuge strategy. Studies conducted by Spanish public institutions suggest that 400 m is a common dispersal distance within which *S. nonagrioides* matings occur at random (Eizaguirre et al., 2006) and that resistance alleles are rare in Spanish populations of *S. nonagrioides* and *O. nubilalis* (Andreadis et al., 2007). Annual monitoring has been conducted with no reported changes in susceptibility of the two stem borer species to the only *Bt* toxin deployed in the field, Cry1Ab (Farinós et al., 2004).

5.4.2.3 Effects on Integrated Pest Management

In spite of the debate on the cultivation of *Bt* maize in Spain, growers have steadily increased their use of *Bt* maize. A survey sponsored by seed companies, revealed that 96% of *Bt* maize growers in Spain were quite satisfied with transgenic varieties to prevent losses due to stem borers. A more in-depth evaluation of the socio-economic impacts of *Bt* maize in Spain used empirical data from on-farm performance for the three-season period 2002–2004 (Gómez-Barbero et al., 2008). In the three main areas growing *Bt* maize growers had 4.7% increases in yield and €85 in gross margins per hectare.

More relevant to IPM practices were the results regarding growers' use of insecticides (Gómez-Barbero et al., 2008). The survey indicated that conventional maize growers were about twice as likely to use insecticides for stem borer suppression (56%, conventional and 30%, *Bt*), and applied on average more than twice as many applications per year as *Bt* maize growers (0.86, conventional and 0.32, *Bt*). The differences between probability of treatment and mean number of applications likely reflects the more common use of multiple applications by conventional growers, among whom 21% used two or more applications compared to 2% for *Bt* maize growers. One likely effect of reduced use of foliar insecticides is conservation of natural enemies, which deter population development of secondary pests, such as aphids and spider mites (Acari: Tetranychidae) (Romeis et al., chapter 4). Other pests that are not currently controlled by *Bt* maize, however, such as *H. armigera* may be more problematic and need to be monitored.

As in the USA, growers may be tempted to use insect-resistant GM maize when it is not justified by economics or IPM principles. Though growing *Bt* maize was profitable across all three main *Bt* maize growing areas, improvement in gross margins ranged from €125/ha in Aragon (northeast) to €7/ha in Castilla La Mancha (central), suggesting that the use of *Bt* maize is not appropriate in all situations and should be used based on the best available economic and ecological data.

5.4.3 China Case Study

China is the second largest producer of maize in the world. In 2004, approximately 24 million hectares of maize were grown, producing a total yield of 125 MMT (average yield ~4.8 t/ha; Wang et al., 2005a). Unlike maize production in the USA, growers in China typically farm relatively small plots with the total production area divided among 100 million maize growers.

5.4.3.1 Major Insect Pests and Their Control

The Asian corn borer, *O. furnacalis*, is the most significant insect pest of maize and occurs in most maize-growing areas from Heilongjiang (northern) to Hainan (southern) provinces. Estimated losses due to this insect range from 6–9 MMT per year (Zhou et al., 1995). Similar to *O. nubilalis* in the USA, direct yield losses come from *O. furnacalis* injury to vegetative stage maize, but the greatest impacts are indirect, from larval feeding on silks and kernels that leads to ear rot, mycotoxin production and reduced grain quality (Zhou et al., 1995; Wang et al., 2005a). Other lepidopteran pests of concern for maize in China include *H. armigera* and *Spodoptera exigua* (Lepidoptera: Noctuidae). In particular, problems with *H. armigera* in maize appear to be increasing as cropping systems from the 1990s have changed (Wang et al., 2001), probably due to the more frequent use of no-till farming and the associated high survival of *H. armigera* pupae in the soil.

Several practical IPM tactics have been developed for *O. furnicalis* including biological, cultural and chemical management. For example, early spring applications of the entomopathogen, *Beauveria bassiana*, over maize stalks can kill ~80% of overwintering larvae, which significantly decreases the number of egg masses in the field and reduces the percentage of infested plants (Wang et al., 2003). Other biological control efforts include mass releases of *Trichogramma dendrolimi* (Hymenoptera: Trichogrammatidae) egg parasitoids on an area of 1.0–1.3 million hectares per year in the northeastern provinces. More recently, the scale of the releases of *T. dendrolimi* has been expanded to 2 million hectares, which includes Huang-Huai-Hai summer maize and northwestern maize regions (Wang et al., 2005a). The program produced 60–85% parasitism of *O. furnacalis* and reduced damage to maize by 65–92% (Piao and Yan, 1996), equal to or better than what is achieved by insecticide-based suppression. With one or two releases, costs are estimated at US$4 or 6/ha, respectively. One remarkable management measure has been an extensive network of light (high intensity mercury-vapor lamp) traps over 320,000 ha; traps reduced *O. furnacalis* plant infestations by ~60% with captured moths used to feed chickens on nearby farms (Yang et al., 1998; Wang et al., 2003). Finally, granular insecticide applications and *B. thuringiensis* insecticides have been used for whorl-stage suppression of *O. furnacalis*.

Though IPM plays an important role in controlling *O. furnacalis* in maize, most growers do not manage Asian corn borer populations because of the costs and required skills, safety and environmental concerns, and uncertainty about the benefits of the management (Zhou et al., 1995).

5.4.3.2 Current Use of GM Maize in China

Although *Bt* maize is not grown in China, its commercialization is currently under consideration by the Chinese government (Wu and Guo, 2005). Extensive laboratory and field trials have been conducted to evaluate the efficiency of transgenic maize on target lepidopteran pests and the potential ecological risks to non-target arthropods (Wang et al., 2005b, c, 2007; Li et al., 2007).

Cry1Ab-expressing maize provided excellent control of *O. furnacalis* in laboratory bioassay and field trials (He et al., 2003a, b, 2004). Neonates of *O. furnacalis* did not survive when fed different tissues of *Bt* maize hybrids that produce Cry1Ab toxins (events MON810 and Bt11) (Wang et al., 2004a). Neonates of *H. armigera* did not survive when fed silk, ear and husk tissues in the laboratory; however, there was low survival of *H. armigera* after artificial infestation of silk-stage maize plants in the field (Chang et al., 2006). Perhaps this was due to cannibalism, which provided a way for some *H. armigera* to avoid the *Bt* toxin. Similarly, laboratory tests demonstrated excellent control of *S. exigua*, but some larvae survived on artificially infested plants in the field. Cry1Ab maize had good control for the less serious lepidopteran pest, oriental armyworm, *Mythimna separata* (Noctuidae) (Wang et al., 2004b, 2005b), but the effects of *Bt* maize on other less serious lepidopteran pests, including the yellow peach borer, *Conogethes punctiferalis* (Pyralidae); sugarcane striped borer, *Proceras venosatus* (Crambidae); and millet borer, *Chilo infuscatellus* (Crambidae), are unknown.

Resistance management research related to *Bt* maize also is in progress. Research on resistance mechanisms and biology of resistant individuals is ongoing for a laboratory-selected *O. furnacalis* strain resistant to Cry1Ab (Xu et al., 2006; He et al., 2007). Additionally, research has started to identify sources of non-*Bt* crops and natural plant refuges for *O. furnacalis* in the different maize growing regions of China. If the variety of crops (e.g., maize, millet, sorghum, wheat, vegetables, soybean, peanut, canola) and natural vegetation produce sufficient numbers of the primary maize pests, then refuge plantings of non-*Bt* maize may not be needed in some areas.

5.4.3.3 Effects on Integrated Pest Management

Bt maize could become a major component of IPM in all the maize-growing areas of China. This assumes *Bt* maize would be affordable, effective, easy for growers to use, and environmentally sound. Potentially the main positive impact would be increased control of *O. furnicalis* in areas (>50%) where currently no control tactics are used. Yet even in areas where control with *T. dendrolimi* has been successful, *Bt* maize could offer an economically viable alternative.

The value of *Bt* maize should be considered a long-term issue. Although preserving the efficacy of *Bt* maize in China using IRM may be challenging. *Ostrinia furnacalis*, *H. armigera* and *S. exigua* will be considered target pests. As in the USA, both *Bt* maize and *Bt* cotton are produced in many of the same areas, which complicates matters since maize is considered a refuge for *H. armigera* (Wu and Guo, 2005). Furthermore, if refuges should be required, given the large number of small farms, it is uncertain whether grower compliance could be assured (or even adequately measured).

It is possible that for some Chinese maize growers, GM insect-resistance would provide considerable benefits in reducing reliance on insecticides, which might also reduce illnesses and deaths related to insecticide use (as has been reported for *Bt* cotton, Pray et al., 2002; Hossain et al., 2004; Qaim et al., chapter 12). However, these benefits are difficult to estimate because of limited data. In addition, reduction in insecticide use will increase opportunities for natural biological control (Romeis et al., chapter 4), especially for the control of secondary pests such as mites, corn leaf aphids and thrips, especially *Frankliniella tenuicornis* (Thysanoptera: Thripidae). At present it is unclear whether traditional IPM practices for lepidopteran maize pests would be enhanced or replaced by *Bt* maize.

5.4.4 Kenya Case Study

Africa grows 26 million hectares of maize, accounting for 18% of global area but only 6.6% of the global production. On average, maize yields within industrial countries are around 8.3 t/ha while for sub-Saharan Africa the average is only 1.3 t/ha (FAOSTAT, 2007). Maize production in Kenya fits the pattern in sub-Saharan Africa; with a production area one-third greater than Canada, total yields from

Kenya are less than one-third that of Canadian maize growers, averaging 1.7 t/ha (FAOSTAT, 2007). Kenyan growers are challenged more by poor soil fertility, drought, and limited funds than by insect pests (De Groote et al., 2004a). As a result, growers usually are unable to make adequate investments in fertilizer or improved maize varieties (Freeman and Omiti, 2003). Biotechnology, however, has the potential to improve agricultural production and sustainability in Kenya and other countries in Africa (Thomson, 2008).

5.4.4.1 Major Insect Pests and Their Control

Key insect pests for Kenyan growers include lepidopteran stem borers and coleopteran storage pests. Maize growers estimate losses from stem borers at 13% (De Groote, 2002); the most important species are the spotted stem borer, *C. partellus*, and African stem borer, *Busseola fusca* (Lepidoptera: Noctudidae) (Ong'amo et al., 2006). Other less common species, including coastal stem borer, *Chilo orichalcociliellus* (Lepidoptera: Crambidae), and pink stem borer, *Sesamia calamistis* (Lepidoptera: Noctuidae), and the African sugarcane borer, *Eldana saccharina* (Lepidoptera: Pyralidae), also occur in Kenya and other maize growing countries in sub-Saharan Africa.

The spotted stem borer was introduced from South Asia and first reported in Kenya in the 1950s (Nye, 1960) and now attacks maize and sorghum at elevations below 1,500 m. In contrast, the African stem borer is prevalent in high- and mid-elevation areas causing at least 10% yield loss (Ong'amo et al., 2006). Both species are attacked by the native parasitoid, *Cotesia sesamiae* (Hymenoptera: Braconidae). *Cotesia flavipes*, which was introduced to help suppress *C. partellus* (Overholt et al., 1997), and the tachinid, *Sturmiopsis parasitica* (Diptera: Tachinidae), is a common parasitoid of *B. fusca* (van Rensburg et al., 1988). Another IPM tactic that combines biologically- and culturally-based pest management is the so called "push-pull strategy" developed by the International Center for Insect Physiology and Ecology (ICIPE) (Khan et al., 1997). In this system, maize is intercropped with grasses such as molasses grass (*Melinis minutiflora*) or desmodium (*Desmodium uncinatum*, *Desmodium intortum*) that repel or push the stem borers *C. partellus* and *B. fusca* away from maize. Though the repellent effect is not absolute, molasses grass also produces a volatile that attracts the parasitoid *C. sesamiae*, increasing the rate of parasitism fourfold. Additional plantings of trap crops around maize, Napier grass (*Pennisetum purpureum*) or Sudan grass (*Sorghum vulgare* var. *sudanense*), help attract or *pull* stem borers out of maize. The push-pull strategy can reduce stem borer populations by 75%. It also addresses other problems of maize growers by helping suppress witchweed (*Striga* spp.), improve soil fertility and provide livestock forage (Khan et al., 2001).

After harvest Kenyan growers must contend with the maize weevil, *Sitophilus zeamais* (Coleoptera: Curculionidae), and the larger grain borer, *Prostephanus truncates* (Coleoptera: Bostrichidae). These beetles can cause losses of 10% or more through consumption of grain, reduced grain quality and contamination with insect body parts. Cultural management of these pests involves storing ears over the cooking area where heat and smoke reduce losses. Open-pollinated and hybrid varieties are being developed with conventional host plant resistance by KARI (Kenya Agricultural Research Institute) and CIMMYT (Centro Internacional de Mejoramiento de Maíz y Trigo) (IRMA, 2002).

5.4.4.2 Current Use of GM Maize in Kenya

Insect-resistant GM maize is not commercially available in Kenya, but pending biosafety legislation may soon allow growers access to GM crops. Comprehensive recommendations for an environmental risk assessment for *Bt* maize in Kenya have been proposed (Hilbeck and Andow, 2004), but many feel they are not appropriate and will delay the adoption. Progress towards approval of GM crops has been facilitated by KARI and CIMMYT through the Insect Resistant Maize for Africa (IRMA) project. The IRMA project holds annual stakeholder meetings to create public awareness on the potential of *Bt* maize and provide policymakers opportunities to visit GM crop improvement research in industrial and developing countries (IRMA, 2000). To date, the IRMA project has conducted the only tests of *Bt* maize in Kenya after material transfer agreements for various *Bt cry* genes (*cry1Ab*, *cry1Ac*, *cry1Ba*, *cry1E*, *cry1Ca*, and *cry2Aa*) were acquired, and new events were developed using a ballistic transformation protocol (Bohorova et al., 2001).

In the laboratory, all events tested produced high mortality of *C. partellus*, but emergence holes noted in field plots raise concerns about the durability of those events (Mugo et al., 2008). The events that provided the best control for *B. fusca* were Event 396 (*Rice Actin* promoter, *cry1Ab*) and Event 127 (*Maize Ubiquitin* promoter, *cry1Ba*), which reduced leaf feeding by 30% in 96-hour bioassays but did not cause a significantly higher mortality than controls (Mugo et al., 2008). In order to release a *Bt* maize variety in Kenya, it is likely that the IRMA project will need to work more closely with the private sector, which has already commercialized *Bt* maize varieties that provide effective control against *B. fusca* and are planted on over 1.2 million hectares in South Africa (James, 2007). KARI and CIMMYT can provide the technical support needed to meet the environmental impact study and testing requirements for a regulatory dossier, and to develop stewardship strategies for smallholder farmers. The commercial event that is most widely used is Monsanto's MON810, which offers good early season control of *B. fusca* in South Africa.

Ongoing research in Kenya also is exploring resistance management for *Bt* maize. Early screening for resistance development in *C. partellus* and *B. fusca* showed no changes in susceptibility to Cry proteins over four generations of selection, increasing hopes that resistance can be effectively managed (Mugo et al., 2005; Tende et al., 2005). However, it is currently unclear how insect resistance management and monitoring for resistance would be conducted. The situation in Kenya shares some features with China, particularly an abundance of very small farms, which may complicate IRM efforts. Consequently, the IRMA project has attempted to determine whether an effective refuge may exist within the existing mixed cropping system. Effective natural refuges appear to exist in some areas and seasons, but alternate hosts such as sorghum may need to be promoted to provide an adequate refuge in arid regions or where maize occupies large areas during the long-rain season from April to June (IRMA, 2005a; Mugo et al., 2005).

5.4.4.3 Effects on Integrated Pest Management

Though insect-resistant GM maize is not commercially grown in Kenya, its potential effects can be examined from two perspectives. The first is to consider whether *Bt* maize resistant to stem borers would improve profitability of maize farming in Kenya. In this case, profitability relates to IPM because of the limited potential for Kenyan farmers to afford expenses associated with improved crop production (e.g., fertilizer) and other pest management efforts (weed suppression and management of other insect pests). The second is the potential for *Bt* maize to impact other IPM practices, particularly insecticide use, which can be estimated.

Information on profitability of *Bt* maize is available from on-farm trials across Kenya. To estimate the potential value of stem borer suppression with *Bt* maize, yields were assessed with and without insecticide use (De Groote et al., 2004b). With overall stem borer losses averaging 13%, current maize production levels equate to about US$80 million. Assuming an effective GM event is found for *B. fusca*, most of these losses (~$10 million) could be preventable. The economic effects of *Bt* maize production elsewhere in sub-Saharan Africa may be informative. Side-by-side plantings of *Bt* and near-isoline maize varieties managed by South African growers showed considerable differences (Gouse et al., 2006). *Bt* maize was perceived to produce greater quality grain and improved yields from 21–62%, depending on location, reinforcing the idea that given affordable seed, *Bt* maize could markedly increase yield and grower profits.

The side-by-side plantings in South Africa also were used to investigate effects on insecticide use (Gouse et al., 2006). No significant differences in insecticide use were found between *Bt* and near-isoline maize, but this may be attributable to a combination of indiscriminate insecticide use and low stem borer numbers. The likelihood of insecticide use by growers varied 20-fold between areas and during two study years; more than half of the growers admitted not observing any stem borers. Other information on the potential impacts of *Bt* maize on IPM include monitoring efforts on non-target species, which indicate that abundance of beneficial non-target arthropods is either unaffected or increased with *Bt* maize (IRMA, 2005a, b).

Collectively, current information suggests that *Bt* or other insect-resistant GM maize could permit greater resources to be committed to pest management and reduce the need for insecticide use. However, this outcome is not assured. The value of *Bt* maize to IPM depends on how maize growers utilize the technology. Ideal use of *Bt* maize would include reduction of insecticide use with maintenance of other traditional IPM practices.

5.5 Summary and Conclusions

Bt maize has revolutionized pest control in a number of countries and may allow growers to expand maize production into regions where high pest populations have made growing maize unprofitable. On balance, benefits of *Bt* maize appear to outweigh possible negative effects. Growers are attracted to convenience of the technology as well as yield protection, reduced use of chemical insecticides and improved grain quality. Some scientists, however, suggest the verdict is still out on *Bt* maize and that more research is needed to sort out issues related to possible non-target effects, gene flow and insect resistance management. With regard to the non-target issue, no surprising negative effects have been found with current *Bt* maize hybrids. Overwhelmingly, experiments have shown toxins produced by *Bt* maize have little if any effects on non-target organisms and, when compared to maize treated with chemical insecticides, *Bt* maize fields usually have higher biodiversity. Gene flow is an important issue, especially related to maize seed producers and organic growers. However, as long as GM material thresholds are reasonable, isolation distances and other measures may effectively limit gene flow. IRM remains a challenge because current high-dose/refuge strategies require growers to plant structured refuges, usually non-*Bt* maize; and often the high-dose criteria for plants are not met for all important pests. However, maize hybrids with genes pyramided against specific lepidopteran and coleopteran pests soon will be available, which should improve resistance management and may allow changes in refuge type and size.

The country-specific case studies indicate there are a variety of ways in which *Bt* maize and future GM maize varieties may affect the practice of IPM by maize growers. Potential benefits, including reduced insecticide use and increased ability to invest in crop production and protection (for growers in developing nations), are considerable. Yet there are challenges, including the possibility that GM maize will displace tools like cultural pest management or conventional host plant resistance. This is most important for resource-limited growers in places such as China and Kenya, who currently rely on a diversity of tactics to manage insect pests. Of course, the worst scenario for the future would include the abandonment of traditional IPM tactics followed by misuse and failure of GM maize due to evolution of pest resistance.

Compared to other IPM practices growing *Bt* maize it is not knowledge intensive because the technology is in the seed. This should be attractive to growers in developing countries where poor infrastructure and inadequate extension services sometimes limit the use of traditional IPM (Shelton, 2007). Growers in developing countries, however, often have other agronomic factors besides pest management to consider before deciding to grow *Bt* maize, as well as social and economic challenges. Nevertheless, *Bt* maize has the potential to reduce extreme yield variability due to lepidopteran pests, which would be an advantage for subsistence growers.

Overall, GM maize should not be considered inherently compatible or incompatible with IPM; rather, like synthetic insecticides developed decades ago, the compatibility of insect-resistant GM crops depends on how they are developed and utilized. As noted, efforts in developing future products may reduce potential problems with secondary pests by broadening activity of GM maize and reducing the chances of resistance evolution by targeted pests. Finally, growers and scientists should understand that GM pest resistance is an important component of maize IPM, but traditional pest management practices must be maintained in order to avoid reliance on a single tactic.

Acknowledgements We would like to thank Tom Sappington, Leslie Lewis and Tom Hunt for useful discussion, and Laura Higgins, Joe Huesing, Tim Nowatski, Nick Storer, and Demetra Vlachos for assistance with rating Cry proteins for controlling various maize pests.

Notes Mention of a proprietary product does not constitute an endorsement or a recommendation for its use by United States Department of Agriculture, University of Lleida, Bill & Melinda Gates Foundation, Chinese Academy of Agricultural Sciences, or Golden Harvest Seeds.

References

Alexander, C., 2007. Insect resistance management plans: The farmers' perspective. AgBioForum 10: 33–43.

Alves, A.P., Spencer, T., Tabashnik, B.E., and Siegfried, B.D., 2006. Inheritance of resistance to the Cry1Ab *Bacillus thuringiensis* toxin in *Ostrinia nubilalis* (Lepidoptera: Crambidae). Journal of Economic Entomology 99: 494–501.

Andow, D.A., Klacan, G.C., Bach, D., and Leahy, T.C., 1995. Limitation of *Trichogramma nubilale* (Hymenoptera: Trichogrammatidae) as an inundative biological control of *Ostrinia nubilalis* (Lepidoptera: Crambidae). Environmental Entomology 24: 1352–1357.

Andreadis, S.S., Álvarez-Alfageme, F., Sánchez-Ramos, I., Stodola, T.J., Andow, D.A., Milonas, P.G., Savopoulou-Soultani, M., and Castánera, P., 2007. Frequency of resistance to *Bacillus thuringiensis* toxin Cry1Ab in Greek and Spanish population of *Sesamia nonagrioides* (Lepidoptera: Noctuidae). Journal of Economic Entomology 100: 195–201.

Anonymous, 1990. Mission Biological Control Laboratory European Corn Borer Parasitoid Survey 1987–1990. Final Report. USDA APHIS PPQ, Mission, TX, USA.

Babendreier, D., Kalberer, N.M., Romeis, J., Fluri, P., Mulligan, E., and Bigler, F., 2005. Influence of Bt transgenic pollen, Bt-toxin and protease inhibitor (SBTI) ingestion on development of the hypopharyngeal glands in honeybees. Apidologie 36: 585–594.

Ball, H.J., and Weekman, G.T., 1962. Insecticide resistance in the western corn rootworm in Nebraska. Journal of Economic Entomology 55: 439–441.

Barry, D., and Darrah, L.L., 1991. Effect of research on commercial hybrid maize resistance to European corn borer (Lepidoptera: Pyralidae). Journal of Economic Entomology 84: 1053–1059.

Bates, S.L., Zhao, J.-Z., Roush, R.T., and Shelton, A.M., 2005. Insect resistance management in GM crops: Past present and future. Nature Biotechnology 23: 57–62.

Baum, J.A., Bogaert, T., Clinton, W., Heck, G.R., Feldmann, P., Ilagan, O., Johnson, S., Plaetinck, G., Munyikwa, T., Pleau, M., Vaughn, T., and Roberts, J., 2007. Control of coleopteran insect pests through RNA interference. Nature Biotechnology 25: 1322–1326.

Baute, T.S., Sears, M.K., and Schaafsma, A.W., 2002. Use of transgenic *Bacillus thuringiensis* Berliner corn hybrids to determine the direct economic impact of the European corn borer (Lepidoptera: Crambidae) on field corn in eastern Canada. Journal of Economic Entomology 95: 57–64.

Beeghly, H.H., Coors, J.G., and Lee, M., 1997. Plant fiber composition and resistance to European corn borer in four maize populations. Maydica 42: 297–303.

Bergvinson, D.J., Arnason, J.T., Hamilton, R.I., Mihm, J.A., and Jewell, D.C., 1994. Determining leaf toughness and its role in maize resistance to the European corn borer (Lepidoptera: Pyralidae). Journal of Economic Entomology 87: 1743–1748.

Bergvinson, D.J., Hamilton, R.I., and Arnason, J.T., 1995. Leaf profile of maize resistance factors to European corn borer, *Ostrinia nubilalis*. Journal of Chemical Ecology 21: 343–354.

Bhatti, M.A., Duan, J., Head, G., Jiang, C., McKee, M.J., Nickson, T.E., Pilcher, C.L., and Pilcher, C.D., 2005a. Field evaluation of the impact of corn rootworm (Coleoptera: Chrysomelidae) protected Bt corn on ground-dwelling invertebrates. Environmental Entomology 34: 1325–1335.

Bhatti, M.A., Duan, J., Head, G., Jiang, C., McKee, M.J., Nickson, T.E., Pilcher, C.L., and Pilcher, C.D., 2005b. Field evaluation of the impact of corn rootworm (Coleoptera: Chrysomelidae) protected Bt corn on foliage-dwelling arthropods. Environmental Entomology 34: 1336–1345.

Bigler, F., 1986. Mass production of *Trichogramma maidis* Pint. et Voeg. and its field application against *Ostrinia nubilalis* Hbn. In Switzerland. Journal of Applied Entomology 101: 23–29.

Blackwood, C., and Buyer, J., 2004. Soil microbial communities associated with Bt and non-Bt corn in three soils. Journal of Environmental Quality 33: 832–836.

Bohorova, N., Frutos, R., Royer, M., Estanol, P., Pacheco, M., Rascon, Q., Mclean, S., and Hoisington, D., 2001. Novel synthetic *Bacillus thuringiensis* cry1B gene and the cry1B-cry1Ab translational fusion confer resistance to southwestern corn borer, sugarcane borer and fall armyworm in transgenic tropical maize. Theoretical and Applied Genetics 103: 817–826.

Bourguet, D., Chaufaux, J., Micoud, A., Delos, M., Naibo, B., Bombarde, F., Marque, G., Eychenne, N., and Pagliari, C., 2002. *Ostrina nubilalis* parasitism and the field abundance of non-target insects in transgenic *Bacillus thuringiensis* corn (*Zea mays*). Environmental Biosafety Research 1: 49–60.

Brookes, G., and Barfoot, P., 2006. GM Crops: The First Ten Years - Global Socio-Economic and Environmental Impacts. ISAAA Brief No. 36, International Services for the Acquisition of Agri-Biotech Applications, Ithaca, NY, USA.

Brookes, G., Barfoot, P., Melé, E., Messeguer, J., Bénétrix, F., Bloc, D., Foueillassar, X., Fabié, A., and Poeydomenge, C., 2004. Genetically Modified Maize: Pollen Movement and Crop Co-existence. PG Economics. http://www.pgeconomics.co.uk/pdf/Maizepollennov2004final.pdf (accessed 18 January 2008).

Butrón, A., Sandoya, G., Santiago, R., Prdás, A., Rial, A., and Malvar, R.A., 2006. Searching for new sources of pink stem borer resistance in maize (*Zea mays* L). Genetic Resources and Crop Evolution 53: 1455–1462.

Byrne, D., 2003. Commission Decision of 24 October 2003 on emergency measures to prevent the spread within the Community of *Diabrotica virgifera* Le Conte. Official Journal of the European Union L275: 49–50.

Caffrey, D.J., and Worthley, L.H., 1927. A progress report on the investigations of the European corn borer. United States Department of Agriculture Bulletin 1476.

Candolfi, M., Brown, K., Reber, B., and Schmidli, H., 2003. A faunistic approach to assess potential side-effects of genetically modified Bt-corn on non-target arthropods under field conditions. Biocontrol Science and Technology 14: 129–170.

Carpenter, J.E., and Gianessi, L.P., 2001. Agricultural Biotechnology: Updated Benefit Estimates. National Center for Food and Agricultural Policy, Washington, DC, USA. http://www.ncfap.org/reports/biotech/updatedbenefits.pdf (accessed 18 January 2008).

Casida, J.E., and Quistad, G.B., 1998. Golden age of insecticide research: Past, present, or future? Annual Review of Entomology 43: 1–16.

Catangui, M.A., and Berg, R.K., 2006. Western bean cutworm, *Striacosta albicosta* (Smith) (Lepidoptera: Noctuidae) as a potential pest of transgenic Cry1Ab *Bacillus thiringiensis* corn hybrids in South Dakota. Environmental Entomology 35: 1439–1452.

Chang, X.Y., He, K.L., Wang, Z.Y., and Bai, S.X., 2006. Evaluation of transgenic Bt maize for resistance to cotton bollworm. Acta Phytophylacica Sinica 33: 374–378.

Chiang, H.C., 1973. Bioeconomics of northern and western corn rootworms. Annual Review of Entomology 18: 47–72.

Chippendale, G.M., and Sorenson, C.E., 1997. Biology and management of the southwestern corn borer. In: Radcliffe's IPM World Textbook, E.B. Radcliffe and W.D. Hutchison, eds., University of Minnesota, St. Paul, MN, USA. http://ipmworld.umn.edu/ (accessed 18 January 2008).

Christou, P., Capell, T., Kohli, A., Gatehouse, J.A., and Gatehouse, A.M.R., 2006. Recent developments and future prospects in insect pest control in transgenic crops. Trends in Plant Science 11: 302–308.

Coors, J.G., 1987. Resistance to the European corn borer, *Ostrinia nubilalis* (Hübner), in maize, *Zea mays* L., as affected by soil silica, plant silica, structural carbohydrates, and lignin. In: Genetics Aspects of Plant Mineral Nutrition, H.W. Gabelman and B.C. Loughman, eds., Martinus Nijhoff, Dordrecht, The Netherlands, pp. 445–456.

Daly, T., and Buntin, G.D., 2005. Effect of *Bacillus thuringiensis* transgenic corn for lepidopteran control on non-target arthropods. Environmental Entomology 34: 1292–1301.

De Groote, H., 2002. Maize yield losses from stemborers in Kenya. Insect Science and Its Application 22: 89–96.

De Groote, H., Okuro, J.O., Bett, C., Mose, L., Odendo, M., and Wekesa, E., 2004a. Assessing the demand for insect resistant maize varieties in Kenya combining Participatory Rural Appraisal into a Geographic Information System. In: Participatory Plant Breeding and Participatory Plant Genetic Resource Enhancement: An Africa-Wide Exchange of Experiences, L. Sperling, J. Lancon and M. Loosvelt, eds., CGIAR System-Wide Program on Participatory Research and Gender Analysis, Cali, Columbia, pp. 148–162.

De Groote, H., Bett, C., Okuro, J.O., Odendo, M., Mose, L., and Wekesa, E., 2004b. Direct estimation of maize crop losses due to stem borers in Kenya, preliminary results from 2000 and 2001. In: Integrated Approaches to Higher Maize Productivity in the New Millennium, D.K. Friesen and A.F.E. Palmer, eds., Proceedings of the Seventh Eastern and Southern Africa Regional Maize Conference, Nairobi, Kenya, 11–15 February 2002, pp. 401–406.

de la Poza, M., Pons, X., Farinós, G.P., López, C., Ortego, F., Eizaguirre, M., Castañera, P., and Albajes, R., 2005. Impact of farm-scale Bt maize on abundance of predatory arthropods in Spain. Crop Protection 24: 677–684.

Demont, M., and Tollens, E., 2004. First impact of biotechnology in the EU: Bt maize adoption in Spain. Annals of Applied Biology 145: 197–207.

Devare, M.H., Jones, C.M., and Thies, J.E., 2004. Effect of Cry3Bb transgenic corn and tefluthrin on the soil microbial community: Biomass, activity, and diversity. Journal of Environmental Quality 33: 837–843.

Devos, Y., Reheul, D., and de Schrijver, A., 2005. The co-existence between transgenic and non-transgenic maize in the European Union: A focus on pollen flow and cross-fertilization. Environmental Biosafety Research 4: 71–87.

Dively, G.P., 2005. Impact of transgenic VIP3A × Cry1Ab Lepidopteran-resistant field corn on the nontarget arthropod community. Environmental Entomology 34: 1267–1291.

Dively, G.P., Rose, R., Sears, M.K., Hellmich, R.L., Stanley-Horn, D.E., Russo, J.M., Calvin, D.D., and Anderson, P.L., 2004. Effects on monarch butterfly larvae (Lepidoptera: Danaidae) after continuous exposure to Cry1Ab-expressing corn during anthesis. Environmental Entomology 33: 1116–1125.

Dow AgroSciences, 2007. Dow and Monsanto to Develop Eight-Way Gene Stack for Corn. Dow News Center, 14 September 2007. http://news.dow.com/dow_news/corporate/2007/20070914a.htm (accessed 18 January 2008).

Dowd, P.F., 2000. Indirect reduction of ear molds and associated mycotoxins in *Bacillus thuringiensis* corn under controlled and open field conditions: Utility and limitations. Journal of Economic Entomology 93: 1669–1679.

Duan, J.J., Marvier, M., Huesing, J., Dively, G., and Huang, Z.J., 2008. A meta-analysis of effects of Bt crops on honey bees (Hymenoptera: Apidae). PLoS ONE 3(1): e1415.

Dutton, A., Klein, H., Romeis, J., and Bigler, F., 2002. Uptake of Bt-toxin by herbivores feeding on transgenic maize and consequences for the predator *Chrysoperla carnea*. Ecological Entomology 27: 441–447.

Dutton, A., Romeis, J., and Bigler, F., 2003. Assessing the risks of insect resistant transgenic plants on entomophagous arthropods: Bt-maize expressing Cry1Ab as a case study. BioControl 48: 611–636.

Dyer, L.E., and Landis, D.A., 1997. Influence of noncrop habitats on the distribution of *Eriborus terebrans* (Hymenoptera: Ichneumonidae) in cornfields. Environmental Entomology 26: 924–932.

Eizaguirre, M., Albajes, R., López, C., Eras, J., Lumbierres, B., and Pons, X., 2006. Six years after the commercial introduction of Bt maize in Spain: Field evaluation, impact and future prospects. Transgenic Research 15: 1–12.

Ellsbury, M.M., Jackson, J.J., Woodson, W.D., Beck, D.L., and Stange, K.A., 1996. Efficacy, application distribution, and concentration by stemflow of *Steinernema carpocapsae* (Rhabditida: Steinernematidae) suspensions applied with a lateral-move irrigation system for corn rootworm (Coleoptera: Chrysomelidae) control in maize. Journal of Economic Entomology 89: 71–81.

Estruch, J.J., Warren, G.W., Mullins, M.A., Nye, G.J., Craig, J.A., and Koziel, A.G., 1996. Vip3A, a novel *Bacillus thuringiensis* vegetative insecticidal protein with a wide spectrum of activity against lepidopteran insects. Proceedings of the National Academy of Sciences of the USA 93: 5389–5394.

European Union, 2003a. Regulation (EC) No. 1829/2003 of the European Parliament and of the Council of 22 September 2003 on genetically modified food and feed. European Parliament and the Council of the European Union, Brussels.

European Union, 2003b. Regulation (EC) No. 1830/2003 of the European Parliament and of the Council of 22 September 2003 concerning the traceability and labelling of genetically modified organisms and the traceability of food and feed products produced from genetically modified organisms and amending. Directive 2001/18/EC. European Parliament and the Council of the European Union, Brussels.

FAOSTAT, 2007. Food and Agriculture Organization of the United Nations. The Statistics Division. http://faostat.fao.org/ (accessed 17 March 2008).

Farinós, G.P., de la Poza, M., Hernández-Crespo, P., Ortego, F., and Castañera, P., 2004. Resistance monitoring of field populations of the corn borers *Sesamia nonagrioides* and *Ostrinia nubilalis* after 5 years of Bt maize cultivation in Spain. Entomologia Experimentalis et Applicata 110: 23–30.

Federici, B.A., 2002. Case study: Bt crops – a novel mode of insect control. In: Genetically Modified Crops: Assessing Safety, K.T. Atherton, ed., Taylor & Francis, New York, USA, pp. 164–200.

Fernandes, O.A., Faria, M., Martinelli, S., Schmidt, F., Carvalho, V.F., and Moro, G., 2007. Short-term assessment of Bt maize on non-target arthropods in Brazil. Scientia Agricola 64: 249–255.

Ferré, J., and Van Rie, J., 2002. Biochemistry and genetics of insect resistance to *Bacillus thuringiensis*. Annual Review of Entomology 47: 501–533.

Freeman, H.A., and Omiti, J.M., 2003. Fertilizer use in semi-arid areas of Kenya: Analysis of smallholder farmers' adoption behavior under liberalized markets. Nutrient Cycling in Agroecosystems 66: 23–31.

Frey, M., Chomet, P., Glawischnig, E., Stettner, C., Grün, S., Winklmair, A., Eisenreich, W., Bacher, A., Meeley, R.B., Briggs, S.P., Simcox, K., and Gierl, A., 1997. Analysis of a chemical plant defense mechanism in grasses. Science 277: 696–699.

Frutos, R., Rang, C., and Royer, M., 1999. Managing insect resistance to plants producing *Bacillus thuringiensis* toxins. Critical Reviews in Biotechnology 19: 227–276.

Gardner, J., Hoffmann, M.P., Cheever, S.A., Seaman, A.J., Westgate, P., and Hazzard, R.V., 2007. Large-scale releases of *Trichogramma ostriniae* to suppress *Ostrinia nubilalis* in commercially grown processing and fresh market sweet corn. Journal of Applied Entomology 131: 432–440.

Georghiou, G.P., 1986. The magnitude of the resistance problem. In: Pesticide Resistance: Strategies and Tactics for Management. National Academy of Sciences, Washington, DC, USA, pp. 14–44.

Gianessi, L.P., Silvers, C.S., Sankula, S., and Carpenter, J.E., 2002. Plant Biotechnology: Current and Potential Impact for Improving Pest Management in U.S. Agriculture, an Analysis of 40 Case Studies. National Center for Food and Agricultural Policy, Washington, DC, USA.

Gillette, C.P., 1912. *Diabrotica virgfera* Lec. as a corn root-worm. Journal of Economic Entomology 5: 364–366.

Goldberger, J., Merrill, J., and Hurley, T., 2005. Bt corn farmer compliance with insect resistance management requirements in Minnesota and Wisconsin. AgBioForum 8: 151–160.

Gómez-Barbero, M., Berbel, J., and Rodríguez-Cerezo, E., 2008. Bt corn in Spain – The performance of the EU's first GM crop. Nature Biotechnology 26: 384–386.

González-Núñez, M., Ortego, F., and Castañera, P., 2000. Susceptibility of Spanish populations of the corn borers *Sesamia nonagrioides* (Lepidoptera: Noctuidae) and *Ostrinia nubilalis* (Lepidoptera: Crambidae) to a *Bacillus thuringiensis* endotoxin. Journal of Economic Entomology 93: 459–463.

Gould, F., 1994. Potential and problems with high-dose strategies for pesticidal engineered crops. Biocontrol Science and Technology 4: 451–461.

Gould, F., 1998. Sustainability of transgenic insecticidal cultivars integrating pest genetics and ecology. Annual Review of Entomology 43: 701–726.

Gouse, M., Pray, C.E., Kirsten, J., and Schimmelpfennig, D., 2005. A GM subsistence crop in Africa: The case of Bt white maize in South Africa. International Journal of Biotechnology 7: 84–94.

Gouse, M., Pray, C., Schimmelpfennig, D., and Kirsten, J., 2006. Three seasons of subsistence insect-resistant maize in South Africa: Have smallholders benefited? AgBioForum 9: 15–22.

Hanley, A.V., Huang, Z.Y., and Pett, W.L., 2003. Effects of dietary transgenic Bt corn pollen on larvae of *Apis mellifera* and *Galleria mellonella*. Journal of Agricultural Research 42: 77–81.

He, K., Wang, Z., Wen, L., Bai, S., Ma, X., and Yao, Z., 2005. Determination of baseline susceptibility to Cry1Ab protein for Asian corn borer (Lepidoptera: Crambidae). Journal of Applied Entomology 129: 407–412.

He, K.L., Wang, Z.Y., Zhou, D.R., Wen, L.P., and Yao, Z.Y., 2003a. Evaluation of transgenic Bt corn for resistance to the Asian corn borer (Lepidoptera: Pyralidae). Journal of Economic Entomology 96: 935–940.

He, K.L., Wang, Z.Y., Wen, L.P., Bai, S.X., Zhou, D.R., and Zhu, Q.H., 2003b. Field evaluation of Asian corn borer control in hybrid of transgenic maize event MON810. Agricultural Sciences in China 2: 1363–1368.

He, K.L., Wang, Z.Y., Wen, L.P., Bai, S.X., and Zhou, D.R., 2004. Transgenic maize evaluated for resistance to the Asian corn borer (Lepidoptera: Pyralidae). Chinese Agricultural Science Bulletin 20: 240–242.

He, K.L., Chang, X., Chang, X.Y., and Wang, Z.Y., 2007. Inheritance and molecular mechanism of the resistance of Asian corn borer to Cry1Ab. Plant Protection 33: 92–93.

Hellmich, R.L., 2006. European corn borer: A diminished insect threat from now on? Illinois Crop Protection Technical Conference: 2006 Proceedings, University of Illinois Extension, Urbana, IL, USA, pp. 62–64.

Hellmich, R.L., Siegfried, B.D., Sears, M.K., Stanley-Horn, D.E., Mattila, H.R., Spencer, T., Bidne, K.G., Daniels, M.J., and Lewis, L.C., 2001. Monarch larvae sensitivity to *Bacillus thuringiensis* purified proteins and pollen. Proceedings of the National Academy of Sciences of the USA 98: 11925–11930.

Hilbeck, A., and Andow, D.A., eds., 2004. Environmental Risk Assessment of Genetically Modified Organisms: Vol. 1. A Case Study of Bt Maize in Kenya. CABI, Wallingford, UK, 281 pp.

Hilbeck, A., Baumgartner, M., Fried, P.M., and Bigler, F., 1998. Effects of transgenic *Bacillus thuringiensis* corn-fed prey on mortality and development time of immature *Chysoperla carnea* (Neuroptera: Chrysiopidae). Environmental Entomology 27: 480–487.

Hossain, F., Pray, C.E., Lu, Y., Huang, J., and Hu, R., 2004. Genetically modified cotton and farmers' health in China. International Journal of Occupational and Environmental Health 10: 296–303.

Huang, F., Buschman, L.L., Higgins, R.A., and McGaughey, W.H., 1999. Inheritance of resistance to *Bacillus thuringiensis* toxin (Dipel ES) in the European corn borer. Science 284: 965–967.

Hunt, T.E., Buschman, L.L., and Sloderbeck, P.E., 2007. Insecticide use in Bt and non-Bt field corn in the Western Corn Belt: As reported by crop consultants in a mail survey. American Entomologist 53: 86–93.

Hurley, T.M., and Mitchell, P.D., 2007. Insect resistance management: Adoption and compliance. In: Insect Resistance Management: Biology, Economics, and Prediction, D.W. Onstad, ed., Elsevier, Amsterdam, The Netherlands, pp. 227–253.

IRMA (Insect Resistant Maize for Africa), 2000. IRMA Updates, June 2000, Vol. 1, Issues 1 + 2. Kenyan Agricultural Research Institute, Nairobi, Kenya. http://www.cimmyt.cgiar.org/english/wpp/gen_res/pdf/IRMA_1&2.pdf (accessed 18 January 2008).

IRMA (Insect Resistant Maize for Africa), 2002. IRMA Updates, August 2002, Vol. 3, Issue 1 + 2. Kenyan Agricultural Research Institute, Nairobi, Kenya. http://www.cimmyt.cgiar.org/english/wpp/gen_res/pdf/IRMA_31&2.pdf (accessed 18 January 2008).

IRMA (Insect Resistant Maize for Africa), 2005a. IRMA Updates, March 2005, Vol. 6, Issue 1. Kenyan Agricultural Research Institute, Nairobi, Kenya. http://www.cimmyt.cgiar.org/english/wpp/gen_res/pdf/IRMA_6&1.pdf (accessed 18 January 2008).

IRMA (Insect Resistant Maize for Africa), 2005b. IRMA Updates, December 2005, Vol. 6, Issue 4. Kenyan Agricultural Research Institute, Nairobi, Kenya. http://www.cimmyt.cgiar.org/english/wpp/gen_res/pdf/IRMA_6&4.pdf (accessed 18 January 2008).

Jackson, J.J., 1996. Field performance of entomopathogenic nematodes for suppression of Western corn rootworm (Coleoptera: Chrysomelidae). Journal of Economic Entomology 89: 366–372.

James, C., 2003. Global Review of Commercialized Transgenic Crops: 2002 Feature: Bt Maize. ISAAA Brief No. 29, International Service for the Acquisition of Agri-Biotech Applications, Ithaca, NY, USA.

James, C., 2007. Global Status of Commercialized Biotech/GM Crops: 2007. ISAAA Brief No. 37, International Services for the Acquisition of Agri-Biotech Applications, Ithaca, NY, USA.

Jesse, L.C.H., and Obrycki, J.J., 2000. Field deposition of Bt transgenic corn pollen: Lethal effects on the monarch butterfly. Oecologia 125: 241–248.

Khan, Z.R., Ampong-Nyarko, K., Chiliswa, P., Hassanali, A., Kimani, S., Lwande, W., Overholt, W.A., Pickett, J.P., Smart, L.E., Wadhams, L.J., and Woodcock, C.M., 1997. Intercropping increases parasitism of pests. Nature 388: 631–632.

Khan, Z.R., Pickett, J.A., Wadhams, L., and Muyekho, F, 2001. Habitat management strategies for the control of cereal stemborers and striga in maize in Kenya. Insect Science and Its Application 21: 375–380.

Klun, J.A., Tipton, C.L., and Brindley, T.A., 1967. 2,4-Dihydroxy-7-methoxy-1,4-benzoxazin-3-one (DIMBOA), an active agent in the resistance of maize to the European corn borer. Journal of Economic Entomology 60: 1529–1533.

Kogan, M., 1998. Integrated pest management: Historical perspectives and contemporary developments. Annual Review of Entomology 43: 243–270.

Krysan, J.L., Foster, D.E., Branson, T.F., Ostlie, K.R., and Cranshaw, W.S., 1986. Two years before the hatch: Rootworms adapt to crop rotations. Bulletin of the Entomological Society of America 32: 250–253.

Kuhar, T.P., Hoffmann, M.P., and Wright, M.G., 2003. Controlling European corn borer in vegetables with a parasitic wasp. Pesticide Outlook 14: 99–101.

Landis, D.A., and Haas, M., 1992. Influence of landscape structure on abundance and within-field distribution of *Ostrinia nubilalis* Hübner (Lepidoptera: Pyralidae) larval parasitoids in Michigan. Environmental Entomology 21: 409–416.

Landis, D.A., Wratten, S.D., and Gurr, G.M., 2000. Habitat management to conserve natural enemies of arthropod pests in agriculture. Annual Review of Entomology 45: 175–201.

Lawo, N.C., and Romeis, J., 2008. Assessing the utilization of a carbohydrate food source and the impact of insecticidal proteins on larvae of the green lacewing, *Chrysoperla carnea*. Biological Control 44: 389–398.

Lee, M.K., Walters, F.S., Hart, H., Palekar, N., and Chen, J.S., 2003. Mode of action of the *Bacillus thuringiensis* vegetative insecticidal protein Vip3A differs from that of Cry1Ab delta-endotoxin. Applied and Environmental Microbiology 69: 4648–4657.
Levine, E., and Oloumi-Sadeghi, H., 1996. Western corn rootworm (Coleoptera: Chrysomelidae) larval injury to corn grown for seed production following soybeans grown for seed production. Journal of Economic Entomology 89: 1010–1016.
Li, L.L., Wang, Z.Y., He, K.L., Bai, S.X., and Hua, L., 2007. Effect of transgenic corn expressing *Bacillus thuringiensis* Cry1Ab toxin on population increase of *Rhopalosiphum maidis* Fitch. Chinese Journal of Applied Ecology 18: 1077–1080.
Linker, H.M., Van Duyn, J.W., Anderson, J.R., and Lewis, W.M., 1990. Scouting Corn in North Carolina. North Carolina Cooperative Extension Service, AG-399, Raleigh, NC, USA.
Losey, J.E., Rayor, L.S., and Carter, M.E., 1999. Transgenic pollen harms monarch larvae. Nature 399: 214.
Marçon, P.C.R.G., Young, L.J., Steffey, K.L., and Siegfried, B.D., 1999. Baseline susceptibility of European corn borer (Lepidoptera: Crambidae) to *Bacillus thuringiensis* toxins. Journal of Economic Entomology 92: 279–285.
Marvier, M., McCreedy, C., Regetz, J., and Kareiva, P., 2007. A meta-analysis of effects of Bt cotton and maize on nontarget invertebrates. Science 316: 1475–1477.
Mason, C.E., Romig, R.F., Wendel, L.E., and Wood, L.A., 1994. Distribution and abundance of larval parasitoids of European corn borer (Lepidoptera: Pyralidae) in the East Central United States. Environmental Entomology 23: 521–531.
Mason, C.E., Rice, M.E., Calvin, D.D., Van Duyn, J.W., Showers, W.B., Hutchinson, W.D., Witkowski, J.F., Higgins, R.A., Onstad, D.W., and Dively G.P., 1996. European corn borer ecology and management. North Central Regional Publication 327. Iowa State University, Ames, IA, USA.
Matten, S.M., Hellmich, R.L., and Reynolds, A., 2004. Current resistance management strategies for Bt corn in the United States. In: Transgenic Crop Production: Concepts and Strategies, O. Koul and G.S. Dhaliwal, eds., Science Publishers, Plymouth, UK, pp. 261–288.
McMullen, M.D., Byrne, P.F., Snook, M.E., Wiseman, B.R., Lee, E.A., Widstrom, N.W., and Coe, E.H., Jr., 1998. Quantitative trait loci and metabolic pathways. Proceedings of the National Academy of Sciences of the USA 95: 1996–2000.
Meinke, L.J., Siegfried, B.D., Wright, R.J., and Chandler, L.D., 1998. Adult susceptibility of Nebraska western corn rootworm (Coleoptera: Chrysomelidae) populations to selected insecticides. Journal of Economic Entomology 91: 594–600.
Messeguer, J., Peñas, G., Ballester, J., Bas, M., Serra, J., Salvia, J., Palaudelmas, L., and Melé, E., 2006. Pollen-mediated gene flow in maize in real situations of coexistence. Plant Biotechnology Journal 4: 633–645.
Meyer, J.D.F., Snook, M.E., Houchins, K.E., Rector, B.G., Widstrom, N.W., and Mcmullen, M.D., 2007. Quantitative trait loci for maysin synthesis in maize (*Zea mays* L.) lines selected for high silk maysin content. Journal of Theoretical and Applied Genetics 115: 119–128.
Mitchell, P.D., and Onstad, D.W., 2007. Valuing pest susceptibility to control. In: Insect Resistance Management: Biology, Economics, and Prediction, D.W. Onstad, ed., Elsevier, Amsterdam, The Netherlands, pp. 17–38.
Moellenbeck, D.J., Peters, M.L., Bing, J.W., Rouse, J.R., Higgins, L.S., Sims, L., Nevshemal, T., Marshall, L., Ellis, R.T., and Bystrak P.G., 2001. Insecticidal proteins from *Bacillus thuringiensis* protect corn from corn rootworms. Nature Biotechnology 19: 668–672.
Mugo, S., De Groote, H., Bergvinson, D., Mulaa, M., Songa, J., and Gichuki, S., 2005. Developing Bt maize for resource-poor farmers - recent advances in the IRMA project. African Journal of Biotechnology 4: 1490–1504.
Mugo, S.N., Gichuki, S.T., Murenga, M.G., Taracha, C.N., Songa, J.M., Bergvinson, D.J., Pellegrineschi, A., and Hoisington, D.A., 2008. Control of stem borers by Bt maize in the first studies in a biosafety laboratory, a biosafety greenhouse, and from the first confined field trials in Kenya. Crop Protection 27 (in press).
Munkvold, G.P., and Hellmich, R.L., 1999. Comparison of fumonisin concentrations in kernels of transgenic Bt maize hybrids and nontransgenic hybrids. Plant Disease 83: 130–138.

Munson, J.D., and Helms, T.J., 1970. Field evaluation of a nematode (DD-136) for control of corn rootworm larvae. Proceedings of the Entomological Society of America, North Central Branch 25: 97–99.

Musser, F.R., Nyrop, J.P., and Shelton, A.M., 2006. Integrating biological and chemical controls in decision making: European corn borer control in sweet corn as an example. Journal of Economic Entomology 99: 1538–1549.

Nye, I.W.R., 1960. The insect pests of graminaceous crops in East Africa. Colonial Research Studies 31: 1–48.

Oberhauser, K.S., Prysby, M., Mattila, H.R., Stanley-Horn, D.E., Sears, M.K., Dively, G.P., Olson, E., Pleasants, J.M., Lam, W.-K.F., and Hellmich, R.L., 2001. Temporal and spatial overlap between monarch larvae and corn pollen. Proceedings of the National Academy of Sciences of the USA 98: 11913–11918.

O'Callaghan, M., Glare, T.R., Burgess, E.P.J., and Malone, L.A., 2005. Effects of plants genetically modified for insect resistance on nontarget organisms. Annual Review of Entomology 50: 271–292.

Omwega, C.O., Kimani, S.W., Overholt, W.A., and Ogol, C.K.P.O., 1995. Evidence of the establishment of *Cotesia flavipes* (Hymenoptera: Bracondiae) in continental Africa. Bulletin of Entomological Research 85: 525–530.

Ong'amo, G.O., Bruno, P., Rü, L., Dupas, S., Moyal, P., Calatayud, P.-A., and Silvain, J.-F., 2006. Distribution, pest status and agro-climatic preferences of lepidopteran stem borers of maize in Kenya. Annales de la Société Entomologique de France 42: 171–177.

Overholt, W.A., Ngi-Song, A.J., Omwega, C.O., Kimani-Njogu, S.W., Mbapila, J., Sallam, M.N., and Ofomata, V., 1997. A review of the introduction and establishment of *Cotesia flavipes* Cameron in East Africa for biological control of cereal stemborers. Insect Science and Its Application 17: 79–88.

Painter, R.H., 1968. Insect Resistance in Crop Plants. University Press of Kansas, Lawrence, KS, USA.

Pedigo, L.P., Hutchins, S.H., and Higley, L.G., 1986. Economic injury levels in theory and practice. Annual Review of Entomology 31: 341–368.

Piao, Y.F., and Yan, S., 1996. Progress of mass production and field application of *Trichogramma dendrolimi*. In: Proceedings of the National Symposium on IPM in China, Z.L. Zhang, Y.F. Piao and J.W. Wu, eds., China Agricultural Scientech Press, Beijing, pp. 1135–1136.

Pilcher, C.D., and Rice, M.E., 2001. Effect of planting dates and *Bacillus thuringiensis* corn on the population dynamics of European corn borer (Lepidoptera: Crambidae). Journal of Economic Entomology 94: 730–742.

Pilcher, C.D., Rice, M.E., Higgins, R.A., Steffey, K.L., Hellmich, R.L., Witkowski, J., Calvin, D.D., Ostlie, K.R., and Gray, M., 2002. Biotechnology and the European corn borer: Measuring historical farmer perceptions and adoption of transgenic Bt corn as a pest management strategy. Journal of Economic Entomology 95: 878–892.

Pilcher, C.D., Rice, M.E., and Obrycki, J.J., 2005. Impact of transgenic *Bacillus thuringiensis* corn and crop phenology on five non-target arthropods. Environmental Entomology 34: 1302–1316.

Piqué, J., Eizaguirre, M., and Pons, X., 1998. Soil insecticide treatments against maize soil pests and corn borers in Catalonia under traditional crop conditions. Crop Protection 17: 557–561.

Pleasants, J.M., Hellmich, R.L., Dively, G.P., Sears, M.K., Stanley-Horn, D.E., Mattila, H.R., Foster, J.E., Clark, P.L., and Jones, G.D., 2001. Corn pollen deposition on milkweeds in and near cornfields. Proceedings of the National Academy of Sciences of the USA 98: 11919–11924.

Pons, X., and Albajes, R., 2002. Control of maize pests with imidacloprid seed dressing treatment in Catalonia (NE Iberian Peninsula) under traditional crop conditions. Crop Protection 21: 943–950.

Pray, C.E., Huang, J., Hu, R., and Rozelle, S., 2002. Five years of Bt cotton in China - The benefits continue. The Plant Journal 31: 423–430.

Prischmann, D.A., Dashiell, K.E., Schneider, D.J., and Hibbard, B.E., 2007. Field screening maize germplasm for resistance and tolerance to western corn rootworms (Coleoptera: Chrysomelidae). Journal of Applied Entomology 131: 406–415.

Pruess, K.P., Witkowski, J.F., and Raun, E.S., 1974. Population suppression of western corn rootworm by adult control with ULV malathion. Journal of Economic Entomology 67: 651–655.

Rasmann, S., Köllner, T.G., Degenhardt, J., Hiltpold, I., Toepfer, S., Kuhlmann, U., Gershenzon, J., and Turlings, T.C.J., 2005. Insect-induced attraction of entomopathogenic nematodes protects maize roots from an invasive pest. Nature 434: 732–737.

Rice, M.E., 2004. Transgenic Rootworm Corn: Assessing Potential Agronomic, Economic, and Environmental Benefits. Plant Health Progress. doi:10.1094/PHP-2004-0301-01-RV. http://www.ent.iastate.edu/Dept/Faculty/rice/rice_plant_health.pdf (accessed 18 January 2008).

Rodrigo-Simón, A., de Maagd, R.A., Avilla, C., Bakker, P.L., Molthoff, J., Gonzalez-Zamora, J.E., and Ferré, J., 2006. Lack of detrimental effects of *Bacillus thuringiensis* Cry toxins on the insect predator *Chrysoperla carnea*: A toxicological, histopathological, and biochemical approach. Applied and Environmental Microbiology 72: 1595–1603.

Romeis, J., Dutton, A., and Bigler, F., 2004. *Bacillus thuringiensis* toxin (Cry1Ab) has no direct effect on larvae of the green lacewing *Chrysoperla carnea* (Stephens) (Neuroptera: Chrysopidae). Journal of Insect Physiology 50: 175–183.

Romeis, J., Meissle, M., and Bigler, F., 2006. Transgenic crops expressing *Bacillus thuringiensis* toxins and biological control. Nature Biotechnology 24: 63–71.

Romeis, J., Meissle, M., Raybould, A., and Hellmich, R.L., 2008. Impact of insect-resistant genetically modified crops on non-target arthropods. In: Environmental Impact of Genetically Modified/Novel Crops, N. Ferry and A.M.R. Gatehouse, eds., CABI, Wallingford, UK (in press).

Rose, R., Dively, G.P., and Pettis, J., 2007. Effects of Bt corn pollen on honey bees: Emphasis on protocol development. Apidologie 38: 368–377.

Rosi-Marshall, E.J., Tank, J.L., Royer, T.V., Whiles, M.R., Evans-White, M., Chambers, C., Griffiths, N.A., Pokelsek, J., and Stephen, M.L., 2007. Toxins in transgenic crop byproducts may affect headwater stream ecosystems. Proceedings of the National Academy of Sciences of the USA 104: 16204–16208.

Roush, R.T., 1998. Two-toxin strategies for management of insecticidal transgenic crops: Can pyramiding succeed where pesticide mixture have not? Philosophical Transactions Royal Society of London B 353: 1777–1786.

Saeglitz, C., Bartsch, D., Eber, S., Gathmann, A., Priesnitz, K.U., and Schuphan, I., 2006. Monitoring the Cry1Ab susceptibility of European corn borer in Germany. Journal of Economic Entomology 99: 1768–1773.

Sankula, S., 2006. Quantification of the Impacts on US Agriculture of Biotechnology-Derived Crops Planted in 2005: Executive Summary. National Center for Food and Agricultural Policy, Washington, DC, USA. http://www.ncfap.org/whatwedo/pdf/2005biotechExecSummary.pdf (accessed 18 January 2008).

Sanvido, O., Widmer, F., Winzeler, M., Streit, B., Szerencsits, E., and Bigler, F., 2008. Definition and feasibility of isolation distances for transgenic maize cultivation. Transgenic Research 17: 317–335.

Schnepf, E., Crickmore, N., van Rie, J., Lereclus, D., Baum, J., Feitelson, J., Zeigler, D.R., and Dean, D.H., 1998. *Bacillus thuringiensis* and its pesticidal crystal proteins. Microbiology and Molecular Biology Reviews 62: 775–806.

Sears, M.K., Hellmich, R.L., Siegfried, B.D., Pleasants, J.M., Stanley-Horn, D.E., Oberhauser, K.S., and Dively, G.P., 2001. Impact of Bt corn pollen on monarch butterfly populations: A risk assessment. Proceedings of the National Academy of Sciences of the USA 98: 11937–11942.

Shelton, A.M., 2007. Considerations on the use of transgenic crops for insect control. Journal of Developmental Studies 43: 890–900.

Siegfried, B.D., Spencer, T., and Nearman, J., 2000. Baseline susceptibility of the corn earworm (Lepidoptera: Noctuidae) to the Cry1Ab toxin from *Bacillus thuringiensis*. Journal of Economic Entomology 93: 1265–1268.

Siegfried, B.D., Spencer, T., and Vaughn, T.T., 2005. Baseline susceptibility of western corn rootworm (Coleoptera: Chrysomelidae) to the Cry3Bb1 *Bacillus thuringiensis* toxin. Journal of Economic Entomology 98: 1320–1324.

Siegfried, B.D., Spencer, T., Crespo, A.L., Storer, N.P., Head, G.P., Owens, E.D., and Guyer, D., 2007. Ten years of Bt resistance monitoring in the European corn borer: What we know, what we don't know, and what we can do better. American Entomologist 53: 208–214.
Stamm, D.E., Mayo, Z.B., Campbell, J.B., Witkowski, J.F., Andersen, L.W., and Kozub, R., 1985. Western corn rootworm (Coleoptera: Chrysomelidae) beetle counts as a means of making larval control recommendations in Nebraska. Journal of Economic Entomology 78: 794–798.
Stanley-Horn, D.E., Dively, G.P., Hellmich, R.L., Mattila, H.R., Sears, M.K., Rose, R., Jesse, L. C.H., Losey, J.E., Obrycki, J.J., and Lewis, L.C., 2001. Assessing the impact of Cry1Ab-expressing corn pollen on monarch butterfly larvae in field studies. Proceedings of the National Academy of Sciences of the USA 98: 11931–11936.
Steffey, K.L., Rice, M.E., All, J., Andow, D.A., Gray, M.E., and Van Duyn, J.W., eds., 1999. Handbook of Corn Insects. Published by Entomological Society of America, Lanham, MD, USA, 164 pp.
Stern, V.M., Smith, R.F., van den Bosch, R., and Hagen, K.S., 1959. The integrated control concept. Hilgardia 29: 81–101.
Tabashnik, B.E., 1994. Evolution of resistance to *Bacillus thuringiensis*. Annual Review of Entomology 39: 47–79.
Tabashnik, B.E., and Croft, B.A., 1982. Managing pesticide resistance in crop-arthropod complexes interactions between biological and operational factors. Environmental Entomology 11: 1137–1144.
Tabashnik, B.E., Cushing, N.L., Finson, N., and Johnson, M.W., 1990. Field development of resistance to *Bacillus thuringiensis* in diamondback moth (Lepidoptera: Plutellidae). Journal of Economic Entomology 83: 1671–1676.
Tabashnik, B.E., Carrière, Y., Dennehy, T.J., Morin, S., Sisterson, M.S., Roush, R.T., Shelton, A.M., and Zhao, J.Z., 2003. Insect resistance to transgenic Bt crops: Lessons from the laboratory and field. Journal of Economic Entomology 96: 1031–1038.
Tams, W.H.T., 1932. New species of African Heterocera. The Entomologist (London) 65: 1241–1249.
Tende, R.M., Nderitu, J.H., Mugo, S., Songa, J.M., Alubayo, F., and Bergvinson, D., 2005. Screening for development of resistance by the spotted stem borer *Chilo partellus* Swinhoe (Lepidoptera: Pyralidae) to Bt-maize delta-endotoxins. Paper presented at the 7th Conference of the African Science Society, Kampala, Uganda, 5–10 December.
Thies, J.E., and Devare, M.H., 2007. An ecological assessment of transgenic crops. Journal of Development Studies 43: 97–129.
Thomson, J.A., 2008. The role of biotechnology for agricultural sustainability in Africa. Philosophical Transactions of the Royal Society B: Biological Sciences 363: 905–913.
Tollefson, J.J., and Calvin, D.D., 1994. Sampling arthropod pests in field corn. In: Handbook of Sampling Methods for Arthropods in Agriculture, L.P. Pedigo and G.D. Buntin, eds., CRC, Boca Raton, FL, USA, pp. 433–473.
Trigo, E.J., and Cap, E.J., 2006. Ten Years of Genetically Modified Crops in Argentine Agriculture. Argentine Council for Information and Development of Biotechnology – ArgenBio, 52 pp. http://www.inta.gov.ar/ies/docs/otrosdoc/Ten_Years_GM_Crops_Argentine_Agriculture.pdf (accessed 18 January 2008).
Umeozor, O.C., van Duyn, J.W., Bradley, J.R., Jr., and Kennedy, G.G., 1985. Comparison of the effect of minimum-tillage treatments on the overwintering emergence of European corn borer (Lepidoptera: Pyralidae) in cornfields. Journal of Economic Entomology 78: 937–939.
USDA-ERS (United States Department of Agriculture, Economic Research Service), 2002. Adoption of Bioengineered Crops. Agricultural Economic Report No. 810. http://www.ers.usda.gov/publications/aer810/aer810.pdf (accessed 18 January 2008).
USDA-ERS (United States Department of Agriculture, Economic Research Service), 2007. Adoption of Genetically-Engineered Crops in the U.S., Washington, DC, USA. http://www.ers.usda.gov/Data/BiotechCrops/alltables.xls (accessed 5 November 2007).
USDA-NASS (United States Department of Agriculture, National Agricultural Statistics Service), 2007. Acreage. Washington, DC, USA. http://www.usda.gov/nass/PUBS/TODAYRPT/acrg0607.pdf (accessed 16 January 2008).

USEPA (United States Environmental Protection Agency), 2007. Current Plant Incorporated Protectant (PIP) Experimental Use Permits. Washington, DC, USA. http://www.epa.gov/pesticides/biopesticides/pips/current_pip_eups.htm (accessed 5 November 2007).

van Rensburg, J.B.J., Walters, M.C., and Giliomee, J.H., 1988. Mortality in natural populations of the maize stalk borer, *Busseola fusca* (Fuller) (Leptidoptera: Noctuidae) in South Africa. Phytophylactica 20: 17–19.

Vaughn, T., Cavato, T., Brar, G., Coombe, T., DeGooyer, T., Ford, S., Groth, M., Howe, A., Johnson, S., Kolacz, K., Pilcher, C., Purcell, J., Romano, C., English, L., and Pershing, J., 2005. A method of controlling corn rootworm feeding using a *Bacillus thuringiensis* protein expressed in transgenic maize. Crop Science 45: 931–938.

Waiss, A.C., Jr., Chan, B.G., Elliger, C.A., Wiseman, B.R., McMillian, W.W., Widstrom, N.W., Zuber, M.S., and Keaster, A.J., 1979. Maysin, a flavone glycoside from corn silks with antibiotic activity toward corn earworm. Journal of Economic Entomology 72: 256–258.

Wang, D.Y., Wang, Z.Y., He, K.L., Cong, B., Bai, S.X., and Wen, L.P., 2004a. Temporal and spatial expression of Cry1Ab toxin in transgenic Bt corn and its effects on Asian corn borer, *Ostrinia furnacalis* (Guenée). Scientia Agricultura Sinica 37: 1155–1159.

Wang, D.Y., Wang, Z.Y., He, K.L., Cong, B., Wen, L.P., and Bai, S.X., 2004b. Food consumption and utilization of the fifth instar larvae of *Mythimna separata* (Walker) feeding on the leaves of transgenic *Bacillus thuringiensis* corn expressing Cry1Ab protein. Acta Entomologica Sinica 47: 141–145.

Wang, Z.Y., He, K.L., Wen, L.P., Zhang, G.Y., and Zheng, L., 2001. Spatial-temporal distributions and cotton bollworm eggs on summer corn seeded at different times in North China. Scientia Agricultura Sinica 34: 153–156.

Wang, Z.Y., He, K.L., Zhao, J.Z., and Zhou, D.R., 2003. Integrated pest management in China. In: Integrated Pest Management in the Global Arena, K.M. Maredia, D. Dakouo and D. Mota-Sanchez, eds., CABI, Wallingford/Oxon, UK, pp. 197–207.

Wang, Z.Y., He, K.L., and Yan, S., 2005a. Large-scale augmentative biological control of Asian corn borer using *Trichogramma* in China: A success story. In: Proceedings of the Second International Symposium on Biological Control of Arthropods, Davos, Switzerland, 12–16 September 2005, M.S. Hoddle, ed., USDA Forest Service, Forest Health Technology Enterprise Team, Morgantown, WV, USA, pp. 487–494.

Wang, Z.Y., Wang, D.Y., He, K.L., Bai, S.X., and Cong, B., 2005b. Evaluation the control effects of the transgenic *Bacillus thuringiensis* corn expressing Cry1Ab protein on the larvae of *Mythimna separate* (Walker) in laboratory. Acta Phytophylacica Sinica 32: 153–157.

Wang, Z.Y., Wang, D.Y., He, K.L., Bai, S.X., Liu, H., and Cong, B., 2005c. Effects of transgenic corn hybrids expressing *Bacillus thuringiensis* Cry1Ab toxin on survival and growth of the beet armyworm, *Spodoptera exigua* (Hübner). Acta Entomologica Sinica 48: 214–220.

Wang, Z.Y., Wu, Y., He, K.L., and Bai, S.X., 2007. Impact of transgenic *Bacillus thuringiensis* Bt maize pollen on longevity and fecundity of *Trichogramma ostriniae* Chen et Pang (Hymenoptera: Trichogrammatidae) in laboratory condition. Bulletin of Insectology 60: 49–55.

Warnock, D.F., Hutchison, W.D., Tong, C.B.S., and Davis, D.W., 2001. Evaluating maize for allelochemicals restricting European corn borer (Lepidoptera: Crambidae) larval development. Crop Science 41: 1761–1771.

Weber, W.E., Bringezu, T., Broer, I., Eder, J., and Holz, F., 2007. Coexistence between GM and non-GM maize crops - Tested in 2004 at the field scale level (Erprobungsanbau 2004). Journal of Agronomy and Crop Science 193: 79–92.

Widstrom, N.W., and Snook, M.E., 2001. Registration of EPM6 and SIM6 maize germplasm, high silk-maysin sources of resistance to corn earworm. Crop Science 41: 2009–2010.

Wiseman, B.R., Snook, M.E., Isenhour, D.J., Mihm, J.A., and Widstrom, N.W., 1992. Relationship between growth of corn earworm and fall armyworm (Lepidoptera: Noctuidae) and maysin concentration. Journal of Economic Entomology 85: 2473–2477.

Wolt, J.D., Peterson, R.K.D., Bystrak, P., and Meade, T., 2003. A screening level approach for non-target insect risk assessment: Transgenic Bt corn pollen and the monarch butterfly (Lepidoptera: Danaidae). Environmental Entomology 32: 237–246.

Wright, R.J., Witkowski, J.F., Echtenkamp, G., and Georgis, R., 1993. Efficacy and persistence of *Steinernema carpocapsae* (Rhabditida: Steinernematidae) applied through a center-pivot irrigation system against larval corn rootworms (Coleoptera: Chrysomelidae). Journal of Economic Entomology 86: 1348–1354.

Wu, F., 2006a. An analysis of Bt corn's benefits and risks for national and regional policymakers considering Bt corn adoption. International Journal of Technology and Globalisation 2: 115–136.

Wu, F., 2006b. Mycotoxin reduction in Bt corn: Potential economic, health and regulatory impacts. Transgenic Research 15: 277–289.

Wu, F., Miller, J.D., and Casman, E.A., 2004. Bt corn and mycotoxin reduction: Economic impacts in the United States and the developing world. Journal of Toxicology - Toxin Reviews 23: 397–424.

Wu, K., and Guo, Y., 2005. The evolution of cotton pest management practices in China. Annual Review of Entomology 50: 31–52.

Xie, Y.S., Arnason, J.T., Philogene, B.J.R., Lambert, J.D.H., Atkinson, J., and Morand, P., 1990. Role of 2,4-dihydroxy-7-methoxy-1,4-benzoxazin-3-one (DIMBOA) in the resistance of maize to western corn rootworm, *Diabrotica virgifera virgifera* (LeConte) (Coleoptera: Chrysomelidae). Canadian Entomologist 122: 1177–1186.

Xu, Y.L., Wang, Z.Y., He, K.L., and Bai, S.X., 2006. Comparative analysis of the detoxification enzyme and midgut proteinase activities between the *Bacillus thuringiensis* resistant and susceptible Asian corn borer, *Ostrinia furnacalis* (Lepidoptera: Crambidae). Journal of Agricultural Biotechnology 14: 889–893.

Yang, G.H., Li, J.P., and Li, M.H., 1998. Investigation on the effect for controlling the Asian corn borer by high intensity mercury-vapor lamp. Journal of Jilin Agricultural Sciences 23: 70–72.

Yorobe, J.M., and Quicoy, C.B., 2006. Economic impact of Bt corn in the Philippines. The Philippine Agricultural Scientist 89: 258–267.

Zangerl, A.R., McKenna, D., Wraight, C.L., Carroll, M., Ficarello, P., Warner, R., and Berenbaum, M.R., 2001. Effects of exposure to event 176 *Bacillus thuringiensis* corn pollen on monarch and black swallowtail caterpillars under field conditions. Proceedings of the National Academy of Sciences of the USA 98: 11908–11912.

Zhou, D.R., He, K.L., Wang, Z.Y., Ye, Z.H., Wen, L.P., Gao, Y.X., and Song, Y.Y., 1995. Asian Corn Borer and Its Integrated Management. Golden Shield Press, Beijing, China.

Zhou, G., and Overholt, W.A., 2001. Spatial-temporal population dynamics of *Cotesia flavipes* (Hymenoptera: Braconidae) in Kenya. Environmental Entomology 30: 869–876.

Chapter 6
The Present and Future Role of Insect-Resistant Genetically Modified Cotton in IPM

Steven E. Naranjo[1,*], John R. Ruberson[2], Hari C. Sharma[3], Lewis Wilson[4], and Kongming Wu[5]

Abstract Transgenic cottons producing Cry toxins from *Bacillus thuringiensis* (*Bt*) provide for control of lepidopteran pests and were first commercially grown in Australia, Mexico and the USA in 1996. As of 2007, a total of six additional countries (Argentina, Brazil, China, Colombia, India, and South Africa) now grow *Bt* cotton on a total production area of 14 million hectares. The technology primarily provides highly selective and effective control of bollworms, which are the most damaging pests of cotton worldwide. It is estimated that between 1996 and 2005 the deployment of *Bt* cotton has reduced the volume of insecticide active ingredient used for pest control in cotton by 94.5 million kilograms and increased farm income through reduced costs and improved yields by US$7.5 billion, with most of the benefit accrued by farmers in developing nations. Reductions in insecticide use have broadened opportunities for biological control of all cotton pests but most other pest management tactics have remained largely unchanged by the use of *Bt* cotton. However, several non-target pests have become more problematic in *Bt* cotton fields in some countries largely due to reductions in insecticide use for target pests. After 11 years of *Bt* cotton cultivation, control failures due to resistance have not been detected under field conditions. This success can be largely credited to pre-emptive resistance management based on mandated refuges and monitoring programs as well as non-mandated refuge crops and natural refuges which collectively act to dilute any resistant alleles in pest populations. New products are in the pipeline to improve the effectiveness of genetically modified cotton cultivars for resistance to lepidopteran pests, and to address other pest problems in cotton. Debate over food and environmental safety, regulatory oversight, and farming community welfare are likely to continue as the technology moves forward with new crops and new adopting countries.

[1] USDA-ARS, Arid-Land Agricultural Research Center, Maricopa, AZ, USA

[2] Department of Entomology, University of Georgia, Tifton, GA, USA

[3] ICRISAT, Patancheru, Andhra Pradesh, India

[4] CSIRO Plant Industry, Narrabri, New South Wales, Australia

[5] Institute of Plant Protection, Chinese Academy of Agricultural Sciences, Beijing, China

* To whom correspondence should be addressed. E-mail: Steve.Naranjo@ars.usda.gov

J. Romeis, A.M. Shelton, G.G. Kennedy (eds.), *Integration of Insect-Resistant Genetically Modified Crops within IPM Programs.*

6.1 Introduction

Cotton was grown in more than 75 countries with a total production of 26.6 billion kilograms in 2006 (National Cotton Council, http://www.cotton.org) and supplied almost 40% of total worldwide demand for fiber. Four species of cotton (*Gossypium*) are cultivated worldwide, including *G. herbaceum* and *G. arboreum*, which are mainly grown in Asia, *G. barbadense* or long staple cotton cultivated in Egypt, India, the West Indies, and parts of the western USA and South America. *Gossypium hirsutum* (upland cotton), the most common species, is cultivated throughout the world. Cotton is a perennial plant, but through manipulation of irrigation, defoliants and cultivation it is grown as an annual crop. The harvestable portions of the plant are found in the cotton fruit where the primary product, lint fiber, arises from the growth of single cells on the seed surface. Cotton seed is used as animal feed or in the production of oil used in some food products.

Cotton is inhabited by a large diversity of arthropods throughout the world. Hargreaves (1948) cataloged >1,300 herbivorous insects on cotton worldwide. Although very few of these are of economic importance, cotton production has a long history of employing insecticides for crop protection. It is estimated that cotton accounts for about 22.5% of total insecticide use worldwide (Anonymous, 1995). Advances in various pest management technologies and practices in cotton are reducing the use and impact of insecticides (Wilson et al., 2004; Brookes and Barfoot, 2006; Naranjo and Luttrell, 2008); most prominent among these has been the use of biotechnology to enhance protection against some of the most severe pests of this crop.

The adoption and use of genetically-modified (GM) crops continues to grow rapidly worldwide. As of 2007, 23 countries were producing GM crops on a total of 114.3 million hectares (James, 2007). Cottons genetically modified to produce the selective toxin proteins of *Bacillus thuringiensis* (*Bt*; Cry1 and Cry2 proteins) have been grown commercially since 1996. All current commercial lines of *Bt* cotton are cultivars of *G. hirsutum*, however, transgenic *G. barbadense* producing *Bt* toxins was approved for limited field evaluation in Australia in late 2007. The USA, Australia and Mexico were the first countries to permit commercial cultivation of *Bt* cotton, followed by China and South Africa in 1997, Argentina in 1998, Colombia and India in 2002, and Brazil in 2005 (Benedict and Ring, 2004; James, 2007). Widespread cultivation of *Bt* cottons in Burkina Faso, West Africa is expected in the near future. Indonesia commercialized *Bt* cotton in 2001 but ceased cultivation 2 years later when regulatory permits expired. Commercial production is expected to resume there in the future. The adoption rate of *Bt* cotton in India has been unprecedented. Production there grew from about 50,000 ha in 2002 to 6.2 million hectares in 2007, a 12,300% increase in 5 years (James, 2007). In 2007, 131 hybrids were approved for planting in India compared with 4 in 2002, the initial year of cultivation. India now grows more *Bt* cotton than any other country in the world. Pakistan, one of the five largest producers of cotton in the world, has developed several indigenous *Bt* cotton varieties and wide-scale testing is on-going.

Worldwide, about 10.8 and 3.2 million hectares of *Bt* cotton and cotton containing both insect resistance and tolerance to herbicides were grown commercially in 2007, respectively (James, 2007). One commercial *Bt* cotton genotype in China also expresses the cowpea trypsin inhibitor in addition to a fusion Cry1A *Bt* protein, but it is used only on a small scale. Cotton production statistics and *Bt* cotton adoption in these nine countries is summarized in Table 6.1.

Many issues underpin the complex nature of developing robust and sustainable pest management strategies for the cotton system. This chapter will expand upon the opportunities and challenges of integrating *Bt* cotton into current and developing IPM systems throughout cotton production regions of the world.

Table 6.1 Summary production statistics and events for *Bt* cotton adopting countries, 2006

Country	Yield (million kg)	Total ha (1,000s)	% *Bt*	First *Bt* production	Events	Insect toxins
China	7,729	6,000	65	1997	MON531	Cry1Ac
					GK12	Cry1A[a]
					SGK321	Cry1A[a] + CpTI
India	4,746	9,166	41	2002	MON531	Cry1Ac,
					MON15985	Cry1Ac + Cry2Ab2
					GFM	Cry1Ab + Cry1Ac
					Event-1	Cry1Ac
USA	4,700	5,152	66	1996	MON531	Cry1Ac
					MON15985	Cry1Ac + Cry2Ab2
					MON531 × 1445[b]	Cry1Ac
					MON15985 × 1445[b]	Cry1Ac + Cry2Ab2
					MON15985 × 88913[b]	Cry1Ac + Cry2Ab2
					DAS21023 × 24236	Cry1Ac + Cry1F
Brazil	1,524	1,094	11	2005	MON531	Cry1Ac
Australia	294	145[c]	83	1996	MON531	Cry1Ac
					MON15985	Cry 1Ac + Cry2Ab2
					MON15985 × 1445[b]	Cry 1Ac + Cry2Ab2
					MON15985 × 88913[b]	Cry 1Ac + Cry2Ab2
Argentina	174	400	67	1998	MON531	Cry1Ac
Mexico	141	115[d]	19	1996	MON531,	Cry1Ac
					MON531 × 1445[b]	Cry1Ac
Colombia	41	53	42	2002	MON531	Cry1Ac
South Africa	12	16	82	1997	MON531,	Cry1Ac
					MON15985	Cry1Ac + Cry2Ab2

Summarized from National Cotton Council (www.cotton.org), James (2006), and the AgBios Database (www.agbios.com). Countries are ranked in order of their total cotton yield. SGK321 also produced a cowpea trypsin inhibitor (CpTI)

[a] This Cry1A toxins represents a fusion of Cry1Ac and Cry1Ab expressed as a single event

[b] MON1445 and MON88913 represent events conferring tolerance to the herbicide glyphosate

[c] Production in Australia was depressed in 2006 due to continued drought; the adoption rate is consistent with recent years

[d] GM cotton seed supply was disrupted in 2006 due to regulatory issues; adoption in 2005 was >90%

6.2 IPM of Target and Non-Target Pests

6.2.1 General Considerations

Modern pest control is guided by the principles of integrated pest management (IPM) that have been articulated by numerous authors for more than 50 years. Kogan (1998) defined IPM as "a decision support system for the selection and use of pest control tactics, singly or harmoniously coordinated into a management strategy, based on cost/benefit analyses that take into account the interests of and impacts on producers, society, and the environment." The use of GM crops, which have biological activity against select insect pests, qualifies as one of the many tactics that can be integrated into pest management strategies for cotton (Fig. 6.1). Caterpillars, particularly the various species referred to as bollworms, are among the most damaging insect pests of cotton in most parts of the world. Within this context, *Bt* cottons are virulent and selective forms of host plant resistance that represent both opportunities and challenges. Aside from the obvious opportunities for improved caterpillar control, the interactions and synergies from a single tactic such as host plant resistance may provide benefits beyond its narrow range of direct control on a specific group of pest species. Thus, although *Bt* cotton directly controls only lepidopteran pests, the associated reduction in insecticide use for these pests may facilitate or enhance the effectiveness of other tactics such as biological control which in turn may directly contribute to control of other pests in the system (e.g., Naranjo, 2001; Wu and Guo, 2003). Further, the large scale adoption of *Bt*

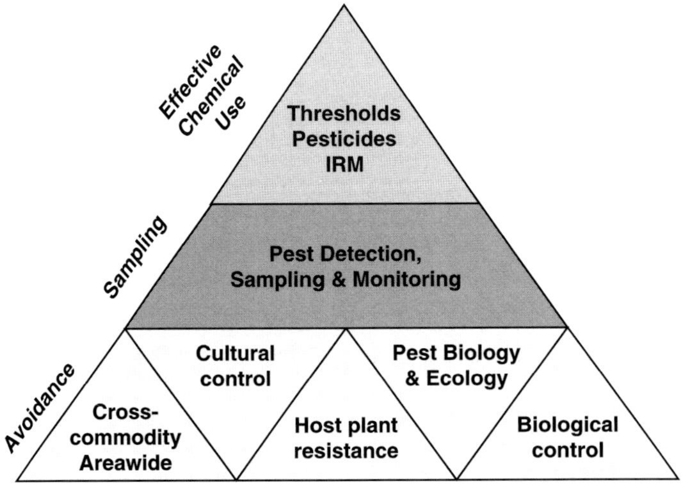

Fig. 6.1 Conceptual diagram of IPM emphasizing the importance of the underlying foundation of pest avoidance through components such as insect-resistant GM varieties and biological control (Modified from Naranjo, 2001; with permission from Elsevier)

cotton in many countries underpins an area-wide suppression of target pests that greatly reduces overall regional populations that can have positive "halo" effects in conventional cotton (e.g., Carrière et al., 2003; Wilson et al., 2004; Storer et al., chapter 10). At the same time, there are significant challenges with using *Bt* cotton, the foremost of which is the potential for the development of resistance to *Bt* toxins in populations of the target pests leading to failure of this control tactic. Other challenges are manifested through the wide diversity of pests affecting cotton worldwide. For example, *Bt* cotton may indirectly cause existing primary pests or secondary pest problems to increase (Wu et al., 2002b; Lei et al., 2003; Williams, 2006). Still more challenges center on the continuing debate on ecological effects and environmental safety of *Bt* cotton and other GM crops (Marvier et al., 2007).

6.2.2 Target Pests

As previously noted, the *Bt* Cry toxins produced in all current commercialized GM cottons have specific activity against various lepidopteran pests (Table 6.2). In countries adopting *Bt* cottons there are roughly 30 species or groups of caterpillar pests of concern. The primary targets of the first generation of single gene transgenic cottons producing Cry1 or Cry2 were the species of the bollworm/budworm complex (*Heliothis* and *Helicoverpa* spp.), the pink bollworm (*Pectinophora gossypiella*), and various spiny and spotted bollworms (*Earias* spp.). These cottons also have good activity against various other pests such as leafworms, leaf perforators, semiloopers and other bollworms. However, single gene cottons have limited activity against pests such as *Spodoptera* spp., *Trichoplusia ni*, *Pseudoplusia includens* and cutworms (Benedict and Ring, 2004). The recently introduced pyramided or dual gene transgenic cottons (e.g., Bollgard II, WideStrike) produce two different Cry toxins and have a broader spectra of activity within the Lepidoptera, including efficacy against many of the pests not previously controlled effectively by single gene constructs, and improved efficacy of the initial primary targets, most notably *H. zea* and *H. armigera* (Fitt and Wilson, 2000; Adamczyk et al., 2001; Chitkowski et al., 2003; Adamczyk and Gore, 2004). Since the 2004/05 season, dual gene *Bt* cottons are now grown exclusively in place of single gene constructs in Australia and use of dual gene *Bt* cottons in the USA grew from about 14% of all *Bt* cottons in 2006 to nearly 34% in 2007 (USDA, 2006, 2007). In Australia, Mexico and the USA, single or double *Bt* gene events are now frequently stacked with genes conferring tolerance to certain herbicides (see Table 6.1).

6.2.3 Non-Target Pests and Changing Pest Problems

A relatively large number of pest species that are not susceptible to the *Bt* toxins expressed in transgenic cottons affect cotton production worldwide (Table 6.3). In

Table 6.2 Lepidoptera pest species affected by *Bt* cotton in adopting countries

Species	Family	Common name	*Bt*[a] efficacy	Argentina	Australia	Brazil	China	Colombia	India	Mexico	South Africa	USA
Alabama argillacea	Noctuidae	Cotton leafworm	4	X		X		X				
Agrotis spp.	Noctuidae	Cutworm	1		X							
Amsacta moorei	Arctiidae	Red hairy caterpillar	3						X			
Anomis flava	Noctuidae	Cotton looper	4		X		X		X			
Bucculatrix gossypii	Tineidae	Cotton leaf perforator	4		X							
Bucculatrix thurberiella	Tineidae	Cotton leaf perforator	4					X				X
Crocidosema plebejana	Tortricidae	Cotton tip worm	4		X							
Diparopis castanea	Noctuidae	Red bollworm	3								X	
Earias biplaza	Noctuidae	Spiny bollworm	3								X	
Earias huegeliana	Noctuidae	Rough bollworm	3		X							
Earias insulana	Noctuidae	Spiny bollworm	3						X		X	
Earias vittella	Noctuidae	Spotted bollworm	3						X			
Earias spp.	Noctuidae	Spiny bollworm	3				X					
Estigmene acrea	Arctiidae	Saltmarsh caterpillar	2					X				X
Helicoverpa gelotopoeon	Noctuidae	Bollworm	3	X								
Helicoverpa armigera	Noctuidae	Old world bollworm	3		X		X		X		X	
Helicoverpa punctigera	Noctuidae	Native budworm	3		X							
Helicoverpa zea	Noctuidae	Cotton bollworm	3					X		X		X
Heliothis virescens	Noctuidae	Tobacco budworm	4	X		X		X		X		X
Ostrinia furnacalis	Crambidae	Corn borer	3				X					
Ostrinia nubilalis	Crambidae	European corn borer	3									X
Pectinophora gossypiella	Gelechiidae	Pink bollworm	4	X	X	X	X	X	X	X	X	X
Pectinophora scutigera	Gelechiidae	Pink spotted bollworm	4		X							
Pseudoplusia includens	Noctuidae	Soybean looper	2					X				X
Sacadotes pyralis	Noctuidae	S.A. bollworm	4					X				
Spodoptera spp.	Noctuidae	Armyworm	2	X	X	X	X	X	X	X	X	X
Sylepta derogata	Pyralidae	Cotton leaf roller	2						X		X	
Trichoplusia ni	Noctuidae	Cabbage looper	2			X		X				X

[a] General efficacy rating for transgenic cottons producing Cry1Ac; 1 = no effect, 4 = completely effective

Table 6.3 Major pest species not susceptible to *Bt* cotton in adopting countries

Order/species	Family	Common name	Argentina	Australia	Brazil	China	Colombia	India	Mexico	South Africa	USA
Hemiptera											
Acrosternum hilare	Pentatomidae	Green stinkbug									X
Adelphocoris spp.	Miridae	Plant bug				X					
Amrasca biguttula	Cicadellidae	Cotton leafhopper						X			
Aphis gossypii	Aphidae	Cotton Aphid	X	X	X	X	X	X			X
Bemisia tabaci	Aleyordidae	Sweetpotato whitefly	X	X	X	X	X	X	X	X	X
Chlorochroa ligata	Pentatomidae	Conchuela stinkbug							X		
Creontiades dilutus	Miridae	Green mirid		X							
Dysdercus spp.	Pyrrhocoridae	Cotton stainer	X		X		X	X		X	
Empoasca spp.	Cicadellidae	Leafhopper				X					
Euschistus spp.	Pentatomidae	Brown stink bug			X						X
Horcias nobilellus	Miridae	Plant bug			X						
Jacobiasca lybica	Cicadellidae	Jassid, leafhopper								X	
Lygus spp.	Miridae	Lygus bug				X			X		X
Neurocolpus nubilus	Miridae	Clouded plant bug									X
Nezara viridula	Pentatomidae	Green stinkbug	X	X	X			X	X		X
Oxycaraenus leatus	Lygaeidae	Dusky cotton bug						X			
Piezodorus guildinii	Pentatomidae	Small green stinkbug			X						
Pseudatomoscelis seriatus	Miridae	Cotton fleahopper							X		X
Scaptocoris castanea	Cydnidae	Burrower bug			X						
Taylorilygus vosseleri	Miridae	Lygus bug								X	
Trialeurodes abutilonea	Aleyordidae	Banded-wing whitefly									X

(continued)

Table 6.3 (continued)

Order/species	Family	Common name	Argentina	Australia	Brazil	China	Colombia	India	Mexico	South Africa	USA
Coleoptera											
Alcidodes affaber	Curculionidae	Cotton stem weevil						X			
Anthonomus grandis	Curculionidae	Boll weevil	X		X		X		X		X
Apion soleatum	Curculionidae	Weevil								X	
Costalimaita ferruginea	Chrysomelidae	Leaf beetle			X						
Cyclocephala sp.	Scarabaeidae	Chafer					X				
Eutinobothrus brasiliensis	Curculionidae	Cotton stem borer			X						
Myllocerus sp.	Curculionidae	Grey weevil						X			
Pempherulus affinis	Curculionidae	Cotton stem weevil						X			
Thysanoptera											
Frankliniella spp.	Thripidae	Thrips		X	X	X					X
Scirtothrips dorsalis	Thripidae	Chili thrips						X			
Thrips spp.	Thripidae	Thrips	X	X	X	X		X	X	X	X
Acari											
Oligonychus spp.	Tetranychidae	Spider mite								X	
Polyphagotarsonemus latus	Tarsonemidae	Broad mite			X						
Tetranychus spp.	Tetranychidae	Spider mites	X	X	X	X	X	X	X	X	X

general, most of these species exhibit the same pest status and continue to be managed identically in *Bt* and conventional cotton systems. However, the use of *Bt* cottons has led to indirect effects on some of these non-target pest species in some productions systems and this is generally thought to be primarily due to reduced insecticide use for caterpillars which previously provided collateral control. In Australia, the reduced use of insecticides for bollworms has allowed some pests to become more prominent. These include the green mirid (*Creontiades dilutus*) (Lei et al., 2003), green vegetable bug (*Nezara viridula*), leaf hoppers (*Austroasca viridigrisea* and *Amrasca terraereginae*), and thrips (*Thrips tabaci*, *Frankliniella schultzei* and *F. occidentalis*) (Wilson et al., 2006). Of these the green mirid is most significant and is now sprayed as many as three times per season (Doyle et al., 2006) with broad spectrum insecticides (Khan et al., 2006). These products are disruptive to a wide range of natural enemies and their use has in turn been linked with increased risk of spider mites (*Tetranychus urticae*), aphids (*Aphis gossypii*), and whitefly (*Bemisia tabaci* B biotype) outbreaks (Wilson et al., 1998; Farrell et al., 2006). This pattern of increased importance of sucking pests, particularly mirid plant bugs, in association with reduced insecticide use for caterpillar pests has played out in other countries. In northern China a complex of mirid plant bugs (*Adelphocoris suturalis*, *A. lineolatus*, *A. fasciaticollis*, *Lygus lucorum*, and *L. pratensis*) have become key insect pests in *Bt* cotton fields in recent years (Wu et al., 2002b). Leafhoppers (*Empoasca biguttula*), cotton aphids (*Aphis gossypii*) and spider mites (*Tetranychus cinnabarinus*) have been observed to occur at higher levels in *Bt* cotton in Henan Province (Deng et al., 2003; Men et al., 2005). Likewise, mirid plant bugs (*Lygus* spp., *Neurocolpus nubilus*) and stinkbugs (e.g., *Nezara viridula*) have increased in pest status since the adoption of *Bt* cottons in the USA, particularly in the mid-southern and southeastern production areas (Williams, 2006). Plant bugs also have become more problematic in South Africa (Gouse et al., 2004). Finally, the reduction in insecticide sprays, especially during both the flowering and boll formation phases, in India has been associated with resurgence of some minor pests such as tobacco caterpillar (*Spodoptera litura*), mealy bugs (*Pseudococcus corymbatus*, *Pulvinaria maxima*, and *Saissetia nigra*), thrips (*Thrips tabaci*) and leafhoppers (*Amrasca biguttula biguttula*) (Sharma et al., 2005). Many of these emergent pests are easily controlled with insecticides and other pest management tactics. In contrast, lepidopteran pests, particularly those feeding within fruiting structures, are among the most difficult to control and *Bt* cotton is a key tactic for their suppression.

The reduction of insecticide use in *Bt* cotton is the likely factor explaining resurgence in some non-target pests, but other factors may be involved as well. Reduced competition from target species may enable non-target pest populations to thrive. Negative effects of *Bt* cotton on natural enemy populations might also lead to enhanced non-target pest problems. However, the bulk of evidence to date suggest that *Bt* crops are highly selective and that negative effects, if any, are relatively minor in magnitude (Naranjo et al., 2005; Romeis et al., 2006; Marvier et al., 2007). Thus, it is more likely that problematic non-target pests are not under good biological control even in conventional systems.

6.2.4 Decision-Making

Sampling and use of economic thresholds for determining the need for control actions are fundamental components of even the most basic IPM programs. The decision to employ *Bt* cotton for caterpillar control is made at planting time before pest populations can be assessed. Thus, the use of *Bt* cotton is most often associated with production areas where caterpillar pests are always a threat. For example, cotton producers in the San Joaquin Valley of California, USA do not use *Bt* cotton because there are no significant lepidopteran pests in this region. Its use also may be associated with risk aversion (insurance), or the simple decision to eliminate any potential need for insecticidal control of caterpillars. However, the grower must weigh the potential benefits against the additional cost of the technology. Excepting greater vigilance and awareness of plant bugs and other pests that have shown a pattern of increasing in *Bt* cotton, identical decision-making protocols generally apply to non-target pests in both *Bt* and conventional production fields. However, mainly because *Bt* cottons producing only Cry1Ac toxins are not completely effective against *Helicoverpa* spp., growers must continue to monitor populations of these pests. In most *Bt* adopting countries there have been slight modifications to sampling protocols and thresholds for the major target pests in *Bt* cotton fields. For example, monitoring eggs to estimate the abundance of *Helicoverpa* spp., which is commonly employed in conventional cotton, is not useful because only the larval stages are susceptible to *Bt* toxins. In Australia, the threshold for conventional cotton is 2 larvae or 1 large larvae (>8 mm)/meter-row, while in *Bt*-cotton the threshold is 2 larvae (>3 mm)/meter-row on two consecutive checks of a *Bt* field or 1 large larvae/m (Farrell et al., 2006). The 'consecutive sample' threshold for larvae >3 mm accounts for the fact that these smaller larvae will usually quickly succumb to the toxin. If larvae were still present on the second sampling date, usually 2–3 days later, this may indicate poor efficacy of the *Bt*-cotton, probably due to poor expression (but also potentially due to resistance) and additional control measures may be required. Similarly, in the USA sampling concentrates on older (2–3 days old) larvae and use of a lower threshold level than conventional cotton. Focus on slightly older larvae helps to identify populations not being effectively controlled by the *Bt* toxins but still of a size amenable to control with available insecticides (Farm Press, 2006). The lower thresholds in both cases reflect the fact that larger larvae, which are the focus of sampling, are capable of greater damage if left untreated. A similar strategy is employed for *H. armigera* in China (Wu and Guo, 2005). Another key target of *Bt* cotton, the pink bollworm, is very effectively controlled by the technology. Nonetheless, scouting for this pest is still recommended, for example, in the southwestern USA. In this instance, the standard methods of sampling adults with pheromone traps and assessment of bolls for larval damage are modified to sampling for later instar larvae (3rd and 4th) within slightly older bolls (Ellsworth, 1997). It is also recommended that growers use monitoring of non-*Bt* refuge fields (see below) as indicators of if and when sampling should be conducted in nearby *Bt* fields.

6.2.5 Biological Control

Worldwide, cotton supports large and diverse arthropod natural enemy communities (Whitcomb and Bell, 1964; Bishop and Blood, 1977; Zhao, 1984; Romeis and Shanower, 1996; Dippenaar-Schoeman et al., 1999) and there is ample evidence to suggest that these natural enemies can have a significant impact on cotton pest population dynamics (e.g., Eveleens et al., 1973; Bottrell and Adkisson, 1977; Abdelrahman and Munir, 1989; Trichilo and Wilson, 1993; Devine et al., 1998; Wilson et al., 1998; Sharma et al., 2007). Reduced insecticide use due to adoption of *Bt*-transgenic cotton, the growing availability and use of selective insecticides, and improvements in other pest management tactics, have created significant opportunities for biological control in the cotton system.

Classical or introductory biological control has been attempted for various exotic pests of cotton but impact on pest populations has been generally minimal (King et al., 1996). With the exception of China, which has active programs in mass production and release of egg and larval parasitoids for control of bollworms (Wu and Guo, 2005), and India where limited releases of *Trichogramma* spp. (Hymenoptera: Trichogrammatidae) and *Chrysoperla carnea* (Neuroptera: Chrysopidae) are practiced (Sithanantham et al., 2005), augmentation has also seen limited success for control of cotton pests in other parts of the world. Conservation biological control, which relies on the preservation, manipulation and/or enhancement of existing natural enemies in the system, is the most widely practiced and successful form of biological pest control in cotton worldwide. For the most part, conservation biological control is achieved with relatively little or no overt action by growers other than consideration of the types of insecticides used, when necessary, to achieve pest control. Use of *Bt* cotton along with other changes in pest management practices have led to generally higher populations of natural enemies in cotton systems in many parts of the world (e.g., Sharma and Ortiz, 2000; Wilson et al., 2004; Naranjo et al., 2004; Wu and Guo, 2005). A reduction in natural enemies that may specialize on pests targeted by *Bt* cotton or other selective control methods is expected (Sisterson and Tabashnik, 2005) and this phenomenon has been exemplified for a specialist parasitoid in the corn system (Pilcher et al., 2005).

The tangible benefits of improved conservation of natural enemy populations in *Bt* cotton have been demonstrated in several systems. In northern China, Wu and Guo (2003) have shown that cotton aphids that are resistant to various insecticides used to control bollworms in cotton are effectively suppressed by natural enemies in *Bt* cotton fields where such sprays are unnecessary. In contrast, insecticides used to control bollworms in non-*Bt* cotton fields disrupt natural enemies leading to outbreaks of aphids. In the western USA and Australia, whiteflies (*Bemisia tabaci*) can be suppressed long-term in cotton fields with a single initial application of selective insecticides in comparison with fields sprayed with broad-spectrum insecticides where repeated applications are necessary. *Bt* cotton underpins this successful whitefly management strategy by reducing or eliminating sprays needed to control caterpillar pests (Naranjo, 2001; Ellsworth and Martinez-Carrillo, 2001). In the mid-southern USA, a cotton aphid threshold has been developed that incorporates the naturally occurring fungus *Neozygites fresenii* (Steinkraus et al., 1996) and aphid predators and parasitoids, allowing growers to actively and effectively integrate natural enemies into their treatment decisions (Chappell et al., 2005). This threshold is feasible because of the insecticide reductions resulting from boll weevil eradication and adoption of *Bt* cotton.

6.2.6 Other Tactics

A wide variety of additional IPM tactics such as cultural control through manipulation of production practices like intercropping, irrigation, fertilization, and cultivar and planting date selection; behavioral control utilizing semiochemicals; and chemical control continue to play important roles in management of pests in both *Bt* and non-*Bt* cotton (see Naranjo and Luttrell, 2008, for recent review). Over many decades, traditional host plant resistance has produced cotton traits with resistance to insects including crop earliness, morphological traits such as nectariless, glabrous or pilose leaf surface, okra-shaped leaf, frego bract, red plant color, and higher concentrations of plant secondary compounds such as gossypol and tannins. However, with the exception of okra leaf shape, for example, which provides moderate resistance to spider mites (Wilson, 1994), few traits have been moved into commercial cultivars and the development time of doing so pales in comparison to the modern transgenic methods that have produced *Bt* cotton (Jenkins and Wilson, 1996). In any case, the widespread adoption of *Bt* cottons has altered some pest management practices. For example, the emergence of plant bugs as more problematic pests in *Bt* cotton in various countries has placed greater emphasis on the use of control options for these pests that are efficacious but more selective. In Australia, effort has focused on reduced rates of insecticides combined with salt (Khan et al., 2006) or petroleum spray oils (Mensah et al., 2004) for control of mirids. These mixtures have shown considerable promise as they are almost as efficacious, more selective and less expensive than the full rate of the insecticide. In the USA, several new putatively selective, foliar insecticides with reduced-risk profiles (e.g., flonicamid, metaflumizone) are being evaluated for control of *Lygus* spp. (Ellsworth and Barkley, 2005; Cook et al., 2007) and continued efforts to develop area-wide IPM programs for *Lygus* spp. based on weed control (Snodgrass et al., 2006) and spatial arrangement of affected crops (Carrière et al., 2006) are underway. Likewise, in China increased focus on weed control and on better management of source crops such as alfalfa are being employed to better manage mirid plant bugs (Wu and Guo, 2005). In India, intercropping cotton with sorghum and pigeonpea is practiced by farmers to help reduce insect damage, and small areas use applications of *H. armigera* NPV and *Bt*- and neem-based insecticides for control of target and non-target pests (Sharma, 2005).

The use of *Bt* cotton may influence other pest control tactics or production issues. For example, in Australia the higher fruit retention of dual gene *Bt* cotton has led to earlier maturity (2–3 weeks) with similar yields to conventional crops. This has allowed growers to plant later into warmer conditions that are more favorable to plant growth and reduce risks from disease (Anderson and Nehl, 2006) and early season pests such as thrips. Another example is the opportunity for more 'site-specific' management. *Helicoverpa* spp. numbers often necessitate a whole farm insecticide application, which decimates natural enemy populations and increases risk of secondary pest outbreaks in Australia. However, the reduced need for control of *Helicoverpa* spp. in *Bt* cotton has allowed other pests to be controlled often on smaller portions of the farm resulting in unsprayed fields serving as refuges for natural enemies that can consequently re-colonize sprayed fields. Finally, *Bt* cotton may be used to support conventional cotton crops. Mensah and Macpherson (2006) showed that use of the moth attractant 'Magnet®' applied to *Bt* cotton with an insecticide (attract and kill) can reduce insecticide costs for *Helicoverpa* spp. control on nearby conventional cotton. Overall, the use of *Bt* cotton has permitted growers to allot more time and effort into other crop production issues such as nutrition and irrigation optimization leading to improved yields in both *Bt* and conventional fields.

6.2.7 Resistance Management

Despite the high adoption rate of *Bt* crops in general, there have not been any reported cases of field failures due to resistance by insects to the Cry proteins since their initial commercial deployment in 1996 (Tabashnik et al., 2003; Mahon et al., 2007; but see Matten et al., chapter 2 for a recent case of putative field resistance to *Bt* maize). Nonetheless there is broad agreement that the sustainability of transgenic *Bt* cottons is one of the most important issues facing cotton insect management and alleles for resistance, and potential resistance mechanisms have been identified in laboratory cultures or field-derived strains of several target pests (e.g., Tabashnik et al., 2003; Gunning et al., 2005; Li et al., 2007; Mahon et al., 2007; Tabashnik et al., 2008). The topic of resistance management is covered in detail by Ferré et al. (chapter 3) for all transgenic crops but some specific information relative to cotton will be highlighted here.

The sustained efficacy of *Bt* toxins against target pests in transgenic cotton can be partly attributed to mandated resistance management programs based around refuge crops and other mitigation approaches in some countries. These programs primarily focus on the key targets (*Helicoverpa*/*Heliothis* spp. and *P. gossypiella*) but potentially delay and mitigate development of resistance in all *Bt*-susceptible pests. The principle behind all refuge strategies is that non-*Bt* cottons or other suitable crop or wild hosts produce susceptible target pests that can readily interbreed with any resistant pests that may arise from *Bt* fields, thereby diluting incipient resistant alleles from the population.

In most countries, growers must abide by legally binding agreements imposed when purchasing the *Bt* cotton seed. In the USA, Mexico, Colombia, and South Africa growers can adopt several options including planting 5% non *Bt* that generally cannot be sprayed for caterpillar pests, or 20% non-*Bt* that can be sprayed for caterpillars (but no *Bt* sprays can be used). These percentages are based on the total amount of all cotton planted. In the USA two additional options for embedding 5% non-*Bt* within a *Bt* cotton field are available where the whole field can be sprayed as needed for any pests (USEPA, 2007). Most requirements also call for the separate 5% and 20% refuge plantings to be within 0.5–1.6 linear km of the *Bt* crop. These distance requirements are based on an understanding of the dispersal behavior of the target pest and attempt to ensure that mating will occur between resistant and susceptible moths (e.g., Tabashnik et al., 1999). The 5% refuge option generally dominates in all these countries. Just recently, the USEPA (US Environmental Protection Agency) approved the use of natural refuges (weeds, alternate crops) hosting bollworm/budworm for growers of dual gene Bollgard II cotton in place of the structured refuges outlined above for single gene *Bt* cottons. This approach is much like that practiced in China (see below) and is based on the rationale that sufficient susceptible moths of the polyphagous *Helicoverpa/Heliothis* spp. can be generated from nearby native vegetation rather than from non-*Bt* cotton planted as a refuge. These natural refuges are allowed from Texas (excluding far-west and northern tip of the panhandle) to the east coast (but excluding southern Florida), but not in excluded areas of Texas and west to California where the monophagous pink bollworm is the main target. In Argentina and Brazil, where only single gene cottons are available, only the 20% non-*Bt* refuge is an option and in Brazil this refuge must be within 0.8 km of the *Bt* cotton. *Bt* maize was provisionally approved in Brazil in mid-2007, but commercial production of the crop awaits review by the Brazilian National Biosafety Council. Further changes in refuge requirements may occur in Brazil if *Bt* maize is eventually approved for commercial cultivation (CTNBio, 2005).

In Australia, the use of cottons containing a single Cry protein (Ingard) was highly restricted with a cap of 30% of the cotton hectarage. With the introduction and now exclusive use of dual gene Bollgard II, a formal plan with a variety of options was initiated (Farrell, 2006). Growers can choose from 100 ha of non-*Bt* that can be sprayed for bollworms (but no *Bt* sprays), 10 ha of non-*Bt* that cannot be sprayed for bollworms, or the use of varying amounts (5–20 ha) of other refuge crops that host bollworms such as pigeon pea, sorghum or maize for every 100 ha of *Bt* cotton planted. Refuge crops must be within 2 km of the *Bt* cotton they are meant to protect. The use of 5 ha of unsprayed pigeon pea for every 100 ha of *Bt* cotton is the dominant option employed by most growers as this minimizes the area lost to cotton production. The Australian program also calls upon four additional elements to lessen the risk of resistance including defined planting windows to reduce the period of exposure of *H. armigera* to Bollgard II by avoiding late planted crops, mandatory cultivation of crop residues to destroy diapausing *H. armigera* pupae which may carry resistance between seasons, defined spray thresholds for *Helicoverpa* spp. (see above) to control potentially resistant survivors in the crops, and control of volunteers of *Bt* cotton in conventional crops and vice versa to reduce the risk of in-field mosaics which increase resistance risk (Fitt and Wilson, 2000).

Resistance management programs are more difficult to implement in China and India because of the challenges associated with educating and monitoring the compliance of the millions of small hectarage cotton farmers in these countries. In

China, wheat is the main host of first-generation *H. armigera* larvae, and cotton, maize, peanut, vegetables and soybean are the major host plants of subsequent generations. The presence of these crops provides a refuge for susceptible bollworm moths over the entire cotton-growing season, and forms the basis of the recommended resistance management strategy. Nonetheless, the size and type of these natural refuges in different provinces is highly variable and provide differing levels of efficacy (Wu et al., 2002a, 2004). In India, permission for environmental release of *Bt* cotton was predicated on farmers providing a "belt" of surrounding non-*Bt* cotton as a refuge. The size of the belt is 20% of the *Bt* crop or five rows of non-Bt cotton on the edge of the *Bt* field, whichever is larger. Adherence to these guidelines is generally poor. Manjunath (2005) and Dhillon and Sharma (2007) argue that, much like the situation in China, there are sufficient alternate bollworm hosts present (chickpeas, pigeon pea, sorghum, tomato) to fulfill the refuge requirement without active participation by growers. Likewise, the large numbers of wild alternate host plants for some of the key target pests in South Africa may contribute to resistance management in that country's small hectarage of *Bt* cotton (Green et al., 2003).

In addition to use of refuge and other tactics for pre-empting resistance, several countries have active programs to monitor for the development of resistance in target pests to *Bt* toxins and/or have developed baseline toxicity to various Cry proteins (Dennehy et al., 2004; Kranthi et al., 2005; Wu et al., 2006; Ali and Luttrell, 2007; Blanco et al., 2007; Mahon et al., 2007; Wu, 2007a). Although resistance to Cry toxins has been selected for in laboratory cultures of all major target species, the frequency of resistant individuals or known resistance alleles have been mostly rare throughout the world and there have been no instances of control failures due to resistance to *Bt* cottons to date. In Australia, the estimated frequency for alleles conferring resistance to Cry1Ac is <0.0003 with a 95% credibility interval (CI) between 0 and 0.0009. In contrast, the *R* frequency for alleles conferring resistance to Cry2Ab2 is over 10x higher at 0.0033 with a 95% CI between 0.0017 and 0.0055 (Mahon et al., 2007). Resistance to Cry2Ab2 appears to be recessive but more work is needed to define its inheritance pattern. As a result, the Australian system is strongly dependent on dual gene *Bt* cottons for resistance management. Likewise, the resistance recently documented in field populations of *H. zea* to Cry1Ac in parts of the USA point to the important role of dual gene *Bt* cottons and other management tactics for effective control and resistance management of this pest (Tabashnik et al., 2008).

Remediation plans to deal with potential resistance episodes have been developed in several countries. For example, in the southwestern USA the industry and scientific community have developed a detailed, multi-pronged action plan for the pink bollworm (Arizona Bt Cotton Working Group, 2002). The first step involves routine monitoring by the scientific community or the grower as noted above. If a ≥3% boll infestation with large larvae is detected the plant is then tested to verify it is producing Cry toxins and the larvae are bioassayed to determine if their susceptibility is below a baseline standard (Simmons et al., 1998). If these burdens are met then alternative pink bollworm control measures (e.g., insecticides) are implemented immediately in the target field and further actions such as crop termination, and early plow-down with shredding and discing of stalks are implemented to reduce the number of insects potentially surviving to the next season. Surrounding fields are then monitored more carefully and if the same criteria are met in these fields then remedial control actions are implemented as well. The perimeter for remedial action expands accordingly. Only non-*Bt* cotton is subsequently allowed in the remediation zone until resistance has declined to an acceptable level defined by the collective experience of the working group.

Similarly, in Australia reports of larvae in commercial *Bt* cotton fields are followed up with collections of larvae, which are tested for resistance to both Cry1Ac and Cry2Ab2 proteins, and tissues from the plant of origin and from adjacent plants are evaluated for levels of *Bt* protein expression. Each year the results from the monitoring program are extensively reviewed by the Transgenic and Insecticides Management Strategies Committee (consisting of researchers and industry representatives) along with information on refuges and *Bt* cotton use rates, so that the effectiveness of the resistance management plan can be assessed. A change in resistance levels would lead to a revision of the resistance management plan which could result in reinstatement of a cap on the proportion of *Bt* cotton, an increase in refuge requirements, or both.

6.2.8 Farm and Landscape Scale Management

Many IPM programs developed for cotton pests focus on what Kogan (1998) refers to as Level I, or approaches aimed at a single pest species or species complexes on a single crop. To some extent, this situation in cotton has been brought about by the need to manage one or a few key pests, often for which selective insecticides or other tactics were few or lacking. This makes development of practical IPM systems challenging. However, more recent IPM strategies are being developed and practiced that focus on higher levels of integration targeting multiple pests within a whole farm context (e.g., Ellsworth et al., 1998; Wilson et al., 2004). In countries like Australia and the USA the advent of more selective insecticides for control of the *Helicoverpa/Heliothis* complex has allowed development of more multi-pest based approaches (see above). The availability of *Bt* cottons has provided a further tool on which to build more inclusive IPM strategies by dramatically and selectively reducing the need to control these primary pests.

However, as emphasized before, the use of selective insecticides or *Bt* cotton represents only a few of many tactics that can be melded into a robust IPM strategy. More resilient IPM systems also entail looking at farm operations and management on a year round basis (Wilson et al., 2004). For example, cultivation can kill diapausing resistant *Helicoverpa* spp. pupae. Cropping rotations can improve soil fertility and soil structure, but also may increase risks of crop diseases and provide overwintering hosts for key pests. Thus, a longer-range view accounts for the effects of farm operations on subsequent pest risks.

It also is important to consider the links between agronomic management and IPM. For instance, use of high fertilizer rates may increase yield (up to a point), but also may delay crop maturity, increasing the crop's exposure to pests and thereby increase costs and risks. Similarly, late irrigation, beyond that required to finish maturation of bolls, also risks excessive late growth, making crops more attractive to pests and more difficult to defoliate.

Another consideration is the landscape in which Bt cotton is planted (Storer et al., chapter 10). For example, it is sensible to use *Bt* cotton in more sensitive areas, such as near neighbors or rivers, where the reduced need for insecticides can lower the risk of off-site movement. Similarly, but at the larger scale, it is important to consider the activities and layout of neighboring farms with regard to issues like spray drift that may disrupt beneficial populations or cultivation of crops that produce large numbers of a key cotton pests (e.g., safflower and mirids in Australia, alfalfa and *Lygus* spp. in the USA). At this scale, formation of IPM or area wide management groups may help growers to communicate and reduce risks from such challenges. At even higher scales, coordinated efforts across a region may be effective in reducing pest populations. For example, in the mixed cropping regions of the Darling Downs region of southern Queensland, Australia, an area-wide strategy was established to manage *H. armigera* populations (Ferguson and Miles, 2002). This approach provided a framework to coordinate efforts to manage the pest across the region by using an understanding of pest ecology to reduce their abundance. Within these frameworks, the deployment of *Bt* cotton plays a critical role by providing broad-scale and selective control over a wide area. Further evidence of the area-wide benefits of *Bt* cotton production is exemplified in the southwest growing areas of the USA where the wide scale use of *Bt* cotton since 1996 has dramatically reduced regional populations of pink bollworm, the main *Bt* cotton target in this area (Carrière et al., 2003). In contrast, several area-wide programs implemented for this pest over the past decades, based on pheromones and insecticides, have never approached such broad and significant population reductions (Henneberry and Naranjo, 1998). In addition, the dramatic reductions in insecticide use for all pests in Arizona cotton (Ellsworth et al., 2007) is in part due to adoption of *Bt* cotton by Arizona growers. *Bt* cotton also is the center-piece technology of an on-going program to eradicate the pink bollworm from its current range in the western USA and northern Mexico (El-Lissy and Grefenstette, 2006).

6.3 Environmental Considerations

Various environmental issues are associated with the use of genetically modified crops including changing patterns of pesticide use (see Fitt, chapter 11 for more detail) and potential effects on animal and plant communities where these crops are grown. Additional detail on these latter topics is provided by Storer et al. (chapter 10) and Romeis et al. (chapter 4).

6.3.1 Non-Target Arthropods

Despite the long history of safety associated with the topical use of *Bt* endotoxins (Glare and O'Callaghan, 2000; Federici, 2003), the season-long expression of these toxins in crop plants through genetic transformation has prompted considerable research to address ecological concerns such as effects on non-target organisms, particularly arthropods. Three countries (Australia, Mexico, and USA) also grow herbicide-tolerant cotton either solely or in combination with *Bt* toxins and these occupy a relatively large share of the GM cotton planted in these countries. There has been relatively little non-target research on herbicide-tolerant cotton and that topic will not be addressed here. However, the vast majority of herbicide-tolerant crops have been modified to be resistant to glyphosate, and there is very little evidence to indicate direct adverse effects of this herbicide on arthropods (Franz et al., 1997).

The effects of pesticide-incorporated crop plants on non-target arthropods have recently been reviewed by O'Callaghan et al. (2005) and Romeis et al. (2006), and the interested reader is referred there for a broader discussion. *Bt* cotton exerts varying effects on non-target organisms, but the results of a recent meta-analysis indicates that overall impacts on arthropod communities appear to be significantly less than those of the insecticide applications that the GM cotton has reduced (Marvier et al., 2007). Numerous studies have examined the potential effects of *Bt*-transgenic crops on natural enemy communities (see Romeis et al., 2006). Not surprisingly, *Bt*-transgenic crops generally tend to adversely affect specialist parasitoid populations by reducing host abundance (a goal of all pest management tactics) and/or reducing individual fitness through indirect host-mediated effects within *Bt* susceptible hosts. Studies utilizing *Bt*-resistant hosts as a means of delivering Cry toxins to parasitoids indicate no direct effect of *Bt* on these natural enemies (Schuler et al., 2004).

Predators are much less affected by *Bt*-transgenic crops. Numerous long-term, large-scale field studies have indicated that there are no meaningful impacts of *Bt* cotton on predator populations (Naranjo et al., 2005) and detailed assessments of predator life histories and function underscore these population-level results, indicating no evidence for direct adverse effects of *Bt* toxin on individual predators (see Romeis et al., 2006). *Bt* toxin can be acquired by predators from *Bt* cotton through prey, but this has not been found to translate into direct negative effects (Torres et al., 2006; Torres and Ruberson, 2006, 2008), and this pattern holds true for other *Bt*-transgenic crops (Harwood et al., 2005; Obrist et al., 2006a, b). Thus, unlike specialist parasitoids, arthropod predators, most of which are generalists, appear to be little hindered by *Bt*-transgenic crops.

Few studies have examined the relative biological control capacities of *Bt* and non-*Bt* cotton fields. However, in no case to date has biological control capacity been reduced in *Bt* cotton fields compared to non-*Bt* fields, and biological control has been improved relative to conventionally-managed fields when these have been included in the comparison (Obrycki et al., 2004; Sisterson et al., 2004; Naranjo, 2005; Head et al., 2005; Romeis et al., chapter 4).

There have been very few studies that have examined impacts of *Bt* cotton production on soil communities. However, Lachnicht et al. (2004) found that decomposition of *Bt* cotton and non-*Bt* cotton residues were similar, and Head et al. (2002) reported that Cry1Ac proteins did not accumulate in biologically active amounts in cotton fields after multiple years of *Bt* cotton production. Further, studies by Shen et al. (2006) indicate that microbial communities in the rhizosphere of soils amended with *Bt* and non-*Bt* cotton residue have the same diversity and functional activity. These results are largely consistent with a larger number of soil residue studies that have been completed for *Bt* maize (Sanvido et al., 2007). Overall, impacts of *Bt* cotton on non-target organisms appear to be limited or negligible, depending on the species and guild, and from a community perspective appear to be significantly less detrimental than most insecticides widely used in cotton pest management. Recent meta-analyses based on the extant literature suggest that these results are consistent among broad taxonomic groups and feeding guilds and should apply to ecologically equivalent taxa worldwide (Marvier et al., 2007; Wolfenbarger et al., 2008).

6.3.2 Gene Flow

An additional environmental concern with genetically modified crops is the flow of genes via pollen to non-transgenic counterparts or other closely related species. Cotton is primarily self-pollinated and the pollen is sticky, relatively heavy, and flowers are only open for a single day (Poehlman, 1987). Thus the opportunity for the spread of transgenes to other plants is thought to be limited and existing research seems to bear this out. Llewellyn and Fitt (1996) found that cross pollination of cotton in Eastern Australia was about 10% or less in adjacent rows 1 m apart. Insect vectored movement in pollen is possible and honey bees have been implicated in moving pollen, however this is only considered a risk if bee numbers are high (Llewellyn et al., 2007). In Northern Australia, movement of pollen was higher than in Eastern Australia, which was attributed to higher bee density. These authors concluded that buffers of 20 m of conventional cotton are adequate to contain cross pollination unless bee numbers are high. The risk of transgenes escaping into wild cottons is extremely low (Brubaker and Brown, 2002; Constable et al., 2007). This is largely because most wild *Gossypium* are diploid, while cultivated cotton is a tretraploid. Cotton volunteers can be found outside of cotton farms, however, Eastick and Hearnden (2006) found that *Bt* cotton does not have the potential to be weedier than conventional cotton. The risk of transgenic cotton becoming a roadside weed, or a weed on dairy farms where raw seed is used as feed also was found to be extremely low (Addison et al., 2007) because the main factors limiting establishment and growth of cotton in these environments were environmental (frost, water, grazing) and the transgenes offered no advantage.

Autoimmunity acts as a potential gene flow barrier in several plant species, including the diploid *G. arboreum*, and the tetraploid, *G. hirsutum* and *G. barbadense* cottons (Bomblies and Weigel, 2007). Thus, no interspecific hybrids have been observed between cultivated cottons and closely related wild relatives in India. Even if a small amount of gene flow occurs, there is little chance that the resulting interspecific hybrids will survive. Similarly, there is little risk of gene flow between *Bt* cotton and wild cotton in China because it is not a native country of wild cottons (Jia, 2001). Recent studies in the USA have suggested that gene flow from *Bt* cottons may occur at low levels resulting in the production of small amounts of fruit in adjacent non-*Bt* cotton plants producing *Bt* toxins (Heuberger, 2006). Patterns appear to be independent of distance from potential sources at least over relatively small scales, but the implications of this for resistance management are currently poorly understood.

6.3.3 Change in Insecticide Use

As highlighted throughout this chapter, one of the most obvious changes in production practices with the introduction of *Bt* cotton has been insecticide use patterns in most adopting countries (also see Fitt, chapter 11 and Qaim et al., chapter 12). On a global scale, Brookes and Barfoot (2006) estimated that *Bt* cotton production during the period 1996–2005 has reduced the total volume of insecticide active ingredient use by 94.5 million kilograms. This represents a 19.4% reduction and was the largest reduction in pesticide use afforded by any GM crop. Moreover, if aspects of the environmental toxicity of the insecticides used are accounted for through the environmental impact quotient (EIQ; Kovach et al., 1992), there was a 24.3% reduction in this quotient over the 10 year period indicating that not only was overall insecticide use reduced but so was their overall environmental impact by an even larger margin. The USA (23%), China (28%) and Australia (22%) realized the largest reductions in EIQ, while countries like Argentina and India had reductions of <4% during the period 1996–2005. In developed nations such as Australia and the USA, changing insecticide use patterns are closely followed by the industry. In the USA there has been an overall reduction in insecticides used for all pests (Fig. 6.2), which has resulted from a combination of factors including the introduction of newer insecticides, the near eradication of the boll weevil (historically one of the most significant pest of cotton in the USA), better adherence to IPM practices, and use of *Bt* cotton. Even with *Bt* cotton, the bollworm/budworm complex remains dominant. Emerging pests like plant bugs and stinkbugs are becoming more of a focus of pest management, and while there has been a significant decline in insecticides for bollworm/budworm there has been a slight but correlated increase in insecticides for these bug pests (Fig. 6.2).

Likewise, in Australia there has been a steady decline in insecticide use since the introduction of *Bt* cotton (Ingard), and insecticide use in Bollgard II cotton is about 75% less than in conventional cotton (Fig. 6.3). This has had obvious benefits in terms of risks of off-farm movement and human health. An additional benefit has been that levels of resistance in *H. armigera*, spider mites and aphids to conventional insecticides in non-*Bt* cotton have also declined (Herron and Wilson, 2006; Rossiter and Kauter, 2006), presumably due to less spraying and the huge sink effect of large areas of Bollgard II cotton, combined with prolonged drought ensuring few alternative hosts.

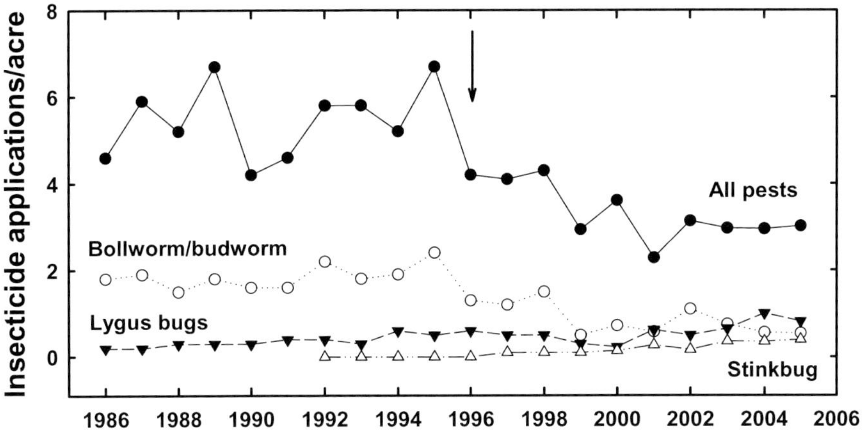

Fig. 6.2 Insecticide use patterns in the USA 1986–2005 relative to all pests, major caterpillar pests and two *Bt* cotton non-target pests that have increased in importance with the production of *Bt* cotton. Arrow denotes the beginning of *Bt* cotton production (Compiled from data of the National Cotton Council, http://www.cotton.org/tech/pest/index.cfm)

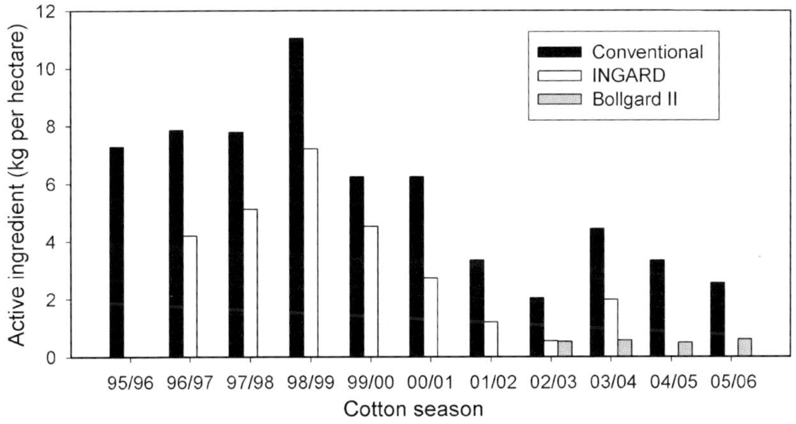

Fig. 6.3 Insecticide use patterns in Australia 1995–2006 for all pests; *Bt* cotton production started in the 1996/97 season (Derived from Pyke, 2008 using data extracted from the Cotton Consultants Australia Market Audit Reports 1998–2006)

Large reductions in insecticide use have also been observed in various developing nations that have adopted *Bt* cotton (Raney, 2006). Based on recent data compiled over two to three growing seasons it is estimated that insecticide costs have been reduced by 47%, 67%, 41%, 77% and 58% in Argentina, China, India, Mexico and South Africa, respectively. These reductions in insecticide usage also have been accompanied by gains in yield, revenue and overall profit in these same countries (see below). In China, resistance in *H. armigera* to commonly used insecticides led to instances in which more than 20 applications were made per season to control this pest in the early 1990s. The adoption of *Bt* cotton shifted this pattern dramatically despite a trend for increasing insecticide use for plant bug control (Pray et al., 2001; Wu, 2007b).

6.4 Adoption and Use

It is often noted that GM crops have been one of the most widespread and rapidly adopted agricultural technologies in history. However, each adopting country has had to struggle with fundamental issues of regulatory authority, environmental and food safety, intellectual property rights, societal demands and other factors governing the production and economics of such crops within their borders. These issues are covered in some detail in other chapters of this book. Here we briefly highlight some of these issues relative to *Bt* cotton.

6.4.1 Regulation

Each country growing GM crops has had to grapple with regulatory issues, with some countries adapting existing governmental organizations to serve these needs and others developing new organizations and authorities (see Matten et al., chapter 2). For example, in the USA various aspects of GM crop regulation are handled by a coordinated framework of three government entities, US Environmental Protection Agency (USEPA), United States Department of Agriculture (USDA) and the Food and Drug Administration (FDA), which oversee pesticide safety, food safety, and movement and transport of seed, respectively (Herdt, 2006). In India, the Ministry of Environment & Forests deals with regulatory issues but calls upon other scientific and policy based entities for safety testing, approval for commercial release, and monitoring. In China, the Ministry of Agriculture oversees the implementation of regulations dealing with safety assessment, production, import, labeling, and processing of GM cotton. In Australia, research and commercial release of transgenic cotton is handled by the Office of the Gene Technology Regulator. However, other agencies also are involved in granting registration, most significantly Food Standards Australia and New Zealand and the Australian Pesticide and Veterinary Medicines Authority. The Brazilian National Biosafety Technical Commission within the Brazilian Federal Science and Technology Department oversees approval of GM crops in that country and the GMO Act in South Africa enabled the establishment of institutions for evaluation and approval. The fact that cotton is primarily a non-food crop has simplified regulatory issues in some countries.

6.4.2 Economics

The economic costs and benefits of *Bt* cotton have been reported by a number of sources over the past decade of adoption (reviewed in Shelton et al., 2002; Brookes and Barfoot, 2006; Smale et al., 2006; Qaim et al., chapter 12) and, much like the general topic of GM crops, the subject is tangled in some controversy (e.g., Benbrook, 2003; Shiva and Jafri, 2003; Pschorn-Strauss, 2005). Nonetheless the bulk of published information shows positive economic benefits associated with the production of *Bt* cotton across a broad cross-section of the industry. A recent synopsis by Brookes and Barfoot (2006), which compiled data over a large number of country-specific, farm-scale studies, indicates that over the 10 year period from 1996–2005, *Bt* cotton production has increased farm income globally by US$7.51 billion. This represents approximately 6.7% of the value of all cotton production worldwide. The largest benefactors, in absolute terms, have not surprisingly been the nations with the greatest adoption and production capacity like China, India and the USA (Table 6.4). Also not surprising given this distribution, in 2005 nearly 80% ($1.38 billion) of the income benefits were garnered by farmers in developing nations. Using a multi-region, equilibrium model approach, Frisvold and Reeves (2007) have suggested a global economic benefit of almost $1.4 billion from world *Bt* cotton production based on 2005 adoption rates despite an estimated 3% decline in world cotton prices due to increased production. Several studies indicate that a significant portion of the overall economic benefits of *Bt* cotton accrue to producers while only a relatively small portion benefits biotechnology firms and seed companies, especially in developing nations (Price et al., 2003; Gouse et al., 2004; Traxler and Godoy-Avila, 2004).

Raney (2006) suggests that GM crops in general, and *Bt* cotton in particular, may benefit the small, resource-poor farmer more than it does large growers through enhancement of yield and reductions in costs and health-related issues associated with insecticide use. However, such associations may not always be clear cut. For example, Shankar and Thirtle (2005) studied small farmers in the Makhathini flats of South Africa and found that the bulk of the positive gains through the adoption of *Bt* cotton has been realized through gains in yield and not insecticide cost saving since relatively few insecticides are used by growers in this region even in the face of heavy insect pest pressure. Such effects are masked in country-wide analyses that pool results from large and small farm operations. Substantial yield gains for *Bt* cotton in other developing countries may follow a similar pattern because pest control in conventional cotton is relatively poor. On the contrary, the relatively small gains in yield observed by adopters of *Bt* cotton in Australia, China and the USA (Table 6.4) arise because growers in these countries engage in aggressive and effective pest control in conventional cotton. In these instances, *Bt* cotton is simply a substitute for insecticides targeting lepidopteran pests. In any case, it is important to consider that not all yield effects may be tied to *Bt* cotton *per se*. The varieties and hybrid producing *Bt* toxins may be better adapted than non-*Bt* cottons grown in the same area (e.g., Qaim et al., 2006), growers

Table 6.4 Estimated economic impacts of *Bt* cotton production 1996–2005

	Farm income benefits (US$ million)	Yield effects (%)	Technology fee (US$/ha)	Cost savings fee excluding (US$/ha)
Argentina	29	24	40–86	18
Australia	150	0	138–250	151–553
China	5,168	8–10	46	261–438
India	463	45–64	62–66[a]	31–66[a]
Mexico	55	3–37	49–65[a]	90–121[a]
South Africa	14	24	53[a]	18[a]
USA	1,627	9–11	58–68	63–74

From Brookes and Barfoot (2006); ranges represent varying estimates over time. Data not currently available for Colombia and Brazil

[a] Values converted from local currency to US dollars based on exchange rates on 31 July 2007

adopting *Bt* cottons may be more efficient farmers overall, and they may grow *Bt* cotton (for which a premium is paid) on their best land and under more optimal conditions (Kambhampati et al., 2006).

The adoption of *Bt* cotton in all countries is closely tied with the perceived or actual benefit of its production balanced against the technology costs imposed universally by providers of the seed and GM traits. These technology fees can vary widely across the world and can even vary within individual countries depending on the level of government involvement and/or other marketing forces (Table 6.4). Australian growers pay the largest fees to access *Bt* technology, but they also reap the largest rewards on a per hectare basis and have among the highest adoption rates in the world. On the contrary, the relatively high technology fee in Argentina has been credited with keeping adoption rates very low in this country (Qaim and DeJanvry, 2005). In China, the government has invested heavily in development of GM crops and this competition helps to keep technology fees low (Raney, 2006). A dualistic system in South Africa which provided credit to resource-poor farmers to acquire the technology initially led to high adoption rates, but this was not sustained when credit became difficult to secure due to changes in local cotton cooperatives (Raney, 2006). In the USA, technology fees vary widely across cotton producing states based on various market forces with fees generally highest in western states (Frisvold et al., 2006).

Additional economic and environmental benefits have been accrued via reductions in fuel costs for growers and reductions in greenhouse gas emissions because of less frequent use of tractors for pesticide application and other cultivation operations (Brookes and Barfoot, 2006).

6.4.3 Social Issues

Bt cotton was the first GM crop commercialized in many parts of the world and as of late 2007 remains the sole insect resistant transgenic crop commercially grown in Australia, Brazil, China, Colombia, India and Mexico (James, 2007), although

Brazil is currently considering the commercial cultivation of *Bt* maize. Its relatively ready acceptance in most countries was related to its primary use as a non-food crop, but its adoption has not been without controversy. There has been strong and vocal opposition to GM crops in general by such non-governmental organizations as Greenpeace, the Union of Concerned Scientists, and more localized opposition exemplified in countries like India which despite the capacity to produce *Bt* cotton for several years did not approve commercial production until 2002 (Jayaraman, 2001, 2002). Critics claim that adoption of *Bt* cotton in India benefits multi-national biotechnology companies while hurting small farmers because of the additional investments needed to grow the new Indian *Bt* hybrids successfully, and the risk of resistance to *Bt* in major pest species. Proponents counter that farms of all size benefit from the technology and note that compliance with refuge requirements to thwart resistance is not critical because adoption is likely to be far from universal. Proponents point to the large success of *Bt* cotton in small-farmer dominated China, while critics offer the example of the short and unsuccessful experience of *Bt* cotton in Indonesia. The overall controversy has been further exacerbated by unscrupulous seed companies and over-eager farmers selling and growing *Bt* cotton before commercial approval had been given in countries such as India, Pakistan and Brazil (Jayaraman, 2001; Ilyas, 2004; Ewing, 2005). The extent of use of so called "stealth seed" prior to official release of *Bt* cotton in India, and its current use in the country is very difficult to gauge, but some estimates put the figure at >50% and it could be higher in certain Indian states (Herring, 2007). Sales of illegal *Bt* cotton seeds have misled and confused farmers and resulted in problems with seed purity, performance, and bio-safety. Bennett et al. (2005) compared the performance of officially released and unofficially cultivated hybrid varieties of *Bt* cotton and conventional hybrids in Gujarat, India by 622 farmers, and observed that the officially released *Bt* varieties outperformed the unofficial varieties. However, unofficially produced *Bt* hybrids also performed better than the non-*Bt* hybrids, although the second generation (F_2) *Bt* seed had no yield advantage over the non-*Bt* hybrids. F_2 seed is regarded as 'GM' by the farmers (and is sold as such), even though its yield performance is little better than the non-*Bt* hybrids. The issue of illegal seed is being aggressively addressed in India and elsewhere, but there seems little doubt that significant debate will continue to embroil the development and adoption of many aspects of GM crops throughout the world.

6.5 Future Developments

Many tactics within the cotton IPM toolbox will continue to be developed, improved, refined, and integrated into strategies that address the interests of producers and society, and positively impact the environment. Host plant resistance, now more and more the product of biotechnological advances, will continue to be a key element in future IPM systems for cotton and many other crops. The promise of transgenes to help solve long-standing pest problems and perhaps other agronomic issues is great and with proper forethought, and scientific, governmental and public guidance and scrutiny should continue to revolutionize IPM and crop production.

6.5.1 New Events

As noted, commercial insect-resistant transgenic cottons are currently limited to various transgenes from *B. thuringiensis* producing δ-endotoxins (see Table 6.1). VipCot cotton (Syngenta Biotech) will express the Vip3A vegetative protein from *B. thuringiensis*, probably in combination with a Cry protein, and is expected to be commercialized in the USA and Australia shortly (Malone et al., chapter 13). As with current Cry endotoxins in cotton, the range of effects of ViP3A is restricted to Lepidoptera and non-target effects appear minimal (Whitehouse et al., 2007). The dual gene cotton WideStrike (Dow AgroScience), which expresses Cry1Ac plus Cry1F and has been available in the USA since 2005, will likely be introduced into Australia in the near future. Bollgard II was commercialized in the USA in 2003, in Australia in 2004 and has recently been approved for use in India. In general, the dual gene constructs have improved efficacy against the primary targets of single gene events and improved efficacy against troublesome pests such as *Spodoptera* spp. among other caterpillars. Technology fees are greater but production will continue to expand so long as the farmers see a significant economic gain. In late 2007, Australia approved limited field evaluation of transgenic *G. barbadense* (long-staple cotton) that produces Cry1Ac and Cry2Ab with or without additional genes for herbicide tolerance. Bollgard III (Hybrid Cry1Ac and Cry1F plus Cry 2Ab) from Monsanto aims to further increase the spectrum of lepidopteran control and provide better late season control through the use of different promoters and proteins (Monsanto, 2007). Cottons based on this technology are in an early product development phases and are several years away from potential commercialization.

The introduction of other insect-resistant GM crops could have important consequences for the production of *Bt* cotton. As noted, *Bt* cotton is currently the only insect-resistant GM crop grown in many adopting nations. The potential commercialization of *Bt* maize is under consideration in countries such as China and Brazil and could have important implications for resistance management. China in particular relies heavily on conventional maize and other crops to act as refuges for susceptible bollworms to ameliorate resistance in *Bt* cotton. The adoption of *Bt* corn could accelerate bollworm resistance in that country without further planning and oversight.

6.5.2 Increasing Pest Spectra of GM Cotton

Lepidopteran pests have been the focus of GM cotton development to date and with good reason given the serious nature of this pest complex worldwide. Nonetheless, a number of other significant pests impact cotton production globally (see Table

6.3) and improvements in host plant resistance through transgenic approaches may help to manage these pests within an overall IPM framework as well. Glare and O'Callaghan (2000) catalog over 170 distinct δ-endotoxins as well as many other toxins from *B. thuringiensis* that could be mined for future transgenic plant development. Lectins and protease inhibitors are being examined in cotton and other crop species for sucking pests and as additional avenues for control of lepidopteran pests (e.g., Carlini and Grossi-de-Sa, 2002; Christou et al., 2006; Malone et al., chapter 13). Monsanto (2007) is in the very early stages of development of transgenic cottons targeting *Lygus* spp. based on *Bt* and non-*Bt* approaches and the publicly-funded Center for Plant Molecular Biology in India has developed a lectin gene for control of sucking pests and is set to license the technology to a private Indian company for use in rice and cotton (Jayaraman, 2004; see also Malone et al., chapter 13).

Efforts are underway to improve many other characteristics of the cotton plant to improve production efficiency and market value. For example, the improvement of several agronomic issues such as tolerance to drought and the herbicide dicamba are in the early stages of development (Monsanto, 2007). Advances are also being made in reducing the concentration of the terpenoid gossypol in cotton seed through RNA interference technology so that the protein in this abundant by-product of cotton lint production may have additional food value for humans (Sunilkumar et al., 2006).

6.6 Summary and Conclusions

The rate and scale of adoption of *Bt* cotton is unprecedented relative to other advances in production of this crop. In 1996, its initial year of commercial production, approximately 1.1 million hectares were grown in three countries. By 2007, the rate of production had increased 12.7-fold to 14 million hectares in nine countries, and several other countries are likely to adopt or re-adopt the technology in the near future. This rapid adoption by a wide cross-section of growers, large and small, has been largely driven by the significant economic benefits of the technology in reducing production costs while improving yield and quality. The technology also has had dramatic positive impacts on the environment globally through the reduction in insecticide usage and even reductions in fuel consumption and associated greenhouse gas emissions in farm operations in a system that has historically been associated with insecticide over-reliance and misuse.

Although the deployment of *Bt* cotton represents only one of a myriad of tactics that can be integrated into efficient and effective pest management strategies, its contribution via reduction in insecticide usage has the potential to cascade through the system and enhance other integrated pest management tactics such as biological control. Numerous non-target studies both in the laboratory and the field have definitively shown the selective nature of this pest control technology and indicate that enhanced biological control should be possible. Indeed, evidence from systems in China and USA has demonstrated the key role of biological control in managing non-target pests in *Bt* cotton. This high degree of selectivity may, however, have some drawbacks. In some systems the use of *Bt* cotton has been associated with increased pressure from other primary or secondary pests not susceptible to *Bt* toxins that may have been previously suppressed by insecticides used for target lepidopteran pest species. In these instances it is likely that effective biological control was absent initially and the selective action of *Bt* has not improved the situation. The adoption of *Bt* cotton has only minor effects on other basic pest management practices such as sampling and use of economic thresholds. Growers may need to be more vigilant of non-target pest dynamics and use modified sampling and threshold protocols for target pests incompletely controlled by current *Bt* cottons. The deployment of *Bt* cotton has the potential to have suppressive area-wide effects on pest populations leading to reduced risk and greater predictability for growers.

To date, resistance to transgenic *Bt* cotton has not been an issue, but constant vigilance and stewardship by all members of the agricultural community will be required to maintain the sustainability of this valuable technology. The use of structured refuges in most countries has been credited with delaying or mitigating resistance but even in countries such as China where resistance management depends on unstructured, natural refuges the technology has been resilient. Sustained pest control also will benefit from future advances in the development of new transgenic events that will improve control efficacy of lepidopteran pests as well as target other key pest species in cotton worldwide. Such advances will rely on continued partnerships between growers, seed and technology industries, and public research institutions. Debate over issues of food and environmental safety, regulatory oversight, and welfare of the farming community as a whole are also likely to continue as the technology moves forward with new crops and new adopting countries.

Acknowledgements We would like to thank Muhammadarshad Awan (Pakistan), Cristina Schetino Bastos (Brazil), Cesar Cardona (Colombia), Jose Martinez-Carrillo (Mexico), Greg Constable (Australia), Eugen Eulitz (South Africa), Graham Head (USA), Joe Huesing (USA), Miguel Serrano (Colombia), Eduardo Trumper (Argentina), and Maurice Vaissayre (France) for their valuable assistance in providing information about *Bt* cotton production in their respective countries and elsewhere.

References

Abdelrahman, A.A., and Munir, B., 1989. Sudanese experience in integrated pest management of cotton. Insect Science and Applications 10: 787–794.

Adamczyk, J.J., Jr., and Gore, J., 2004. Laboratory and field performance of cotton containing Cry1ac, Cry1f, and both Cry1ac and Cry1f (Widestrike™) against beet armyworm and fall armyworm larvae (Lepidoptera: Noctuidae). Florida Entomologist 87: 427–432.

Adamczyk, J.J., Jr., Adams, L.C., and Hardee, D.D., 2001. Field efficacy and seasonal expression profiles for terminal leaves or single and double *Bacillus thuringiensis* toxin cotton genotypes. Journal of Economic Entomology 94: 1589–1593.

Addison, S., Farrell, T., Roberts, G., and Rogers, D., 2007. Roadside surveys support predictions of negligible naturalisation potential for cotton in north-east Australia. Weed Research 47: 192–201.

Ali, M.A., and Luttrell, R.G., 2007. Monitoring Bt susceptibility in *Helicoverpa zea* and *Heliothis virescens*: Results of 2006 studies. In: Proceedings of the Beltwide Cotton Conferences, National Cotton Council, Memphis, TN, USA, pp. 1062–1072.

Anderson, C., and Nehl, D., 2006. Delayed sowing as a best-bet approach to minimise the impacts of *Fusarium* wilt. In: Proceedings of the 13th Australian Cotton Conference, Australian Cotton Growers Research Association, Narrabri, New South Wales, Australia, pp. 443–446.

Anonymous, 1995. Cotton - the crop and its pesticide market. Pesticide News 30: 1.

Arizona Bt Cotton Working Group, 2002. A Remedial Action Plan for Responding to Pink Bollworm Resistance in Bt Cotton in Arizona. Cooperative Extension, University of Arizona, Tucson, AZ, USA.

Benbrook, C.M., 2003. Impact of genetically engineered crops on pesticide use in the United States: The first eight years. BioTech InfoNet, Technical Paper No. 6.

Benedict, J.H., and Ring, D.R., 2004. Transgenic crops expressing Bt proteins: Current status, challenges and outlook. In: Transgenic Crop Protection: Concepts and Strategies, O. Koul and G.S. Dhaliwal, eds., Science Publishers, Enfield, NH, USA, pp. 15–84.

Bennett, R., Ismael, Y., and Morse, S., 2005. Explaining contradictory evidence regarding impacts of genetically modified crops in developing countries. Varietal performance of transgenic cotton in India. Journal of Agricultural Science 143: 35–41.

Bishop, A.L., and Blood, P.R.B., 1977. A record of beneficial arthropods in South East Queensland cotton. Pest Articles and News Summaries 23: 384–386.

Blanco, C.A., Perera, O.P., Boykin, D., Abel, C., Gore, J., Matten, S.R., Ramirez-Sagahon, J.C., and Teran-Vargas, A.P., 2007. Monitoring *Bacillus thuringiensis*-susceptibility in insect pests that occur in large geographies: How to get the best information when two countries are involved. Journal of Invertebrate Pathology 95: 201–207.

Bomblies, K., and Weigel, D., 2007. Hybrid necrosis: Autoimmunity as a potential gene-flow barrier in plant species. Nature Reviews Genetics 8: 382–393.

Bottrell, D.G., and Adkisson, P.L., 1977. Cotton insect pest management. Annual Review of Entomology 22: 451–481.

Brookes, G., and Barfoot, P., 2006. Global impact of biotech crops: Socio-economic and environmental effects in the first ten years of commercial use. AgBioForum 9: 139–151.

Brubaker, C., and Brown, A., 2002. An Evaluation of the Potential for Gene Flow Between Commercial Cotton Cultivars and Wild Australian Cotton Species. CSIRO Plant Industry Report.

Carlini, C.R., and Grossi-de-Sa, M.F., 2002. Plant toxic proteins with insecticidal properties. A review on their potentialities as bioinsecticides. Toxicon 40: 1515–1539.

Carrière, Y., Ellers-Kirk, C., Sisterson, M., Antilla, L., Whitlow, M., Dennehy, T.J., and Tabashnik, B.E., 2003. Long-term regional suppression of pink bollworm by *Bacillus thuringiensis* cotton. Proceedings of the National Academy of Sciences of the USA 100: 1519–1523.

Carrière, Y., Ellsworth, P.C., Dutilleul, P., Ellers-Kirk, C., Barkley, V., and Antilla, L., 2006. A GIS-based approach for areawide pest management: The scales of *Lygus hesperus* movements to cotton from alfalfa, weeds, and cotton. Entomologia Experimentalis et Applicata 118: 203–210.

Chappell, A.S., Kring, T.J., Lorenz, G.M., Greene, J.K., and Studebaker, G.E., 2005. Reliance on predators in making cotton aphid treatment decisions. In: Proceedings of the Beltwide Cotton Conferences, National Cotton Council, Memphis, TN, pp. 1610–1612.

Chitkowski, R.L., Turnipseed, S.G., Sullivan, M.J., and Bridges, W.C., Jr., 2003. Field and laboratory evaluations of transgenic cottons expressing two proteins compared with one of *Bacillus thuringiensis* var. *kurstaki* Berliner for management of noctuid (Lepidoptera) pests. Journal of Economic Entomology 96: 755–762.

Christou, P., Capell, T., Kohli, A., Gatehouse, J.A., and Gatehouse, A.M.R., 2006. Recent developments and future prospects in insect pest control in transgenic crops. Trends in Plant Science 11: 302–308.

Constable, G., Preston, C., and Gupta, V., 2007. Genomic age series Part 3: GM cotton - benefits, risks and opportunities. Agricultural Science 20: 28–32.

Cook, D.R., Leonard, B.R., Burris, E., and Burns, D.R., 2007. Tarnished plant bug management in Northeast Louisiana. In: Proceedings of the Beltwide Cotton Conferences, National Cotton Council, Memphis, TN, USA, pp. 1704–1710.

CTNBio, 2005. Commercial use of resistant insect cotton. Proceedings of the Brazilian Federal Biosafety Technical Commission (CTNBio). http://www.ctnbio.gov.br/index.php/content/view/3663.html (accessed 12 March 2008).

Deng, S.D., Xu, J., Zhang, Q.W., Zhou, S.W., and Xu, G.J., 2003. Effect of transgenic *Bacillus thuringiensis* cotton on population dynamics of the non-target pests and natural enemies. Acta Entomologica Sinica 46: 1–5.

Dennehy, T.D., Unnithan, G.C., Brink, S.A., Wood, B.D., Carrière, Y., Tabashnik, B.E., Antilla, L., and Whitlow, M., 2004. Update on pink bollworm resistance in Bt cotton in the southwest. In: Cotton, A College of Agriculture Report, Series P-138, University of Arizona, Tucson, AZ, USA, pp. 213–223.

Devine, G.J., Ishaaya, I., Horowitz, A.R., and Denholm, I., 1998. Effects of piperonyl butoxide on *Bemisia tabaci* Genn. (Homoptera: Aleyrodidae): Mortality, development, parasitism and predation in Israeli cotton fields. Crop Protection 17: 717–726.

Dhillon, M.K., and Sharma, H.C., 2007. Survival and development of *Campoletis chlorideae* on various insect and crops hosts. Implications for Bt-transgenic crops. Journal of Applied Entomology 131: 179–185.

Dippenaar-Schoeman, A.S., van den Berg, A.M., and van den Berg, A., 1999. Spiders in South African cotton fields: Species diversity and abundance (Arachnida: Araneae). African Plant Protection 5: 93–103.

Doyle, B., Reeve, I., and Coleman, M., 2006. The CCA 2005 Bollgard Comparison Report: A Survey of Cotton Growers' and Consultants' Experience with Bollgard in the 2004–2005 Season. The Cotton Research and Development Corporation and The Cotton Catchment Community Cooperative Research Centre. Institute for Rural Futures, Armidale, New South Wales, Australia.

Eastick, R., and Hearnden, M., 2006. Potential for weediness of Bt-cotton in Northern Australia. Weed Science 54: 1142–1151.

El-Lissy, O.A., and Grefenstette, W.J., 2006. Progress of pink bollworm eradication in the U.S and Mexico, 2005. In: Proceedings of the Beltwide Cotton Conferences, National Cotton Council, Memphis, TN, USA, pp. 1313–1319.

Ellsworth, P.C., 1997. Bt Cotton in Arizona. What Will Change? University of Arizona, Cooperative Extension No. 1 (revised), Tucson, AZ, USA.

Ellsworth, P.C., and Barkley, V., 2005. Transitioning *Lygus* chemical controls to more selective options for Arizona Cotton. In: Cotton, A College of Agriculture Report, Series P-142, University of Arizona, Tucson, AZ, USA, pp. 165–178.

Ellsworth, P.C., and Martinez-Carrillo, J.L., 2001. IPM for *Bemisia tabaci*: A case study from North America. Crop Protection 20: 853–869.

Ellsworth, P.C., Naranjo, S.E., Castle, S.J., Hagler, J., and Henneberry, T.J., 1998. Whitefly management in Arizona: Looking at whole systems. In: Cotton, A College of Agriculture Report, Series P-112, University of Arizona, Tucson, AZ, USA, pp. 311–318.

Ellsworth, P.C., Fournier, A., and Smith, T.D., 2007. Arizona cotton insect losses. Publ. No. AZ1183. University of Arizona, College of Agriculture and Life Sciences, Cooperative Extension, Tucson, AZ, USA. Based on P.C. Ellsworth, J.S. Jones and T.D. Smith. 2000. http://cals.arizona.edu/crops/cotton/insects/cil/cil.html (accessed 12 March 2008).

Eveleens, K.G., van den Bosch, R., and Ehler, L.E., 1973. Secondary outbreak induction of beet armyworm by experimental insecticide application in cotton in California. Environmental Entomology 2: 497–503.

Ewing, R., 2005. End of Brazil GMO ban to curb rampant black market. Grain, Barcelona, Spain. http://www.grain.org/research/contamination.cfm?id = 286 (accessed 15 January 2008).

Farm Press, 2006. Delta Agricultural Digest 2006. Farm Press, Clarksdale, MS, USA.

Farrell, T., ed., 2006. Cotton Pest Management Guide 2006–07. New South Wales Department of Primary Industries, Australia.

Farrell, T., Mensah, R., Sequeira, R., Wilson, L., and Dillon, M., 2006. Key insect and mite pests of Australian Cotton. In: Cotton Pest Management Guide 2006–07, T. Farrell, ed., New South Wales Department of Primary Industries, Australia, pp. 1–17.

Federici, B.A., 2003. Effects of Bt on non-target organisms. Journal of New Seeds 5: 11–30.

Ferguson, J., and Miles, M., 2002. Area wide management on the Darling Downs - has it worked? In: Proceedings of the Australian Cotton Conference, Australian Cotton Growers Research Association, Narrabri, New South Wales, Australia, pp. 711–719.

Fitt, G.P., and Wilson, L.J., 2000. Genetic engineering in IPM: Bt cotton. In: Emerging Technologies for Integrated Pest Management: Concepts, Research and Implementation, G.G. Kennedy and T.B. Sutton, eds., APS Press, St. Paul, MN, USA, pp. 108–125.

Franz, J.E., Mao, M.K., and Sikorski, J.A., 1997. Glyphosate: A unique global herbicide. American Chemical Society Monograph 189, Washington, DC, USA, 653 pp.

Frisvold, G., and Reeves, J.M., 2007. Economy-wide impacts of Bt cotton. In: Proceedings of the Beltwide Cotton Conferences, National Cotton Council, Memphis, TN, USA, pp. 1–7.

Frisvold, G., Reeves, J.M., and Tronstad, R., 2006. Bt cotton adoption in the United State and China: International trade and welfare effects. AgBioForum 9: 69–78.

Glare, T.R., and O'Callaghan, M., 2000. *Bacillus Thuringiensis*: Biology, Ecology and Safety. Wiley, NY, USA.

Gouse, M., Pray, C., and Schimmelpfennig, D., 2004. The distribution of benefits from Bt cotton adoption in South Africa. AgBioForum 7: 187–194.

Green, W.M., deBillott, M.C., Joffe, T., vanStaden, L., Bennet-Nel, A., duToit, C.L.N., and vanderWesthuizen, L., 2003. Indigenous plants and weeds on the Maktathini Flats as refuge hosts to maintain bollworm population susceptibility to transgenic cotton (Bollgard™). African Entomology 11: 21–29.

Gunning, R.V., Dang, H.T., Kemp, F.C., Nicholson, I.C., and Moores, G.D., 2005. New resistance mechanism in *Helicoverpa armigera* threatens transgenic crops expressing *Bacillus thuringiensis* Cry1Ac toxin. Applied and Environmental Microbiology 71: 2558–2563.

Hargreaves, H., 1948. List of the Recorded Cotton Insects of the World. Commonwealth Institute of Entomology, London, UK.

Harwood, J.D., Wallin, W.G., and Obrycki, J.J., 2005. Uptake of Bt endotoxins by non-target herbivores and higher order arthropod predators: Molecular evidence from a transgenic corn agroecosystem. Molecular Ecology 14: 2815–2823.

Head, G., Surber, J.B., Watson, J.A., Martin, J.W., and Duan, J.J., 2002. No detection of Cry1Ac protein in soil after multiple years of transgenic Bt cotton (Bollgard) use. Environmental Entomology 31: 30–36.

Head, G., Moar, M., Eubanks, M., Freeman, B., Ruberson, J., Hagerty, A., and Turnipseed, S., 2005. A multiyear, large-scale comparison of arthropod populations on commercially managed Bt and non-Bt cotton fields. Environmental Entomology 34: 1257–1266.

Henneberry, T.J., and Naranjo, S.E., 1998. Integrated management approaches for pink bollworm in the southwestern United States. Integrated Pest Management Reviews 3: 31–52.

Herdt, R.W., 2006. Biotechnology in agriculture. Annual Review of Environment and Resources 31: 265–295.

Herring, R.J., 2007. Stealth seeds: Bioproperty, biosafety, biopolitics. Journal of Developmental Studies 43: 130–157.

Herron, G., and Wilson, L.J., 2006. Insecticide resistance in cotton aphid and two-spotted spider mite: Seasons 2004–2005 and 2005–2006. In: Proceedings of the 13th Australian Cotton Conference, Australian Cotton Growers Research Association, Narrabri, New South Wales, Australia, pp. 491–495.

Heuberger, S.M., 2006. Contamination of refuges by transgenic Bt cotton: Implications for pink bollworm resistance. Master's thesis, University of Arizona, Tucson, AZ, USA, 84 pp.

Ilyas, M., 2004. Researchers employ genetic engineering: Cotton varieties. Grain, Barcelona, Spain. http://www.grain.org/research/btcotton.cfm?id = 155 (accessed 12 March 2008).

James, C., 2006. Global Status of Commercialized Biotech/GM Crops: 2006. ISAAA Brief No. 35, International Service for the Acquisition of Agri-Biotech Applications, Ithaca, NY, USA.
James, C., 2007. Global Status of Commercialized Biotech/GM Crops: 2007. ISAAA Brief No. 37, International Service for the Acquisition of Agri-Biotech Applications, Ithaca, NY, USA.
Jayaraman, K.S., 2001. Illegal Bt-cotton in India haunts regulators. Nature Biotechnology 19: 1090.
Jayaraman, K.S., 2002. India approves GM cotton. Nature Biotechnology 20: 415.
Jayaraman, K.S., 2004. India produces homegrown GM cotton. Nature Biotechnology 22: 255–256.
Jenkins, J.N., and Wilson, F.D., 1996. Host plant resistance. In: Cotton Insects and Mites: Characterization and Management, E.G., King, J.R. Phillips and R.J. Coleman, eds., The Cotton Foundation Publisher, Memphis, TN, USA, pp. 563–600.
Jia, S., 2001. Transgenic Cotton. Science Press, Beijing, China.
Kambhampati, U., Morse, S., Bennett, R., and Ismael, Y., 2006. Farm-level performance of genetically modified cotton: A frontier analysis of cotton production in Maharashtra. Outlook on Agriculture 35: 291–297.
Khan, M., Quade, A., and Murray, D., 2006. Mirid management - effect of salt rate when mixed with reduced rates of chemical. In: Proceedings of the 13th Australian Cotton Conference, Australian Cotton Growers Research Association, Narrabri, New South Wales, Australia, pp. 537–542.
King, E.G., Coleman, R.J., Morales-Ramos, J.A., Summy, K.R., Bell, M.R., and Snodgras, G.L., 1996. Biological control. In: Cotton Insects and Mites: Characterization and Management, E.G., King, J.R. Phillips and R.J. Coleman, eds., The Cotton Foundation Publisher, Memphis, TN, pp. 511–538.
Kogan, M., 1998. Integrated pest management: Historical perspectives and contemporary developments. Annual Review of Entomology 43: 243–270.
Kovach, J., Petzoldt, C., Degni, J., and Tette, J., 1992. A method to measure the environmental impact of pesticides. New York Food and Life Sciences Bulletin, New York State Agricultural Experiment Station, Cornell University, Geneva, NY, USA. http://www.nysipm.cornell.edu/publications/eiq/default.asp (accessed 12 March 2008).
Kranthi, M., Dhawd, C.S., Naidu, K., Mate, K., Patil, E., and Kranthi, S., 2005. Bt cotton seed as a source of *Bacillus thuringiensis* insecticidal toxin for bioassays to detect and monitor bollworm resistance to Bt cotton. Current Science 88: 796–800.
Lachnicht, S.L., Hendrix, P.F., Potter, R.L., Coleman, D.C., and Crossley, D.A., 2004. Winter decomposition of transgenic cotton residue in conventional-till and no-till systems. Applied Soil Ecology 27: 135–142.
Lei, T., Khan, M., and Wilson, L., 2003. Boll damage by sucking pests: An emerging threat, but what do we know about it? In: World Cotton Research Conference III: Cotton for the New Millennium, A. Swanepoel, ed., Agricultural Research Council - Institute for Industrial Crops, Cape Town, South Africa, pp. 1337–1344.
Li, G.P., Wu, K.M., Gould, F., Wang, J.K., Miao, J., Gao, X.W., and Gao, Y.Y., 2007. Increasing tolerance to Cry1Ac cotton from cotton bollworm, *Helicoverpa armigera*, was confirmed in Bt cotton farming area of China. Ecological Entomology 32: 366–375.
Llewellyn, D., and Fitt, G., 1996. Pollen dispersal from two field trials of transgenic cotton in the Namoi Valley, Australia. Molecular Breeding 2: 157–166.
Llewellyn, D., Tyson, C., Constable, G., Duggan, B., Beale, S., and Steel, P., 2007. Containment of regulated genetically modified cotton in the field. Agriculture Ecosystems and Environment 121: 419–429.
Mahon, R., Olsen, K., Downes, S., and Addison, S., 2007. Frequency of alleles conferring resistance to the Bt toxins Cry1Ac and Cry2Ab in Australian populations of *Helicoverpa armigera* (Hübner) (Lepidoptera: Noctuidae). Journal of Economic Entomology 100: 894–902.
Manjunath, T.M., 2005. Safety of Bt cotton: Facts allay fear. http://www.agbioworld.org/biotech-info/articles/biotech-art/safety-bt-cotton.html (accessed 12 March 2008).

Marvier, M., McCreedy, C., Regetz, J., and Kareiva, P., 2007. A meta-analysis of effects of Bt cotton and maize on non-target invertebrates. Science 316: 1475–1477.
Men, X., Ge, F., Edwards, C.A., and Yardim, E.N., 2005. The influence of pesticide applications on *Helicoverpa armigera* Hubner and sucking pests in transgenic Bt cotton and non-transgenic cotton in China. Crop Protection 24: 319–324.
Mensah, R., and Macpherson, I., 2006. Attracting and killing moths on Bollgard?II ® cotton crops: A new strategy for managing *Helicoverpa* spp. on conventional cotton crops. In: Proceedings of the 13th Australian Cotton Conference, Australian Cotton Growers Research Association, Narrabri, New South Wales, Australia, pp. 555–565.
Mensah, R., Coates, R., and Hoque, Z., 2004. Petroleum spray oils - lubricating the path to IPM. Part 4. Use of synthetic insecticides and petroleum spray oil combinations for improved efficacy against *Helicoverpa* spp. and green mirids on cotton crops. In: Proceedings of the 12th Australian Cotton Conference, Australian Cotton Growers Research Association, Narrabri, New South Wales, Australia, pp. 699–710.
Monsanto, 2007. Putting technology to work in the field 2007. R&D pipeline at a glance. http://www.monsanto.com/pdf/products/productPipeline.pdf (accessed 12 March 2008).
Naranjo, S.E., 2001. Conservation and evaluation of natural enemies in IPM systems for *Bemisia tabaci*. Crop Protection 20: 835–852.
Naranjo, S.E., 2005. Long-term assessment of the effects of transgenic Bt cotton on the function of the natural enemy community. Environmental Entomology 34: 1211–1223.
Naranjo, S.E., and Luttrell, R.G., 2008. Cotton arthropod IPM. In: Integrated Pest Management, E. B. Radcliff and W.D. Hutchison, eds., Cambridge University Press, Cambridge, UK (in press).
Naranjo, S.E., Ellsworth, P.C., and Hagler, J.R., 2004. Conservation of natural enemies in cotton: Role of insect growth regulators in management of *Bemisia tabaci*. Biological Control 30: 52–72.
Naranjo, S.E., Head, G., and Dively, G.P., 2005. Field studies assessing arthropod non-target effects in Bt transgenic crops: Introduction. Environmental Entomology 34: 1178–1180.
Obrist, L.B., Dutton, A., Albajes, R., and Bigler, F., 2006a. Exposure of arthropod predators to Cry1Ab toxin in Bt maize fields. Ecological Entomology 31: 143–154.
Obrist, L.B., Dutton, A., Romeis, J., and Bigler, F., 2006b. Biological activity of Cry1Ab expressed by Bt maize following ingestion by herbivorous arthropods and exposure of the predator *Chrysoperla carnea*. BioControl 51: 31–48.
Obrycki, J.J., Ruberson, J.R., and Losey, J.E., 2004. Interactions between natural enemies and transgenic insecticidal crops. In: Genetics, Evolution and Biological Control, L.E. Ehler, R. Sforza and T. Mateille, eds., CABI, Wallingford/Oxon, UK, pp. 183–206.
O'Callaghan, M., Glare, T.R., Burgess, E.P.J., and Malone, L.A., 2005. Effects of plants genetically modified for insect resistance on non-target organisms. Annual Review of Entomology 50: 271–292.
Pilcher, C.D., Rice, M.E., and Obrycki, J.J., 2005. Impact of transgenic *Bacillus thuringiensis* corn and crop phenology on five non-target arthropods. Environmental Entomology 34: 1302–1316.
Poehlman, J.M., 1987. Breeding cotton. In: Breeding Field Crops, 3rd edition, J.M. Poehlman, ed., Van Nostrand Reinhold Publisher, New York, USA, pp. 556–591.
Pray, C., Huang, J., Ma, D., and Qiao, F., 2001. Impact of Bt cotton in China. World Development 29: 813–825.
Price, G.K., Lin, W., Falck-Zepeda, J.B., and Fernandez-Cornejo, J., 2003. The size and distribution of market benefits from adopting agricultural biotechnology. United States Department of Agriculture, Economic Research Service, Technical Bulletin #1906.
Pschorn-Strauss, E., 2005. Bt cotton in South Africa: The case of the Makhathini farmers. Grain. http://www.grain.org/seedling/?id = 330 (accessed 12 March 2007).
Pyke, B., 2008. The impact of high adoption of BollgardII cotton on pest management in Australia. In: Proceedings of the World Cotton Research Conference, Lubbock, TX, USA, 9–15 September 2007 (in press).
Qaim, M., and DeJanvry, A., 2005. Bt cotton and pesticide use in Argentina: Economic and environmental effects. Environmental and Developmental Economics 10: 179–200.

Qaim, M., Subramanian, A., Naik, G., and Zilberman, D., 2006. Adoption of Bt cotton and impact variability: Insights from India. Review of Agricultural Economics 28: 48–58.

Raney, T., 2006. Economic impact of transgenic crops in developing countries. Current Opinion in Biotechnology 17: 174–178.

Romeis, J., and Shanower, T.G., 1996. Arthropod natural enemies of *Helicoverpa armigera* (Hübner) (Lepidoptera: Noctuidae) in India. Biocontrol Science and Technology 6: 481–508.

Romeis, J., Meissle, M., and Bigler, F., 2006. Transgenic crops expressing *Bacillus thuringiensis* toxins and biological control. Nature Biotechnology 24: 63–71.

Rossiter, L., and Kauter, G., 2006. *Helicoverpa armigera* insecticide resistance management - key considerations for 2006–2007. In: Proceedings of the 13th Australian Cotton Conference, Australian Cotton Growers Research Association, Narrabri, New South Wales, Australia, pp. 521–524.

Sanvido, O., Romeis, J., and Bigler, F., 2007. Ecological impacts of genetically modified crops: Ten years of field research and commercial cultivation. Advances in Biochemical Engineering and Biotechnology 107: 235–278.

Schuler, T.H., Denholm, I., Clark, S.J., Stewart, C.N., and Poppy, G.M., 2004. Effects of Bt plants on the development and survival of the parasitoid *Cotesia plutellae* (Hymenoptera: Braconidae) in susceptible and Bt-resistant larvae of the diamondback moth, *Plutella xylostella* (Lepidoptera: Plutellidae). Journal of Insect Physiology 50: 435–443.

Shankar, B., and Thirtle, C., 2005. Pesticide productivity and transgenic cotton technology: The South African smallholder case. Journal of Agricultural Economics 56: 97–115.

Sharma, H.C., ed., 2005. *Heliothis/Helicoverpa* Management: Emerging Trends and Strategies for Future Research. Oxford/IBH, New Delhi, India.

Sharma, H.C., and Ortiz, R., 2000. Transgenics, pest management and the environment. Current Science 79: 421–437.

Sharma, O.P., Bambawale, O.M., Dhandapani, A., Tanwar, R.K., Bhosle, B.B., Lavekar, R.C., and Rathod, K.S., 2005. Assessment of severity of important diseases of rainfed *Bt* transgenic cotton in Southern Maharashtra. Indian Phytopathology 58: 483–485.

Sharma, H.C., Arora, R., and Pampapathy, G., 2007. Influence of transgenic cottons with *Bacillus thuringiensis cry1Ac* gene on the natural enemies of *Helicoverpa armigera*. BioControl 52: 469–489.

Shelton, A.M., Zhao, J.Z., and Roush, R.T., 2002. Economic, ecological, food safety, and social consequences of the deployment of Bt transgenic plants. Annual Review of Entomology 47: 845–881.

Shen, R.F., Cai, H., and Gong, W.H., 2006. Transgenic Bt cotton has no apparent effect on enzymatic activities or functional diversity of microbial communities in rhizosphere soil. Plant and Soil 285: 149–159.

Shiva, V., and Jafri, A.J., 2003. Failure of the GMO's in India. http://www.mindfully.org/GE/2003/India-GMO-Failure-Shiva31may03.htm (accessed 12 March 2008).

Simmons, A.L., Dennehy, T.J., Tabashnik, B.E., Antilla, L., Bartlett, A., Gouge, D., and Staten, R., 1998. Evaluation of Bt cotton deployment strategies and efficacy against pink bollworm in Arizona. In: Proceedings of the Beltwide Cotton Conferences, National Cotton Council, Memphis, TN, USA, pp. 1025–1030.

Sisterson, M.S., and Tabashnik, B.E., 2005. Simulated effects of transgenic Bt crops on specialist parasitoids of target pests. Environmental Entomology 34: 733–742.

Sisterson, M.S., Biggs, R.W., Olson, C., Carrière, Y., Dennehy, T.J., and Tabashnik, B.E., 2004. Arthropod abundance and diversity in Bt and non-Bt cotton fields. Environmental Entomology 33: 921–929.

Sithanantham, S., Singh, S.P., and Romeis, J., 2005. Biological control of *Helicoverpa*: Research status, constraints and opportunities. In: *Heliothis/Helicoverpa* Management: Emerging Trends and Strategies for Future Research, H.C. Sharma, ed., Oxford/IBH, New Delhi, India, pp. 329–369.

Smale, M., Zambrano, P., and Cartel, M., 2006. Bales and balances: A review of the methods used to assess the economic impact of Bt cotton on farmers in developing economies. AgBioForum 9: 195–212.

Snodgrass, G.L., Scott, W.P., Abel, C.A., Robbins, J.T., Gore, J., and Hardee, D.D., 2006. Suppression of tarnished plant bugs (Heteroptera: Miridae) in cotton by control of early season wild host plants with herbicides. Environmental Entomology 35: 1417–1422.

Steinkraus, D.C., Hollingsworth, R.G., and Boys, G.O., 1996. Aerial spores of *Neozygites fresenii* (Entomophthorales: Neozygitaceae): Density, periodicity, and potential role in cotton aphid (Homoptera: Aphididae) epizootics. Environmental Entomology 25: 48–57.

Sunilkumar, G., Campbell, L.M., Puckhaber, L., Stipanovic, R.D., and Rathore, K.S., 2006. Engineering cottonseed for use in human nutrition by tissue-specific reduction of toxic gossypol. Proceedings of the National Academy of Sciences of the USA 103: 18054–18059.

Tabashnik, B.E., Patin, A.L., Dennehy, T.J., Liu, Y.B., Miller, E., and Staten, R.T., 1999. Dispersal of pink bollworm (Lepidoptera: Gelechiidae) males in transgenic cotton that produces a *Bacillus thuringiensis* toxin. Journal of Economic Entomology 92: 772–780.

Tabashnik, B.E., Carrière, Y., Dennehy, T.J., Morin, S., Sisterson, M.S., Roush, R.T., Shelton, A.M., and Zhao, J.Z., 2003. Insect resistance to transgenic Bt crops: Lessons from the laboratory and field. Journal of Economic Entomology 96: 1031–1038.

Tabashnik, B.E., Gassmann, A.J., Crowder, D.W., and Carriere, Y., 2008. Insect resistance to *Bt* crops: evidence versus theory. Nature Biotechnology 26: 199–202.

Torres, J.B., and Ruberson, J.R., 2006. Interactions of Bt-cotton and the omnivorous big-eyed bug *Geocoris punctipes* (Say), a key predator in cotton fields. Biological Control 39: 47–57.

Torres, J.B., and Ruberson, J.R., 2008. Interactions of *Bacillus thuringiensis* Cry1Ac toxin in genetically engineered cotton with predatory heteropterans. Transgenic Research 17: 345–354.

Torres, J.B., Ruberson, J.R., and Adang, M.J., 2006. Expression of *Bacillus thuringiensis* Cry1Ac protein in cotton plants, acquisition by pests and predators: A tritrophic analysis. Agricultural and Forest Entomology 8: 191–202.

Traxler, G., and Godoy-Avila, S., 2004. Transgenic cotton in Mexico. AgBioForum 7: 57–62.

Trichilo, P.J., and Wilson, L.T., 1993. An ecosystem analysis of spider mite outbreaks: Physiological stimulation or natural enemy suppression. Experimental and Applied Acarology 17: 291–314.

USDA (United States Department of Agriculture), 2006. Cotton Varieties Planted: 2006 Crop. U.S. Department of Agriculture, Agricultural Marketing Service – Cotton Program, Memphis, TN, USA.

USDA (United States Department of Agriculture), 2007. Cotton Varieties Planted: 2007 Crop. U.S. Department of Agriculture, Agricultural Marketing Service – Cotton Program, Memphis, TN, USA.

USEPA (United States Environmental Protection Agency), 2007. Insecticide resistance management fact sheet for *Bacillus thuringiensis* (Bt) cotton products. http://www.epa.gov/pesticides/biopesticides/pips/bt_cotton_refuge_2006.htm (accessed 12 March 2008).

Whitcomb, W.H., and Bell, K., 1964. Predaceous insects, spiders and mites of Arkansas cotton fields. Arkansas Agricultural Experiment Station Bulletin 690: 1–84.

Whitehouse, M., Wilson, L., and Constable, G., 2007. Target and non-target effects on the invertebrate community of Vip cotton, a new insecticidal transgenic. Australian Journal of Agricultural Research 58: 273–285.

Williams, M.R., 2006. Cotton insect losses 2005. In: Proceedings of the Beltwide Cotton Conferences, National Cotton Council, Memphis, TN, USA, pp. 1151–1204.

Wilson, L.J., 1994. Resistance of okra-leaf cotton genotypes to two-spotted spider mites (Acari: Tetranychidae). Journal of Economic Entomology 87: 1726–1735.

Wilson, L.J., Bauer, L.R., and Lally, D.A., 1998. Effect of early season insecticide use on predators and outbreaks of spider mites (Acari: Tetranychidae) in cotton. Bulletin of Entomological Research 88: 477–488.

Wilson, L.J., Mensah, R.K., and Fitt, G.P., 2004. Implementing integrated pest management in Australian cotton. In: Novel Approaches to Insect Pest Management in Field and Protected Crops, A.R. Horowitz and I. Ishaaya, eds., Springer, Berlin, pp. 97–118.

Wilson, L., Hickman, M., and Deutscher, S., 2006. Research update on IPM and secondary pests. In: Proceedings of the 13th Australian Cotton Research Conference, Broadbeach, Queensland, Australia, pp. 249–258.

Wolfenbarger, L.L., Naranjo, S.E., Lundgren, J.G., Bitzer, R.J., and Watrud, L.S. 2008. Bt Crop effects on functional guilds of non-target arthropods: A meta-analysis. PLoS ONE 3(5): e2118. doi:10.137/journal.pone.0002118.

Wu, K., 2007a. Monitoring and management strategy of *Helicoverpa armigera* resistance to Bt cotton in China. Journal of Invertebrate Pathology 95: 220–223.

Wu, K., 2007b. Environmental impacts of *Bt* cotton commercialization and strategy for risk management. Journal of Agricultural Biotechnology 15: 1–4.

Wu, K., Guo, Y., and Gao, S., 2002a. Evaluation of the natural refuge function for *Helicoverpa armigera* (Hübner) within Bt transgenic cotton growing areas in north China. Journal of Economic Entomology 95: 832–837.

Wu, K., Li, W., Feng, H., and Guo, Y., 2002b. Seasonal abundance of the mirids, *Lygus lucorum* and *Adelphocoris* spp. (Hemiptera: Miridae) on Bt cotton in northern China. Crop Protection 21: 997–1002.

Wu, K., Feng, H., and Guo, Y., 2004. Evaluation of maize as a refuge for management of resistance to Bt cotton by *Helicoverpa armigera* (Hubner) in the Yellow River cotton farming region of China. Crop Protection 23: 523–530.

Wu, K., Guo, Y., and Head, G., 2006. Resistance monitoring of *Helicoverpa armigera* (Lepidoptera: Noctuidae) to Bt insecticidal protein during 2001–2004 in China. Journal of Economic Entomology 99: 893–896.

Wu, K.M., and Guo, Y.Y., 2003. Influences of *Bacillus thuringiensis* Berliner cotton planting on population dynamics of the cotton aphid, *Aphis gossypii* Glover, in northern China. Environmental Entomology 32: 312–318.

Wu, K.M., and Guo, Y.Y., 2005. The evolution of cotton pest management practices in China. Annual Review of Entomology 50: 31–52.

Zhao, J.Z., 1984. Species distribution and population fluctuation of predacious spiders in cotton fields in China. Natural Enemies of Insects 6: 1–12.

Chapter 7
The Present and Future Role of Insect-Resistant Genetically Modified Potato Cultivars in IPM

Edward J. Grafius[1,*], and David S. Douches[2]

Abstract Potato, *Solanum tuberosum* L., is one of the world's principal food crops. Important potato insect pests include Colorado potato beetle, *Leptinotarsa decemlineata* Say, potato tuberworm, *Phthorimaea operculella* (Zeller), and aphids, especially as they transmit potato leafroll virus and potato virus Y. Management of insect pests of potato relies almost entirely on chemical insecticides. Potato breeding is complicated by the potato's tetraploidy. Numerous *Solanum* spp. have resistance to insects but these properties have not been transferred into commercially desirable cultivars. Insect-resistant cultivars are generally not available. GM potatoes expressing resistance to *L. decemlineata*, potato virus Y, and potato leaf roll virus were registered and marketed in the USA from 1995–2000, but were withdrawn from the market in response to marketing concerns about GM crops. *Bacillus thuringiensis* (*Bt*) genes for resistance to various insect pests are likely candidates for inclusion into potatoes using genetic engineering. Other resistance factors, including glandular trichomes, leptine glycoalkaloids, and other genes encoding for insecticidal proteins also show promise, especially if pyramided with appropriate *Bt* genes. Re-introduction of GM potatoes in the USA and elsewhere awaits changes in consumer preferences.

7.1 Introduction

Potato, *Solanum tuberosum* L., is an important crop worldwide and ranks fourth in production among food crops after maize (*Zea mays* L.), rice (*Oryza sativa* L.), and wheat (*Triticum aesitvum* L.) (FAOSTAT data, 2006). The importance of potatoes is increasing due to the rising world population, the capability of potatoes to grow well in adverse conditions, and its high nutritional value (Anonymous, 1984). By itself, potato is not a good source of energy and is comparatively lower in calories

[1] Department of Entomology, Michigan State University, East Lansing, MI, USA

[2] Department of Crop and Soil Sciences, Michigan State University, East Lansing, MI, USA

* To whom correspondence should be addressed. E-mail: grafius@msu.edu

J. Romeis, A.M. Shelton, G.G. Kennedy (eds.), *Integration of Insect-Resistant Genetically Modified Crops within IPM Programs.*

than other cooked foods such as cassava, plantain, or beans (Woolfe, 1987). However, the protein content of a cooked potato is comparable to that of rice (Paul and Southgate, 1978). Potatoes are rich in antioxidants that are associated with many health benefits, including lower incidences of heart disease, and reductions in some types of cancers, macular degeneration, and cataracts (Brown, 2005).

The potato presents unique challenges and advantages to plant breeders. Because it is propagated vegetatively by tuber cuttings, potato cultivars don't need to be bred to produce homogenous plants from true seed. A major disadvantage of potatoes for breeders is that *S. tuberosum* is tetraploid making it difficult to transfer desirable traits between cultivars and have them expressed in progeny. There are numerous species of *Solanum*, several of which are cultivated in Peru and central America. These provide a rich source for potential traits to breed into *S. tuberosum*, including tuber qualities (e.g., colors ranging from white to deep purple skin and flesh) and resistance to insect pests and diseases. Unfortunately, many of these wild *Solanum* relatives are diploid, greatly complicating the breeding process. Thus, insertion of candidate genes by genetic engineering is a particularly valuable process for developing new potato cultivars, but the commercialization of genetically modified (GM) potato cultivars is difficult because of the current unwillingness of Japanese and European markets to accept them. Largely because of marketing concerns, genetic engineering is not a significant part of most potato breeding programs. Potato cultivars expressing the *Bacillus thuringiensis* var. *tenebrionis* Cry 3A toxin for resistance to the Colorado potato beetle, *Leptinotarsa decemlineata* Say (Coleoptera: Chrysomelidae) (cv. NewLeaf, Monsanto Corp.) were the first GM food crop approved for human consumption and commercially produced in the USA (1995). Because of consumer concern *Bt* potato cultivars were taken off the market in 2000.

7.2 Potato Insect Pests

Potatoes are attacked by a wide range of insect pests, including insects that attack foliage (e.g., *L. decemlineata*; potato leafhopper, *Empoasca fabae* L. [Homoptera: Cicadellidae]), insects that damage tubers (e.g., wireworms, Coleoptera: Elateridae), insects that attack both foliage and tubers (e.g., potato tuber moth, *Phthorimaea operculella* [Zeller] [Lepidoptera: Gelechiidae]; black cutworm, *Agrotis ipsilon* Hufnagel [Lepidoptera: Noctuidae]), and insects that are important mainly because they transmit virus diseases (e.g., green peach aphid, *Myzus persicae* [Sulzer] [Hemiptera: Aphidiidae]).

7.2.1 *Leptinotarsa decemlineata*

L. decemlineata is one of the most economically significant pests of potato in North America, Europe, and western Asia (Fig. 7.1). *L. decemlineata* led to the

Fig. 7.1 Colorado potato beetle adults feeding on potato (Photo by Joseph Coombs, Michigan State University Potato Breeding and Genetics)

first large-scale use of insecticides on any crop in 1864 (Gauthier et al., 1981) and insecticides have long been the primary means of *L. decemlineata* control (Casagrande, 1987).

L. decemlineata has shown a remarkable ability to develop resistance to every insecticide used extensively for its control (Bishop and Grafius, 1996) and has done so at an increasing rate with each new insecticide (Forgash, 1985; Heim et al., 1990; Ioannidis et al., 1991). Up to 100-fold resistance to insecticides can appear in as few as three generations (Ioannidis et al., 1992). Resistance to neonicotinoids appeared within 2 years on Long Island, New York (Zhao et al., 2000) and became widespread throughout the eastern USA by 2000, although neonicotinoids have remained effective in the Midwestern USA, except for Michigan (see discussion below). *L. decemlineata* has developed resistance to over 40 different insecticides (Whalon et al., 2007). In some areas, prior to the introduction of the neonicotinoids, the only products remaining effective were *Bt* var. *tenebrionis* (Cry 3A) foliar sprays and sodium fluoraluminate (cryolite). However, both of these materials are generally effective on only newly hatched larvae and require repeated applications. Growers in Michigan reported crop losses of 20% or higher from 1990–1994, despite intensive use of multiple tank-mixed insecticides and control costs of >(US)$200–300/ha (Grafius, 1997).

The first neonicotinoid insecticide (imidacloprid) was registered for control of *L. decemlineata* in the USA in 1995 (coincidental with the first introduction of *Bt* potatoes) and was widely adopted immediately (>90% of the potatoes in Michigan were treated in 1995; Grafius, 1997), because of the severity of the problem. Imidacloprid is a systemic insecticide that can be applied once at planting, to control first generation *L. decemlineata* and to provide season-long control of aphids. Growers in the USA and Canada now rely heavily on the neonicotinoid insecticides imidacloprid and thiamethoxam for control of *L. decemlineata*, with 65–91% of the potato acreage in the eastern and midwestern USA and 51% of potatoes in the USA treated with imidacloprid or thiamethoxam (National Agricultural Statistics Service, 2006).

Imidacloprid resistance first occurred on Long Island, New York in 1997 (Zhao et al., 2000) and has since been detected in Pennsylvania in 2002 (Bishop et al., 2003), and Michigan in 2004 (Grafius et al., 2005) and appears to be becoming more widespread. Cross resistance within the neonicotinoid class is common (Grafius et al., 2005; Mota-Sanchez et al., 2006) and *L. decemlineata* resistant to imidacloprid also express some level of resistance to other neonicotinoid insecticides (Byrne et al., 2004; Mota-Sanchez et al., 2006).

During the early 1990s, when resistance problems in *L. decemlineata* were so severe in the northeastern and midwestern USA and Canada, growers adopted multiple techniques for control. These included propane flamers targeted to control adults on young potato plants (Moyer et al., 1992), crop vacuums (Boiteau et al., 1992), plastic-lined trenches to limit dispersal of adults between fields (Misener et al., 1993), and increased crop scouting. Also, many growers increased distances between potato crops in rotational systems to reduce dispersal of adults to new fields in the spring (Wright, 1984; Weisz et al., 1994). Since the adoption of neonicotinoid treatments at planting, propane flamers, crop vacuums, and trenches are no longer being used because they are costly and difficult to implement, and because *L. decemlineata* populations are generally not high enough to encourage their use.

7.2.2 *Phthorimaea operculella*

One of the most common and destructive insect pests of potato in tropical and subtropical areas worldwide is *P. operculella* (Visser, 2005). It causes damage in both field and storage (Westedt et al., 1998), reduces the quality of produce, and increases the risk of pathogen infection. Furthermore, because it attacks both the foliage and the tuber, damage caused by *P. operculella* can reduce the potato yield tremendously (Capinera, 2001). In warmer climates, losses in storage can reach 100% (Lagnaoui et al., 2001).

P. operculella spends its larval stages either in the foliage or tuber. The larvae mine both the foliage and the petiole creating transparent leaf blisters and may move from the foliage to the tubers. At high infestation levels, damage to the foliage may be sufficient to kill the plant. Additionally, larvae hatching from eggs laid on the exposed potato tubers or in cracks in the soil make slender tunnels throughout the tubers and introduce bacterial rots. Mounds of frass at the tunnel entrances can indicate an infested tuber. *P. operculella* can infest tubers in storage, when tubers infested prior to harvest are placed in storage or when moths enter storage facilities from nearby infested fields. The potato crop may have to be treated multiple times with insecticides in the field and in storage to limit damage by *P. operculella* (Douches et al., 2004). Because of its ability to develop resistance to chemical insecticides, *P. operculella* is an increasing agricultural problem in tropical and subtropical areas.

7.2.3 Aphids and Virus Diseases

Potato aphid, *Macrosiphum euphorbiae* (Thomas), and green peach aphid, *Myzus persicae* (Sulzer), are the two most common aphids colonizing potatoes (Radcliffe and Ragsdale, 2002). *M. persicae* numbers can build up to levels where significant damage and yield losses occur (Sexson et al., 2005). However, the most significant effect of aphids on potatoes is transmission of virus diseases, among which potato virus Y and potato leafroll virus are of particular importance. Numerous aphid species, including *M. persicae* and *M. euphorbiae*, are capable of transmitting virus diseases to potatoes (Difonzo et al., 2007). However, the most common vectors of potato virus Y are aphids that do not colonize and reproduce on potatoes, such as *Rhopalosiphum padi* (L.) and *Aphis glycines* Matsumura in the USA (Radcliffe and Ragsdale, 2002). These aphids acquire the virus as the result of casual feeding on infected potatoes or other hosts and disperse widely, transmitting the virus as they travel. Because transmission of potato virus Y can occur in just a few seconds of feeding after the aphid's arrival in the potato field and because most of the aphid species involved in transmission do not colonize potatoes, insecticides are ineffective for control of potato virus Y. However, insecticide treatments may be useful for control of potato leaf roll virus, since it is transmitted mainly by *M. persicae* and requires 12 hours or more for the acquisition and transmission process (Difonzo et al., 2007). Although GM potato cultivars resistant to aphids have not been available, GM cultivars resistant to potato virus Y and potato leafroll virus along with resistance to *L. decemlineata* were produced and marketed briefly until they were removed from the market in 2000. Potato virus Y continues to be a serious problem for which there is no effective control (Davis et al., 2007).

7.3 Pest Management Systems in Potatoes

Potato pest management has relied heavily on the use of insecticides and fungicides for control of insect pests and diseases. Current IPM recommendations for potatoes include the use of scouting and economic thresholds (e.g., for *L. decemlineata*,

E. fabae, and aphid species; Sexson et al., 2005) and the use of pheromone or black light trap monitoring (e.g., for *P. operculella* [Visser, 2005] and European corn borer, *Ostrinia nubilalis* [Hübner] [Lepidoptera: Crambidae] [Sexson et al., 2005]). Resistant cultivars are generally not available. Likewise, biological control has not played a major role in management of insect pests of potatoes. The high value of the potato crop means that treatment thresholds are low for pests such as *L. decemlineata*, *E. fabae*, aphid species, and *P. operculella*. The frequent use of insecticides often has adverse effects on natural enemies such as coccinellids and chrysopids (Reed et al., 2001). Numerous studies have been conducted on endemic and introduced natural enemies of *L. decemlineata* (e.g., Harcourt, 1971; Groden et al., 1990; Hough-Goldstein et al., 1993; Arpaia et al., 1997; Hilbeck et al., 1997) but none have proven reliable enough at maintaining pest populations below economic levels.

Throughout most of its current range in North America and in Europe and western Asia, *L. decemlineata* is an introduced pest. Ancestral *L. decemlineata* populations exist in the southwestern USA and Mexico. However, searches for natural enemies in ancestral locations have not resulted in identification and introduction of natural enemies that can effectively regulate *L. decemlineata* populations (Cappaert et al., 1991; Hazzard and Ferro, 1991; Hough-Goldstein and Keil, 1991; Biever and Chauvin, 1992).

Thus, growers continue to rely heavily on insecticides and fungicides for insect and disease management in potatoes. While they have been able to manage pests in this manner overall, pesticide costs continue to rise and availability of products continues to be a concern, especially for *L. decemlineata* control. Pesticide resistance has resulted in major control failures and remains a constant threat. Pesticide resistance has been an especially severe problem with *L. decemlineata*, but resistance problems have also occurred with *M. persicae* and *E. fabae* (Whalon et al., 2007). Pesticide resistance has become such a serious problem that the U.S. National Potato Council has developed specific recommendations for management of resistance to insecticides in *M. persicae* and *L. decemlineata* (Anonymous, 2005). These recommendations include some cultural controls but emphasize avoiding repeated use of pesticides from the same chemical groups.

For control of aphids and virus disease transmission, insecticides have limited effectiveness; especially for non-persistent viruses like potato virus Y. Vegetatively propagated seeds in potatoes makes it extremely difficult to maintain virus-free plant material. In these situations, resistant cultivars seem to be the only reliable management option. Unfortunately, current cultivars do not show significant resistance. Cultivars that are symptomless carriers (carry potato virus Y but do not show visible foliar symptoms, although yield is affected) are especially problematic because disease levels cannot be kept low by normal seed screening, and asymptomatic plants provide a large reservoir of the disease in areas where potato may be grown both for seed and for tablestock or processing.

In spite of progress toward diversifying pest management techniques in the 1990s, following the introduction of cultural controls for *L. decemlineata* and insect and disease resistant GM cultivars, none of these practices are in use commercially at this time, except for crop rotation. They have been displaced by neonicotinoid insecticides, which offer low cost, effectiveness, and ease of application. The potato industry today continues to rely heavily on insecticides and fungicides. In developing countries, intensive pesticide use creates an especially severe economic burden and potential risk to the farmers, applicators, and farm workers. Without additional non-chemical options for control of insect and disease pests, especially resistant cultivars, pesticide use will continue to be high, with high costs and potential environmental and human health risks and repeated occurrence of pesticide resistance. Only with a more balanced management program that includes multiple control options can pesticide resistance and its associated costs and crop losses be reduced or prevented. Resistant cultivars may be developed using traditional plant breeding methods, but the benefits of molecular biological techniques for rapidly inserting resistance genes into existing cultivars could be large.

7.4 Potato Breeding

The three main targets of potato breeding are cultivars for fresh food market, for the processing industry, and for non-food industrial uses. At present, potato breeders are giving emphasis to meeting the quality requirements of processors and fresh market buyers, maximizing yield potential, and developing cultivars that are resistant to insects and diseases. Potato is clonally propagated and is therefore vulnerable to many diseases and insect pests that colonize tubers and can be spread in potato "seed."

There are 199 wild and 7 cultivated species within *Solanum* section *Petota* (Solanaceae) that can be called "potatoes;" most are tuber-bearing (Spooner and Hijmans, 2001). Spooner et al. (2004) describe two major epicenters of potato germplasm diversity. The first stretches from the southwestern USA to the Mexican central highlands, and the second region is the Andes of South America. The latter claims the majority of diversity found in potato.

Given the significance of the potato, research on the genetic improvement of this crop is important. Potato breeders are challenged by an autotetraploid genome and asexual propagation; consequently, the breeding principles, and practices are quite different from those employed for the majority of diploid (or allopolyploid) seed-propagated crops. Potatoes also have numerous market limiting standards, such as color, shape, and dry matter and sugar content (important for processing potatoes) that must be met by new cultivars. Moreover, Tarn et al. (1992) have identified 17 pathogen and 6 pest resistance traits that need to be considered in a potato breeding program.

7.5 Host Plant Resistance

The genus *Solanum* has tremendous natural diversity, including a large number of natural traits that confer resistance to insect pests. The two most commonly exploited host plant resistance factors in *Solanum* are glycoalkaloids and glandular trichomes.

7.5.1 Glycoalkaloids

The cultivated potato naturally produces glycoalkaloid compounds, which can deter insect feeding (Sinden et al., 1980, 1986). High glycoalkaloid levels are useful host plant resistance factors, but they impart a bitter taste in the tuber and at high concentrations induce nausea and vomiting in mammals and are cholinesterase inhibitors (Sinden and Webb, 1972; Van Gelder, 1990; Lachman et al., 2001). North Dakota State University recently released a cultivar, Dakota Diamond (ND5822C-7), with insect resistance attributed to glycoalkaloids (A. Thompson, University of North Dakota, personal communication). Most glycoalkaloids are distributed throughout the potato plant, in tubers and foliage. Glycoalkaloid concentrations in potato plants can be strongly affected by light, which explains why these protective glycoalkaloids are normally present at higher concentration in the aerial parts of the potato plant (leaves, stems, and sprouts) than in the tubers (Lachman et al., 2001).

The two most common glycoalkaloids found in potatoes are α-chaconine and α-solanine, which together comprise as much as 95% of the total glycoalkaloids present in the potato (Lachman et al., 2001). Chemically, α-chaconine and α-solanine, are classified as steroidal glycoalkaloids. *Solanum* steroidal glycoalkaloids are large, biologically-active, secondary metabolites that have been isolated from more than 350 plant species (Ripperger and Schreiber, 1981; Roddick, 1986; Lawson et al., 1993). These are toxins that occur naturally in many edible and non-edible members of the *Solanaceae*, such as eggplant (*S. melongena*) and potato. Presence of steroidal glycoalkaloids in potato could both be a benefit and a concern (Lawson et al., 1993). The fatal oral dose for an adult human would be 420 mg (Lachman et al., 2001). Due to human health issues associated with glycoalkaloids, the industry has limited tuber glycoalkaloids levels to 20 mg/100 g of fresh tissue for newly released cultivars (Van Gelder, 1990).

S. chacoense Bitter produces novel glycoalkaloids called leptines that are expressed only in the foliage (Lorenzen et al., 2001). Leptines are acetylated analogs of the common potato steroid glycoalkaloids, solanine and chaconine. Acetylated glycoalkaloids are the most active form of steroid glycoalkaloids present in potato. Leptines such as those found in the breeding line USDA8380-1, and other acetylated steroid glycoalkaloids are only reported to be synthesized by some accessions of *S. chacoense* and are synthesized only in leaves and not the tubers (Sanford et al., 1996). Although leptines have not been introgressed into any current commercial cultivars, they could provide protection from foliar pests and alleviate the human health concern associated with high glycoalkaloid content in the tuber (Sinden et al., 1986).

7.5.2 Glandular Trichomes

Some wild species of potato have glandular trichomes that confer resistance to small insects, such as aphids and leafhoppers (Lapointe and Tingey, 1986; Tingey,

1991; Yencho and Tingey, 1994). Glandular trichomes also affect small natural enemies. Three wild *Solanum* species, *S. berthaultii, S. polyadenium, and S. tarijense*, have high densities of glandular trichomes that have been bred into cultivated potato (Tingey et al., 1984). Breeding line NYL235-4 has glandular trichomes derived from *S. berthaultii* and is available for further research and breeding (Plaisted et al., 1992).

7.6 Genetic Engineering

Potatoes were among the first successful GM crop plants (An et al., 1986). Potato transformation is achieved by a number of methods and can be directed to either the nuclear or plastid genome. Electroporation and *Agrobacterium tumefaciens*-mediated techniques can be used to incorporate genes into the nuclear genome; biolistic methods can be used for the incorporation of genes into both the nuclear and plastid genome (Daniell et al., 1998; Huang et al., 2002; Maliga, 2004). *A. tumefaciens*-mediated transformation is the predominant method currently used in potato. This procedure offers the advantage of mobilizing defined regions of DNA into the nuclear genomes of the plant. Using these procedures, individual genes or a cassette of genes can be inserted into the genome of a plant.

Genetic engineering allows for the introduction of foreign genes or the reintroduction of individual potato genes to elite cultivars. *Solanum* has immense diversity with many beneficial traits, including natural resistance to pests. Unfortunately, many resistant species are not readily accessible to breeders using traditional breeding techniques due to issues with endosperm balance numbers and incompatibility. Additionally, many of the beneficial genes possessed by wild, weedy potato relatives are often masked and/or difficult to remove from the wild background. Even if crosses can be performed between domesticated potato and wild relatives, multiple generations of backcrossing are usually required to remove undesirable traits.

Although a great deal of work is required, important genes can be identified and cloned from a wild *Solanum* species. If an important gene is cloned, it can be inserted and expressed into a number of elite potato lines more easily via genetic engineering than by traditional breeding methods. For example, *S. bulbocastanum* is highly resistant to late blight. Recently, a late blight resistant gene from *S. bulbocastanum* was cloned (Ballvora et al., 2003; Song et al., 2003). The resistant gene, *RB*, was inserted into a susceptible cultivar, cv. Katahdin, conferring resistance to late blight in the transformed plants (Song et al., 2003). Prior to the use of genetic engineering techniques, plant breeders did not readily have access to this resistance source because *S. bulbocastanum* is a diploid (2x = 24) species with an endosperm balance number of 1 compared to *S. tuberosum*, which is a tetraploid species with an endosperm balance number of 4; therefore *S. bulbocastanum* must be crossed with a bridging species before the genome can be introgressed into *S. tuberosum* (Hawkes, 1994). Even in cases where sexual incompatibility is not an issue, genetic engineering can shorten breeding time because extensive backcrossing is not needed to remove unwanted traits of the wild parent or to pyramid and combine traits for durable host plant resistance or for resistance to multiple pests.

7.7 Genes of Interest: Insect-Resistance

7.7.1 *Bacillus thuringiensis* Genes

Strains of *Bt* have been formulated for use as a foliar spray for many decades, but *Bt* sprays provide limited protection because the toxins are photosensitive and degrade quickly compared to most chemical insecticides (Whalon and Wingerd, 2003). To increase efficiency, genes coding for *Bt* Cry proteins have been inserted into many agricultural crops. The resulting plants express Cry proteins constantly in their tissue, which alleviates problems of toxin distribution that are associated with foliar applications. The major advantages to this delivery system are increased efficacy, reduced application costs and minimal scouting needs compared with conventional insecticide sprays (Lambert and Peferoen, 1992). The specificity of Cry proteins allows plant breeders to target a single insect pest and not kill beneficial insects; on the other hand the specificity does not provide a wide range of protection (Ferré and Van Rie, 2002).

Bt cry3A, from *B. thuringiensis* subsp. *tenebrionis*, targets coleopteran pests and is effective against *L. decemlineata*. GM potato cultivars expressing the Cry3A toxin provide good control of *L. decemlineata* and were commercially available in the USA from 1996–2000 (see discussion section 7.9 of this chapter). *Bt cry1Ia1*, from *B. thuringiensis* subsp. *kurstaki*, targets lepidopteran pests and is effective against *P. operculella* (Perlak et al., 1993; Douches et al., 2004). Davidson et al. (2005) demonstrated potato tuber moth control when a *cry1Ac* gene was expressed in tubers of some lines of cvs. Russet Burbank and Red Rascal. Davidson et al. (2004) also examined the efficacy of *cry1Ac9* expressing potato plants for control of potato tuber moth. Using a growth index, they showed that field grown foliage fed to potato tuber moth larvae inhibited larval growth over three seasons of testing. Meiyalaghan et al. (2006) expressed independently *cry1Ac9* and *cry9Aa2* using the *Lhca3* light-inducible promoter. They identified potato lines with potato tuber moth larvae resistance in the foliage but minimal or no expression of the gene in the tubers. Recently, chimeric *Bt* genes have been engineered to broaden the range of pests affected (Naimov et al., 2003; Singh et al., 2004; Chen et al., 2006). Using the SN19 hybrid gene, Naimov et al. (2003) showed transgenic plants to be resistant to larvae and adults of *L. decemlineata*, and larvae of *P. operculella* and *O. nubilalis*.

7.7.2 *Inhibitors of Insect Digestive Enzymes*

In 1993, Michaud et al. demonstrated the use of two cysteine proteinase inhibitors, oryzacystatins I and II, as a means to control coleopteran insect pests. Since then, transformation and expression of proteinaceous cysteine inhibitors in potato have been tested to control *L. decemlineata*. Cloutier et al. (1999) studied the growth of *L. decemlineata* on potato foliage expressing oryzacystatin I. Interestingly, L1 to L3 larvae consumed more foliage of the transgenic line than of the non-transgenic line, apparently to compensate for poorer nutrition. Gatehouse et al. (1997) evaluated bean chitinase, wheat α-amylase and cowpea trypsin inhibitor genes expressed in potato. Expression of bean chitinase had no effect on *Lacanobia oleracea* (L.) (Lepidoptera: Noctuidae), while the wheat α-amylase gene expression was not detected in the potato plants. The expression of the cowpea typsin inhibitor gene in potato reduced insect biomass and survival, but did not reduce foliage damage in the potato. The Soya bean Kunitz trypsin inhibitor in potato caused only marginal insect control that decreased over time (Gatehouse et al., 1999). In contrast, Lecardonnel et al. (1999) reported that potato lines, transformed with the rice cysteine proteinase inhibitor, caused substantial mortality of *L. decemlineata* larvae reared on the foliage. The authors suggest that this resistance factor be used in combination with different resistance genes. Brunelle et al. (2005) state that more appropriate inhibitors will be key to the effective use of extracellular protease inhibition as a pest control strategy.

7.7.3 Plant Derived Lectin Genes

The snowdrop lectin (see Malone et al., chapter 13) has been examined for its potential as a transgenic insect resistance trait in potato. Down et al. (1996) demonstrated a fourfold decrease in population growth by the aphid *Aulacarthum solani* Kaltenbach (Hemiptera: AA A phididae) on transgenic potato plants expressing the snowdrop lectin relative to control plants in greenhouse trials. Gatehouse et al. (1997) reported that transgenic plants expressing snowdrop lectin showed enhanced resistance to tomato moth (*L. oleracea*), which was attributed in part to an antifeedant effect. Gatehouse et al. (1999) subsequently demonstrated that the constitutive expression of concanavalin A, a lectin from jackbean, in potato reduced the fecundity of the aphid *M. persicae* and larval weight and foliage consumption by *L. oleracea* but did not affect larval survival. Lectin genes may not provide high levels of control but may be useful to control aphids and other pests not susceptible to *Bt* and in combination with other genes to provide a broader base for resistance.

7.7.4 Vegetative Insecticidal Proteins (Vips)

Vegetative insecticidal proteins (Vip) from *B. thuringiensis* or *B. cereus* Frankland and Frankland are less well known than Cry proteins, but are active against a wider range of insects (Estruch et al., 1996; Sharma et al., 2002). Unlike Cry proteins, which are produced during sporulation, the distinctly different Vip proteins are produced by *Bt* or *B. cereus* in the secreted supernatant fluids collected during the vegetative growth stage prior to sporulation (Estruch et al., 1996). Vip3A, from *B. thuringiensis*, is active against Lepidoptera; Vip1 and Vip2, from *B. cereus*, are toxic to Coleoptera (Estruch et al., 1996; Moellenbeck et al., 2001).

Vips result in the lysing of the gut epithelial cells, although the exact mechanism is not known (Yu et al., 1997; Lee et al., 2005). The development of cross-resistance between the two classes of toxins is unlikely, because the membrane receptors and pH requirement for toxicity are different (Lee et al., 2005).

7.8 Resistance Management

The major biological concerns surrounding crops engineered to produce *Bt* toxins are sustainability and management. Sustainability of insect resistance in GM crops depends on the pest and the resistance gene used. *Bt* resistance management strategies include: a high dose/refuge strategy, pyramiding genes for resistance, alternation of crops with different resistance genes, and wound-inducible resistance gene expression (Bates et al., 2005; Ferré et al., chapter 3). For potatoes, only Cry3a *Bt* plants resistant to *L. decemlineata* were available commercially. Research is on-going to develop additional sources of *L. decemlineata* resistance via traditional breeding and GM technology. Potatoes resistant to *P. operculella* have been developed using Cry1Ia1 (Douches et al., 2004), but are not commercially available. Thus, with only single resistance genes available, the high dose/refuge strategy is the only resistance management option available (Shelton et al., 2002). Regulations in place for deployment of *Bt* transgenic crops in the USA are based on this model (USEPA, 2001), but there have been serious concerns about the level of compliance (Jaffe, 2003).

The high dose/refuge model has several critical assumptions: resistance is inherited as a recessive factor, the initial frequency of resistance alleles in the pest population is low, large numbers of pests will be produced in the refuges, and individuals will widely disperse before interbreeding to ensure genetic mixing between susceptible individuals from the refuges and any resistant homozygote individuals that may be produced in the GM crop (Ferré et al., chapter 3). Dispersal and mating behavior of the pest are keys to whether the high dose/refuge strategy will be successful or not.

7.8.1 Resistance Management for L. decemlineata

Unfortunately, *L. decemlineata* has several characteristics that may prove problematic for the resistance management strategies that have been proposed for *Bt* crops (Ferré et al., chapter 3). Larvae pupate in the soil within the crop field and adults often disperse very little and mate within a few days of emergence from pupation (Hare, 1990). Thus, it is unlikely that a refuge, even one adjacent to the potato crop, will produce the level of inter-mating required for resistance management to be successful. Seed mixtures would provide mixing between resistant and susceptible

L. decemlineata, but could allow individuals to receive low doses and move to non-GM plants to complete development, fostering low-dose selection for resistance development. Another complicating factor is that females mate multiple times and randomly allocate sperm from multiple matings to offspring (Boiteau, 1988), which would increase the probability of producing one or more homozygous resistant offspring. Finally, because *L. decemlineata* is so destructive (economic thresholds are 0.5 to 1 large larva or adult per plant, Sexson et al., 2005), difficult to control with conventional insecticides, and prone to development of resistance to conventional insecticides (Bishop and Grafius, 1996), growers may be hesitant to leave untreated refuges to produce large numbers of susceptible individuals.

7.8.2 Resistance Management for P. operculella

P. operculella may be more amenable to resistance management practices than *L. decemlineata*. Adults may disperse 360 m or more and males tend to disperse prior to mating (Cameron et al., 2005), fostering genetic mixing in a high dose/refuge resistance management strategy. In addition, there may be multiple alternative crops, such as tomatoes, peppers, and eggplant, and wild hosts near potato fields in tropical and semi-tropical areas such as Egypt and South Africa, and crops are often grown in very small plots, interspersed with other crop hosts of *P. operculella* (Santos, unpublished data). Time during the year when potatoes are not planted also encourages dispersal of *P. operculella* adults to other host crops and weeds. Implementing resistance management for *P. operculella* infestations in potato tubers and seed potatoes stored without refrigeration will be difficult. Populations can increase rapidly under these conditions, infesting 100% of susceptible potatoes within 3 months (Douches et al., 2004), providing an ideal environment for selection, inbreeding, and development of *Bt* resistance. Planting infested seed into the field (a common practice) provides a ready mechanism for dispersal and reinfestation by resistant individuals.

7.8.3 Developing Lines with Multiple Resistance Genes to Delay/Prevent L. decemlineata or P. operculella Adaption

A key tactic to resistance management for adaptable insects such, as *L. decemlineata* and *P. operculella*, is the development of alternative resistance traits to be used in combination with the expression of *Bt* Cry toxins since combining multiple resistance factors can delay resistance development exponentially (Roush, 1998; Zhao et al., 2005). A resistance management model using two different sources of resistance (e.g., combining GM and traditionally bred host plant resistance) in a single cultivar has the potential to be more widely adopted and more durable in potatoes than the high dose/refuge strategy. Combining host plant resistance factors as a resistance management strategy for the introduced *Bt* Cry protein does not require grower cooperation or regulatory monitoring or enforcement. Multiple toxins can be employed for resistance management by combining GM insecticidal proteins with host plant resistance factors or by stacking host plant resistance factors into plants (Mani, 1985; Roush, 1998; Zhao et al., 2005) (Fig. 7.2). *Solanum* has immense potential genetic diversity for host plant resistance. Since it is also amenable to genetic engineering, potato breeders are in a unique position with the ability to readily access both natural and engineered host plant resistance for plant protection. Use of diverse host plant resistance factors, such as the combination of *Bt* and leptine genes derived from *S. chacoense* (Coombs et al., 2002; Cooper et al., 2004), may provide the most benefit for managing resistance.

A primary effort of the Michigan State University potato breeding program is to combine these two resistance mechanisms into one commercial cultivar. Breeding efforts have been initiated to develop advanced breeding lines that express leptine-based insect resistance for the chip processing and tablestock industry. These lines have been crossed to our most advanced breeding lines that have good agronomic performance along with either chip processing or tablestock qualities. The superior individuals from each cross were selected for specific gravity, tuber appearance rating, chip processing, leptine/total glycoalkaloid content, and *L. decemlineata* resistance. HPLC procedures are being used to measure leptine/total glycoalkaloid concentration in the potato foliage (Sinden et al., 1986). The superior selections from this breeding effort will be candidates for combining with the *Bt cry3A* gene via *Agrobacterium*-mediated transformation (Douches et al., 1998). If there are problems with accumulation of glycoalkaloids in the tubers of clones developed from *S. chacoense*, then antisense total glycoalkaloid technology will be applied (Stapleton et al., 1991). The use of glycoalkaloid-based resistance combined with *Bt cry1Ac* is also being applied to host plant resistance to *P. operculella* (Estrada et al., 2007).

This same strategy of combining natural and engineered traits can also be applied to glandular trichome-mediated resistance (Fig. 7.2). The combination of *Bt cry3A* gene and glandular trichomes may provide a broader-based insect resistance providing control of small-bodied insects such as leafhoppers and aphids, using the glandular trichomes, along with *L. decemlineata* resistance provided by *Bt cry3A*.

7.8.4 Integrated Pest Management and Resistance Management

Finally, to effectively manage resistance, we must not forget the diversity of pest control methods available through integrated pest management practices. For managing resistance to conventional insecticides, alternation of products and product mixtures (analogous to multiple genetic resistance factors presented in alternation or by pyramiding) have repeatedly failed for insects such as *L. decemlineata* and *P. xylostella*. While it is true that many of these practices were not adopted until after significant insecticide resistance problems had occurred, implementation of similar practices for managing GM crops may also be delayed or ignored until after significant problems have occurred. Other mortality factors, such as crop rotation or biological control may

Fig. 7.2 Colorado potato beetle choice field study. Defoliation of susceptible potato clones (left) compared to natural (leptine glycoalkaloids, center) and engineered (*Bt cry3A* in combination with glandular trichomes, right) host plant resistance clones developed by Michigan State University (Photo by Joseph Coombs, Michigan State University Potato Breeding and Genetics)

be highly effective and must be included in the management system, if our experiences with insecticide resistance are any indication. For example, for *L. decemlineata*, one of the most effective management strategies is crop rotation with a non-host crop, locating the new potato field more than 0.5 km from the previous location. This distance requires *L. decemlineata* adults to disperse by flight rather than walking to the new potato fields; densities are greatly reduced in rotated fields and spring adults arrive later, reducing reproductive potential (Weisz et al., 1994). Crop rotation also encourages gene flow. For *P. operculella*, a grandulosis virus is available and can be effectively used in potato storages, in place of conventional insecticides (Sporleder et al., 2005). Incorporation of host plant resistance into an integrated pest management system involving multiple biological, cultural, and chemical controls will greatly increase the sustainability of a pest management system.

7.9 History of GM Potatoes

With the advent of genetic engineering, applications of this technology were quickly targeted to the potato. Being a heterozygous, tetraploid species with a history of slow varietal adoption, the insertion of economically important genes is considered a positive strategic approach to improvement of potato. Moreover, the gene insertion technology is an expedient, knowledge-based approach that is scale neutral. In potato, genetic engineering has focused on insect resistance to *L. decemlineata* and *P. operculella*, disease resistance to late blight, nutritional enhancement, stress tolerance, virus resistance, blackspot bruise resistance, starch modification, cold-induced sweetening resistance and vaccine delivery (reviewed by Bradeen et al., 2008). The use of *Bt* genes offers unique solutions that integrate technology delivery and do not displace traditional breeding methods. The implementation of the *Bt* technology also is environmentally friendly and can lead to more sustainable and stable pest and crop management. Research has shown that *Bt* potatoes reduce insecticide use with benefits for biological control that help control other pests, such as aphids, that are not targeted by the *Bt* toxin (Reed et al., 2001; Duan et al., 2004; Romeis et al., 2006, chapter 4). In some cases it can lead to increasing crop productivity.

The first *Bt* solanaceous plants (tomatoes) were reported in 1987 with partial resistance to lepidopteran insects (Fischoff et al., 1987; Vaeck et al., 1987). Following improvements through truncation of the *Bt* gene and codon modification to optimize protein expression in plants, *Bt* potato plants expressing the *Bt cry3A* gene were developed that provided control of *L. decemlineata* (Adang et al., 1993; Perlak et al., 1993). Monsanto, through its subsidiary, NatureMark, first launched *Bt cry3A* potatoes under the Newleaf® trademark onto the market in North America in 1995. *L. decemlineata* was a pest that North American growers were having increasing problems controlling due to the high levels of insecticide resistance in the population (Simon, 2003). In 1997 the release of the *Bt cry3A* potatoes combined with either resistance to potato virus Y or potato leaf roll virus were referred to NewLeaf Y® and NewLeaf Plus®, respectively (Thomas et al., 1997). When NewLeaf® cultivars were introduced in 1995, 1,500 acres were grown commercially and as seed stocks increased, the commercial acreage reached 50,000 acres. Market success of the NewLeaf, NewLeaf Y® and NewLeaf Plus® potatoes could be attributed to the difficulty in controlling *L. decemlineata* and also high pest populations of aphids and associated virus problems due to mild winters in the Pacific Northwest (Thornton, 2003). With the NewLeaf® potatoes, growers were able to reduce insecticide costs. NewLeaf Plus® was grown mainly in the Pacific Northwest. The added virus resistance benefited seed producers, while commercial growers benefited from higher yields and reduced need for insecticides (Thornton, 2003). The processing industry and consumers benefited from improved quality. Potatoes were one of the first foods from a GM crop that was commonly served in restaurants.

NewLeaf® cultivars were the fastest varietal adoption in the history of the USA potato industry (Thornton, 2003). Although GM insect and virus resistant potatoes were highly effective and growers were increasingly using them, acreage did not increase at the same rate as in maize or cotton, and market growth for GM potatoes was not as rapid as Monsanto would have liked for several reasons (Thornton, 2003). The ability of growers to save potato seed for future crops necessitated that NatureMark devise a system to control the distribution of GM seed potatoes and maintain royalties needed to offset their investment in the technology. Clonal variation that existed in the seed stock of the Russet Burbank variety, the predominant North American variety (Simon, 2003), led to concerns that the transgenic selections of Russet Burbank and other NewLeaf® cultivars did not perform equivalent to their non-transgenic counterparts. Another factor was the registration of a new insecticide, imidicloprid, which when applied in furrow gave excellent control of *L. decemlineata* and aphids and early season control of *E. fabae* (Thornton, 2003). Imidacloprid offered the growers a conventional alternative for *L. decemlineata* control without being limited by varietal selection. Another factor working against NewLeaf® adoption was the requirement that a portion of the acreage be planted with non-*Bt* cultivars for resistance management, whereas imidacloprid could be used on 100% of a grower's acreage.

With public debate about the risks and benefits of biotechnology gaining attention in the media, segregation of the GM potatoes was requested by consumers. In 1999, when the organized anti-biotech campaigns aimed at consumers began, the large end-users of potatoes in the North American quick serve industry became concerned that their market share could be negatively impacted in Europe and Japan because they used *Bt* potatoes in their North American markets (Simon, 2003). Strategies to segregate GM and non-GM potatoes were attempted, but these changes in practice did not add value to the business. McDonald's decision to ban GM crops from its food chain had a major impact. After the 1999 season the processors decided they could not afford the market risk associated with GM potatoes. Potato processors, also under pressure from export markets in Europe, were forced to suspend contracts. International trade barriers were more substantial for GM potatoes than other technology adoptions (Guenthner, 2002). Thus, more than 60% of the USA market was closed to GM potatoes. This led to the processor and commercial grower discontinuing use, hence the loss of a market for NatureMark potatoes. One additional factor that led to the rapid demise was that only 3% of the USA potato acreage was *Bt* potatoes (Guenthner, 2002). Closing this part of the market had little impact on potato supplies and didn't lead to widespread market disruption. NatureMark dissolved after the 2001 season.

"It is ironic that those activists who list reduction in use of pesticides as a major goal are those that have effectively blocked the most successful scientific approach to that end"(Kaniewski and Thomas, 2004).

GM potatoes entered a complex, dynamic market in which consumer acceptance is a powerful force (Guenthner, 2002). Anti-biotech organizations spent and continue to raise money trying to influence societal acceptance of GM crops by creating doubt about the safety of the deregulated *Bt* crops. Hence, GM potato acceptance was slowed by perceived health concerns. The key reason for the withdrawal of GM cultivars from the potato industry was that big brand users of potatoes in the food service industry reacted to the consumer concerns in Asia and Europe (Simon, 2003). In a precautionary stance they instructed their suppliers to stop processing GM potatoes for them. Processors had little to gain by accepting GM raw materials but exposed themselves to significant risk in marketing. Although growers accepted GM potatoes, *Bt* potatoes, which provided benefits to producers but not consumers, did not have a consumer-acceptance accelerator (Guenthner, 2002). Unlike cotton and most maize and soybeans, *Bt* potatoes are directly consumed, which may affect consumer attitudes. Most North American growers are increasing their plantings of GM crops other than potatoes because of the significant advantages they provide to their production system (James, 2007). Unfortunately for *Bt* potatoes, imidacloprid was registered in the same year *Bt* potatoes became available and proved to be a very strong competitor in the potato insect control market. Not only was imidacloprid priced competitively with *Bt* potatoes, but it was more widely available and controlled a wider range of pests. It is not known whether *Bt* potatoes would have become more successful if imidacloprid had not become available. Still, the primary reason that *Bt* potatoes are now off the market is because of trade issues driven by international consumers.

7.10 Future of GM Potatoes

GM potatoes will probably reappear in Eastern Europe, Africa, and Asia before Europe and North America because in the former locations, the producer is more likely to be the consumer and GM technology may have a great attraction. For example, in Africa, the primary potato insect pest is *P. operculella*, which is difficult to control and requires multiple applications of insecticides, including multiple treatments to tubers in storage (Douches et al., 2004). Trials with GM potatoes have been very successful in South Africa and Egypt (Douches et al., 2004). Currently, the only available means to control the potato tuber moth and avoid major crop losses is the use of chemical insecticides.

Michigan State University, funded by the U.S. Agency for International Development (USAID), through its Agricultural Biotechnology Support Project, initiated biotechnology research on the development of *P. operculella* resistant cultivars in 1992. A *Bt cry1Ia1* gene was obtained from ICI Seeds (now Syngenta seed company) and successfully introduced into several potato cultivars, including the Dutch variety Spunta. Transgenic lines were shown to parovide a high level of control of *P. operculella*. The *Bt* potato will be one of the first public sector-developed products to reach farmers in developing countries and will serve as a model for the public-sector deployment of crops that are resistant to insects. A research team from Michigan State University and the South African Agricultural Research Council has been leading this effort in South Africa since 2001. The benefits of this product to the farmer and end-users will be reduced insecticide use, increased marketable yield, improved quality, reduced post-harvest losses, reduced human exposure to pesticides, and less pesticide residues on potato tubers.

The commercialization project includes six components: product development, regulatory file development, obtaining freedom to operate and establishing licensing relationships, marketing and technology delivery, documentation of socio-economic benefits, and public communication. USAID is interested in expanding this effort in other developing countries. The regulatory dossier to address food and environmental safety of the *Bt* potato will be submitted to the South African Biosafety Committee in January 2008. With government approval, the *Bt* potato will continue towards commercialization through farmer participatory trials and the breeding of new potato lines derived from the approved transgenic event. Other countries may be interested in these *P. operculella* resistant potatoes after commercial success is demonstrated in South Africa. Other than at the International Potato Center (CIP) in Peru, no other insect-resistant GM potatoes are being developed elsewhere internationally.

In Western Europe late blight and eelworm, *Heterodera rostochiensis*, resistance could be the primary drivers for market penetration of GM potatoes (Simon, 2003). When the consumer has confidence in the technology, it will spread quickly due to the environmental and production benefits. These two GM opportunities would greatly reduce pesticide use and would be extremely attractive to potato growers. Jiang and Helgeson cloned the late blight resistance gene (*RB*) from *S. bulbocastanum*, a Mexican diploid potato species (Song et al., 2003). Potato lines expressing the *RB* gene show a level of foliar late blight resistance that would have value in commercial potato production (Kuhl et al., 2007). The *RB* gene belongs to a class of characterized resistance genes that encode proteins with nucleotide binding and leucine-rich repeat domains (Song et al., 2003). Recently other late blight resistance genes have been mapped and cloned from S. *bulbocastanum* (van der Vossen et al., 2003), *S. mochiquense* (Smilde et al., 2005), and a complex genomic hybrid (Park et al., 2005). The resistance genes from these *Solanum* species offer race non-specific resistance, unlike those previously utilized from *S. demissum*. The ability to transform major late blight resistance genes into potato provides a unique opportunity to pyramid late blight resistance genes in an analytic manner. Moreover, the pyramided resistance genes in a single genotype should be a better strategy to deploy late blight resistant potato cultivars (Dangl and Jones, 2001).

Virus resistance in GM potatoes will be especially valuable because there are no pesticides for control of viruses, in contrast to insect pests or fungal diseases. Since potatoes are propagated by tuber cuttings, viruses are easily spread during the seed multiplication process. The existence of potato cultivars that do not show visible symptoms means that infected plants cannot be identified visually and removed. These cultivars may serve as inoculation sources for other nearby potatoes. Also, control of aphids is ineffective for controlling most virus diseases of potatoes, except for control of aphid vectors of potato leafroll virus, where long acquisition and infection periods are necessary.

New developments in the USA indicate that GM potatoes may return to the market. J.R. Simplot Co., the Boise Idaho-based frozen potato processing company and major supplier of McDonald's french fries, discontinued its use of GM potatoes in 1999. However, the Simplot Co. has continued to invest in biotechnology research and development. They have developed a transgenic approach they refer to as "Precision Breeding" or an "all-native approach" (Rommens et al., 2004). Precision breeding uses only native potato DNA in the patented transformation system. Target traits at this time include black spot bruise resistance, reduced acrylimide levels, and lower reducing sugar in the tubers (Rommens et al., 2006). The positive public perception of precision breeding may allay some concerns about GM technology.

Another positive development is the public/private effort to establish a Specialty Crops Regulatory program in the USA. The goal of the Specialty Crop Research Initiative is to establish an organization at the federal level that will facilitate the development of biotech-derived specialty crops including potatoes. If this program is put in place, it will assist in generating necessary regulatory data and facilitate obtaining regulatory clearances necessary for a biotech product.

7.11 Conclusions

Potato is a high value crop with low thresholds for damage and multiple pest problems. They have a wide variety of pest, crop quality, and yield problems that cannot be dealt with only using pesticides or traditional breeding methods. Imidacloprid and other pesticides currently are highly effective and economical. Resistance to imidacloprid in *L. decemlineata*, increasing problems with aphid-transmitted viruses, and the presence of *P. operculella* in the Pacific Northwest may rapidly increase the value of GM potatoes in the USA. Globally, insecticide resistant *L. decemlineata* are present throughout Europe and western Asia and *P. operculella* is a major pest in the field and in storage throughout tropical and subtropical regions. *P. operculella* is especially damaging to potatoes in unrefrigerated storages, as are common in developing countries, but can be 100% controlled by resistant GM potatoes (Douches et al., 2004). *L. decemlineata* and *P. operculella* are resistant to multiple insecticides (Whalon et al., 2007) making control with insecticides difficult or impossible in some regions. Presently available control options for growers are limited to crop rotation plus the use of new (and more costly) insecticides, if these are available, and alternating or mixing insecticides. Excessive use of insecticides in response to a resistance problem decreases the effectiveness of biological control agents and increases the risk of environmental and human health impacts, increasing the utility of GM cultivars with resistance to pests such as *L. decemlineata* and *P. operculella*. The advantages of GM technology to improve the potato crop, reduce pesticide use, increase yields, and lower production costs will continue to provide incentive for integration of this technology into potato breeding and commercial crop production. As consumers become more accustomed to other GM crops, concerns about GM potatoes are likely to lessen and markets will accept the new products. Potato, a minor crop by itself, was unable to lead the change to consumer acceptance of GM crops in the USA, but can perhaps follow the widespread acceptance of GM maize, soybeans, and other crops.

References

Adang, M.J., Brody, M.S., Cardineau, G., Eagan, N., Roush, R.T., Shewmaker, C., Jones, A., Oakes, J., and McBride, K., 1993. The reconstruction and expression of a *Bacillus thuringiensis cryIIIA* gene in protoplasts and potato plants. Plant Molecular Biology 21: 1131–1145.

An, G., Watson, B., and Chiang, C., 1986. Transformation of tobacco, tomato, potato, and *Arabidopsis thaliana* using a binary Ti vector system. Plant Physiology 81: 301–305.

Anonymous, 1984. Potatoes for the Developing World. International Potato Center, Lima, Peru, pp. 13–16.

Anonymous, 2005. Neonicotinoid Insecticides–A Grower Approach to Resistance Management for Colorado Potato Beetle and Green Peach Aphid in Potatoes. National Potato Council. Washington, DC, USA.
http://www.irac-online.org/documents/insecticides_press.pdf (accessed 23 January 2008).

Arpaia, S., Gould, F., and Kennedy, G., 1997. Potential impact of *Coleomegilla maculata* predation on adaptation of *Leptinotarsa decemlineata* to Bt transgenic potatoes. Entomologia Experimentalis et Applicata 82: 91–100.

Ballvora, A., Ercolano, M.R., Weiss, J., Meksem, K., Bormann, C.A., Oberhagemann, P., Salamini, F., and Gebhardt, C., 2003. The R1 gene for potato resistance to late blight (*Phytophthora infestans*) belongs to the leucine zipper/NBS/LRR class of plant resistance genes. The Plant Journal 30: 361–371.

Bates, S.L., Zhao, J.Z., Roush, R.T., and Shelton, A.M., 2005. Insect resistance management in GM crops: past, present and future. Nature Biotechnology 23: 57–62.

Biever, K.D., and Chauvin, R.L., 1992. Suppression of the Colorado potato beetle (Coleoptera: Chrysomelidae) with augmentative releases of predaceous stinkbugs (Hemiptera: Pentatomidae). Journal of Economic Entomology 85: 720–726.

Bishop, B.A., and Grafius, E., 1996. Insecticide resistance in the Colorado potato beetle. In: Chrysomelidae Biology, Vol. 1, The Classification, Phylogeney and Genetics, P. Jolivet and M. Cox, eds., SPB Academic Publishing, Amsterdam, The Netherlands, pp. 355–377.

Bishop, B., Grafius, E., Byrne, A., Pett, W., and Bramble, E., 2003. Potato insect biology and management. 2002 Research Report. Michigan Potato Research Report 34: 113–130. Michigan Agricultural Experiment Station, East Lansing, MI, USA.

Boiteau, G., 1988. Sperm utilization and post-copulatory female guarding in the Colorado potato beetle, *L. decemlineata*. Entomologia Experimentalis et Applicata 47: 183–187.

Boiteau, G., Misener, G., Singh, R., and Bernard, G., 1992. Evaluation of a vacuum collector for insect pest control in potato. American Potato Journal 69: 157–166.

Bradeen, J.M., Carputo, D., and Douches, D.S., 2008. Potato. In: Transgenic Series, Vol. 4, Sugar, Tuber and Fiber Crops, C.R. Kole and T.C. Hall, eds., Blackwell, Oxford, UK (in press).

Brown, C.R., 2005. Antioxidants in potato. American Journal of Potato Research 82: 163–172.

Brunelle, F., Girard, C., Cloutier, C., and Michaud, M., 2005. A hybrid, broad-spectrum inhibitor of Colorado potato beetle aspartate and cysteine digestive proteinases. Archives of Insect Biochemistry and Physiology 60: 20–31.

Byrne, A., Grafius, E., Bishop, B., and Pett, W., 2004. Susceptibility of Colorado potato beetle populations to imidacloprid and thiamethoxam. Arthropod Management Tests 29: L12.

Cameron, P.J., Wallace, A.R., Madhusudhan, V.V., Wigley, P.J., Qureshi, M.S., and Walker, G.P., 2005. Mating frequency in dispersing potato tuber moth, *Phthorimaea operculella*, and its influence on the design of refugia to manage resistance in Bt transgenic crops. Entomologia Experimentalis et Applicata 115: 323–332.

Cappaert, D.L., Drummond, F.A., and Logan, P.A., 1991. Incidence of natural enemies of the Colorado potato beetle, *Leptinotarsa decemlineata* (Coleoptera: Chrysomelidae) on a native host in Mexico. Entomophaga 36: 369–378.

Capinera, J.L., 2001. Handbook of Vegetable Insects. Academic, NY, USA, 729 p.

Casagrande, R.A., 1987. The Colorado potato beetle: 125 years of mismanagement. Bulletin of the Entomological Society of America 18: 142–150.

Chen, M., Ye, G.Y., Liu, Z.C., Yao, H.W., Chen, X.X., Shen, S.C., Hu, C., and Datta, S.K., 2006. Field assessment of the effects of transgenic rice expressing fused gene of *cry1Ab* and *cry1Ac* from *Bacillus thuringiensis* Berliner on nontarget planthopper and leafhopper populations. Environmental Entomology 35: 127–134.

Cloutier, C., Fournier, M., Jean, C., Yelle, S., and Michaud, D., 1999. Growth compensation and faster development of Colorado potato beetle (Coleoptera: Chrysomelidae) feeding on potato foliage expressing oryzacystatin I. Archives of Insect Biochemistry and Physiology 40: 69–79.

Coombs, J.J., Douches, D.S., Li, W.B., Grafius, E.J., and Pett, W.L., 2002. Combining engineered (*Bt-cry3A*) and natural resistance mechanisms in potato (*Solanum tuberosum* L.) for control of the Colorado potato beetle. Journal of the American Society of Horticultural Science 127: 62–68.

Cooper, S.G., Douches, D.S., and Grafius, E.J., 2004. Combining genetic engineering and traditional breeding to provide durable resistance in potatoes to Colorado potato beetle. Entomologia Experimentalis et Applicata 112: 37–46.

Dangl, J.L., and Jones, J.D.G., 2001. Plant pathogens and integrated defense responses to infection. Nature 411: 826–833.

Daniell, H., Datta, R., Varma, S., Gray, S., and Lee, S.B., 1998. Containment of herbicide resistance through genetic engineering of the chloroplast genome. Nature Biotechnology 16: 345–348.

Davidson, M.M., Butler, R.C., Wratten, S.D., and Conner, A.J., 2004. Resistance of potatoes transgenic for a *cry* 1Ac9 gene, to *Phthorimaea operculella* (Lepidoptera: Gelichiidae) over field seasons and between plant organs. Annals of Applied Biology 145: 271–277.

Davidson, M.M., Butler, R.C., Wratten, S.J., and Conner, A.J., 2005. Age-specific bioassays and fecundity of *Phthorimaea operculella* (Lepidoptera: Gelichiidae) reared on *cry*1Ac-transgenic potato plants. Annals of Applied Biology 146: 493–499.

Davis, J.A., Radcliffe, E.B., and Ragsdale, D.W., 2007. Resistance to green peach aphid, *Myzus persicae* (Sulzer), and potato aphid, *Macrosiphum euphorbiae* (Thomas), in potato cultivars. American Journal of Potato Research 84: 259–269.

Difonzo, C., Ragsdale, D.W., and Radcliffe, E.B., 2007. Integrated Management of Seed Potato with Emphasis on the Red River Valley of Minnesota and North Dakota. Radcliffe's IPM World Textbook. http://www.ipmworld.umn.edu/chapters/difonzo.htm (accessed 23 January 2008).

Douches, D.S., Westedt, A.L., Zarka, K., Schroeter, B., and Grafius, E.J., 1998. Transformation of CryV-Bt transgene combined with natural resistance mechanisms for resistance to tuber moth in potato (*Solanum tuberosum* L.). HortScience 33: 1053–1056.

Douches, D.S., Pett, W., Santos, F., Coombs, J., Grafius, E., Li, W., Metry, E.A., Nasr El-Din, T., and Madkour, M., 2004. Field and storage testing Bt-potatoes for resistance to potato tuber moth (Lepidoptera: Gelichiidae). Journal of Economic Entomology 97: 1425–1431.

Down, R.E., Gatehouse, A.M.R., Hamilton, W.D.O., and Gatehouse, J.A., 1996. Snowdrop lectin inhibits development and decreases fecundity of the glasshouse potato aphid (*Aulacorthum solani*) when administered in vitro and via transgenic plants both in laboratory and glasshouse trials. Journal of Insect Physiology 42: 1035–1045.

Duan, J.J., Head, G., Jensen, A., and Reed, G., 2004. Effects of transgenic *Bacillus thuringiensis* potato and conventional insecticides for Colorado potato beetle (Coleoptera: Chrysomelidae) management on the abundance of ground-dwelling arthropods in Oregon potato ecosystems. Environmental Entomology 33: 275–281.

Estrada, M.A., Zarka, K., Cooper, S., Coombs, J., Douches, D.S., and Grafius, E., 2007. Potato tuberworm (Lepidoptera: Gelechiidae) resistance in potato lines with the *Bacillus thuringiensis cry1Ac* gene and natural resistance factors. HortScience 42: 1306–1311.

Estruch, J.J., Warren, G.W., Mullins, M.A., Nye, G.J., Craig, J.A., and Koziel, M.G., 1996. Vip3A, a novel *Bacillus thuringiensis* vegetative insecticidal protein with a wide spectrum of activities against lepidopteran insects. Proceedings of the National Academy of Sciences of the USA 93: 5389–5394.

FAOSTAT data, 2006. http://faostat.fao.org/site/408/DesktopDefault.aspx?PageID = 408 (accessed 19 December 2007).

Ferré, J., and Van Rie, J., 2002. Biochemistry and genetics of insect resistance to *Bacillus thuringiensis*. Annual Review of Entomology 47: 501–533.

Fischoff, D., Bowdish, K.S., Perlak, F.J., Marrone, P.G., McCormick, S.M., Niedermeyer, J.G., Dean, D.A., Kusano-Kretzmer, K., Mayer, E.J., Rochester, D.E., Rogers, S.G., and Fraley, R. T., 1987. Insect tolerant transgenic tomato plants. Nature Biotechnology 5: 807–813.

Forgash, A.J., 1985. Insecticide resistance in the Colorado potato beetle. In: Proceedings of the symposium of the Colorado potato beetle, D.N. Ferro and R.H. Voss, eds., XVIIth International Congress of Entomology, Massachusetts Agricultural Experiment Station Bulletin. No. 704. Amherst, MA, USA, pp. 33–52.

Gatehouse, A.M.R., Davison, G.M., Newell, C.A., Merryweather, A., Hamilton, W.D.O., Burgess, E.P.J., Gilbert, R.J.C., and Gatehouse, J.A., 1997. Transgenic potato plants with enhanced resistance to the tomato moth, *Lacanobia oleracea*: growth room trials. Molecular Breeding 3: 49–63.

Gatehouse, A.M.R., Davison, G.M., Steward, J.N., Gatehouse, L.N., Kumar, A., Geoghegan, I.E., Birch, A.N.E., and Gatehouse, J.A., 1999. Concanavalin A inhibits development of tomato moth (*Lacanobia oleracea*) and peach-potato aphid (*Myzus persicae*) when expressed in transgenic potato plants. Molecular Breeding 5: 153–165.

Gauthier, N.L., Hofmaster, R.N., and Semel, M., 1981. History of Colorado potato beetle control. In: Advances in Potato Pest Management, J.H. Lashomb and R.A. Casgrande, eds., Hutchinson Ross, Stroudsburg, PA, USA, pp. 13–33.

Grafius, E., 1997. Economic impact of insecticide resistance in the Colorado potato beetle (Coleoptera: Chrysomelidae) on the Michigan potato industry. Journal of Economic Entomology 90: 1144–1151.

Grafius, E., Pett, W., Bishop, B. Byrne, A., and Bramble, E., 2005. Potato insect biology and management. Michigan Potato Research Report 36: 42–56. Michigan Agricultural Experiment Station, East Lansing, MI, USA.

Groden, E., Drummond, F.A., Casagrande, R.A., and Haynes, D.L., 1990. *Coleomegilla maculata* (Coleoptera: Coccinellidae): its predation upon the Colorado potato beetle (Coleoptera: Chrysomelidae) and its incidence in potatoes and surrounding crops. Journal of Economic Entomology 83: 1306–1315.

Guenthner, J.F., 2002. Consumer acceptance of genetically modified potatoes. American Journal of Potato Research 79: 309–316.

Harcourt, D.G., 1971. Population dynamics of *Leptinotarsa decemlineata* (Say) in eastern Ontario. III. Major population processes. Canadian Entomologist 103: 1049–1061.

Hare, J.D., 1990. Ecology and management of the Colorado potato beetle. Annual Review of Entomology 35: 81–100.

Hawkes, J.G., 1994. Origins of the cultivated potatoes and species relationships. In: Potato Genetics, J. Bradshaw and G. Mackay, eds., CAB International, Wallingford, UK, pp. 3–42.

Hazzard, R.V., and Ferro, D.N., 1991. Feeding response of adult *Coleomegilla maculata* (Coleoptera: Coccinellidae) to eggs of Colorado potato beetle (Coleoptera: Chrysomelidae) and green peach aphids (Homoptera: Aphididae). Environmental Entomology 20: 644–651.

Heim, D.C., Kennedy, G.G., and Vanduyn, J.W., 1990. Survey of insecticide resistance among North Carolina Colorado potato beetle (Coleoptera, Chrysomelidae) populations. Journal of Economic Entomology 83: 1229–1235.

Hilbeck, A., Eckel, C., and Kennedy, G.G., 1997. Predation on Colorado potato beetle eggs by generalist predators in research and commercial potato plantings. Biological Control 8: 191–196.

Hough-Goldstein, J., and Keil, C.B., 1991. Prospects for integrated control of the Colorado potato beetle (Coleoptera: Chrysomelidae) using *Perillus bioculatus* (Hemiptera: Pentatomidae) and various pesticides. Journal of Economic Entomology 84: 1645–1651.

Hough-Goldstein, J.A., Heimpel, G.E., Bechmann, H.E., and Mason, C.E., 1993. Arthropod natural enemies of the Colorado potato beetle. Crop Protection 12: 324–334.

Huang, F.C., Klaus, S., Herz, S., Zou, Z., Koop, H.U., and Golds, T., 2002. Efficient plastid transformation in tobacco using the aphA-6 gene and kanamycin selection. Molecular Genetics and Genomics 268: 19–27.

Ioannidis, P.M., Grafius, E., and Whalon, M.E., 1991. Patterns of insecticide resistance to azinphosmethyl carbofuran and permethrin in Colorado potato beetle (Coleoptera: Chrysomelidae). Journal of Economic Entomology 84: 1417–1423.

Ioannidis, P.M, Grafius, E., Wierenga, J.M., Whalon, M.E., and Hollingworth, R.M., 1992. Selection, inheritance and characterization of carbofuran resistance in Colorado potato beetle (Coleoptera: Chrysomelidae). Pesticide Science 35: 215–222.

Jaffe, G., 2003. Planting Trouble: Are Farmers Squandering Bt Corn Technology. Center for Science in the Public Interest, Washington, DC, USA. http://www.cspinet.org/new/pdf/bt_corn_report.pdf (accessed 23 January 2008).

James, C., 2007. Global Status of Commercialized Biotech/GM Crops: 2007. ISAAA Brief No. 37, International Service for the Acquisition of Agri-Biotech Applications, Ithaca, NY, USA.

Kaniewski, W., and Thomas, P., 2004. The potato story. AgBioForum 7(1&2): 41–46.

Kuhl, J.C., Zarka, K., Coombs, J., Kirk, W.W., and Douches, D.S., 2007. Late blight resistance of *RB* transgenic potato lines. Journal of the American Society of Horticultural Science 132: 783–789.

Lachman, J., Hamouz, K., Orsák, M., and Pivec, V., 2001. Potato glycoalkaloids and their significance in plant protection and human nutrition – review. Series Rostlinná Vyroba 47: 181–191.

Lagnaoui, A., Cañedo, V., and Douches, D.S., 2001. Evaluation of *Bt-cry1Ia1* (*cryV*) transgenic potatoes on two species of potato tuber moth, *Phthorimaea operculella* and *Symmetrischema tangolias* (Lepidoptera: Gelechiidae) in Peru. CIP Program Report 1999–2000, International Potato Center, Lima, Peru, pp. 117–121.

Lambert, B., and Peferoen, M., 1992. Insecticidal promise of *Bacillus thuringiensis*. BioScience 42: 112–122.

Lapointe, S.L., and Tingey, W.M., 1986. Glandular trichomes of *Solanum neocardenasii* confer resistance to green peach aphid (Homoptera: Aphididae). Journal of Economic Entomology 79: 1264–1268.

Lawson, D.R., Veilleux, R.E., and Miller, A.R., 1993. Biochemistry and genetics of *Solanum chacoense* steroidal alkaloids: natural resistance factors to the Colorado potato beetle. Current Topics in Botanical Research 1: 335–352.

Lecardonnel, A., Chauvin, L., Jouanin, L., Beaujean, A., Prevost, G., and Sangwan-Norreel, B., 1999. Effects of rice cystatin I expression in transgenic potato on Colorado potato beetle larvae. Plant Science 140: 71–79.

Lee, M.K., Miles, P., and Chen, J.S., 2005. Brush border membrane binding properties of *Bacillus thuringiensis* Vip3A toxin to *Heliothis virescens* and *Helicoverpa zea* midguts. Biochemical and Biophysical Research Communications 339: 1043–1047.

Lorenzen, J.H., Balbyshev, N.F., Lafta, A.M., Casper, H., Tian, X., and Sagredo, B., 2001. Resistant potato selections contain leptine and inhibit development of the Colorado potato beetle (Coleoptera: Chrysomelidae). Journal of Economic Entomology 94: 1260–1267.

Maliga, P., 2004. Plastid transformation in higher plants. Annual Review of Plant Biology 55: 289–313.

Mani, G.S., 1985. Evolution of resistance in the presence of two insecticides. Genetics 109: 761–783.

Meiyalaghan, S., Jacobs, J.M.E., Butler, R.C., Wratten, S.D., and Conner, A.J., 2006. Expression of *cry* 1Ac9 and *cry*9Aa2 genes under a potato light-inducible *Lhca*3 promoter in transgenic potatoes for tuber moth resistance. Euphytica 147: 297–309.

Michaud, D., Nguyen-Quoc, B., and Yelle, S., 1993. Selective inhibition of Colorado potato beetle cathepsin H by oryzacystatins I and II. Federation of European Biochemical Societies Letters 331: 173–176.

Misener, G.C., Bioteau, G., and McMillan, L.P., 1993. A plastic-lining trenching device for the control of Colorado potato beetle: beetle excluder. American Potato Journal 70: 903–908.

Moellenbeck, D.J., Peters, M.L., Bing, J.W., Rouse, J.R., Higgins, L.S. Sims, L., Nevshemal, T., Marshall, L., Ellis, R.T., Bystrak, P.G., Lang, B.A., Stewart, J.L., Kouba, K., Sondag, V., Gustafson, V., Nour, K., Xu, D., Swenson, J., Zhang, J., Czapla, T., Schwab, G., Jayne, S., Stockhoff, B.A., Narva, K., Schnepf, H.E., Stelman, S.J., Poutre, C., Koziel, M., and Duck, N., 2001. Insecticidal proteins from *Bacillus thuringiensis* protect corn from corn rootworms. Nature Biotechnology 19: 668–672.

Mota-Sanchez, D., Hollingworth, R.M., Grafius, E.J., and Moyer, D.D., 2006. Resistance and cross-resistance to neonicotinoid insecticides and spinosad in the Colorado potato beetle (Coleoptera: Chrysomelidae). Pest Management Science 62: 30–37.

Moyer, D.D., Derksen, R.C., and McLeod, M.J., 1992. Development of a propane flamer for Colorado potato beetle. American Potato Journal 69: 599–600.

Naimov, S., Dukiandjiev, S., and de Maagd, R.A., 2003. A hybrid *Bacillus thuringiensis* delta-endotoxin gives resistance against a coleopteran and a lepidopteran pest in transgenic potato. Plant Biotechnology Journal 1: 51–57.

National Agricultural Statistics Service, 2006. http://usda.mannlib.cornell.edu/MannUsda/viewDocumentInfo.do?documentID = 1560 (accessed April 2007).

Park, T.H., Gros, J., Sikkema, A., Vleeshouwers, V., Muskens, M., Allefs, S., Jacobsen, E., Visser, R.G.F., and van der Vossen, E.A.G., 2005. The late blight resistance locus *Rpi-blb3* from *Solanum bulbocastanum* belongs to a major late blight *R* gene cluster on chromosome 4 of potato. Molecular Plant Microbe Interactions 18: 722–729.

Paul, A.A., and Southgate, D.A.T., 1978. McCance and Widdowson's the composition of foods, 4th ed. MRC Special Report No. 297. Her Majesty's Stationery Office, London, UK.

Perlak, F.J., Stone, T.B., Muskopf, Y.M., Petersen, L.J., Parker, G.B, McPherson, S.A., Wyman, J., Love, S., Reed, G., Biever, D., and Fischhoff, D.A., 1993. Genetically improved potatoes: protection from damage by Colorado potato beetles. Plant Molecular Biology 22: 313–321.

Plaisted, R.L., Tingey, W.M., and Steffens, J.C., 1992. The germplasm release of NYL235-4: a clone with resistance to the Colorado potato beetle. American Potato Journal 69: 843–846.

Radcliffe, E.B., and Ragsdale, D.W., 2002. Aphid-transmitted potato viruses: the importance of understanding vector biology. American Journal of Potato Research 79: 353–386.

Reed, G.L., Jensen, A.S., Riebe, J., Head, G., and Duan, J.J., 2001. Transgenic *Bt* potato and conventional insecticides for Colorado potato beetle management: comparative efficacy and non-target impacts. Entomologia Experimentalis et Applicata 100: 89–100.

Ripperger, H., and Schreiber, K., 1981. *Solanum* steroid alkaloids. In: The Alkaloids, Chemistry and Physiology, R. Manske and R. Rodrigo, eds., Academic, New York, USA, pp. 81–192.

Roddick, J.G., 1986. Steroidal alkaloids of the Solanaceae. In: Solanaceae, Biology and Systematics, W.G. D'Arcy, ed., Columbia University Press, New York, USA, pp. 201–222.

Romeis, J., Meissle, M., and Bigler, F., 2006. Transgenic crops expressing *Bacillus thuringiensis* toxins and biological control. Nature Biotechnology 24: 63–71.

Rommens, C.M., Humara, J.M., Ye, J., Yan, H., Richael, C., Zhang, L., Perry, R., and Swords, K., 2004. Crop improvement through modification of the plant's own genome. Plant Physiology 135: 421–431.

Rommens, C.M., Ye, J., Richael, C., and Swords, K., 2006. Improving potato storage and processing characteristics through all-native DNA transformation. Journal of Agriculture and Food Chemistry 54: 9882–9887.

Roush, R.T., 1998. Two-toxin strategies for management of insecticidal transgenic crops: can pyramiding succeed where pesticide mixtures have not? Philosophical Transactions of the Royal Society London Series B-Biological Science 353: 1777–1786.

Sanford, L.L., Kobayshi, R.S., Deahl, K.L., and Sinden, S.L., 1996. Segregation of leptines and other glycoalkaloids in *Solanum tuberosum* (4X) X *S. chacoense* (4X) crosses. American Potato Journal 73: 21–31.

Sexson, D.L., Wyman, J., Radcliffe, E.B., Hoy, C.J., Ragsdale, D.W., and Dively, G.P., 2005. Potato. In: Vegetable Insect Management, R. Foster and B. Flood, eds., Meister Publishing, Willoughby, OH, USA, pp. 92–107.

Sharma, H.C., Crouch, J.H., Sharma, K.K., Seetharama, N., and Hash, C.T., 2002. Applications of biotechnology for crop improvement: prospects and constraints. Plant Science 163: 381–395.

Shelton, A.M., Zhao, J.Z., and Roush, R.T., 2002. Economic, ecological, food safety and social consequences of the deployment of Bt transgenic plants. Annual Review of Entomology 47: 845–881.

Simon, W., 2003. The agricultural advantages and consumer acceptance or otherwise of genetically modified crops with particular emphasis on potatoes. Nuffield Reports: Biotechnology Reports. Nuffield Farming Scholarships Trust of Market Harborough, UK, 28 p.

Sinden, S.E., and Webb, R.E., 1972. Effect of variety and location on the glycoalkaloid content of potatoes. American Potato Journal 49: 334–338.

Sinden, S.L., Sanford, L.L., and Osman, S.F., 1980. Glycoalkaloid and resistance to the Colorado potato beetle in *Solanum chacoense* Bitter. American Potato Journal 57: 331–343.

Sinden, S.L., Sanford, L.L., Cantelo, W.W., and Deahl, K.L., 1986. Leptine glycoalkaloids and resistance to the Colorado potato beetle (Coleoptera: Chrysomelidae) in *Solanum chacoense*. Environmental Entomology 15: 1057–1062.

Singh, P., Kumar, M., Chaturvedi, C., Yadav, D., and Tuli, R., 2004. Development of a hybrid delta-endotoxin and its expression in tobacco and cotton for control of a polyphagous pest *Spodoptera litura*. Transgenic Research 13: 397–410.

Smilde, W.D., Brigneti, G., Jagger, L., Perkins, S., and Jones, J.D.G., 2005. *Solanum mochiquense* chromosome IX carries a novel late blight resistance gene *Rpi-moc1*. Theoretical and Applied Genetics 110: 252–258.

Song, J., Bradeen, J.M., Naess, S.K., Raasch, J.A., Wielgus, S.M., Haberlach, G., Liu, J., Kuang, H., Austin-Phillips, S., Buell, C.R., Helgeson, J.P., and Jiang, J., 2003. Gene RB cloned from *Solanum bulbocastanum* confers broad spectrum resistance to potato late blight. Proceedings of the National Academy of Sciences of the USA 100: 9128–9133.

Spooner, D.M., and Hijmans, R.J., 2001. Potato systematics and germplasm collecting, 1989–2000. American Journal of Potato Research 78: 237–268.

Spooner, D.M., van den Berg, R.G., Rodriguez, A., Bamberg, J.B., Hijmans, R.J., and Lara-Caberera, S.I., 2004. Wild potatoes (Solanum section Petota; Solanaceae) of the North and Central America. American Society of Plant Taxonomists, Ann Arbor, MI, USA.

Sporleder, M., Kroschel, J., Huber, J., and Lagnaoui, A., 2005. An improved method to determine the biological activity (LC_{50}) of the granulosis PoGV in its host *Phthorimaea operculella*. Entomologia Experimentalis et Applicata 116: 191–197.

Stapleton, A., Allen, P.V., Friedman, M., and Belknap, W.R., 1991. Purification and characterization of solanidine glycosyltransferase from potato (*Solanum tuberosum*). Journal of Agriculture and Food Chemistry 39: 1187–1193.

Tarn, T.R., Tai, G.C.C., De Jong, H., Murphy, A.M., and Seabrook, J.E.A., 1992. Breeding potatoes for long-day, temperate climates. Plant Breeding Reviews 9: 217–332.

Thomas, P.E., Kaniewski, W.K., and Lawson, E.C., 1997. Reduced field spread of potato leafroll virus in potatoes transformed with the potato leafroll virus coat protein gene. Plant Disease 81: 1447–1453.

Thornton, M., 2003. The Rise and Fall of NewLeaf Potatoes. NABC Report 15: Biotechnology: Science and Society at a Crossroad, pp. 235–243. http://nabc.cals.cornell.edu/pubs/nabc_15/chapters/Thornton.pdf (accessed 23 January 2008).

Tingey, W.M., 1991. Potato glandular trichomes: defensive activity against insect attack. In: Naturally Occurring Pest Bioregulators, P.A. Hedin, ed., American Chemical Society Symposium Series 449, American Chemical Society, Books, Washington, DC, USA, pp. 126–135.

Tingey, W.M, Gregory, P., Plaisted, R.L., and Tauber, M.J., 1984. Research progress: Potato glandular trichomes and steroid glycoalkaloids. Report of the XXII planning conference on Integrated Pest Management, International Potato Center, Lima, Peru, pp. 115–124.

USEPA (United States Environmental Protection Agency), 2001. Biopesticide Registration Action Document. *Bacillus thuringiensis* (*Bt*) Plant-Incorporated Protectants. 15 October 2001. http://www.epa.gov/oppbppd1/biopesticides/pips/bt_brad.htm (accessed 15 January 2008).

Vaeck, M., Reynaerts, A., Hofte, H., Jansens, S., De Beuckeleer, M., Dean, C., Zabeau, M., Van Montagu, M.V., and Leemans, J., 1987. Transgenic plants protected from insect attack. Nature 328: 33–37.

van der Vossen, E., Sikkema, A., Hekkert, B.T.L., Gros, J., Stevens, P., Muskens, M., Wouters, D., Pereira, A., Stiekema, W., and Allefs, S., 2003. An ancient R gene from the wild potato species

Solanum bulbocastanum confers broad-spectrum resistance to *Phytophthora infestans* in cultivated potato and tomato. The Plant Journal 37: 867–882.

Van Gelder, W.M.J., 1990. Chemistry, toxicology and occurrence of steroidal glycoalkaloids: potential contaminants of the potato (*Solanum tuberosum* L.). In: Poisonous Plant Contamination of Edible Plants, A.F.M. Rizk, ed., CRC, Boca Raton, FL, USA, pp. 117–156.

Visser, D., 2005. Guide to Potato Pests and Their Natural Enemies in South Africa. Arc-Roodeplaat Vegetable and Ornamental Plant Institute, Pretoria, South Africa, 105 p.

Weisz, R., Smilowitz, Z., and Christ, B., 1994. Distance, rotation, and border crops affect Colorado potato beetle (Coleoptera: Chrysomelidae) colonization and population density and early blight (*Alternaria solani*) severity in rotated potato fields. Journal of Economic Entomology 87: 723–729.

Westedt, A.L., Douches, D.S., Pett, W., and Grafius, E.J., 1998. Evaluation of natural and engineered resistance mechanisms in *Solanum tuberosum* L. for resistance to *Phthorimaea operculella* Zeller. Journal of Economic Entomology 91: 552–556.

Whalon, M.E., and Wingerd, B.A., 2003. Bt: mode of action and use. Archives of Insect Biochemistry and Physiology 54: 200–211.

Whalon, M.E., Mota-Sanchez, D., Hollingworth, R.M., and Duynslager, L., 2007. Arthropod Pesticide Resistance Database. http://www.pesticideresistance.org (accessed December 2007).

Woolfe, J.A., 1987. The Potato in the Human Diet. Cambridge University Press, Cambridge, UK.

Wright, R.J., 1984. Evaluation of crop rotation for control of Colorado potato beetles (Coleoptera: Chrysomelidae) in commercial potato fields on Long Island. Journal of Economic Entomology 77: 1254–1259.

Yencho, G.G., and Tingey, W.M., 1994. Glandular trichomes of *Solanum berthaultii* alter host preference of the Colorado potato beetle, *Leptinotarsa decemlineata*. Entomologia Experimentalis et Applicata 70: 217–225.

Yu, C.G., Mullins, M.A., Warren, G.W., Koziel, M.G., and Estruch, J.J., 1997. The *Bacillus thuringiensis* vegetative insecticidal protein Vip3A lyses midgut epithelium cells of susceptible insects. Applied and Environmental Microbiology 63: 532–536.

Zhao, J.Z., Grafius, E.J., and Bishop, B.A., 2000. Inheritance and synergism of resistance to imidacloprid in the Colorado potato beetle (Coleoptera: Chrysomelidae). Journal of Economic Entomology 93: 1508–1514.

Zhao, J.Z., Cao, J., Collins, H.L., Bates, S.L., Roush, R.T., Earle, E.D., and Shelton, A.M., 2005. Concurrent use of transgenic plants expressing a single and two *Bacillus thuringiensis* genes speeds insect adaptation to pyramided plants. Proceedings of the National Academy of Sciences of the USA 102: 8426–8430.

Chapter 8
Bt Rice in Asia: Potential Benefits, Impact, and Sustainability

Michael B. Cohen[1,*], Mao Chen[2], J.S. Bentur[3], K.L. Heong[4], and Gongyin Ye[5]

Abstract Numerous lines of genetically modified rice expressing *cry* genes from *Bacillus thuringiensis* (*Bt*) have been shown to confer resistance to lepidopterous pests of rice (stem borers and leaffolders) under laboratory and field conditions. Stem borers and leaffolders are chronic pests that generally do not cause high yield losses on an individual field basis, but their feeding damage results in substantial cumulative yield losses in rice growing countries because of their widespread occurrence. The adoption of *Bt* rice will therefore provide modest but consistent yield increases. Understanding farmer decision-making practices will be necessary to achieve the full potential of *Bt* rice to decrease insecticide use. No negative effects of *Bt* rice on predators, parasitoids, non-lepidopterous herbivores, or soil invertebrates have been detected, except when natural enemies are fed *Bt*-intoxicated prey. Effects of *Bt* rice on soil microorganisms have been observed but have not been shown to have negative consequences for soil health. Despite the fact that outcrossing of *Bt* genes to wild and weedy rice will almost certainly occur, possible consequences of outcrossing have received little study. There is a high risk of development of pest resistance to *Bt* rice, because of the challenge of implementing resistance management programs for millions of small farmers and because the major target pests (the stem borers *Scirpophaga incertulas* and *Chilo suppressalis*) do not have important alternative hosts that can provide natural refuges. As of December 2007, *Bt* rice had not been commercialized in any country.

[1] Edmonton, Canada

[2] Cornell University/NYSAES, Geneva, NY, USA

[3] Directorate of Rice Research, Hyderabad, India

[4] International Rice Research Institute, Los Baños, The Philippines

[5] Zhejiang University, Zhejiang, China

* To whom correspondence should be addressed. E-mail: michael.cohen@shaw.ca

J. Romeis, A.M. Shelton, G.G. Kennedy (eds.), *Integration of Insect-Resistant Genetically Modified Crops within IPM Programs.*

8.1 Introduction

Rice (*Oryza sativa*) is the most widely consumed food crop and was grown on over 152 million hectares worldwide in 2004 (FAO, 2007). More than 90% of the world's rice is produced and consumed in Asia, where it is grown in temperate, subtropical, and tropical regions, and under irrigated and rainfed conditions (Maclean et al., 2002). The pest complex of rice in Asia varies substantially with geographic area and production system (Savary et al., 2000). A large diversity of insects feed on the crop, among which species of Hemiptera, Diptera, Lepidoptera, and Coleoptera are pests of regional or local significance (Dale, 1994).

Lepidopteran stem borers are chronic pests in all rice ecosystems. In Asia, the most important tropical species is the yellow stem borer, *Scirpophaga incertulas* (Pyralidae), while the striped stem borer, *Chilo suppressalis* (Crambidae) is generally the most abundant in temperate regions. At the vegetative growth stage of rice, stem borer feeding results in damaged tillers called deadhearts. At the reproductive stage, stem borers cause the production of panicles of unfilled grains called whiteheads. Numerous species of foliage-feeding Lepidoptera also occur in rice, the most important of which are leaffolders, *Cnaphalocrocis medinalis* and *Marasmia* spp. (Pyralidae).

Stem borers generally do not cause large yield losses in individual fields, but result in substantial cumulative yield losses in rice growing countries because of their chronic and widespread occurrence. An empirical analysis of rice production across Asia found that stem borers caused yield losses of 2.3%, the highest of any group of insects (Savary et al., 2000). A Chinese study estimated that stem borers cause a 3.1% loss in the national rice yield (Sheng et al., 2003). Yield losses due to leaffolders are generally small because rice plants at the vegetative growth stage have a large capacity to compensate for damage to foliage. However, leaffolder damage is highly visible to farmers and is often the most important stimulus for insecticide applications (Way and Heong, 1994; Matteson, 2000). In addition to cumulative yield losses resulting from chronic levels of infestation, outbreaks of stem borers and leaffolders can cause higher localized losses.

Other rice insect pests of regional significance in Asia include the brown planthopper, *Nilaparvata lugens*, whitebacked planthopper, *Sogatella furcifera*, and small brown planthopper, *Laodelphax striatellus* (Hemiptera: Delphacidae); the green leafhopper, *Nephotettix virescens* (Hemiptera: Cicadellidae) (the vector of viruses causing rice tungro disease); the rice bug, *Leptocorisa* spp. (Hemiptera: Alydidae); the Asian rice gall midge, *Orseolia oryzae* (Diptera: Cecidiomyiidae); and the rice water weevil, *Lissorhoptrus oryzophilus* (Coleoptera: Curculionidae) (Dale, 1994). Planthoppers, leafhoppers, and the viruses that they vector were the most serious pests of rice from the late 1960s to the mid-1980s. It was eventually determined that these pests were primarily insecticide-induced, and they are now maintained at low levels in most areas by natural biological control and resistant varieties (Gallagher et al., 1994; Way and Heong, 1994; Matteson, 2000). In addition to stem borers and leaffolders, the Asian gall midge (Rajamani et al., 2004) and the rice water weevil (Saito et al., 2005) are current pests of chronic importance in some tropical and temperate areas, respectively.

There is a long history of conventional breeding for insect resistance in rice (Heinrichs, 1994; Khush, 1995). Many genes conferring resistance to planthoppers and leafhoppers and the Asian rice gall midge have been identified in rice germplasm and crossed into in modern rice cultivars. Major gene resistance to lepidopteran rice pests has not been identified, despite the evaluation of thousands of accessions (Heinrichs, 1994). Quantitative resistance to lepidopterans has been identified but no highly resistant varieties with have been released.

Because of the prominent pest status of stem borers and leaffolders, the limited sources of resistance to these pests in rice germplasm, and the success of transgenic *Bt* technology in maize and cotton, there have been extensive investments in *Bt* rice research and development. Fujimoto et al. (1993) reported the first transformation of rice with a *Bt* gene. Many papers reporting the development and evaluation of *Bt* rice lines have since appeared, as reviewed by High et al. (2004) and Chen et al. (2006b). Rice lines expressing *cry1Aa*, *cry1Ab*, *cry1Ac*, a *cry1Ab/cry1Ac* fusion, *cry1B*, *cry1C*, and *cry2A*, and a pyramid of *cry1Ac* with *cry2A*, under control of various promoters, have been shown to confer resistance to stem borers and to leaffolders and other foliage-feeding Lepidoptera (Table 8.1). In addition to cry toxins, several protease inhibitors have been introduced into rice to confer resistance to Lepidoptera (Malone et al., chapter 13). Of these, the cowpea trypsin inhibitor (*CpTI*) has been most extensively studied. Rice lines containing CpTI alone (Huang et al., 2005) and in combination with Cry1Ac have been field tested in China (Zhao et al., 2004) (Table 8.1).

Genetic engineering has also been applied to the control of planthopper and leafhopper pests of rice, with the use of plant lectin genes. The snowdrop lectin gene, *Galanthus nivalus* agglutinin (*gna*), has been transferred to several rice varieties and has been shown to provide partial resistance to planthoppers and leafhoppers in laboratory and field tests (e.g., Foissac et al., 2000; Malone et al., chapter 13). Rice plants containing pyramids of *gna* and cry genes have also been produced (Malone et al., chapter 13). Recently, partial resistance to leafhoppers and planthoppers was demonstrated by rice transformation with a lectin gene from garlic (*Allium sativa* leaf agglutinin gene, *ASAL*) (Saha et al., 2006).

The first field trials of *Bt* rice first took place in China in 1998 (Shu et al., 2000; Tu et al., 2000; Ye et al., 2001a, b). Large field trials (referred to as productive testing) of several *Bt* rice lines have continued in China since then (Chen et al., 2006b; Wang and Johnson, 2007). Field trials have also taken place in Pakistan (Bashir et al., 2005; Mahmood-ur-Rahman et al., 2007), Spain (Breitler et al., 2004), Iran (James, 2005), and India (Bunsha, 2006). Several of the *Bt* lines and a CpTI line have shown high levels of protection from stem borer and leaffolder damage under field conditions (Chen et al., 2006b). In a study in farmers' fields in China, a *Bt* rice line resulted in yield increases of 6% to 9% and substantial reductions in insecticide use, in comparison to conventional varieties (Huang et al., 2005).

Bt rice is not currently in commercial production in any country. Iran grew 4,000 ha of *Bt* rice for seed multiplication in 2005 (James, 2005), but commercial

Table 8.1 Representative rice lines transformed with genes for resistance to Lepidoptera

Toxin gene	Promoter	Cultivar	Peak expression level (% soluble protein or µg/g fresh wt.)	Location of field testing	References
cry1Aa	*Ubiquitin*	Ariete, Senia	0.4–0.5%	Spain	Breitler et al., 2004
cry1Ab	*Ubiquitin*	Xiushui 11	0.12%	China	Shu et al., 2000; Wu et al., 2002
cry1Ab	*PEPC*	Tarom Molaii	0.1%	Iran	Ghareyazie et al., 1998; James, 2005
cry1Ab/cry1Ac fusion	*Actin1*	Minghui 63, Shanyou 63	0.002%	China	Tu et al., 2000; Huang et al., 2005
cry1B	*Ubiquitin*	Ariete, Senia	0.4–0.5%	Spain	Breitler et al., 2004
cry1C	*Ubiquitin*	Minghui 63	2.5–3.5 µg/g	–	Wei et al., 2006
cry2A	*CaMV 35S*	Minghui 63	9.7–12.1 µg/g	–	Chen et al., 2005
cry1Ac + *cry2A*	*CaMV 35S*	Basmati 370	Cry1Ac: 6.1 µg/g Cry2A: 0.7 µg/g	Pakistan	Bashir et al., 2005
cry1Ac + *CpTI*	*Ubiquitin*	Minghui 86	Cry1Ac: 1.2 µg/g	China	Zhao et al., 2004; Han et al., 2006
CpTI	*CaMV 35S*	Youming 86	–	China	Deng et al., 2003; Huang et al., 2005

Abbreviations: CaMV, cauliflower mosaic virus; PEPC, phosphoenolypyruvate carboxylase production in Iran is not currently permitted (F. Alinia, March 2, 2007, personal communication). The only transgenic rice lines currently approved for commercial production are three herbicide (glufosinate)-resistant lines in the United States (AGBIOS, 2007).

In this chapter, we examine some of the biological and sociological factors that will determine the success of *Bt* rice and suggest important topics for future research. We focus on three topics where rice presents unique challenges: compatibility with naturally-occurring biological control, knowledge sharing with farmers, and resistance management. In addition to Asia, rice is grown in Europe, Africa, Australia, and North and South America, and lepidopterous pests occur on rice on all these continents. However, aside from some work on European cultivars (e.g., Breitler et al., 2004), there has been little research outside of Asia on *Bt* rice meant for cultivation. Consequently this chapter will focus on Asia.

8.2 Non-Target Impacts of *Bt* Rice

8.2.1 Compatibility of Bt Rice with Biological Control

The fundamental importance of biological control to insect pest management in Asian rice fields gained wide recognition following the vast brown planthopper outbreaks of the 1970s (Gallagher et al., 1994). It was eventually determined that the complex of natural enemies that had kept this pest in check had been devastated by the overuse of broad-spectrum insecticides. These insecticides were introduced to rice production along with high-yielding varieties and chemical fertilizers as part of the Green Revolution. Biological control of planthoppers and other pests, including stem borers and leaffolders, has since been intensively studied in various regions of Asia and rice production systems (Ooi and Shepard, 1994; Matteson, 2000). In addition, it has been found that the interactions of predators and parasitoids and their herbivorous prey form only one part of the highly diverse and interconnected rice field food web. A food web for irrigated and rainfed rice in the Philippines contains 546 taxa and more than 9,300 linkages among organisms found at and above the water line (Schoenly et al., 1996). Settle et al. (1996) documented 127 species of herbivores and 332 species of predators and parasitoids from lowland irrigated rice fields in Indonesia.

Because of the critical role of biological control in rice production and previous experience with indiscriminate use of another insect control technology (broad-spectrum insecticides), the need to carefully evaluate the ecological effects of *Bt* rice before its release has been generally recognized. Many trials have assessed the potential impacts of *Bt* rice on non-target herbivores, parasitoids, predators, and soil-dwelling detritivores (reviewed in Chen et al., 2006b) and microbial organisms (Wu et al., 2003; Xu et al., 2004) under laboratory and field conditions, some over multiple years and sites. These trials include studies of direct toxicity of purified

Cry toxins or *Bt* rice conducted under laboratory and greenhouse conditions, and ecological studies conducted in the field. Importantly, the field studies have examined not only populations of target herbivores and their natural enemies, but non-target herbivores as well. Various insect sampling methods have been used in field studies, including vacuum sampling, sweep nets, and sticky traps. In general, negative effects of *Bt* rice on non-target organisms have not been observed, as measured by indicators of fitness, population density and dynamics, and biodiversity indices (Chen et al., 2006b). The results from rice are consistent with those from other *Bt* crops, as reviewed by Romeis et al. (2006), who compiled data from more than 100 peer-reviewed studies on the impact of *Bt*-transgenic plants on biological control organisms (see also Romeis et al., chapter 4).

Most field studies of the effects of *Bt* rice on rice ecosystems have been conducted in China, the first country to conduct field trials of *Bt* rice and the only country where large-scale field trials suitable for ecological studies can be routinely planted. Field trials conducted in other countries have generally been small in scale and primarily designed to evaluate pest resistance and agronomic performance. As an alternative to field experiments with *Bt* rice in the Philippines, pending possible future approval of large-scale field plots, Schoenly et al. (2003) used applications of *Bt* sprays to selectively remove foliage-feeding lepidopteran larvae from rice fields and study the effects on the rice field food web.

As expected, some negative effects on natural enemies have been observed when *Bt*-susceptible, sublethally damaged herbivores are used as prey or hosts. Cocoon formation of *Apanteles chilonis* (Hymenoptera: Braconidae) and its parasitism rate were negatively affected when reared on *Bt* rice-fed *C. suppressalis* (Jiang et al., 2004, 2005). In the field, the dispersal dynamics of parasitoids of non-target planthoppers and leafhoppers (Chen et al., 2003) and overall temporal dynamics of species richness, diversity, evenness and dominance indices of the parasitoid communities (Liu et al., 2006; Li et al., 2007a) were very similar in *Bt* rice and non-*Bt* rice. Fourteen parasitic arthropod families were collected from both *Bt* and non-*Bt* rice fields in a 2-year field study and no consistent differences were found in population dominance in two different rice ecosystems (Liu et al., 2003). Schoenly et al. (2003) found that *Bt* sprays directed against lepidopteran larvae had negligible impacts on parasitoid and predator populations.

Several laboratory studies have examined the effects *Bt* rice on the fitness of predators. Bernal et al. (2002) did not detect fitness effects on the predator *Cyrtorhinus lividipennis* (Hemiptera: Miridae) feeding on *N. lugens* that were reared on *Bt* rice. Similarly, Bai et al. (2006a) reported that development of *Propylea japonica* (Coleoptera: Coccinellidae) was not affected when fed on *N. lugens* reared on *Bt* rice. The fitness of *P. japonica* and *Chrysoperla sinica* (Neuroptera: Chrysopidae) was not negatively affected when feeding on *Bt* rice pollen containing Cry1Ab toxin (Bai et al., 2005a, b). Under field conditions, Liu et al. (2002) found that no marked differences between *Bt* and non-*Bt* rice fields were detected for the population dynamics of main non-target insect pests and five common spider species. In a 2-year field study, Liu et al. (2003) reported that 26 different predatory arthropod families belonging to Hemiptera, Neuroptera, Coleoptera, Diptera,

Hymenoptera, and Aranaea were collected both from *Bt* and non-*Bt* rice fields and were not significantly affected by *Bt* rice. Another 2-year field study did not detect effects on population dynamics of the predator *C. lividipennis* (Chen et al., 2007).

8.2.2 Non-Target Herbivores

Studies of the effects of *Bt* rice on non-target herbivores have focused on the planthoppers *N. lugens* and *S. furcifera*, and the leafhoppers *Nephotettix virescens* and *N. cinticeps*. Overuse of insecticides resulted in severe outbreaks of both planthopper species throughout much of Asia in the 1970s, and local outbreaks are still common in China. The leafhoppers are important as vectors of the viruses that cause rice tungro disease, which was a serious problem in the 1970s but now occurs sporadically. In tests of direct toxicity, Bernal et al. (2002) and Bai et al. (2006a) used enzyme-linked immunosorbent assays (ELISA) to demonstrate that *N. lugens* ingest Cry toxins from some *Bt* rice lines, and detected no negative effects on fitness. Tan et al. (2006) reported that *Bt* rice did not significantly affect oviposition behavior of *S. furcifera* adults when detected by an electronic monitor, although developmental time of *S. furcifera* nymphs on *Bt* rice was 1–2 days longer than on non-*Bt* rice. Similarly, *N. lugens* and *N. cincticeps* (Chen et al., 2006a) did not show feeding or oviposition preference for *Bt* rice compared to non-*Bt* rice under laboratory conditions.

In terms of indirect effects, it has been hypothesized that the reduction in foliage-feeding lepidopterans caused by *Bt* rice might lead to reduced populations of generalist natural enemies and consequently to increased populations of non-target herbivores, particularly planthoppers. In a field study conducted at three sites in China over a 2-year period with two *Bt* rice lines, populations of planthoppers and leafhoppers were very similar in *Bt* and non-*Bt* rice fields (Chen et al., 2006b, 2007). In the Philippines, *Bt* sprays did not significantly affect the population trajectories of planthoppers and leafhoppers in rice fields (Schoenly et al., 2003).

8.2.3 Soil Biota

Soil-dwelling detritivores, such as Collembola, play an important role in rice ecosystems, influencing soil structure and nutrient mineralization as well as the activity and composition of the microbial community (Guo et al., 1995; Brussaard, 1998; Haimi, 2000). Bai et al. (2005c) found that Cry1Ab toxin could be detected in *Entomobrya griseoolivata* (Collembola: Entomobryidae) feeding on *Bt* rice tissue litter in the laboratory. Population densities of four collembolan species, *E. griseoolivata*, *Bourletiella christianseni*, *Hypogastrura matura* and *Isotoma monochaeta* did not differ significantly between *Bt* and non-*Bt* rice fields in a 2-year field survey (Bai et al., 2006b). In a 2-year field trial, Liu et al. (2003) found that population levels of three collembolan families (Scatopsidae, Sminthuridae, and Tomoceridae) and four detritivorous dipteran families (Ceratopogonidae, Mycetophilidae, Phoridae, Psychodidae, and Scatopsidae) were very similar in *Bt* and non-*Bt* rice fields.

The environmental behavior and fate of Cry toxins from *Bt* plants have drawn wide attention due to the potential impact on microbial communities and enzyme activities in the rhizosphere. Studies with *Bt* maize and cotton have shown that Cry toxins from crop residues and root exudates are degraded over time but that degradation may be slowed due to toxin binding to soil components (Stotzky, 2004). Effects of *Bt* cultivars on the soil nematode and bacterial communities have been detected in some studies (Stotzky, 2004; Mulder et al., 2006; Griffiths et al., 2007; Sun et al., 2007), but the differences between *Bt* and non-*Bt* isolines have in some cases been smaller than differences among a series of non-*Bt* cultivars (Griffiths et al., 2007) or differences among plant growth stages (Baumgarte and Tebbe, 2005). In the 10 years since *Bt* cotton and *Bt* maize were introduced, there have been no published reports of problems with soil function or fertility in farmers' fields.

In contrast to the aerobic soil conditions under which cotton and maize are grown, most rice is grown under flooded, anaerobic conditions. A few studies have begun to examine the possible impacts of *Bt* rice on the microbial community in flooded soils. Saxena et al. (2004) found that Cry toxin was released in root exudates of *Bt* rice. Bai et al. (2006b, c) observed that Cry toxin was exuded from *Bt* rice tissue into water and maintained its activity for more than 160 days. Wang et al. (2007) found that rapid degradation of purified Cry1Ab from *Bt* rice occurred in soils under aerobic conditions with a half-life ranging from 19.6 to 41.3 days. However, under flooded conditions, the half-life of Cry1Ab was prolonged to 45.9–141 days. Li et al. (2007b) measured the degradation of Cry1Ac in rice stalks and roots remaining in an unplowed field after harvest. Toxin content declined by approximately 80% before winter, and further declined to below levels of detection by the following April. Under laboratory conditions, Xu et al. (2004) found that *Bt* rice straw significantly increased the number of hydrolytic-fermentative and anaerobic nitrogen-fixing bacteria and decreased denitrifying and methanogenic bacteria in flooded paddy soil. However, it is not clear whether this phenomenon could happen in actual rice ecosystems and lead to a subsequent soil structure change. Ren et al. (2004) observed some significant differences in the number of anaerobic fermentative bacteria, denitrifying bacteria, hydrogen producing acetogenic bacteria, and methanogenic bacteria between flooded paddy soils amended with *Bt* and non-*Bt* rice straw during the early stages of incubation. The differences disappeared 11 weeks after inoculation. Wu et al. (2003) found that colony forming units of cultureable bacteria, actinomycetes and fungi were not significantly different between soil amended with *Bt* rice straw and non-*Bt* rice straw.

8.2.4 Outcrossing of Insect Resistance Transgenes

In Asia, transgenes will almost certainly outcross from transgenic rice varieties to non-transgenic rice varieties, weedy rice, and two wild species closely-related to *O. sativa*, the perennial *O. rufipogon* and the annual *O. nivara* (Lu and Snow, 2005). Weedy rices, which include various weedy types of *O. sativa* and *O. rufipogon*, are among the most important weeds of cultivated rice in Asia (Baki et al., 2000). Cultivated rice is primarily self-pollinating, but a low level of outcrossing (ca. 0.01% to 1%) occurs among cultivars under field conditions. It has also been shown that pollen from cultivated rice can fertilize weedy rice (Chen et al., 2004) and *O. rufipogon* (Song et al., 2003; Chen et al., 2004) under field conditions, and that fertile progeny are produced. The rate of outcrossing declines rapidly with distance (Wang et al., 2006), but weedy rice occurs within rice fields and in some areas *O. rufipogon* is abundant in adjacent irrigation ditches and canals. Weedy rice generally flowers synchronously with the rice crop. Overlap in flowering time of wild rice with weedy and cultivated rice occurs at some times of year (Lu and Snow, 2005; Cohen et al., 2008).

The possible consequences of *cry*, *CpTI* or lectin gene outcrossing to weedy rices and *O. rufipogon* have not been assessed in field experiments. Such experiments would require the deliberate release in rice-growing areas of hybrid progeny of transgenic rice with wild or weedy rice. Thus, evaluation of the consequences of outcrossing must rely primarily on indirect methods and inference. Outcrossing of transgenes for insect resistance could have adverse consequences if the fitness and weediness of the recipient plant populations were increased. This could occur if the affected insects constrain the distribution or abundance of the recipient populations. Major genes for resistance to planthoppers, leafhoppers, and gall midges occur in rice germplasm and have been transferred among rice cultivars by conventional breeding for decades. Increased weediness of wild and weedy rices as a result of outcrossing of these genes has not been reported, although the question has not been directly addressed through experimentation. The target pests for *Bt* and CpTI rice are lepidopterous stem borers and leaffolders, taxa for which major genes have not been identified in rice germplasm. Thus, these genes should be subject to additional scrutiny.

Cohen et al. (2008) conducted a 2-year survey of the plant pathogens and insect herbivores on cultivated, weedy, and wild rice (*O. rufipogon*) in three habitats (fresh water, saline water, and acid sulfate soils) of the Mekong Delta, Vietnam. Based on observations of insects or of the damage symptoms caused by pathogens and insects, they concluded that major pathogen and insect pests of cultivated rice also occur on wild and weedy rice. The most common insect damage observed on wild rice was attributable to locusts and leaffolders. The survey was not designed to determine whether pest damage resulted in decreased fitness of *O. rufipogon*. However, if *O. rufipogon* is highly tolerant of vegetative damage, as is cultivated rice, then increased resistance to foliage-feeding insects conferred by transgene outcrossing will probably not have a large impact on the weediness of this wild rice species. Stem borer damage, which at the reproductive stage directly decreases seed production, was found only at low levels on *O. rufipogon* and weedy rice.

8.2.5 Recommendations for Further Research

Rice ecosystems exhibit large spatial and temporal variation. Consequently, it will be necessary to study the post-release ecological impacts of *Bt* rice over large areas and multiple years. The effects of *Bt* rice on non-target herbivores, natural enemies, detritivores, and soil microbial organisms has been evaluated under field conditions in several studies. Most of these studies focused on a few selected arthropod taxa, although Liu et al. (2003, 2004, 2006) and Schoenly et al. (2003) evaluated the potential impact of *Bt* toxin at guild and community levels in *Bt* rice fields and *Bt*-sprayed fields, respectively. To date, no studies longer than 2 years have been carried out in *Bt* rice fields. In *Bt* cotton and *Bt* maize, longer-term (3–6 year) field trials have been conducted to evaluate the impact of *Bt* plants on arthropods and natural enemies (Naranjo, 2005a, b; Bhatti et al., 2005; Hellmich et al., chapter 5; Naranjo et al., chapter 6). Comparable long-term, detailed studies are needed for *Bt* rice in farmers' fields after commercialization. Some effects that might result from the large-scale adoption of *Bt* rice cannot be projected from field trials a few hectares in size.

All the studies of ecological impact of *Bt* rice have been conducted in China. Because of regional variation in rice ecosystems, similar studies should be conducted in other countries prior to the release of *Bt* rice to farmers. Cohen et al. (2008) found differences among three rice habitats in the Mekong Delta in the relative abundance of pathogens and insects on *O. rufipogon*. This finding suggests that geographic variation in pest incidence should be taken into account in future studies of the possible fitness effects of transgene outcrossing and of impacts on ricefield communities.

Research on the impact of *Bt* rice on non-target herbivores has given emphasis to non-lepidopteran herbivores (such as planthoppers and leafhoppers). However, the impact on non-target and target lepidopterans should be carefully evaluated after commercialization. Gao et al. (2006) found that *cry1Ac/CpTI* rice was significantly more resistant to the striped stem borer, *C. suppressalis*, than to the pink stem borer, *Sesamia inferens* (Lepidoptera: Noctuidae), and that over 99% of *C. suppressalis* but only 44–64% of *S. inferens* were controlled in field plots. Differences in the efficacy of *Bt* rice on different lepidopterans and the substantial reductions in populations of the more susceptible species could result in a change in pest status of lepidopteran species with lower toxin susceptibility.

The impact of *Bt* rice on insects living below the water line, such as ephemeropteran and chironomid larvae, has not been extensively evaluated. Many of these insects serve as prey for generalist predators in rice fields (Guo et al., 1995). Future studies on interactions of *Bt* rice, aquatic insects and predators in rice fields would provide a better understanding of possible impacts of *Bt* rice on this portion of the rice field food web. Additional study of impacts on the microbial community would also be valuable, because of the distinctiveness of the anaerobic conditions in which most rice is grown.

8.3 Impact of *Bt* Rice on Farmers' Pest Management Practices

8.3.1 Decision-Making by Rice Farmers

Will Bt rice change farmers' pest management practices? To address this question there is a need to understand how rice farmers make decisions. Many decisions people make are based on behavioral responses to descriptive considerations such as beliefs and preferences, and normative considerations such as logic and rationality. Asian rice farmers are no exceptions and tend to be "anchored" on the belief that insects are yield constraints and they need to respond by "killing" them (Bentley, 1989; Heong and Escalada, 1997). The preferred killing agents are insecticides and many sprays are targeted at highly visible insect damages, such as those caused by leaffolders. Stem borers are also common spray targets because of the visible whitehead symptoms they cause. Farmers tend to overestimate yield loss associated with whitehead symptoms by more than tenfold. In a Philippine study, Heong and Escalada (1999) found that rice farmers generally act to prevent the worst case from occurring and had a high perception of susceptibility. About 59% of farmers believed that a loss of 10 to 15% was "extremely likely" if they did not apply insecticides. Thus insecticide use decisions are often based on their perceived prospects of incurring loss rather than actual need (Tait, 1977; Mumford and Norton, 1984) – loss aversion according to Prospect Theory (Tversky and Kahnemann, 1992). This high loss aversion behavior might also be due to farmers lacking knowledge in identifying stem borer adults and misperceptions that all lepidopteran adults in the vicinity of the rice crop are stem borers. Farmers who participated in an exercise that taught them a simple way to compute yield loss from whitehead counts modified their loss aversion behavior and reduced their loss perceptions and insecticide spraying by 27% and 22%, respectively (Escalada and Heong, 2004).

Farmers' subjective norm attitudes (or peer pressure), which are related to their perceptions of what specific community reference groups expect of them, play a significant role in influencing decisions. Subjective norm measures, quantified using the Theory of Reasoned Action model (Fishbein and Ajzen, 1975), were found to strongly influence farmers' spray decisions (Heong and Escalada, 1999; Heong et al., 2002). Important reference groups in rural communities in Asia are extension technicians, chemical sales representatives, community leaders and neighbors and most farmers tend to believe that they expect them to use pesticides. In the Philippines, an increase in pesticide misuse was found to be strongly associated with visits by chemical company representatives or by agricultural technicians (Tjornhom et al., 1997).

Decisions made under uncertainty, like the likelihood of a high stem borer attack, with limited knowledge and time are often based on bounded rationality (Gigerenzer and Todd, 1999). Unbounded rationality on the other hand assumes that decision making has no constraints of time, knowledge, computation power and reasoning abilities, as with maximization of expected utility and in Bayesian models. There are two forms of bounded rationality: satisficing heuristics (or rules-of-thumb) for searching through available alternatives, and fast and frugal heuristics that use little information, time and computation. Rice farmers' decisions appear to be based of a set of heuristics of both forms developed from experiences and guesswork and tend to have inherent faults and biases (Tversky and Kahnemann, 1974). Understanding these faults and biases will be useful in developing interventions to improve decision making.

8.3.2 Approaches to Achieve a Positive Impact of Bt Rice on Pest Management Practices

To achieve a positive impact on farmers' pest management practices, *Bt* rice will need to change farmers' insecticide use attitudes and to result in increased profits. Huang et al. (2005) found that farmers in China growing *Bt* rice and CpTI rice obtained higher yields, used less insecticide, and had higher profits than farmers growing conventional rice. If similar results are obtained after *Bt* rice is commercialized, then the technology will be rapidly adopted by farmers. However, to substantially reduce unnecessary insecticide use, more information is needed on how best to communicate information about *Bt* rice to farmers. The methodology of Huang et al. (2005) did not include a placebo or blind controls, in which groups of farmers growing *Bt* and conventional rice under different information regimes are compared. The authors were thus unable to determine whether farmers reduced their insecticide use because they observed lower stem borer damage, or because they knew beforehand that they were growing *Bt* rice and decided *a priori* to use less insecticide (Heong et al., 2005).

Naik et al. (2005) found that some *Bt* cotton adopters in Andhra Pradesh, India, did not realize the full benefits of the technology because they did not sufficiently reduce insecticide use, probably due to misinformation and a lack of training. If the potential of *Bt* rice is to be realized, farmers will need to be trained not to use insecticides for stem borer control. Adoption is a learning process and to facilitate this, farmers can be invited to participate in experiments (Huan et al., 2004) where they can learn about the benefits of *Bt* rice and modify their attitudes and decision heuristics. After extensive farmer participatory evaluations, information about the potential benefits of *Bt* rice, including increased profits and reduced insecticide use, might be communicated widely through media campaigns (Escalada and Heong, 2004), extension channels, and seed multiplication programs.

8.3.3 Recommendations for Post-Commercialization Actions

Periodic surveys of farmers' beliefs and practices in stem borer management will provide information for modifying deployment strategies and media campaigns. Most farmers' insecticide use decisions are influenced by their lack of information and loss aversion attitudes. Thus, for farmers to fully benefit from *Bt* rice, their insecticide decisions will need to modified. To achieve and sustain this success might require developing a post release communication strategy to support the adopters and continuously provide information and motivation.

8.4 Resistance Management

8.4.1 Challenges to Resistance Management for Bt Rice

Resistance management for *Bt* crops has received substantial attention from researchers and government regulatory bodies because of the tremendous value of *Bt* toxins as safe and effective insecticides and the novel opportunities for durability enabled by *Bt* crop technology as compared to insecticide sprays. The high dose-refuge strategy for resistance management of *Bt* crops has strong theoretical support (Gould, 1998; Ferré et al., chapter 3) and has been implemented, with apparent success, for *Bt* maize and/or cotton in the USA, Australia, and Canada (Bates et al., 2005). In these countries, growers are required to maintain refuges consisting of fields of non-*Bt* host plants. The refuges serve as sources of homozygous susceptible adults, and must be of sufficient size and proximity such that resistant adults from *Bt* fields rarely mate with each other (Ferré et al., chapter 3; Hellmich et al., chapter 5; Naranjo et al., chapter 6). On-farm refuges are not required in the cotton industry in China, where *Bt* cotton now accounts for large proportions of some cotton production areas (James, 2007). Cotton in China is produced by millions of small farmers, and it is thus impractical for farmers to maintain refuges or for authorities to enforce a refuge requirement. However, the principal target pest for *Bt* cotton, *Helicoverpa armigera* (Lepidoptera: Noctuidae), is highly polyphagous, and it has been shown that non-*Bt* crops such as soybean and peanut can function as refuges for this pest (Wu and Guo, 2005).

Rice production systems in tropical Asia present several unique challenges to resistance management for *Bt* rice. These challenges relate to difficulties in establishing adequate refuges for the most important target pests, the yellow stem borer (*S. incertulas*) and striped stem borer (*C. suppressalis*), and to ensuring that stem borer larvae are exposed to a consistently high dose of toxin. As is the case with *Bt* cotton in China and India, rice in tropical Asia is grown by millions of small farmers, making it unlikely that a requirement for on-farm refuges can be enforced. An important contrast to the situation with *H. armigera* in cotton, however, is that there are no significant alternative wild or cultivated host plants for *C. suppressalis* or *S. incertulas* in most rice-growing regions (Cuong and Cohen, 2002).

Khan et al. (1991) compiled a list of 41 alternative hosts of *C. suppressalis* in six plant families, as cited in 33 publications. A critical review of these papers found only nine with firsthand field observations of alternate host use, none of which included quantitative data (Cuong and Cohen, 2002). Eight of the nine papers were reports of unusual incidents of alternative host use by *C. suppressalis*. Only one paper (Hachiya, 1981) described a alternative host on which *C. suppressalis* is often found, *Zizania latifolia* (Poaceae). *Z. latifolia* occurs in some rice-growing areas of temperate China and Japan, where it grows wild and is also cultivated as a vegetable. However, there is evidence of reproductive isolation between *C. suppressalis* populations from *O. sativa* and *Z. latifolia* (Konno and

Tanaka, 1996), which would diminish its value as a refuge. Cuong and Cohen (2002) also conducted a field survey of five grass species that are purported alternative *C. suppressalis* hosts in the Philippines. Only four *C. suppressalis* larvae were found in almost 44,000 tillers collected in Laguna Province over a 13-month period, all of which had likely moved into the alternative hosts from adjacent rice plants. All five grasses were poor hosts for *C. suppressalis* in greenhouse experiments. Khan et al. (1991) also list numerous reports of alternative hosts for *S. incertulas*, but these reports are likely to be spurious as well. *Scirpophaga incertulas* is generally considered to be monophagous on *O. sativa* (Dale, 1994).

The history of farmer adoption of new rice varieties, such as the semi-dwarf rices of the green revolution (Khush, 1995), suggests that farmers will rapidly adopt *Bt* rice should it become available. Consequently, it is possible that, in some major rice-growing regions, the only significant refuges that will exist will be the rice fields of the few farmers who choose to grow non-*Bt* varieties. No comparable situation exists for any commercialized *Bt* crop.

As defined in the high dose-refuge strategy, a high dose is a level of toxin that is sufficient to kill almost all insects heterozygous at the resistance locus (Gould, 1998; Ferré et al., chapter 3). High heterozygote mortality in combination with the presence of an appropriate refuge can substantially delay the increase in frequency of resistance alleles in a population. There is a risk that the toxin dose of *Bt* rice lines will be reduced in farmers' fields due to within-field mixtures of *Bt* and non-*Bt* plants. Within-field mixtures can decrease the dose to which pests are exposed, and decrease the size of refuges, if the feeding stages of the pest move between *Bt* and non-*Bt* plants (Mallet and Porter, 1992; Gould, 1998; Ferré et al., chapter 3). For such pests, seed mixtures are less effective than field-to-field refuges. Because most larvae of *C. suppressalis* and *S. incertulas* move from plant to plant at least once during development (Cohen et al., 2000b; Dirie et al., 2000), seed mixtures would not be a recommended approach to maintaining refuges for *Bt* rice. However, it is likely that deliberate or inadvertent mixing of *Bt* and non-*Bt* seeds will occur in rice-producing areas. In most rice-growing areas, the great majority of farmers plant inbred varieties. These farmers usually replant their own seed or obtain seed from other farmers, and only occasionally purchase certified seed. These practices will result in gradual mixing of *Bt* and non-*Bt* seed during storage and handling. In addition, unscrupulous seed dealers selling inbred or hybrid seed may dilute *Bt* seed with non *Bt*-seed to boost profits, while farmers may dilute these *Bt* seeds as a way to stretch their investment.

8.4.2 Simulation Modeling

We used a deterministic simulation model (Gould et al., 2006) to examine the durability of *Bt* rice under some of the unique circumstances that are likely to prevail for *Bt* rice in Asia, and to evaluate the effectiveness of actions that might be implemented to increase durability. The model calculates the frequency of the resistance allele (R) after each insect generation. The pest population is considered to be resistant once the frequency of R exceeds 0.5. We examined the effect of two variables: refuge size and refuge type (i.e., separate *Bt* and refuge fields, and seed mixtures).

The simulations were conducted assuming a high-dose cultivar with a single *Bt* toxin. We further assumed that resistance to the *Bt* toxin is inherited as a monogenic, recessive trait with two alleles; that the allele for resistance (R) is recessive to that for susceptibility (S); that there is a recessive 10% fitness cost to the R allele; that the initial frequency of the R allele is 0.001; and that random mating occurs between insects from *Bt* and refuge fields.

In simulations involving seed mixtures, we assumed that 25% of larvae remain on the same plant throughout development, while the remaining 75% of larvae change plants once and do not discriminate between *Bt* and non-*Bt* plants when moving. This is a simplification of the complex movement dynamics of *C. suppressalis* and *S. incertulas* larvae (Cohen et al., 2000b; Dirie et al., 2000), but incorporates the key behavioral attribute that most larvae feed on more than one plant during development.

We assigned average fitness values to the SS, RS, and RR genotypes under pure stand and mixed-stand scenarios as shown in Table 8.2, based on the following assumptions. In pure stands, the fitness values of larvae that move are the same as those that do not move. In seed mixtures, larvae that do not move have the same fitness values as larvae in pure stands, for both *Bt* plants and non-*Bt* plants. The average fitness of SS and RS larvae that move from *Bt* to non-*Bt* plants is slightly higher than those of larvae of these genotypes that feed only on *Bt* plants, as some may move before consuming a lethal dose of toxin. The fitness of RS larvae in mixed stands is higher than that of SS larvae because when larvae feed on both *Bt* and non-*Bt* plants the dose of toxin is reduced, and under these conditions we assume that the R allele is not completely recessive. SS and RS larvae that move from non-*Bt* to *Bt* plants have slightly higher fitness than larvae of these genotypes that feed only on *Bt* plants or move from *Bt* to non-*Bt* plants. Most stem borer larvae move between 2 and 15 days after eclosion, and these older larvae may have a greater probability of survival on *Bt* plants than would neonate larvae.

Table 8.2 Fitness values for insect genotypes used in simulations of *Bt* rice durability[a]

	Genotype		
Scenario	SS	RS	RR
Pure stand of *Bt* plants	0.001	0.01	0.8
Pure stand of non-*Bt* plants	1.0	1.0	0.8
Seed mixture; larvae moving from *Bt* to non-*Bt* plants	0.002	0.05	0.8
Seed mixture; larvae moving from non-*Bt* to *Bt* plants	0.005	0.1	0.8

[a]We assume a 10% recessive fitness cost to the R allele, and that 75% of larvae move once between plants during development and 25% do not move. In pure stands, the fitness values of larvae that move are the same as those that do not move. In seed mixtures, larvae that do not move have the same fitness values as larvae in pure stands, for both *Bt* plants and non-*Bt* plants. See text for additional explanation of fitness values.

Under the conditions described above, and with a landscape consisting of 20% refuge fields and 80% *Bt* fields and no mixing of seeds in either refuge or *Bt* fields, 73 generations are required before the frequency of R exceeds 0.5. This is a highly favorable scenario, such as that mandated for *Bt* maize in North America. Within a few years of the release of *Bt* rice, however, the proportion of refuge fields in some rice-growing areas will probably be much lower than 20%. With a landscape consisting of 5% refuge fields and 95% *Bt* fields, R exceeds a frequency of 0.5 after 36 generations (Table 8.3). Durability of the *Bt* toxin declines to 18 generations when refuges account for only 2% of rice fields.

Inadvertent mixing of *Bt* and non-*Bt* rice seeds will probably occur during planting, harvest and storage, and as a result of outcrossing. Rice is primarily a self-pollinating crop, but a low level of outcrossing occurs under field conditions (Lu and Snow, 2005). After a few cropping seasons, inadvertent mixing might result in levels of 5% non-*Bt* seed in *Bt* seed, and vice-versa. Under these conditions, and with a landscape consisting of 5% refuge fields and 95% *Bt* fields, the frequency of R would exceed 0.5 after 31 generations. This is a small decrease in durability compared with pure stands of *Bt* and non-*Bt* fields and a 5% refuge. A similarly small decrease occurs when comparing the 2% refuge scenarios with pure stands and 5% seed contamination (Table 8.3).

Intentional dilution of *Bt* seed by dealers or farmers could result in *Bt* fields that consist of 50–50 mixtures of *Bt* and non-*Bt* seed. Under such conditions, and assuming that 5% of fields are refuges consisting of non-*Bt* seed contaminated with 5% *Bt* seed, we found that the frequency of R would exceed 0.5 after 19 generations. With 2% of the landscape consisting of similar refuge fields, durability declines to nine generations.

8.4.3 Options to Increase Durability

In a tropical irrigated rice ecosystem with continuous rice cropping, such as that found in the most productive rice bowls in Asia, there are approximately six to eight

Table 8.3 Durability of a single-toxin *Bt* rice variety under different refuge and seed contamination scenarios

Refuge type	Percent of refuge fields in landscape	No. generations until frequency of R >0.5
Pure stands	20	73
	5	36
	2	18
Seed mixture, 5% contamination[a]	20	117
	5	31
	2	16
Seed mixture 50% contamination[b]	20	65
	5	19
	2	9

[a] *Bt* fields have 5% non-*Bt* plants and non-*Bt* fields have 5% *Bt* plants
[b] *Bt* fields have 50% non-*Bt* plants and non-*Bt* fields have 5% *Bt* plants

generations of stem borers per year. Therefore, under some of the scenarios examined above, a *Bt* toxin could become ineffective within a few years of its deployment in *Bt* rice. Given the socio-economic characteristics of rice production in Asia and the biology of rice stem borers, there are few practical interventions that governments can make to prolong the effectiveness of *Bt* toxins.

One practical option is to deploy only rice varieties with two pyramided toxins, both at a high dose. Two-toxin varieties require smaller refuges than do one-toxin varieties (Roush, 1997; Bates et al., 2005) and are more durable (Zhao et al., 2003). They are most effective when both toxins are at a high dose and there is no cross-resistance between toxins. Biochemical studies with rice stem borers have identified *Bt* toxins that appear to bind to different midgut receptors and that therefore can be recommended for use in combination. These combinations include Cry1Ab or Cry1Ac with Cry1C or Cry2A (Fiuza et al., 1996; Alcantara et al., 2004).

A second resistance management option for governments is to restrict the number of popular varieties that are available in *Bt* form, and to ensure that adequate seed supplies of popular non-*Bt* varieties are maintained. This kind of seed supply management could encourage greater numbers of farmers to grow non-*Bt* varieties, e.g., if they have a favored variety that is only available in non-*Bt* form.

Finally, it may be possible to maintain refuge fields through community participatory mechanisms, in which members would make collective decisions to implement resistance management plans. Such approaches have been successfully used in implementing natural resource management (Pound et al., 2003) and sustainable land use (Neef, 2005) in Asia.

8.4.4 Recommendations for Further Research and Cultivar Development

It is uncertain whether random mating between insects from *Bt* rice and refuge fields will occur with refuge levels as low as 2% and 5%, such as we have simulated above. This will depend in part on the distance moved by adults before mating. Insufficient movement would result in assortative mating and a more rapid increase in the frequency of R. If the size of each rice field is 1 ha and refuge fields have a regular spatial distribution, then we calculate that the maximum distance from the edge of a *Bt* field to the edge of a non-*Bt* field would be 105 and 100 m, for the 2% and 5% refuge scenarios, respectively. Cuong and Cohen (2003) found that all *S. incertulas* observed (male and female), and >85% of *C. suppressalis*, flew away from the site of eclosion before mating. However, no data are available on the distance moved before mating, and such studies should be conducted.

Because *Bt* rice will probably be grown under conditions that pose a high risk for the evolution of pest resistance, implementation of well-designed resistance monitoring programs is of great importance. In addition to providing early warning of possible crop resistance failures, data from monitoring programs could be used to design improved strategies for the release of future insect-resistant rice varieties in rice production systems in Asia. The first step in developing a monitoring program is to collect baseline data on the susceptibility of the target pest populations. Meng et al. (2003) obtained baseline data on *C. suppressalis* susceptibility to Cry1Ab and Cry1Ac in the major rice growing regions of China. Similar studies are needed in other countries and with *S. incertulas*. Resistance monitoring would be made more efficient if discriminating doses (to distinguish susceptible from resistant insects) were determined for target stem borer species.

The F_2 screen has also been suggested as a monitoring tool (Andow and Alstad, 1998), although it is recognized to be labor intensive and to have other methodological problems (Zhao et al., 2002; Ferré et al., chapter 3). Bentur et al. (2000) applied the F_2 screen to estimate the frequency of resistance genes in an *S. incertulas* population in the Philippines. They noted that difficulties in rearing inbred lines of *S. incertulas* under *Bt* selection and the large area of greenhouse space required for the procedure were practical limitations for the F_2 screen with this species.

In addition to monitoring for resistance, it will also be useful to implement monitoring of stem borer damage in farmers' fields in areas growing *Bt* rice. Alstad and Andow (1995) suggested that extensive adoption of *Bt* crops might result in a halo effect, whereby non-*Bt* fields experience decreased pest damage because of area-wide reductions in pest populations. This effect would be most likely to occur for pests with a restricted host range, and has apparently occurred with *Pectinophora gossypiella* (Lepidoptera: Gelichiidae) in Arizona cotton fields (Carrière et al., 2003). If area-wide suppression is documented to occur with rice stem borers, then more farmers might be convinced to grow non-*Bt* varieties and thereby increase the proportion of refuge fields in the area.

High-dose cultivars that kill almost all insects heterozygous at resistance loci are an essential component of the high-dose refuge strategy. There are five different imperfect ways to demonstrate that a *Bt* cultivar produces a high dose relative to a particular insect population (Ferré et al., chapter 3). None of these methods has been applied to any rice pest for any line of *Bt* rice available today, and thus no *Bt* rice line has been shown to express a high dose. Two of the methods that have been widely used on pests of other crops require the availability of *Bt*-resistant pest colonies or reliable LC_{99} values. No resistant colonies of rice stem borers or leaffolders have been established, nor are there any published LC_{99} estimates for these species. High control mortality and other difficulties in rearing and handling rice stem borers are obstacles to developing resistant colonies and conducting dose-response experiments. Some *Bt* rice lines described in the literature (Table 8.1) have doses of toxin that are comparable to high-dose lines of *Bt* cotton and maize varieties in commercial production, i.e., >0.2% soluble protein or >2 μg/g of fresh leaf tissue (Cohen et al., 2000a), and thus may function as high-dose lines. In the absence of experimental data demonstrating whether rice lines produce a high dose, lines with each *Bt* toxin produced at a level comparable to those of high-dose *Bt* cotton or maize cultivars could be given priority for commercial release.

Development of additional *Bt* rice cultivars with appropriate pyramids of two Cry toxins is needed. A basmati rice with a pyramid of *cry1Ab* and *cry2A* has been field tested in Pakistan (Bashir et al., 2005; Mahmood-ur-Rahman et al., 2007). Several single-toxin parental lines of the popular Chinese hybrid Minghui 63 have been produced (Table 8.1). Some of these lines could be crossed to produce pyramided varieties. A rice line with a pyramid of *cry1Ab* and *CpTI* has been produced (Zhao et al., 2004). However, no plant of any crop species transformed with a protease inhibitor gene has been shown to perform as a high dose plant, and thus the effectiveness of pyramids of *cry* and protease inhibitor genes remains to be demonstrated.

8.5 Conclusions

Bt maize and *Bt* cotton have been highly successful crop protection technologies. Similarly, *Bt* rice has the potential to be a beneficial innovation in rice production, although there are several challenges and constraints to its success. Preliminary results indicate that *Bt* rice is compatible with biological control and soil health, but there is a need for studies of larger scale and longer duration and in additional environments. Products derived from *Bt* maize and *Bt* cotton have been approved for human consumption in several countries, and studies on the food safety of *Bt* rice have not revealed any concerns (High et al., 2004). Resistance management for *Bt* rice cannot rely on alternative hosts to provide refuges, as has been the case for *Bt* cotton in Asia, nor will it be possible in most rice-growing areas to implement grower requirements for refuge fields as has been done for *Bt* cotton and *Bt* maize in some countries. It is likely that the durability of *Bt* toxins in rice will be lower than in other crops. Finally, most farmers who grow *Bt* rice will obtain higher yields and will use less insecticide, although the yield increases will not be as dramatic as those obtained for *Bt* cotton in many countries (see Naranjo et al., chapter 6) and full reduction of unnecessary insecticide use will require investment in grower education.

Given the potential benefits of *Bt* rice, the favorable data on safety to the environment and human health, and the advanced state of development of single-toxin *Bt* cultivars, why have no countries released *Bt* rice to farmers? The reasons appear to be concerns about consumer acceptance and international trade. Rice is of great cultural importance throughout Asia and is the predominant staple food. It is not surprising that there is some reluctance on the part of consumers in Asia to accept the introduction of *Bt* rice, and caution on the part of their governments about approving commercial release of the crop to farmers. Countries that export substantial amounts of rice must also consider the acceptance of *Bt* rice by their foreign markets. (In 2004, the largest rice exporters in Asia were, in descending order, Thailand, India, Vietnam, Pakistan and China (FAO, 2007).) Once a transgenic rice is approved for cultivation, it will be difficult to maintain its segregation from conventional rice within the country of production. Thus, the release of any transgenic variety may put at risk all rice exports from the country of production to some trading partners.

It is possible that "Golden Rice," which produces provitamin A, will be the first type of transgenic rice released to farmers in Asia because of the important benefits it can provide to some rice consumers (Al-Babili and Beyer, 2005). If so, the acceptance of Golden Rice by consumers and international markets will help to determine the time course for the release of *Bt* rice.

Acknowledgements We thank Fred Gould for providing the simulation model and guidance on its use. The opinions expressed in this article are those of the authors and not of the institutions with which they are affiliated.

References

AGBIOS, 2007. GM crop database. http://www.agbios.com/dbase.php (accessed 11 October 2007).

Al-Babili, S., and Beyer, P., 2005. Golden rice – five years on the road – five years to go? Trends in Plant Science 10: 565–573.

Alcantara, E.P., Aguda, R.M., Curtiss, A., Dean, D.H., and Cohen, M.B., 2004. *Bacillus thuringiensis* δ-endotoxin binding to brush border membrane vesicles of rice stem borers. Archives of Insect Biochemistry and Physiology 55: 169–177.

Alstad, D.N, and Andow, D.A., 1995. Managing the evolution of insect resistance to transgenic plants. Science 268: 1894–1896.

Andow, D.A., and Alstad, D.N., 1998. F_2 screen for rare resistance alleles. Journal of Economic Entomology 91: 572–578.

Bai, Y.Y., Jiang, M.X., and Cheng, J.A., 2005a. Effects of transgenic *cry1Ab* rice pollen on the oviposition and adult longevity of *Chrysoperla sinica* Tjeder. Acta Phytophylacica Sinica 32: 225–230.

Bai, Y.Y., Jiang, M.X., and Cheng, J.A., 2005b. Effects of transgenic cry1Ab rice pollen on fitness of *Propylea japonica* (Thunberg). Journal of Pest Science 78: 123–128.

Bai, Y.Y., Jiang, M.X., and Cheng, J.A., 2005c. Impacts of transgenic *cry1Ab* rice on two collembolan species and predation of *Microvelia horvathi* (Hemiptera: Veliidae). Acta Entomologica Sinica 48: 42–47.

Bai, Y.Y., Jiang, M.X., Cheng, J.A., and Wang, D., 2006a. Effects of Cry1Ab toxin on *Propylea japonica* (Thunberg) (Coleoptera: Coccinellidae) through its prey, *Nilaparvata lugens* Stål (Homoptera: Delphacidae), feeding on transgenic Bt rice. Environmental Entomology 35: 1130–1136.

Bai, Y.Y., Jiang, M.X., Cheng, J.A., and Wang, D., 2006b. Effects of transgenic Bt *cry1Ab* rice on collembolan population in paddy field. Chinese Journal of Applied Ecology 17: 903–906.

Bai, Y.Y., Jiang, M.X., and Cheng, J.A., 2006c. Residues of Cry1Ab insecticidal protein in water released from Bt rice tissue litter and surveying of three aquatic insect groups in Bt rice fields. Journal of Sichuan Agricultural University 24: 25–29.

Baki, B.B., Chin, D.V., and Mortimer, A.M., eds., 2000. Wild and weedy rice in rice ecosystems in Asia – a review. International Rice Research Institute, Los Baños, The Philippines.

Bashir, K., Husnain, T., Fatima, T., Riaz, N., Makhdoom, R., and Riazuddin, S., 2005. Novel indica basmati line (B-370) expressing two unrelated genes of *Bacillus thuringiensis* is highly resistant to two lepidopteran insects in the field. Crop Protection 24: 870–879.

Bates, S.L., Zhao, J.-Z., Roush, R.T., and Shelton, A.M., 2005. Insect resistance management in GM crops: past, present, and future. Nature Biotechnology 23: 57–62.

Baumgarte, S., and Tebbe, C.C., 2005. Field studies on the environmental fate of the Cry1Ab Bt-toxin produced by transgenic corn (MON810) and its effect on bacterial communities in the maize rhizosphere. Molecular Ecology 14: 2539–2551.

Bentley, J.W., 1989. What farmers don't know can't help them: the strengths and weaknesses of indigenous technical knowledge in Honduras. Agriculture and Human Values 6: 25–31.

Bentur, J.S., Andow, D.A., Cohen, M.B., Romena, A.M., and Gould, F., 2000. Frequency of alleles conferring resistance to a *Bacillus thuringiensis* toxin in a Philippine population of *Scirpophaga incertulas* (Lepidoptera: Pyralidae). Journal of Economic Entomology 93: 1515–1521.

Bernal, C.C., Aguda, R.M., and Cohen, M.B., 2002. Effect of rice lines transformed with *Bacillus thuringiensis* toxin genes on the brown planthopper and its predator *Cyrtorhinus lividipennis*. Entomologia Experimentalis et Applicata 102: 21–28.

Bhatti, M.A., Duan, J., Head, G., Jiang, C.J., McKee, M.J., Nickson, T.E., Pilcher, C.L., and Pilcher, C.D., 2005. Field evaluation of the impact of corn rootworm (Coleoptera: Chrysomelidae)–protected *Bt* corn on ground-dwelling invertebrates. Environmental Entomology 34: 1193–1210.

Breitler, J.C., Vsssal, J.M., Catala, M.D.M., Meynard, D., Marfa, V., Mele, E., Royer, M., Murillo, I., Segundo, B.S., Guiderdoni, E., and Messeguer, J., 2004. Bt rice harbouring *cry* genes controlled by a constitutive or wound-inducible promoter: protection and transgene expression under Mediterranean field conditions. Plant Biotechnology Journal 2: 417–430.

Brussaard, L., 1998. Soil fauna, guilds, functional groups and ecosystem processes. Applied Soil Ecology 9: 123–135.

Bunsha, D., 2006, November 18. Crops on trial. Frontline 23(23). http://www.hinduonnet.com/fline/fl2323/stories/20061201003603000.htm (accessed 15 January 2008).

Carrière, Y., Ellers-Kirk, C., Sisterson, M., Antilla, L., Whitlow, M., Dennehy, T.J., and Tabashnik, B.E., 2003. Long-term regional suppression of pink bollworm by *Bacillus thuringiensis* cotton. Proceedings of the National Academy of Sciences of the USA 100: 1519–1523.

Chen, H., Tang, W., Xu, C., Li, X., Lin, Y., and Zhang, Q., 2005. Transgenic indica rice plants harboring a synthetic cry2A* gene of *Bacillus thuringiensis* exhibit enhanced resistance against lepidopteran rice pests. Theoretical and Applied Genetics 111: 1330–1337.

Chen, L.J., Lee, D.S., Song, Z.P., Suh, H.S., and Lu, B.R., 2004. Gene flow from cultivated rice (*Oryza sativa*) to its weedy and wild relatives. Annals of Botany 93: 67–73.

Chen, M., Ye, G.Y., Hu, C., and Datta, S.K., 2003. Effect of transgenic *Bt* rice on the dispersal of planthoppers and leafhoppers as well as their egg parasitic wasps. Journal of Zhejiang University (Agriculture and Life Science) 29: 29–33.

Chen, M., Ye, G.Y., Liu, Z.C., Yao, H.W., Chen, X.X., Sheng, Z.C., Hu, C., and Datta, S.K., 2006a. Field assessment of the effects of transgenic rice expressing a fused gene of *cry1Ab* and *cry1Ac* from *Bacillus thuringiensis* Berliner on non-target planthoppers and leafhoppers. Environmental Entomology 35: 127–134.

Chen, M., Zhao, J.Z., Ye, G.Y., Fu, Q., and Shelton, A.M., 2006b. Impact of insect-resistant transgenic rice on target insect pests and non-target arthropods in China. Insect Science 13: 409–420.

Chen, M., Liu, Z.C., Ye, G.Y., Shen, Z.C., Hu, C. Peng, Y.F., Altosaar, I., and Shelton, A.M., 2007. Impacts of transgenic *cry1Ab* rice on non-target planthoppers and their main predator *Cyrtorhinus lividipennis* (Hemiptera: Miridae) - A case study of the compatibility of Bt rice with biological control. Biological Control 42: 242–250.

Cohen, M.B., Gould, F., and Bentur, J.S., 2000a. Bt rice: practical steps to sustainable use. International Rice Research Notes 25(2): 4–10.

Cohen, M.B., Romena, A.M., and Gould, F., 2000b. Dispersal by larvae of the stem borers *Scirpophaga incertulas* (Lepidoptera: Pyralidae) and *Chilo suppressalis* (Lepidoptera: Crambidae) in plots of transplanted rice. Environmental Entomology 29: 958–971.

Cohen, M.B., Arpaia, S., Lan, L.P., Chau, L.M., and Snow, A.A., 2008. Shared flowering phenology, insect pests, and pathogens among wild, weedy, and cultivated rice in the Mekong Delta, Vietnam: implications for transgenic rice. Environmental Biosafety Research 7: in press.

Cuong, N.L., and Cohen, M.B., 2002. Field surveys and greenhouse evaluation of non-rice host plants of the striped stem borer, *Chilo suppressalis* (Lepidoptera: Pyralidae), as refuges for resistance management of rice transformed with *Bacillus thuringiensis* toxin genes. Bulletin of Entomological Research 92: 265–268.

Cuong, N.L., and Cohen, M.B., 2003. Mating and dispersal behaviour of *Scirpophaga incertulas* and *Chilo suppressalis* (Lepidoptera: Pyralidae) in relation to resistance management for rice transformed with *Bacillus thuringiensis* toxin genes. International Journal of Pest Management 49: 275–279.

Dale, D., 1994. Insect pests of the rice plant–their biology and ecology. In: Biology and Management of Rice Insects, E.A. Heinrichs, ed., Wiley Eastern/New Age International, New Delhi, India, pp. 363–485.

Deng, C., Song, G., Xu, J., and Zhu, Z., 2003. Increasing accumulation level of foreign protein in transgenic plants through protein targeting. Acta Botanica Sinica 45: 1084–1089.

Dirie, A.M., Cohen, M.B., and Gould, F., 2000. Larval dispersal and survival of two stem borer (Lepidoptera: Crambidae) species on *cry1Ab*-transformed and non-transgenic rice. Environmental Entomology 29: 972–978.

Escalada, M.M., and Heong, K.L, 2004. A participatory exercise for modifying rice farmers' beliefs and practices in stem borer assessment. Crop Protection 23: 11–17.

FAO, 2007. FAOSTAT. http://faostat.fao.org/site/567/default.aspx (accessed 1 September 2007).

Fishbein, M., and Ajzen, I., 1975. Belief, Attitude, Intention and Behavior: An Introduction to Theory and Research. Addison-Wesley, Reading, MA, USA.

Fiuza, L.-M., Nielson-Leroux, C., Goze, E., Frutos, R., and Charles, J.-F., 1996. Binding of *Bacillus thuringiensis* Cry1 toxins to the midgut brush border membrane vesicles of *Chilo suppressalis* (Lepidoptera: Pyralidae): evidence of shared binding sites. Applied and Environmental Microbiology 62: 1544–1549.

Foissac, X., Nguyen, T.L., Christou, P., Gatehouse, A.M.R., and Gatehouse, J.A., 2000. Resistance to green leafhopper (*Nephotettix virescens*) and brown planthopper (*Nilaparvata lugens*) in transgenic rice expressing snowdrop lectin (*Galanthus nivalis* agglutinin; GNA). Journal of Insect Physiology 46: 573–583.

Fujimoto, H., Itoh, K., Yamamoto, M., Kyozuka, J., and Shimamoto, K., 1993. Insect resistant rice generated by introduction of a modified endotoxin gene of *Bacillus thuringiensis*. BioTechnology 11: 1151–1155.

Gallagher, K.D., Kenmore, P.E., and Sogawa, K., 1994. Judicious use of insecticides deter planthopper outbreaks and extend the life of resistant varieties in Southeast Asian rice. In: Planthoppers: Their Ecology and Management, R.F. Denno and J.T. Perfect, eds., Chapman & Hall, New York, USA, pp. 599–614.

Gao, Y.L., Fu, Q., Wang, F., Lai, F.X., Luo, J., Peng, Y.F., and Zhang, Z.T., 2006. Effects of transgenic rice harboring *cry1Ac* and *CpTI* Genes on survival of *Chilo suppressalis* and *Sesamia inferens* and field composition of rice stem borers. Chinese Journal of Rice Science 20: 543–548.

Ghareyazie, B., Alinia, F., Menguito, C.A., Rubia, L.G., de Palma, J.M., Liwanag, E.A., Cohen, M.B., Khush G.S., and Bennett, J., 1997. Enhanced resistance to two stem borers in aromatic rice containing a synthetic *cryIAb* gene. Molecular Breeding 3: 401–414.

Gigerenzer, G., and Todd, P.M., 1999. Adaptive Behavior and Cognition research group. Simple Heuristics That Make Us Smart. Oxford University Press, New York, USA.

Gould, F., 1998. Sustainability of transgenic insecticidal cultivars-integrating pest genetics and ecology. Annual Review of Entomology 43: 701–726.

Gould, F., Cohen, M.B., Bentur, J.S., Kennedy, G., and Van Duyn, J., 2006. Impact of small fitness costs on pest adaptation to crop varieties with multiple toxins: a heuristic model. Journal of Economic Entomology 99: 2091–2099.

Griffiths, B.S., Heckmann, L.H., Caul, S., Thompson, J., Scrimgeour, C., and Krogh, P.H., 2007. Varietal effects of eight paired lines of transgenic Bt maize and near-isogenic non-Bt maize on soil microbial and nematode community structure. Plant Biotechnology Journal 5: 60–68.

Guo, Y.J., Wang, N.Y., Jiang, J.W., Chen, J.W., and Tang, J., 1995. Ecological significance of neutral insects as nutrient bridge for predators in irrigated rice arthropod community. Chinese Journal of Biological Control 11: 5–9.

Hachiya, K., 1981. Notes on host plants of the genus *Chilo* (Lepidoptera, Pyralidae) in Hokkaido. Bulletin of the Hokkaido Prefecture Agricultural Experiment Station 45: 47–52.

Haimi, J., 2000. Decomposer animals and bioremediation of soils. Environmental Pollution 107: 233–238.

Han, L., Wu, K., Peng, Y., Wang, F., and Guo, Y., 2006. Evaluation of transgenic rice expressing Cry1Ac and CpTI against *Chilo suppressalis* and intrapopulation variation in susceptibility to Cry1Ac. Environmental Entomology 35: 1453–1459.

Heinrichs, E.A., 1994. Host plant resistance. In: Biology and Management of Rice Insects, E.A. Heinrichs, ed., Wiley Eastern/New Age International, New Delhi, India, pp. 517–547.

Heong, K.L., and Escalada, M.M., eds., 1997. Pest Management of Rice Farmers in Asia. International Rice Research Institute, Los Baños, The Philippines.

Heong, K.L., and Escalada, M.M., 1999. Quantifying rice farmers' pest management deicisions – beliefs and subjective norms in stem borer control. Crop Protection 18: 315–322.

Heong, K.L., Escalada, M.M., Sengsoulivong, V., and Schiller, J., 2002. Insect management beliefs and practices of rice farmers in Laos. Agriculture, Ecosystems and Environment 92: 137–145.

Heong, K.L., Chen, Y., Johnson, D.E., Jahn, G.C., Hossain, M., and Sackville-Hamilton, R., 2005. Letter to editor: debate over a GM rice trial in China. Science 310: 231.

High, S.M., Cohen, M.B., Shu, Q.Y., and Altosaar, I., 2004. Achieving successful deployment of *Bt* rice. Trends in Plant Science 9: 286–292.

Huan, N.H., Thiet, L.V, Chien, H.V., and Heong, K.L., 2004. Farmers' evaluation of reducing pesticides, fertilizers and seed rates in rice farming through participatory research in the Mekong Delta, Vietnam. Crop Protection 24: 457–464.

Huang, J.K., Hu, R.F., Rozelle, S., and Pray, C., 2005. Insect-resistant GM rice in farmers' fields: assessing productivity and health effects in China. Science 308: 688–690.

James, C., 2005. Global Status of Commercialized Biotech/GM Crops: 2005. ISAAA Brief No. 34, International Service for the Acquisition of Agri-Biotech Applications, Ithaca, New York, USA.

James, C., 2007. Global Status of Commercialized Biotech/GM Crops: 2007. ISAAA Brief No. 37, International Service for the Acquisition of Agri-Biotech Applications, Ithaca, New York, USA.

Jiang, Y.H., Fu, Q., Cheng, J.A., Ye, G.Y., Bai, Y.Y., and Zhang, Z.T., 2004. Effects of transgenic *Bt* rice on the biological characteristics of *Apanteles chilonis* (Munakata) (Hymenoptera: Braconidae). Acta Entomologica Sinica 47: 124–129.

Jiang, Y.H., Fu, Q., Cheng, J.A., Zhu, Z.R., Jiang, M.X., Ye, G.Y., and Zhang, Z.T., 2005. Effect of transgenic *sck* + cry1Ac on the survival and growth of *Chilo Suppressalis* (Walker) and its parasitoid *Apanteles chilonis* (Munakata). Acta Entomologica Sinica 48: 554–560.

Khan, Z.R., Litsinger, J.A., Barrion, A.T., Villanueva, F.F.D., Fernandez, N.J., and Taylo, L.D., 1991. World Bibliography of Rice Stem Borers, 1794–1990. International Rice Research Institute, Los Baños, The Philippines.

Khush, G.S., 1995. Breaking the yield frontier of rice. GeoJournal 35: 329–332.

Konno, Y., and Tanaka, F., 1996. Mating time of the rice-feeding and water-oat-feeding strains of the rice stem borer, *Chilo suppressalis* (Walker) (Lepidoptera: Pyralidae). Japanese Journal of Applied Entomology and Zoology 40: 245–247.

Li, F.F., Ye, G.Y., Wu, Q., Peng, Y.F., and Chen, X.X., 2007a. Arthropod abundance and diversity in Bt and non-Bt rice fields. Environmental Entomology 36: 646–654.

Li, Y., Wu, K., Zhang, Y., and Yuan, G., 2007b. Degradation of Cry1Ac protein within transgenic *Bacillus thuringiensis* rice tissues under field and laboratory conditions. Environmental Entomology 36: 1275–1282.

Liu, Y.F., He, L., Wang, Q., Hu, S.Q., Liu, W.H., Chen, K.G., and You, M.S., 2006. Evaluation of the effects of insect-resistant *cry1Ac/sck* transgenic rice on the parasitoid communities in paddy fields. Acta Entomologica Sinica 49: 955–962.

Liu, Z.C., Ye, G.Y., Hu, C., and Datta, S.K., 2002. Effects of Bt transgenic rice on population dynamics of main non-target insect pests and dominant spider species in rice paddies. Acta Phytophylacica Sinica 29: 138–144.

Liu, Z.C., Ye, G.Y., Hu, C., and Datta, S.K., 2003. Impact of transgenic indica rice with a fused gene of *cry1Ab/cry1Ac* on the rice paddy arthropod community. Acta Entomologica Sinica 46: 454–465.

Liu, Z.C., Ye, G.Y., and Hu, C., 2004. Effects of *Bacillus thuringiensis* transgenic rice and chemical insecticides on arthropod communities in paddy-fields. Chinese Journal of Applied Ecology 15: 2309–2314.

Lu, B.R, and Snow, A.A., 2005. Gene flow from genetically modified rice and its environmental consequences. BioScience 55: 669–678.

Maclean, J.L., Dawe, D.C., Hardy, B., Hettel, G.P., eds., 2002. Rice Almanac, 3rd edition. CABI, Wallingford/Oxon, UK.

Mallet, J., and Porter, P., 1992. Preventing insect adaptation to insect-resistant crops: are seed mixtures or refugia the best strategy? Proceedings of the Royal Society B, Biological Sciences 250: 165–169.

Mahmood-ur-Rahman, Rashid, H., Shahid, A.A., Bashir, K., Husnain, T., and Riazuddin, S., 2007. Insect resistance and risk assessment studies of advanced generations of basmati rice expressing two genes of *Bacillus thuringiensis*. Electronic Journal of Biotechnology (online) 10(2). 15 April 2007. Available from: http://www.ejbiotechnology.info/content/vol10/issue2/full/3/ (accessed 3 January 2008).

Matteson, P.C., 2000. Insect pest management in tropical Asian irrigated rice. Annual Review of Entomology 45: 549–574.

Meng, F., Wu, K., Gao, X., Peng, Y., and Guo, Y., 2003. Geographic variation in susceptibility of *Chilo suppressalis* (Lepidoptera: Pyralidae) to *Bacillus thuringiensis* toxins in China. Journal of Economic Entomology 96: 1838–1842.

Mulder, C., Wouterse, M., Raubuch, M., Roelofs, W., and Rutgers, M., 2006. Can transgenic maize affect soil microbial communities? PLoS Computational Biology 2(9): e128, 1165–1173.

Mumford, J.D., and Norton, G.A., 1984. Economics of decision making in pest management. Annual Review Entomology 29: 157–174.

Naik, G., Qaim, M., Subramaniam, A., and Zilberman, D., 2005. Bt cotton controversy – some paradoxes explained. Economics and Political Weekly, April 9: 1515–1517.

Naranjo, S.E., 2005a. Long-term assessment of the effects of transgenic *Bt* cotton on the abundance of nontarget arthropod natural enemies. Environmental Entomology 34: 1193–1210.

Naranjo, S.E., 2005b. Long-term assessment of the effects of transgenic *Bt* cotton on the function of the natural enemy community. Environmental Entomology 34: 1211–1223.

Neef, A., ed., 2005. Participatory Approaches for Sustainable Land Use in Southeast Asia. White Lotus, Bangkok, Thailand.

Ooi, P.A.C., and Shepard, B.M., 1994. Predators and parasitoids of rice insect pests. In: Biology and Management of Rice Insects, E.A. Heinrichs, ed., Wiley Eastern/New Age International, New Delhi, India, pp. 585–612.

Pound, B., Snapp, S., McDougall, C., and Braun, A., eds., 2003. Managing Natural Resources for Sustainable Livelihoods – Uniting Science and Participation. Earthscan, UK.

Rajamani, S., Pasalu, I.C., Mathur, K.C., and Sain, M., 2004. Biology and ecology of rice gall midge. In: New Approaches to Gall Midge Resistance in Rice, J. Bennett, J.S. Bentur, I.C. Pasalu and K. Krishnaiah, eds., International Rice Research Institute/Indian Council of Agricultural Research, Los Baños, The Philippines/New Delhi, India, pp. 7–15.

Ren, X., Wu, W.X., Ye, Q.F., Chen, Y.X., and Janice, E.T., 2004. Effect of Bt transgenic rice straw on the bacterial community composition in a flooded paddy soil. Acta Scientiae Circumstantiae 24: 871–878.

Romeis, J., Meissle, M., and Bigler, F., 2006. Transgenic crops expressing *Bacillus thuringiensis* toxins and biological control. Nature Biotechnology 24: 63–71.

Roush, R.T., 1997. Two-toxin strategies for management of insect resistant transgenic crops: can pyramiding succeed where pesticide mixtures have not? Philosophical Transactions of the Royal Society of London B, Biological Sciences 353: 1777–1786

Saha, P., Majumder, P., Dutta, I., Ray, T., Roy, S.C., and Das, S., 2006. Transgenic rice expressing *Allium sativum* leaf lectin with enhanced resistance against sap-sucking insect pests. Planta 223: 1329–1343.

Saito, T., Hirai, K., and Way, M.O., 2005. The rice water weevil, *Lissorhoptrus oryzophilus* Kuschel (Coleoptera: Curculionidae). Applied Entomology and Zoology 40: 31–39.

Savary, S., Willocquet, L., Elazegui, F.S., Castilla, N.P., and Teng, P.S., 2000. Rice pest constraints in tropical Asia: quantification of yield losses due to rice pests in a range of production situations. Plant Disease 84: 357–369.

Saxena, D., Stewart, C.N., Altosaar, I., Shu, Q., and Stotzky, G., 2004. Larvicidal Cry proteins from *Bacillus thuringiensis* are released in root exudates of transgenic *B. thuringiensis* corn, potato, and rice but not of *B. thuringiensis* canola, cotton, and tobacco. Plant Physiology and Biochemistry 42: 383–387.

Schoenly, K., Cohen, J.E., Heong, K.L., Litsinger, J.A., Aquino, G.B., Barrion, A.T., and Arida, G. 1996. Food web dynamics of irrigated rice fields at five elevations in Luzon, Philippines. Bulletin of Entomological Research 86: 451–466.

Schoenly, K.G., Cohen, M.B., Barrion, A.T., Zhang, W.J., Gaolach, B., and Viajante, V.D., 2003. Effects of *Bacillus thuringiensis* on non-target herbivore and natural enemy assemblages in tropical irrigated rice. Environmental Biosafety Research 2: 181–206.

Settle, W.H., Ariawan, H., Astuti, E.T., Cahyana, W., Hakim, A.L., Hindayana, D., Lestari, A.S., and Pajarningsih, 1996. Managing tropical rice pests through conservation of generalist enemies and alternative prey. Ecology 77: 1975–1988.

Sheng, C.F., Wang, H.T., Gao, L.D., and Xuan, W.J., 2003. The current status on large scale occurrence of rice stem borers, their loss estimation and control and protection strategies in China. Plant Protection 29: 37–39.

Shu, Q.Y., Ye, G.Y., Cui, H.R., Cheng, X.Y., Xiang, Y.B., Wu, D.X., Gao, M.W., Xia, Y.W., Hu, C., Sardana, R., and Altosaar, I., 2000. Transgenic rice plants with a synthetic *cry1Ab* gene from *Bacillus thuringiensis* were highly resistant to eight lepidopteran rice pest species. Molecular Breeding 6: 433–439.

Song, Z.P., Lu, B.R., Zhu, Y.G., and Chen, J.K., 2003. Gene flow from cultivated rice to the wild species *Oryza rufipogon* under experimental field conditions. New Phytologist 157: 657–665.

Stotzky, G., 2004. Persistence and biological activity in soil of the insecticidal proteins from *Bacillus thuringiensis*, especially from transgenic plants. Plant and Soil 266: 77–89.

Sun, C.X., Chen, L.J., Wu, Z.J., Zhou, L.K., and Shimizu, H., 2007. Soil persistence of *Bacillus thuringiensis* (Bt) toxin from transgenic Bt cotton tissues and its effect on soil enzyme activities. Biology and Fertility of Soils 43: 617–620.

Tait, E.J., 1977. A method for comparing pesticide usage patterns between farmers. Annals of Applied Biology 86: 229–240.

Tan, H., Ye, G.Y., Shen, J.H., Pengy, Y.F., and Hu, C., 2006. Effects of transgenic indica rice expressing a gene of cry1Ab with insect resistance on the development and reproduction of nontarget pest, *Sogatella furcifera* (Homoptera: Delphacidae). Acta Phytophylacica Sinica 33: 251–256.

Tjornhorm, J.D., Norton, G.W., Heong, K.L., Talekar, N.S., and Gapud, V., 1997. Socioeconomic characteristics and pesticide misuse in Philippine onion production. Philippines Entomologist 11: 139–149.

Tu, J., Zhang, G., Datta, K., Xu, C., He, Y., Zhang, Q., Khush, G.S., and Datta, S.K., 2000. Field performance of transgenic elite commercial hybrid rice expressing *Bacillus thuringiensis* d-endotoxin. Nature Biotechnology 18: 1101–1104.

Tversky, A., Kahnemann, D., 1992. Advances in prospect theory: cumulative representation of uncertainty. Journal of Risk and Uncertainty 5: 297–323.

Tversky, A., and Kahnemann, D., 1974. Judgement under uncertainty: heuristics and biases. Science 185: 1124–1131.

Way, M.J., and Heong, K.L., 1994. The role of biodiversity in the dynamics and management of insect pests of tropical irrigated rice–a review. Bulletin of Entomological Research 84: 567–587.

Wang, H.Y., Ye, Q.F., Gan, J., and Wu, L.C., 2007. Biodegradation of Cry1Ab protein from Bt transgenic rice in aerobic and flooded paddy soils. Journal of Agricultural and Food Chemistry 55: 1900–1904.

Wang, F., Yuan, Q.-H., Shi., L., Qian, Q., Liu, W.-G., Kuang, B.-G., Zeng, D.-L., Liao, Y.-L., Cao, B., and Jia, S.-R., 2006. A large-scale field study of transgene flow from cultivated rice (*Oryza sativa*) to common wild rice (*O. rufipogon*) and barnyard grass (*Echinochloa crusgalli*). Plant Biotechnology Journal 4: 667–676.

Wang, Y., and Johnson, S., 2007. The status of GM rice R&D in China. Nature Biotechnology 25: 717–718.

Wei, T., Chen, H., Xu, C.G., Li, X.H., Lin, Y.J., and Zhang, Q.F., 2006. Development of insect-resistant transgenic indica rice with a synthetic cry1C* gene. Molecular Breeding 18: 1–10.

Wu, G., Cui, H., Ye, G., Xia, Y., Sardana, R., Cheng, X., Li, Y., Altosaar, I., and Shu, Q., 2002. Inheritance and expression of the *cry1Ab* gene in *Bt* (*Bacillus thuringiensis*) transgenic rice. Theoretical and Applied Genetics 104: 727–734.

Wu, K.M., and Guo, Y.Y., 2005. The evolution of cotton pest management practices in China. Annual Review of Entomology 50: 31–52.

Wu, W.X., Ye, Q.F., Min, H., and Chen, H.E., 2003. Effect of cry1Ab toxin released from straw of Bt-transgenic rice on microflora and enzymatic activities in upland soil. Acta Pedologica Sinica 40: 606–612.

Xu, X.Y., Ye, Q.F., Wu, W.X., and Min, H., 2004. Incorporation of the straw of transgenic Bt rice into the soil on anaerobic microbial populations and enzyme activity in paddy soil. Plant Nutrition and Fertilizer Science 10: 63–67.

Ye, G.Y., Shu, Q.Y., Yao, H.W., Cui, H.R., Cheng, X.Y., Hu, C., Xia, Y.W., Gao, M.W., and Altosaar, I., 2001a. Field evaluation of resistance of transgenic rice containing a synthetic *cry1Ab* gene from *Bacillus thuringiensis* Berliner to two stem borers. Journal of Economic Entomology 94: 271–276.

Ye, G.Y., Tu, J.M., Cui, H., Datta, K., and Datta, S.K., 2001b. Transgenic IR72 with fused *Bt* gene *cry1Ab/cry1Ac* from *Bacillus thuringiensis* is resistant against four lepidopteran species under field conditions. Plant Biotechnology 18: 125–133.

Zhao, H.Y., Zhang, Y.J., Wu, K.M., Zhao, K.J., Peng, Y.F., and Gu, Y.Y., 2004. Expression of Cry1Ac/CpTI transgenic rice and its resistance in different stages to *Chilo suppressalis*. Journal of Agricultural Biotechnology 12: 76–79.

Zhao, J.-Z., Li, Y., Collins, H.L., and Shelton, A.M., 2002. Examination of the F2 screen for rare resistance alleles to *Bacillus thuringiensis* toxins in the diamondback moth. Journal of Economic Entomology 95: 14–21.

Zhao, J.-Z., Cao, J., Li, Y., Collins, H.L., Roush, R.T., Earle, E.D., and Shelton, A.M., 2003. Plants expressing two *Bacillus thuringiensis* toxins delay insect resistance compared to single toxins used sequentially or in a mosaic. Nature Biotechnology 21: 1493–1497.

Chapter 9
Transgenic Vegetables and Fruits for Control of Insects and Insect-Vectored Pathogens

Anthony M. Shelton[1,*], Marc Fuchs[2], and Frank A. Shotkoski[3]

Abstract Fruits and vegetables are major components of a healthy diet, but are subject to severe pest pressure. Approximately 30% of all insecticides applied worldwide are used to control insects affecting vegetables and fruits. Transgenic, or more commonly referred to as genetically modified (GM), vegetables and fruits offer unique opportunities for controlling insects and the pathogens they transmit. Aphid transmitted viruses have been particularly difficult to manage by tactics aimed at reducing aphid populations and in many cases there has not been virus resistant plant germplasm. Farmers in the USA have benefited from having GM virus resistant squash and papaya available to them as tools in their overall IPM programs. In the USA, *Bt* sweet corn has proven effective for control of Lepidoptera and continues to be accepted in the fresh market. However, the best opportunities for GM vegetables and fruits are in developing countries where 83% of the world's population lives, the majority of vegetables and fruits are produced and pest problems are most acute.

9.1 Introduction

Vegetables and fruits are essential for well-balanced diets since they supply many of the essential nutrients not found in many of the staple crops. Additionally, there is compelling evidence that a diet rich in vegetables and fruits can lower the risk of heart disease, strokes and several forms of cancer, as well as improve gastrointestinal health and vision (HSPH, 2007).

Of the total worldwide production of vegetables and fruits in 2004, China (36.6%) and India (9.2%) produce the largest shares, with the USA (5.0%) a distant third (FAOSTAT, 2007). In 2006, world production of vegetables was 903,405,299

[1] Department of Entomology, Cornell University/NYSAES, Geneva, NY 14456, USA

[2] Department of Plant Pathology, Cornell University/NYSAES, Geneva, NY 14456, USA

[3] International Programs/ABSP II, Cornell University, Ithaca, NY 14853, USA

* To whom correspondence should be addressed. E-mail: ams5@cornell.edu

J. Romeis, A.M. Shelton, G.G. Kennedy (eds.), *Integration of Insect-Resistant Genetically Modified Crops within IPM Programs.*

metric tonnes (MT), while production of fruit was 526,496,050 MT. China was the leading producer of vegetables (49.6% of the world's vegetables or > 5-fold the production of the next leading country, India). China was also the world's leading producer of fruits, with 17.7% of the world's production, followed by India with 8.3%. The expansion of vegetable production in China has been particularly significant at almost 6% per year over the last 20 years. Expansion of vegetable production has exceeded 3% per annum in other developing countries in Asia and in developed countries. Worldwide, the area of arable land devoted to vegetables is expanding at 2.8%; higher than fruits (1.75%), oil crops (1.47%), root crops (0.44%) and pulses (0.39%), and at the expense of cereals (–0.45%), and fiber crops (−1.82%).

Many vegetables and fruits are consumed near where they are produced, especially in China and India. Besides local marketing, there is considerable movement of vegetables on the world market, and this includes fresh and processed vegetables. The European Union (EU), followed by North America and Japan, are the world's most important fresh produce import regions. To some extent, countries vary in their standards of acceptable pest management practices and this can affect imports, especially of GM products.

Production of fruits and vegetables is becoming attractive for many poor farmers worldwide because it is profitable. Farmers involved in production of fruits and vegetables usually earn much higher farm incomes compared to cereal producers, with per capita farm income up to 5-fold higher (Lumpkin et al., 2005). Horticultural crops are generally more knowledge and capital intensive than cereal crops and they suffer from many biological stresses including insects, diseases and weeds. Because of their diversity within and between plant families, their pest complexes are far more varied and complex compared to field crops. Considerably fewer resources have been directed at improving their production and pest management options compared to staple crops such as rice, wheat and maize (Lumpkin et al., 2005). Vegetables and fruits are high value commodities with high cosmetic standards, and the main method for control has been the frequent use of pesticides. In the case of insecticides, nearly 30% of the worldwide $8.1 billion annual insecticide market is applied to fruits and vegetables (Krattiger, 1997). Insecticides are regularly applied to control a complex of insects that cause damage by feeding directly on the plant or by transmitting pathogens that harm plants.

Genetic modification of vegetables and fruits for management of insects and insect-transmitted pathogens has provided some successes, and several examples are worth noting besides those covered under potatoes (Grafius and Douches, chapter 7) and maize (Hellmich et al., chapter 5). The literature suggests that more GM vegetables are commercialized (or soon to be so) than GM fruits. This is understandable since vegetables are annuals and require the purchase of new seed, ensuring a continuing market for seeds. Consequently, in this chapter more emphasis will be placed on GM vegetables. What follows is not meant to be a comprehensive review, but an effort to provide insights into some important projects and the issues they represent. Unlike other examples of GM insect-resistant plants discussed in other chapters of this book (i.e. cotton and maize), GM insect-resistant vegetables and fruits have often been developed through partnerships between the private and public sector. This is likely the result of vegetables and fruits being "minor crops" that are not grown on such extensive areas and generally have limited research and extension resources compared to field crops. Private-public partnerships help leverage resources and create new opportunities for moving much needed technologies forward. In this chapter, several projects on insect-resistant GM plants are described and the ways in which these technologies are being incorporated into IPM programs are discussed.

9.2 Insect Transmitted Virus Protected Vegetables and Fruits

9.2.1 Plant Viruses and Their Vectors

Viruses can substantially reduce production and quality of vegetable and fruit crops and are becoming increasingly problematic worldwide. Many plant viruses are vectored by insects, including aphids, whiteflies, thrips and leafhoppers. Aphid-vectored viruses are particularly problematic because many are transmitted in a non-circulative and non-persistent manner (Zitter et al., 1996; Gonsalves, 1998). This means that a very short time, i.e. a few seconds or minutes, is sufficient for aphids to acquire virus particles when probing on infected plants. A similarly short time period is enough for aphids to release virus particles when probing on healthy plants. The primary injury caused by aphid-vectored viruses arises not from the direct feeding damage of the aphids but from their ability to allow the virus to enter the plant and initiate the disease. Control has focused on using insecticides to control the vectors (aphids, whiteflies, thrips) or tactics such as mulches or barrier crops that lessen the likelihood of the vector landing on the crop (Hooks and Fereres, 2006). Successes with both strategies have been variable.

Host plant resistance should be the foundation of IPM (Kennedy, chapter 1; Naranjo et al., chapter 6). However, virus-resistant germplasm has not been available for many important vegetable and fruit crops. In addition to developing plants that directly resist insect feeding or development, another successful application of agriculture biotechnology is the development of plants that resist insect-transmitted viruses.

Virus resistance can be achieved by applying the concept of pathogen-derived resistance (Sanford and Johnston, 1985). The insertion of a virus gene fragment into a susceptible plant can activate RNA silencing, a potent defense mechanism against viruses (Voinnet, 2005). This mechanism confers virus resistance, providing a high degree of nucleotide sequence homology between the virus-derived transgene and the challenge virus. Resistance can be engineered against multiple viruses if gene fragments from different viruses are fused within a single expression cassette or pyramided within a single T-DNA region of a binary plasmid. The coat protein (CP) is commonly used to engineer virus resistance (Fuchs and Gonsalves, 2007). Described below are two examples of virus-resistant transgenic crop cultivars expressing CP genes.

9.2.2 Papaya

Papaya ringspot virus (PRSV) is a major virus affecting papaya (*Carica papaya* L.) worldwide that causes foliar mosaic, distortion, and plant stunting. PRSV also causes ringspots on fruits and affects fruit yield and quality. No useful resistance is known in the *Caricaceae* family. In Hawaii, PRSV caused production to fall from 58 million pounds in 1992 to 24 million pounds by 1998 (Fuchs and Gonsalves, 2007). Orchard scouting and elimination of PRSV-infected trees is routinely used to limit the spread of the virus, but these approaches are only moderately successful (Gonsalves, 1998). In the early 1980s, a team of scientists characterized PRSV, used rDNA to isolate and clone a gene for the CP of the virus, introduced the gene into plant cells using particle bombardment and created the first transgenic fruit for virus resistance (Fig. 9.1). Genetically modified papaya resistant to PRSV was commercially released in Hawaii in 1998 (Gonsalves, 1998) and has had a tremendous socio-economic impact. The adoption rate of virus-resistant transgenic papaya was rapid and widespread. Transgenic papaya cultivars were planted on more than half of the total papaya production area (480 ha) in 2004 in Hawaii (Fuchs and Gonsalves, 2007; Shankula, 2006). Since the release of PRSV-resistant papaya cultivars, papaya production in Hawaii has reached a level similar to that before PRSV became epidemic in the 1980s (Shankula, 2006).

In addition to Hawaii, China recommended the commercialization of PRSV-resistant transgenic papaya in late 2006 (James, 2006). Efforts are underway to commercialize PRSV-resistant papaya cultivars for the Philippines where the majority of the crop is consumed locally (Hautea et al., 1999; ABSP II, 2007). Through a partnership between Philippine public institutions, Monsanto and the Malaysian Agricultural Research Development Institute, a local Philippine variety was made resistant to PRSV (Hautea et al., 1999; ABSP II, 2007). Contained trials were completed in 2006 and confined trials are underway to assess the safety and efficacy of the new variety. In addition to the Philippines, field tests with PRSV-resistant papaya cultivars have been conducted in Thailand (Gonsalves et al., 2006), Brazil (Souza et al., 2005) and Jamaica (Tennant et al., 2005).

9.2.3 Squash

Yield losses due to viruses in the USA often range from 20% to 80% in summer squash (*Cucurbita pepo* L.) with a reported $2.6 million economic loss in the state of Georgia in 1997 (Gianessi et al., 2002). Three of the most important viruses affecting squash production are *Zucchini yellow mosaic virus* (ZYMV), *Watermelon mosaic virus* (WMV), and *Cucumber mosaic virus* (CMV) (Zitter et al., 1996). No summer squash cultivar with satisfactory resistance to CMV, ZYMV and WMV has yet been developed by conventional breeding (Gaba et al., 2004; Munger, 1993). Control of squash viruses has focused on cultural practices, including delayed transplanting relative to aphid flights, use of reflective film mulch to repel aphids,

Fig. 9.1 Papaya plant on the left was infected by papaya ringspot virus while plant on right was genetically engineered to be resistant to the virus (Photo by J. Ogrodnick)

and application of stylet oil (used to reduce virus transmission) in combination with insecticides to reduce aphid vector populations (Perring et al., 1999). In the state of Georgia, it is estimated that ten applications of stylet oils and insecticides are made routinely to control aphids and, hence, limit virus incidence and transmission (Gianessi et al., 2002).

Squash expressing the CP gene of ZYMV and WMV was exempted from regulation in the US in 1994 and was released thereafter (Tricoli et al., 1995; Acord, 1996). In addition, squash expressing the CP gene of ZYMV, WMV and CMV was deregulated and commercialized in 1995 (Medley, 1994). Subsequently, numerous squash types and cultivars have been developed by crosses and back crosses with the two initially deregulated lines. This material is highly resistant to infection by one, two or three of the target viruses, i.e. CMV, ZYMV and WMV (Arce-Ochoa et al., 1995; Clough and Hamm, 1995; Fuchs and Gonsalves, 1995; Tricoli et al., 1995; Fuchs et al., 1998; Schultheis and Walters, 1998). The adoption of virus-resistant squash cultivars is steadily increasing in the USA. In 2005, the adoption rate was estimated at 12% (approximately 3,100 ha) across the country with the highest rates in New Jersey (25%), Florida (22%), Georgia (20%), South Carolina (20%) and Tennessee (20%) (Shankula, 2006). Virus-resistant transgenic squash has allowed growers to achieve yields comparable to those obtained in the absence of viruses with a net benefit of $22 million in 2005 (Shankula, 2006).

9.2.4 Other Crops

In China, tomato and pepper resistant to CMV through expression of the viral CP gene have also been released (Shotkoski, personal communication). However, limited information is available on their adoption rate. There is an ongoing project on CMV resistant tomato in Indonesia and the Philippines (ABSP II, 2007). Since research on the development of virus-resistant transgenic plants is making substantial progress, it is anticipated that more crops will be released in the future (Fuchs and Gonsalves, 2007).

9.2.5 Virus-Resistant Plants and IPM

Virus-resistant transgenic squash and papaya are facilitating the implementation of IPM practices because insecticides directed to control arthropod vector populations are reduced or eliminated. This is particularly true in squash for which applications of stylet oil and insecticides are made routinely to control aphids in an effort to limit virus incidence and transmission (Gianessi et al., 2002).

Virus-resistant plants can be a major tool in IPM. Virus-resistant transgenic squash limits virus infection rates by restricting challenge viruses, reducing their titers, or inhibiting their replication and/or cell-to-cell or systemic movement. Therefore, lower virus levels reduce the frequency of acquisition by vectors and subsequent transmission within and between fields. Consequently, virus epidemics are substantially limited. Recently, it has been shown that commercial transgenic squash resistant to ZYMV and WMV does not serve as a virus source for secondary (i.e. within field) spread (Klas et al., 2006). Virus-resistant transgenic plants are particularly valuable if no genetic source of resistance has been identified or if host resistance is difficult to transfer into elite cultivars by traditional breeding approaches due to genetic incompatibility or links to undesired traits. In such cases, engineered resistance may be the only viable option to develop virus-resistant cultivars. This is well illustrated by PRSV-resistant papaya. Engineered resistance may also be the only approach to develop cultivars with multiple sources of resistance. This has been the case for squash resistant to CMV, ZYMV and WMV.

In Hawaii, an additional IPM benefit developed because of PRSV-resistant transgenic papaya. Prior to the introduction of PRSV-resistant transgenic papaya, growing papaya was no longer viable despite area-wide efforts to eradicate infected trees in order to limit the propagation of the virus. The cultivation of PRSV-resistant transgenic papaya cultivars dramatically reduced the incidence of PRSV in many areas allowing some growers to return to growing non-GM papaya, which is important for the high value Japanese market in which GM papaya is not allowed. Growers have been able to use the PRSV-resistant transgenic papaya cultivars as a trap crop (Shelton and Badenes-Perez, 2006) by growing it as a border around the non-GM crop and allowing it to cleanse viruliferous aphids of PRSV (Fuchs and Gonsalves, 2007). Thus, the Hawaian papaya industry can now produce and market both transgenic and conventional papaya in the same field, and even organic papaya in adjacent fields if other organic practices are performed. This is a case in which organic agriculture directly benefits from GM crops, which are not allowed as part of the organic production philosophy.

9.3 *Bt* Vegetables and Fruits

Bacillus thuringiensis (*Bt*) genes that encode for Cry proteins that control insect pests have been transferred into a wide variety of crop plants, but only *Bt* maize and cotton are grown commercially on a large scale (James, 2007). However, there are tremendous opportunities to use *Bt* for controlling insect pests in several fruit and vegetable crops. Potatoes have been discussed previously (Grafius and Douches, chapter 7).

9.3.1 Sweet Corn

Presently the only *Bt* vegetable crop commercially available in the USA is *Bt* sweet corn. Of the 262,196 ha of sweet corn (fresh and processing) grown in the USA in 2006 (NASS, 2007), it is estimated that <5% is *Bt* sweet corn expressing Cry1Ab endotoxin (Event Bt11). Processors have avoided growing *Bt* sweet corn because of concerns about export markets, so it has been grown only as a fresh market crop. Studies in New York have shown it to be very effective against the European corn borer, *Ostrinia nubilalis* (Hübner) (Lepidoptera: Crambidae), providing 100% clean ears when no other lepidopteran species were present and >97% when the two noctuids, *Spodoptera frugiperda* (Smith) (fall armyworm), and *Helicoverpa zea* (Boddie) (corn earworm), were also present (Musser and Shelton, 2003). Studies in other states have shown that *Bt* sweet corn provided consistently excellent control of the lepidopteran pest complex (Lynch et al., 1999; Sorenson and Holloway, 1999; Burkness et al., 2001; Hassell and Shepard, 2002; Speese et al., 2005). However, experiences indicate that under very high pressure by *H. zea*, supplemental sprays of synthetic insecticides often are required and that removal of broad-spectrum sprays targeting Lepidoptera have resulted in damage from other species. Studies conducted by Dively and colleagues (unpublished) under high *H. zea* pressure in Maryland have indicated superior control, compared to Bt11, with sweet corn expressing both Cry1Ab endotoxin (Bt11 event) and the vegetative insecticidal protein VIP3A (MIR 162 event). By using appropriately timed insecticide applications with *Bt* sweet corn varieties, fresh market sweet corn growers in North Carolina have been able to extend their production later into the season when populations of *H. zea* and *S. frugiperda* are generally too high to control satisfactorily with insecticide applications alone (G.G. Kennedy, personal communication). Similar findings have been reported in South Carolina (Hassell and Shepard, 2002). Even when two insecticide sprays are required on Bt11 sweet corn (e.g., for late-season control of *H. zea*), an economic assessment in Virginia found a gain of $1,777/ha for fresh-market sweet corn vs. non-*Bt* sweet corn sprayed up to six times with pyrethroid insecticides (Speese et al., 2005).

Similar to results described by Hellmich et al. (chapter 5) on maize, *Bt* sweet corn has proven to be soft on the major predators of *O. nubilalis*, including the ladybeetles *Coleomegilla maculata* (DeGeer) and *Harmonia axyridis* (Pallas), and the hemipteran *Orius insidiosus* (Say) (Musser and Shelton, 2003; Hoheisel and Fleischer, 2007) and a complex of epigeal coleopterans (Leslie et al., 2007). Overall, *Bt* sweet corn was much better at preserving these predators while controlling *O. nubilalis* than were the commonly used insecticides lambda-cyhalothrin, indoxacarb and spinosad. Results from these studies led to the development of a decision guide for sweet corn growers that uses information on biological control and can advise them on the economic return of using various options, including *Bt* sweet corn (Musser et al., 2006).

While *Bt* sweet corn can replace the traditional method of controlling Lepidoptera with broad-spectrum insecticides, it may also allow secondary pests to arise. In Florida, the corn silk fly, *Euxesta stigmatias* Loew (Diptera: Ulidiidae), has become problematic on *Bt* sweet corn and requires treatment (Nuessly and Hentz, 2004). In some states dusky sap beetles, *Carpophilus lugubris* Murray (Coleoptera: Nitidulidae), normally controlled by foliar insecticides, have become more problematic (Dowd, 2000). In Minnesota and Pennsylvania, corn rootworm beetles or Japanese beetles clipping silks have recently become more prevalent, requiring at least one insecticide application for ear protection (W. Hutchison and S. Fleischer, personal communication). Nevertheless, the use of *Bt* sweet corn has proven to be very effective against the targeted, key pests (Lepidoptera) and plantings of *Bt* sweet corn continue to rise in the USA, with new *Bt* fresh-market hybrids being released each year.

9.3.2 Brassica Vegetables, the Diamondback Moth and Other Lepidoptera

Brassica vegetables include cabbage, cauliflower, broccoli, Brussels sprout (*Brassica oleracea*); turnip, Chinese cabbage, pak choi (*B. rapa*) and mustards (*B. nigra, B. juncea, B. carinata*). In 2005, the area of cabbages harvested worldwide was 3,136,540 ha with an additional 983,730 ha of cauliflower and broccoli (FAOSTAT, 2007). Of this total, 80% was grown in developing countries. Cabbages and cauliflower are important vegetable cash crops for low-income farmers throughout Asia, Africa, Latin America and the Caribbean. They serve as important staple dietary items and are high in folate, vitamins B and C and other micronutrients (HSPH, 2007).

Lepidopteran larvae are the most problematic insect pests of vegetable brassicas on a worldwide basis. One species in particular, the diamondback moth, *Plutella xylostella* (L.) (Lepidoptera: Plutellidae), is considered the most destructive insect pest and has severely limited brassica production, especially in resource-poor regions (Talekar and Shelton, 1993) (Fig. 9.2). *Plutella xylostella* now occurs wherever brassicas are grown and causes losses to the world economy of about US$1 billion yearly (Talekar and Shelton, 1993). Losses of cabbage and cauliflower due to *P. xylostella* frequently reach 90% without the use of insecticides (CIMBAA, 2007). Even with frequent use of insecticides, substantial losses occur and threaten food security. In tropical areas where pest pressure is high, it is not uncommon to apply insecticides every other day. Such intense use of insecticides poses hazards to farmers, consumers and the environment and has caused populations of this insect to become resistant to most of the major insecticides.

9.3.3 Bt Brassica Vegetables, a Model System

Cry1 *Bt* genes have been introduced into several *Brassica* species, conferring resistance to *P. xylostella* and other Lepidoptera (Earle et al., 2004; Paul et al., 2005). Early work in a collaborative program (Earle and Shelton at Cornell University) on

Fig. 9.2 Cauliflower in India devastated by the Diamondback moth, *Plutella xylostella*, despite nearly 50 sprays of insecticides (Photo courtesy of Nunhems India, Inc.)

Bt brassica vegetables used cytoplasmic male sterile broccoli, *Brassica oleracea* L. subsp. *italica*, transformed using *Agrobacterium tumefaciens* strain LBA4404 containing the binary vector pMON10517-1 (Metz et al., 1995). The latter carried the neomycin phosphotransferase gene and a full-length, synthetic, *B. thuringiensis* Cry1Ac-like gene, derived from HD-73. Progeny were produced by pollinating transformed plants with Green Comet hybrid broccoli and were used in our experiments. In addition to introducing *cry1Ac* into plants, similar methods were employed to produce plants expressing a *cry1C* gene (Cao et al., 1999) and pyramided plants expressing both *cry1A* and *cry1C* genes (Cao et al., 2002). Strains of *P. xylostella* that had developed resistance in the field to Cry1A, Cry1C or both were used with the three types of *Bt* broccoli to conduct the studies highlighted below.

Much of our initial interest in *Bt* brassicas was as a research tool to study Insecticide Resistance Management (IRM) strategies (Metz et al., 1995) because *P. xylostella* is the only insect to have developed resistance to Cry proteins under field conditions (Tabashnik et al., 2003). This resistance evolved from foliar sprays of *Bt* products and not to *Bt*-transgenic brassicas, which are not yet commercially available. However, field collected Cry1-resistant populations of *P. xylostella,* when combined with *Bt* brassicas, resulted in a model system to study IRM strategies for other *Bt* crops such as maize and cotton (Bates et al., 2005b; Ferré et al., chapter 3).

Use of this model system over the last 15 years has led to the following key findings that have implications for commercialized *Bt* crops and those yet to come. Studies have confirmed the importance of refuges (Ferré et al., chapter 3) in maintaining susceptible alleles in the population (Shelton et al., 2000; Tang et al., 2001); demonstrated the superiority of using dual gene (pyramided) *Bt* plants compared to introducing single gene plants in a mosaic or sequential fashion (Zhao et al., 2003); demonstrated the increased speed of resistance evolution to dual gene *Bt* plants if they are grown in association with single *Bt* gene plants that express one of the same proteins (Zhao et al., 2005); demonstrated the potential usefulness of inducible promoters in plants for creating a refuge in time or space (Bates et al., 2005a); and demonstrated the lack of toxicity by a Cry1 toxin to a hymenopteran endoparasitoid (Chen et al., 2008). This last finding supports the idea that previous reports showing harm to parasitoids by Cry1 toxins were likely due to poor host quality (sick or dying insects) rather than toxicity to the parasitoid (Romeis et al., 2006). While these findings from this model insect-*Bt* plant system have been helpful for understanding IRM and biological control in the currently commercialized *Bt* crops (maize and cotton), they have also prepared the way for the introduction of commercialized *Bt* brassica vegetables. In fact, there is far more information available about IRM and non-target effects for *Bt* brassicas prior to their commercialization than was the case for *Bt* maize or *Bt* cotton.

9.3.4 Commercializing Bt Cauliflower and Cabbage

It was the ability of the high expressing pyramided *Bt* plants to delay the evolution of *Bt* resistance in *P. xylostella* populations that led to the formation of a private-public partnership called the Collaboration on Insect Management for Brassicas in Asia and Africa (CIMBAA, 2007) in 2003. This partnership involves Nunhems, a major vegetable breeding company located in The Netherlands, and the following public partners: the Asian Vegetable Research and Development Center in Taiwan (AVRDC), the Centre for Environmental Stress and Adaptation Research at the University of Melbourne in Australia, Cornell University in the USA, and the Natural Resources Institute, University of Greenwich – UK. CIMBAA includes additional international institutions to address specific research aspects of the project. The initial goal of CIMBAA is to make the dual-*Bt* technology available in varieties that are optimally adapted to growing conditions for cabbage and cauliflower in India. Once the developed material meets the regulatory standards for efficacy and environmental and human health, the bioengineered plants will be submitted to the regulatory system of India. Although India is the first country of interest, other countries have also expressed interest in joining CIMBAA. The ownership of the material, the regulatory dossier and any intellectual property rights (IPR) owned or jointly licensed by the collaborating parties will, to the extent possible, be transferred into public hands for dissemination, without technology fees, to brassica breeders in the developing world. These breeders will then be free to grow the material directly or breed the trait into their own varieties for sale and consumption. Nunhems' commercial interest in CIMBAA is to be the first company to produce the high value hybrids while the public side is focused on helping disseminate the technology into areas where it is most needed.

The initial technology discussions focused on which proteins should be expressed in the plants. Because some populations of *P. xylostella* are reported to have already evolved resistance to Cry1A in sections of India (Mittal et al., 2007), other genes were selected. After considerable discussion, it was decided that the CIMBAA plants would use *cry1C* and *cry1B* genes, because they had been shown to be effective against *P. xylostella* and cross-resistance between the two toxins was not detected (Zhao et al., 2001). Additionally, studies had shown that resistance to Cry1C in *P. xylostella* is polygenic (Zhao et al., 2000), making it more difficult for the insect to evolve resistance. Because the material would be transferred to brassica breeders, it was important that the integrity of the material be maintained. Therefore, Nunhems focused on placing the two genes so closely linked on the chromosome that they would not be separated in conventional breeding programs, thus ensuring that lines originating from them would contain the pyramided genes.

As of January, 2008, over 25 field populations of *P. xylostella* have been collected throughout India and several other countries and tested in the laboratory and found to be very susceptible to both proteins (Gujar and Shelton, unpublished). Likewise, breeding efforts by Nunhems have produced cabbage and cauliflower lines that express both proteins at levels sufficient to control not only populations of susceptible *P. xylostella* but also populations of *P. xylostella* that are resistant to Cry1C (Shelton et al., unpublished). No populations of *P. xylostella* have developed resistance to Cry1B so they are not able to be tested. Although *P. xylostella* is the chief target of the CIMBAA plants, other Lepidoptera may also be problematic in

India and other future locations for CIMBAA products, so it is appropriate that they also be tested. Laboratory studies, conducted with the purified proteins, confirmed that the following Lepidoptera are very susceptible to at least one of the proteins: *Pieris rapae* (L.) (Pieridae), *P. brassicae* (L.) (Pieridae), *Crocidolomia binotalis* Zeller (Pyralidae), *Hellula undalis* (Fabricius) (Pyralidae). On the other hand, laboratory studies have shown that two other Lepidoptera species that may feed on cabbage are much less susceptible: *Spodoptera litura* (Fabricius) (Noctuidae) and *Helicoverpa armigera* (Hübner) (Noctuidae). Laboratory and greenhouse studies with CIMBAA breeding lines have shown excellent control of *P. xylostella* and *P. rapae*, and there are ongoing studies with other pest species as well as non-target organisms.

9.3.5 *Bt* Cauliflower and Cabbage Within IPM

CIMBAA is committed to introducing dual *Bt* gene plants into an overall Integrated Pest Management (IPM) Program. While Lepidoptera, especially *P. xylostella*, are the primary pests, aphids can also be problematic and must be controlled in a fashion that does not compromise the use of the *Bt* plants. Therefore, selective neonicitinoid insecticides, applied as a seed treatment or drench, are being considered as part of the overall strategy for the CIMBAA plants. Use of more broad-spectrum insecticides could potentially disrupt biological control agents such as predators and parasitoids that help control the different pests, as well as affecting other non-target arthropods. Studies on an important predator, *Chrysoperla carnea* (Neuroptera: Chrysopidae), have shown that it is not harmed by either Cry1B or Cry1C when it is fed either protein directly (J. Romeis, unpublished). However, testing the effects of either protein against a hymentopteran parasitoid is more problematic because the parasitoid feeds internally on the host's tissues. Thus, rigorous studies require populations of the host that are resistant to the toxin so host-quality mediated effects can be excluded. Using Cry1C-resistant *P. xylostella*, our studies have shown that its important parasitoid, *Diadegma insulare* (Hymenoptera: Ichneumonidae), is not harmed when it parasitizes *Bt*-fed *P. xylostella* larvae (Chen et al., 2008). This is in stark contrast to parallel studies that demonstrated the parasitoid was harmed by commonly used insecticides.

A major concern about introducing *Bt* brassicas for control of *P. xylostella* is the potential for the evolution of resistance. While models (Roush, 1997a) and greenhouse studies (Zhao et al., 2003) have shown the wisdom of the dual gene approach compared to single *Bt* genes, there should be extra caution about using *Bt* plants for *P. xylostella* management since it has demonstrated its ability to develop resistance to Cry toxins in the field, albeit when applied as foliar sprays. However, an alternative approach to using *Bt* plants would be to spray conventional insecticides, but these have generally failed even after only 3 years of use because of resistance evolution (Bates et al., 2005b). Spraying formulated *Bt* insecticides is another option, but the risk of resistance evolution is greater with *Bt* sprays than with *Bt* plants because sprays create a mosaic of toxin concentrations on plants, which is not the case with *Bt* plants that more uniformly express a high dose of the Cry toxins. On sprayed plants, the insect population is subjected to a mosaic of doses, many of which are sublethal to heterozygotes in the insect population. It is the survival of the heterozygotes that drives resistance evolution and this can explain why *P. xylostella* resistance to a *Bt* protein developed faster with foliar sprays than when the insects were exposed to high dose *Bt* plants (Roush, 1997b).

As the CIMBAA plants are being developed, it will be important to verify that they are expressing season-long high doses of both toxins. It will also be important that other strategies be included in the overall management of *P. xylostella*. These strategies include the conservation of natural enemies, crop destruction at the end of the season to reduce population spread, and regular monitoring of susceptibility to the toxins. In many places where CIMBAA products will be grown, it will be difficult to promote the idea of planting refuges of non-*Bt* brassicas, although this is one of the major requirements for IRM in some countries and may account for the present lack of resistance evolution to *Bt* crops (Tabashnik et al., 2003; Ferré et al., chapter 3, but see Matten et al., chapter 2 for a recent case of putative field resistance). While it has been demonstrated that *Bt* plants can, when used over multiple generations, drive down populations of *P. xylostella* (Shelton et al., 2008), other IPM tactics should also be utilized. As an overall strategy, growers should be encouraged to grow non-brassica crops close by that will not serve as hosts for *P. xylostella*. Long-term studies in the broccoli production area of Mexico have shown the value of crop diversity in the landscape for managing *P. xylostella* over the long term (Hoy et al., 2007). *Bt* brassicas will be a tremendous tool for farmers worldwide, but they should be incorporated into a larger IPM program.

9.3.6 Eggplant

Eggplant (*Solanum melongena* L.) is a popular vegetable crop grown in many countries throughout the subtropics and tropics on a total of 1,857,230 ha in 2006 (FAOSTAT, 2007). It is commonly known as *brinjal* in India (510,000 ha) and Bangladesh (64,208 ha) and is the most popular vegetable grown in the Philippines (20,000 ha). The crop is often considered a "poor man's vegetable" and is mainly cultivated on small family farms. It is an important source of nutrition and cash income for many resource-poor farmers. Eggplant is an annual plant attacked by a number of devastating diseases (*Phomopsis* blight, *Verticillium* wilt, and several viruses [Chen et al., 2002]), and insects (including thrips, cotton leafhopper, jassids and aphids); however, the most damaging is the eggplant fruit and shoot borer (FSB), *Leucinodes orbonalis* Guenée (Lepidoptera: Crambidae) (Fig. 9.3). Infestation is caused by adults migrating from neighboring fields, from eggplant seedlings, or from previously grown eggplants in the same planting area. Damage from *L. orbonalis* starts at the nursery stage and continues after crop transplanting until harvest. Losses have been estimated to be 54–70% in India and Bangladesh and up to 50% in the Philippines (ABSP II, 2007).

Fig. 9.3 Eggplant infested by the eggplant fruit and shoot borer, *Leucinodes orbonalis* (Photo by A.M. Shelton)

Recommended insect pest management practices include the prompt manual removal of wilted shoots, trapping male moths using pheromones to prevent mating, ensuring regular crop rotation and using nylon net barriers. These methods, however, are not widely adopted by farmers because of time and resource constraints or lack of awareness (K. Vijayraghavan, personal communication). There are no known eggplant varieties resistant to the borer, so the use of insecticide sprays continues to be the most common control method used by farmers. Fruit and shoot borer are only vulnerable to sprays for a few hours before they bore into the plant. Therefore, farmers spray insecticides as many as 80 times over a 7-month cropping season (AVRDC, 2001). Farmers may even spray every other day, particularly during the fruiting stage (K. Vijayraghavan, personal communication). In Asia, chemical spraying for this insect accounts for 24% of the total cost of production (ABSP II, 2007). Intensive use of insecticides raises serious concerns for environmental and human health. A study conducted in the Jessore District of Bangladesh found that "98% of farmers felt sickness and more than 3% were hospitalized due to various complexities related to pesticide use" (AVRDC, 2003).

9.3.7 Bt Eggplant

Transformation of eggplant with *cry1Ac* was done by the Maharashtra Hybrid Seeds Company Limited (Mahyco) under a collaborative agreement with Monsanto, and the first *Bt* transgenic eggplant with resistance to *L. orbonalis* (FSB) (FSBR-eggplant) was produced in 2000. The first contained trial for the elite event was undertaken in 2002, and in 2003 a collaboration was initiated with the Cornell University-led Agricultural Biotechnology Support Program II (ABSP II) funded by the United States Agency for International Development (USAID). ABSP II developed a consortium of private and public sector partners to develop FSBR-eggplant for resource-poor farmers. The strategy is that Mahyco will make its profit by selling hybrids while the public sector institutions in India, Bangladesh, the Philippines and other countries will distribute open pollinated lines at a much reduced cost. The potential economic, social and environmental benefits are discussed in detail by Qaim et al. (chapter 12).

9.3.8 Bt Eggplant Within IPM

From an IPM standpoint, there are many benefits as well as some concerns about the use of FSBR-eggplant. The first concern is the potential for the insect to develop resistance to FSBR-eggplant (for a detailed discussion on the factors that influence resistance evolution, see Ferré et al., chapter 3). In the case of *L. orbonalis*, an IRM plan was developed and submitted to the Indian Genetic Engineering Approval Committee (GEAC), which is responsible for environmental approval of activities involving transgenic products. The plan consists of a high dose-refuge strategy in which 5% of an area should be planted to non-*Bt* eggplant at the same time as the main planting. The refuge can be treated with another insecticide but not with *Bt*, and the crop in the refuge must not be destroyed at harvest so that susceptible alleles will be maintained in the *L. orbonalis* population. An intended IRM strategy is that seeds of *Bt* and non-*Bt* plants will be distributed together but in different packets, thus facilitating the refuge strategy. Training guides for *Bt* eggplant have been developed and will distributed when the product is commercialized. These training guides emphasize the need for growers to pay attention to any secondary insect pests, such as aphids and leafhoppers, as well as pathogens that will not be controlled by *Bt* eggplant, thus reconfirming the idea that the *Bt* eggplant is a component in the overall IPM program. In the future it may be possible to incorporate other types of insect-resistance genes into *Bt* eggplant. For example, Ribeiro et al. (2006) has described a genetically modified eggplant line expressing oryzacystatin, an inhibitory protein of cystein proteinases, that has a negative impact on population growth and mortality rates of the aphids, *Myzus persicae* (Sulzer) and *Macrosiphum euphorbiae* (Thomas).

In each country, designated agencies have overall responsibility for IRM (in India it is the Department of Biotechnology in collaboration with State Agricultural Universities and other agencies). Plans are being developed on how they will coordinate a monitoring program and who will pay for it. One proposal is that the assays will be performed in the universities, agricultural research institutes and by the company (Mahyco) with each using the same methods. Multi-location field trials of FSBR-eggplant are being conducted in 2007 and 2008 and commercial release of FSBR-eggplant is anticipated to be in 2009 in India.

9.3.9 *Biosafety and Food Safety Issues*

As the first transgenic food crop to be commercialized in India, *Bt* eggplant has had to undergo several regulatory tests to ensure food and environmental safety. In the developed world, conducting such tests is expensive and time-consuming. As eggplant is not a major crop in North America or Europe, India's regulatory agencies have insisted on the generation of in-country data. This has included both laboratory and field tests to demonstrate field efficacy, safety to beneficial insects, sheep and humans. For additional details see: http://www.envfor.nic.in/divisions/csurv/geac/macho.htm.

Developing in-country safety data requires appropriate laboratories and skills. India, because of its early work with *Bt* cotton, has been able to apply its own expertise to develop such data. Despite such developments, an NGO in India challenged the commercialization via a Supreme Court case that was later dismissed. Such hurdles can cause major delays and substantially increase the cost for product commercialization.

9.3.10 *Warranites, Indemnity and Damages*

These may become a major issue in the transfer of any proprietary gene based technologies. In the case of *Bt* eggplant, it was essential to include stewardship warranty and infringement claims. Many public supported programs in India and in other developing countries are unaware of how to structure such agreements, and this may delay the development of much needed technologies. In the *Bt* eggplant project, ABSP II was able to facilitate these agreements so that both parties (the donor of the technology and recipient) understood clearly their roles and responsibilities.

9.4 Challenges and Opportunities for Transgenic Vegetables and Fruits for Control of Insects and Insect-Vectored Pathogens

9.4.1 *Plant Development*

Vegetables and fruits are considered minor crops and traditionally have had fewer resources channeled to them compared to the staple crops. While it is becoming less expensive to create GM crops for pest management, developing a marketable product and a regulatory package remains costly. Development and regulatory costs can be more readily recouped if the product is grown on an extensive area, as would be done with staple crops, but which is not generally the case for individual fruit and vegetable crops. For example, the large agriculture biotechnology companies have for the most part abandoned the development of GM vegetable and fruit crops because of the high costs associated with product development and deregulation. For vegetables, there are many varieties of the same crop and the half-life of a particular variety can be quite limited. Introducing a GM trait into a breeding program can be complicated and cost prohibitive, especially in crops where backcrossing is difficult or impossible (e.g. potatoes). In most countries, deregulation of a GM trait is event specific. For many vegetable crops, it is not possible to develop a single GM event that can be converted into many different varieties of a single vegetable species via conventional breeding. For example, *Brassica* contains about 100 species, including rapeseed, cabbage, cauliflower, broccoli, Brussels sprouts, turnip and various mustards. No single parent exists that can be used to backcross the transgene into the many different types of *Brassica* species. Individual events would have to be developed for most of the different crop types and deregulation of more than one event for a single protein is problematic for most business models. For the few transgenic vegetable crops that are being developed, novel or unconventional strategies have been employed to bring the crops to market.

For GM papaya, Gonsalves and his colleagues undertook much of the work without large financial backing from industry. They and Cornell (Gonsalves' institution) were able to develop freedom to operate (FTO) policies with little or no cost because the companies that held patents believed there was little financial incentive for them in papaya. Another approach is for a company to piggy-back its vegetable work with larger scale crops. This is essentially what was done with *Bt* sweet corn and *Bt* maize. A third approach is to develop the private-public partnerships in which the private sector would focus on selling hybrids to higher end producers while the public sector would focus on resource-poor farmers. The roles and the financial responsibilities of each partner need to be clearly defined and the eggplant model serves as a good example of a private-public partnership (for a detailed description, see Medakker and Vijayaraghavan, 2007).

9.4.2 Stewardship and IPM

Production of vegetables and fruits in industrialized and developing countries tends to be on smaller areas and in more diversified holdings than staple crops like rice and maize. Thus, they often operate in more complex agricultural systems in which insects may move from one crop to the next within the same farm. How this will impact the use and effects of GM plants in the agricultural landscape can be complex (Storer et al., chapter 10). If multiple GM insect-resistant plants are grown within the same area and if a polyphagous insect is exposed to the same *Bt* protein expressed in the different species, this will challenge the conventional IRM strategies developed for cotton and maize. Thoughtful consideration will be needed before choosing which toxins vegetable plants should express, and the selection should be based not only on what will be an effective toxin against the target insect but what toxins are already in use in other crops that may be hosts for the target insect. Additionally, the difficulty of sampling insect populations for resistant alleles will take on a higher level of complexity in a diversified vegetable system. Further consideration should also have to be given to the effects on non-target organisms within diversified GM plantings. In a study conducted in the northeastern USA, Hoheisel and Fleischer (2007) investigated the seasonal dynamics of coccinellids and their food (aphids and pollen) in a farm system containing plantings of *Bt* sweet corn, *Bt* potato and GM insect-resistant squash. Their results indicated that the transgenic vegetable crops provided conservation of cocinellids and resulted in a 25% reduction in insecticides. In a similar study with these same crops, Leslie et al. (2007) compared the soil surface dwelling communities of Coleopetera and Formicidae in the transgenic crops and their isolines and found no differences in species richness and species composition, but did find the transgenic vegetables required fewer insecticide applications. Such results bode well for GM plants within vegetable IPM systems.

In small, diversified vegetable plantings typical of those found throughout developing countries, the challenges for regulatory oversight of GM plants are immense. Farmers will likely save GM seed, move GM seed between locations, and some GM products may move into markets that do not permit these products. These concerns will be lessened if GM plants are consumed locally and in accordance with national biosafety regulatory policies. However, it is likely that violations will occur and this will challenge legal systems.

It is clear that GM vegetables and fruits can offer novel and effective ways of controlling insects and the pathogens they transmit. It is equally clear that such technology must be introduced within the context of IPM. While each vegetable and fruit has its own set of one of more key pests, other pests can also be problematic. Traditional broad-spectrum insecticides often controlled a suite of pest insects. Thus, when *Bt* (or other GM) vegetables and fruits are introduced into production systems, other methods of control will have to be applied or developed for secondary pests. Because the present GM technologies have proven to be less harmful to natural enemies, biological control of secondary pests may be more achievable but other tactics such as the use of selective insecticides (applied either as seed treatments or foliar sprays) may be necessary (Romeis et al., chapter 4).

9.5 Conclusions

GM vegetables and fruits can have a major role in the management of insects and the diseases they transmit. However, to date they have largely played a secondary role compared to the large areas planted to cotton and maize and have generally been under the radar of those opposed to biotechnology. In the USA where labeling of GM products is not required, virus resistant squash and papaya and *Bt* sweet corn are consumed by the public with little or no thought about the role these crops have played in managing difficult pests. When the markets have allowed the production of GM plants, farmers have readily adopted the technology as part of their pest management practices and this is likely to continue with GM vegetables and fruits.

What will be of great interest and importance for the future of GM vegetables and fruits will be the course set by developing countries. In 2007 about 83% of the world's 6.3 billion population lived in developing countries, and this proportion is expected to increase rapidly in the next several decades. Nearly 46% of the world's vegetables and fruits are grown in China and India, two countries that account for nearly 40% of the world's population and where pest problems are severe. Both countries have readily adopted *Bt* cotton (Naranjo et al., chapter 6) and it is likely that *Bt* rice will be commercialized in China in the near future (Cohen et al., chapter 8). Acceptance of GM crops in these two countries will make it more likely they will adopt GM vegetables and fruits. This in turn will likely hasten their adoption in other parts of the world and allow farmers to use this technology in their overall IPM programs. With the eventual acceptance of GM technology, it is expected that the costs associated with deregulation will become more affordable and that the biotech industry will become more interested in developing GM vegetables and fruits, especially for the developing countries.

References

ABSP II (Agricultural Biotechnology Support Project II), 2007. http://www.absp2.cornell.edu/ (accessed 1 December 2007).

Acord, B.D., 1996. Availability of determination of non-regulated status for a squash line genetically engineered for virus resistance. Federal Register 61: 33484–33485.

Arce-Ochoa, J.P., Dainello, F., Pike, L.M., and Drews, D., 1995. Field performance comparison of two transgenic summer squash hybrids to their parental hybrid lineage. HortScience 30: 492–493.

AVRDC (Asian Vegetable Research and Development Center), 2001. Fact Sheet: Eggplant Fruit and Shoot Borer. www.avrdc.org/LC/eggplant/borer.html (accessed 1 December 2007).

AVRDC (Asian Vegetable Research and Development Center), 2003. Technical Bulletin #29: Socio-economic Parameters of Eggplant Pest Control in Jessore District of Bangladesh. http://www.avrdc.org/pdf/TB29.pdf (accessed 1 December 2007).

Bates, S.L., Cao, J., Zhao, J.-Z., Earle, E.D., Roush, R.T., and Shelton, A.M., 2005a. Evaluation of a chemically inducible promoter for developing a within-plant refuge for resistance management. Journal of Economic Entomology 98: 2188–2194.

Bates, S.L., Zhao, J.-Z., Roush, R.T., and Shelton, A.M., 2005b. Insect resistance management in GM crops: Past, present and future. Nature Biotechnology 23: 57–62.

Burkness, E.C., Hutchison, W.D., Bolin, P.C., Bartels, D.W., Warnock, D.F., and Davis, D.W., 2001. Field efficacy of sweet corn hybrids expressing a *Bacillus thuringiensis* toxin for management of *Ostrinia nubilalis* (Lepidoptera: Crambidae) and *Helicoverpa zea* (Lepidoptera: Noctuidae). Journal of Economic Entomology 94: 197–203.

Cao, J., Tang, J.D., Strizhov, N., Shelton, A.M., and Earle, E.D., 1999. Transgenic broccoli with high levels of *Bacillus thuringiensis* Cry1C protein control diamondback moth larvae resistant to Cry1A or Cry1C. Molecular Breeding 5: 131–141.

Cao, J., Zhao, J.-Z., Tang, J.D., Shelton, A.M., and Earle, E.D., 2002. Broccoli plants with pyramided cry 1Ac and cry1C Bt genes control diamondback moths resistant to Cry1A and Cry1C proteins. Theoretical and Applied Genetics 105: 258–264.

Chen, M., Zhao, J.-Z., Colins, H.L., Earle, E.D., Cao, J., and Shelton, A.M., 2008. A critical assessment of the effects of Bt transgenic plants on parasitoids. PLoS ONE 3(5): e2284. doi:10.1371/journal.pone.0002284.

Chen, N.C., Kalb, T., Talekar, N.S., Wang, J.F., and Ma, C.H., 2002. AVRDC Training Guide: Suggested Cultural Practices for Eggplant. http://www.avrdc.org/LC/eggplant/practices.pdf (accessed 1 December 2007).

CIMBAA (Collaboration on Insect Management for Brassicas in Asia and Africa), 2007. http://www.cimbaa.org/ (accessed 1 December 2007).

Clough, G.H., and Hamm, P.B., 1995. Coat protein transgenic resistance to watermelon mosaic and zucchini yellow mosaic virus in squash and cantaloupe. Plant Disease 79: 1107–1109.

Dowd, P.F., 2000. Dusky sap beetles and other kernel damaging insects in Bt and non-Bt sweet corn in Illinois. Journal of Economic Entomology 93: 1714–1720.

Earle, E.D., Cao, J., and Shelton, A.M., 2004. Insect-resistant transgenic Brassicas. In: Biotechnology in Agriculture and Forestry, Vol. 54, E.C. Pua and C.J. Douglas, eds., Springer, Berlin/Heidelberg/New York, pp. 227–252.

FAOSTAT, 2007. http://faostat.fao.org/ (accessed 1 December 2007).

Fuchs, M., and Gonsalves, D., 1995. Resistance of transgenic squash Pavo ZW-20 expressing the coat protein genes of zucchini yellow mosaic virus and watermelon mosaic virus 2 to mixed infections by both potyviruses. BioTechnology 13: 1466–1473.

Fuchs, M., and Gonsalves, D., 2007. Safety of virus-resistant transgenic plants two decades after their introduction: Lessons from realistic field risk assessment studies. Annual Review of Phytopatholology 45: 173–202.

Fuchs, M., Tricoli, D.M., McMaster, J.R., Carney, K.J., Schesser, M., McFerson, J.R., and Gonsalves, D., 1998. Comparative virus resistance and fruit yield of transgenic squash with single and multiple coat protein genes. Plant Disease 82: 1350–1356.

Gaba, V., Zelcer, A., and Gal-On, A., 2004. Cucurbit biotechnology - the importance of virus resistance. In vitro Cellular & Developmental Biology - Plant 40: 346–358.

Gianessi, L.P., Silvers, C.S., Sankula, S., and Carpenter, J.E., 2002. Virus resistant squash. In: Plant Biotechnology: Current and Potential Impact for Improving Pest Management in U.S. Agriculture. An Analysis of 40 Case Studies. National Center for Food and Agricultural Policy, Washington, DC, USA.

Gonsalves, D., 1998. Control of papaya ringspot virus in papaya: A case study. Annual Review of Phytopathology 36: 415–437.

Gonsalves, D., Vegas, A., Prasartsee, V., Drew, R., Suzuki, J., and Tripathi, S., 2006. Developing papaya to control papaya ringspot virus by transgenic resistance, intergeneric hybridization, and tolerance breeding. Plant Breeding Reviews 26: 35–78.

Hassell, R., and Shepard, B.M., 2002. Insect populations on *Bacillus thuringiensis* transgenic sweet corn. Journal of Entomological Science 37: 285–292.

Hautea, R., Chan, Y.K., Attathom, S., and Krattiger, A.F., 1999. The Papaya Biotechnology Network of Southeast Asia: Biosafety Considerations and Papaya Background Information. ISAAA Brief No. 11, International Service for the Acquisition of Agri-Biotech Applications, Ithaca, NY, USA.

Hoheisel, G.A., and Fleischer, S.J., 2007. Coccinellids, aphids, and pollen in diversified vegetable fields with transgenic and isoline cultivars. Journal of Insect Science 7: 64. http://www.insectscience.org/7.61/.

Hooks, C.R., and Fereres, A., 2006. Protecting crops from non-persistently aphid-transmitted viruses- a review on the use of barrier plants as a management tool. Virus Research 120: 1–16.

Hoy, C.W., McCully, J.E., Laborde, J.A., Vargas-López, A., Bujanos-Muñiz, R., and Rangel, E., 2007. The linkage between integrated pest management and agroecosystem management- a case study in the Bajio, Mexico. American Entomologist 53: 174–183.

HSPH (Harvard School of Public Health), 2007. Fruits and Vegetables. www.hsph.harvard.edu/nutritionsource/fruits.html (accessed 1 December 2007).

James, C., 2007. Global Status of Commercialized Biotech/GM Crops: 2007. ISAAA Brief No. 37, International Service for the Acquisition of Agri-Biotech Applications, Ithaca, NY, USA.

Klas, F.E., Fuchs, M., and Gonsalves, D., 2006. Comparative spatial spread overtime of Zucchini yellow mosaic virus (ZYMV) and Watermelon mosaic virus (WMV) in fields of transgenic squash expressing the coat protein genes of ZYMV and WMV, and in fields of non-transgenic squash. Transgenic Research 15: 527–541.

Krattiger, A.F., 1997. Insect Resistance in Crops: A Case Study of *Bacillus thuringiensis* (Bt) and Its Transfer to Developing Countries. ISAAA Brief No. 2, International Service for the Acquisition of Agri-Biotech Applications, Ithaca, NY, USA.

Leslie, T.W., Hoheisel, G.A., Biddinger, D.J., Rohr, J.R., and Fleisher, S.J., 2007. Transgenes sustain epigeal insect biodiversity in diversified vegetable farm systems. Environmental Entomology 36: 234–244.

Lumpkin, T.A., Weinberger, K., and Moore, S., 2005. Increasing income through fruit and vegetable production opportunities and challenges. Draft Note for Consultative Group on International Agriculture Research Science Forum, CGIAR Priorities: Science for the Poor. http://www.globalhort.org/downloads/pdf/forum-agm2005.pdf (accessed 1 December 2007).

Lynch, R., Wiseman, B., Sumner, H., Plaisted, D., and Warnick, D., 1999. Management of corn earworm and fall armyworm (Lepidoptera: Noctuidae) injury on a sweet corn hybrid expressing a cryIA(b) gene. Journal of Economic Entomology 92: 1217–1222.

Medakker, A., and Vijayaraghavan, V., 2007. Successful commercialization of insect-resistant eggplant by a public–private partnership: Reaching and benefiting resource-poor farmers. In: Intellectual Property Management in Health and Agricultural Innovation: A Handbook of Best Practices, A. Krattiger, R.T. Mahoney, L. Nelsen et al., eds., MIHR, Oxford, UK/PIPRA, Davis, CA, USA. www.ipHandbook.org (accessed 1 December 2007).

Medley, T.L., 1994. Availability of determination of non-regulated status for virus resistant squash. Federal Register 59: 64187–64189.

Metz, T.D., Roush, R.T., Tang, J.D., Shelton, A.M., and Earle, E.D., 1995. Transgenic broccoli expressing a *Bacillus thuringiensis* insecticidal crystal protein: Implications for pest resistance management strategies. Molecular Breeding 1: 309–317.

Mittal, A., Kumari, A., Kalia, V., Singh, D., and Gujar, G.T., 2007. Spatial and temporal baseline susceptibility of Diamondback Moth, *Plutella xylostella*, to *Bacillus thuringiensis* spore crystal mixture, purified crystal toxins and mixtures of cry toxins in India. Biopesticides International 3: 58–70.

Munger, H.M., 1993. Breeding for viral resistance in cucurbits. In: Resistance to Viral Diseases of Vegetables: Genetics and Breeding, M.M. Kyle, ed., Timber Press, Portland, OR, USA, pp. 8–43.

Musser, F.R., and Shelton, A.M., 2003. Bt sweet corn and selective insecticides: Their impacts on sweet corn pests and predators. Journal of Economic Entomology 96: 71–80.

Musser, F.R., Nyrop, J.P., and Shelton, A.M., 2006. Integrating biological and chemical controls in decision-making: European corn borer control in sweet corn as an example. Journal of Economic Entomology 99: 1538–1549.

NASS (National Agricultural Statistics Service), 2007. Vegetables: 2006 Annual Summary. http://usda.mannlib.cornell.edu/usda/nass/VegeSumm//2000s/2007/VegeSumm-01-26-2007.pdf (accessed 1 December 2007).

Nuessly, G.S., and Hentz, M.G., 2004. Contact and leaf residue activity of insecticides against the sweet corn pest *Euxesta stigmatias*. Journal of Economic Entomology 97: 496–502.

Paul, A., Sharma, S.R., Sresty, T.V.S., Devi, S., Bala, S., Kumar, P.S., Saradhi, P.P., Frutos, R., and Altosaar, I., 2005. Transgenic cabbage resistant to diamondback moth. Indian Journal of Biotechnology 4: 72–77.

Perring, T.M., Gruenhagen, N.M., and Farrar, C.A., 1999. Management of plant viral diseases through chemical control of insect vectors. Annual Review of Entomology 44: 457–481.

Ribeiro, A.P.O., Pereira, E.J.G., Galvan, T.L., Picano, M.C., Picoli, E.A.T., da Silva, D.J.H., and Fari, M.G., 2006. Effect of eggplant transformed with oryzacystatin gene on *Myzus persicae* and *Macrosiphum euphorbiae*. Journal of Applied Entomology 130: 84–90.

Romeis, J., Meissle, M., and Bigler, F., 2006. Transgenic crops expressing *Bacillus thuringiensis* toxins and biological control. Nature Biotechnology 24: 63–71.

Roush, R.T., 1997a. Managing resistance to transgenic crops. In: Advances in Insect Control, N. Carozzi and M. Koziel, eds., Taylor & Francis, London, UK, pp. 271–294.

Roush, R.T., 1997b. Managing risk of resistance in pest to insect-tolerant transgenic crops. In: Commercialization of Transgenic Crops: Risk, Benefit and Trade Consideration. Proceedings of a workshop, Canberra, 11–13 March 1997, G.D. McLean, P.M. Waterhouse, G. Evans and M.J. Gibbs, eds., Cooperative Research Center for Plant Sciences and Bureau of Statistics, Canberra, Australia, pp. 259–271.

Sanford, J.C., and Johnston, S.A., 1985. The concept of parasite-derived resistance - deriving resistance genes from the parasites own genome. Journal of Theoretical Biology 113: 395–405.

Schultheis, J.R., and Walters, S.A., 1998. Yield and virus resistance of summer squash cultivars and breeding lines in North Carolina. HortScience 8: 31–39.

Shankula, S., 2006. Quantification of the Impacts on US Agriculture of Biotechnology-Derived Crops Planted in 2005. http://www.ncfap.org/ (accessed 1 December 2007).

Shelton, A.M., Tang, J.D., Roush, R.T., Metz, T.D., and Earle, E.D., 2000. Field tests on managing resistance to Bt-engineered plants. Nature Biotechnology 18: 339–342.

Shelton, A.M., and Badenes-Perez, F.R., 2006. Concept and applications of trap cropping in pest management. Annual Review of Entomology 51: 285–308.

Shelton, A.M., Hatch, S.L., Zhao, J.-Z., Chen, M., Earle, E.D., and Cao, J., 2008. Suppression of diamondback moth using Bt-transgenic plants as a trap crop. Crop Protection 27: 403–409.

Sorenson, K.A., and Holloway, C.W., 1999. Corn earworm and European corn borer control with Bt treated corn, Brunswick and Sampson Co., NC, 1998. Arthropod Management 24: 429–430.

Souza, M.T., Jr., Nickel, O., and Gonsalves, D., 2005. Development of virus resistant transgenic papayas expressing the coat protein gene from a Brazilian isolate of *Papaya ringspot virus*. Fitopatologia Brasileira 30: 357–365.

Speese III, J., Kuhar, T.P., Bratsch, A.D., Nault, B.A., Barlow, V.M., Cordero, R.J., and Shen, Z., 2005. Efficacy and economics of fresh-market Bt transgenic sweet corn in Virginia. Crop Protection 24: 57–64.

Tabashnik, B.E., Carriere, Y., Dennehy, T.J., Morin, S., Sisterson, M.S., Roush R.T., Shelton, A.M., and Zhao, J.-Z., 2003. Insect resistance to transgenic Bt crops: Lessons from the laboratory and field. Journal of Economic Entomology 96: 1031–1038.

Talekar, N.S., and Shelton, A.M., 1993. Biology, ecology and management of the diamondback moth. Annual Review of Entomology 38: 275–301.

Tang, J.D., Collins, H.L., Metz, T.D., Earle, E.D., Zhao, J.-Z., Roush, R.T., and Shelton, A.M., 2001. Greenhouse tests on resistance management of Bt transgenic plants using refuge strategies. Journal of Economic Entomology 94: 240–247.

Tennant, P., Ahmad, M.H., and Gonsalves, D., 2005. Field resistance of coat protein transgenic papaya to *Papaya ringspot virus* in Jamaica. Plant Disease 89: 841–847.

Tricoli, D.M., Carney, K.J., Russell, P.F., McMaster, J.R., Groff, D.W., Hadden, K.C., Himmel, P.T., Hubbard, J.P., Boeshore, M.L., and Quemada, H.D., 1995. Field evaluation of transgenic squash containing single or multiple virus coat protein gene constructs for resistance to cucumber mosaic virus, watermelon mosaic virus 2, and zucchini yellow mosaic virus. BioTechnology 13: 1458–1465.

Voinnet, O., 2005. Induction and suppression of RNA silencing: Insights from viral infections. Nature Reviews Genetics 6: 206–221.

Zhao, J.-Z., Collins, H.L., Tang, J.D., Cao, J., Earle, E.D., Roush, R.T., Herrero, S., Escriche, B., Ferre, J., and Shelton, A.M., 2000. Development and characterization of diamondback moth resistance to transgenic broccoli expressing high levels of Cry1C. Applied Environmental Microbiology 66: 3784–3789.

Zhao, J.-Z., Li, Y., Collins, H.L., Cao, J., Earle, E.D., and Shelton, A.M., 2001. Different cross-resistance patterns in the diamondback moth resistant to *Bacillus thuringiensis* toxin Cry1C. Journal of Economic Entomology 94: 1547–1552.

Zhao, J.-Z., Cao, J., Li, Y., Collins, H.L., Roush, R.T., Earle, E.D., and Shelton, A.M., 2003. Plants expressing two *Bacillus thuringiensis* toxins delay insect resistance compared to single toxins used sequentially or in a mosaic. Nature Biotechnology 21: 1493–1497.

Zhao, J.-Z., Cao, J., Collins, H.L., Bates, S.L., Roush, R.T., Earle, E.D., and Shelton, A.M., 2005. Concurrent use of transgenic plants expressing a single and two Bt genes speeds insect adaptation to pyramided plants. Proceedings of the National Academy of Sciences of the USA 102: 8426–8430.

Zitter, T.A., Hopkins, D.L., and Thomas, C.E., 1996. Compendium of Cucurbit Diseases. APS Press, St. Paul, MN, USA.

Chapter 10
Landscape Effects of Insect-Resistant Genetically Modified Crops

Nicholas P. Storer[1,*], Galen P. Dively[2], and Rod A. Herman[1]

Abstract Putative effects of incorporating GM crops into agricultural landscapes have long been proposed. Here we examine the evidence for such effects within dynamic and heterogeneous agroecosystems, based on widespread commercial deployment insect-resistant *Bt* crops for over 10 years. While there is good evidence for changes in the population sizes of several target pest populations and for the increasing importance of some secondary pests, there is no evidence of landscape-level effects on non-target species. These findings were anticipated by laboratory and field characterization of the high specificity of action of the *Bacillus thuringiensis* proteins currently deployed, and the equivalence of the GM crops to their non-transformed conventional counterparts. Indirect effects of the insect-resistant GM crops on the agricultural ecosystems due to multitrophic exposure, loss of prey, or reduction of prey quality, are generally negligible compared with the direct effects of other more dramatic environmental manipulations that are standard agricultural practices.

10.1 Introduction

Insect-resistant genetically modified (GM) crops are becoming widely deployed around the world. In 2007, crops genetically modified to express insecticidal proteins were grown on more than 42 million hectares globally (James, 2007). Since the advent of this technology, many ecological consequences of the long-term intensive use have been postulated (e.g., Betz et al., 2000; Cannon, 2000; Obrycki et al., 2001). These proposed effects include: pest adaptation to the insecticidal proteins; pest population suppression; non-target population reduction through direct effects or indirect trophic effects; increase in populations of non target species (including pest and beneficial species) through insecticide reduction; alteration

[1] Dow AgroSciences LLC, Indianapolis, IN, USA

[2] University of Maryland, College Park, MD, USA

* To whom correspondence should be addressed. E-mail: nstorer@dow.com

J. Romeis, A.M. Shelton, G.G. Kennedy (eds.), *Integration of Insect-Resistant Genetically Modified Crops within IPM Programs.*

of soil biota; alteration of cropping systems due to changes to the economics of growing different crops; changes in the land area used for crop cultivation; reduction of use of cultural pest control techniques such as crop rotation; a decline in the practice of integrated pest management (IPM) programs to manage the insect pests targeted by the crops; or an increase IPM programs to manage other pests once the key pest populations are controlled. The assessment of any ecological effects should be conducted from the perspective of sustainability of agricultural ecosystems and in the context of existing agricultural practices. Sustainability in agriculture requires melding of three key aspects: meeting human food and fiber needs, environmental stewardship, and economic profitability at the farm and community level. From a sustainability perspective, effects of insect-resistant crops may be beneficial, harmful, or neutral. To investigate environmental sustainability, one can ask a series of questions. How do the crops affect long-term trends in productivity of the agricultural system? What are the measurable ecological effects? What are the effects on long-term trends in environmental quality across the broader agricultural and non-agricultural landscape?

Annual cropping systems are by their nature temporally and spatially unstable (Kennedy and Storer, 2000) and therefore become colonized by organisms with biological traits that are suited to exploiting such environments – polyphagy, mobility, and high reproductive capacity in suitable environments being three such key traits (Southwood, 1962; Stinner et al., 1983; Wallner, 1987; Way, 1977). Conventional agricultural processes create instability within and across years. Across growing seasons, the choice of which crops are grown (in many cases non-native plants), and the varieties of each crop affects the community of herbivores that can be supported (Kennedy and Storer, 2000). Crop rotation creates enormous temporal flux in habitat availability, often with the explicit goal of disrupting pest populations, but with the consequence of disrupting other aspects of the ecological community. Irrigation, especially in arid regions, and fertilizer applications dramatically alter the ability of the ecosystem to support a community of plants and arthropods. Within a season, annual crops rapidly develop from seed to maturity, providing qualitatively different habitats at each phenological stage (vegetative, reproductive, maturity). Pre-planting preparation of fields, such as tillage and plowing, disturbs the habitat of resident species. Planting choices, such as planting date and plant density affect the ability of different species to successfully colonize the crop (e.g., Bradley et al., 1986; Terry et al., 1987). Weed management practices, including use of pre-emergent and post-emergent herbicide, or mechanical cultivation, removes an important habitat for arthropod populations (e.g., Hawes et al., 2003). Insecticide applications clearly make the habitat transiently unsuitable for affected insect species. Crop harvest (also cotton defoliation) causes instantaneous reduction in habitat suitability, which can lead to efflux of herbivores (e.g., Shelton and North, 1986; Schaber et al., 1990).

Additional broader scale instability in the agricultural ecosystem can derive from landscape-level changes in agricultural practices driven by broader economic and technological opportunities – rapid increase in crop acreage may follow elimination of a key pest (e.g., Carlson et al., 1989). Dramatic shifts in cropping patterns are seen as a result of the growing thirst for biofuels. Herbicide tolerance traits (GM or conventionally derived), which often are available in combination with insect resistance traits, dramatically alter weed management practices (Champion et al., 2003; Gianessi, 2005). Development of new crop technology, such as tolerance to low soil fertility or drought, may enable changes in land use patterns, such as increasing the productivity of marginal land. Economic or social factors drive changes in land usage, for example from arable to rangeland, reforestation of agricultural land, or suburbanization.

There are clearly huge landscape-level impacts of all aspects of agriculture; GM crop cultivation should be viewed in the context of the diverse and often ecologically disruptive practices that characterize agriculture. In looking for landscape-level ecological effects of insect-resistant GM crop cultivation that are qualitatively different from effects of other, "conventional", agricultural practices, one should first assess the properties of the GM crops that make them unique, and then focus on how or whether those unique properties alter the suitability of the habitat for colonization in a manner that is qualitatively different from other agricultural practices. Furthermore, any landscape-level effects must be mediated by agents that encompass more than a single field – mobile insect (and other animal) populations; pollen and seed dispersal; and human activities. This chapter considers the potential landscape-level effects of insect-resistant GM crops in the context of the innate instability of modern agricultural systems. We examine evidence for and against such effects of GM crops expressing insecticidal proteins from *Bacillus thuringiensis* strains (*Bt* crops) as the only types of insect-resistant GM crops currently grown on a commercial scale. We focus on cropping systems in the United States of America since this country has the largest area and longest history of use of these crops (Hellmich et al., chapter 5; Naranjo et al., chapter 6). *Bt* maize and *Bt* cotton have been commercially cultivated in the USA for 12 years, and in 2007 almost 50% of maize and 60% of cotton expressed *Bt* proteins; in several states, *Bt* crop adoption levels exceed 50% of maize hectares and 80% of cotton hectares (USDA, 2007). The extent that conclusions drawn based on this experience can be extrapolated to future GM crops using different sources for insect resistance is discussed.

10.2 Unique Properties of GM Crops Within Broader Agricultural Context

The insect-resistant GM crops currently cultivated are substantially equivalent to their conventional counterparts, both agronomically and compositionally, with the exception of the addition of the specific insecticidal trait(s) (and in many cases the addition also of specific herbicide-tolerance traits) (Smith, 2000; Cellini et al., 2004; König et al., 2004). Indeed, unintended changes that are due to the transformation process would likely be manifested as undesirable phenotypic traits and would require additional regulatory assessment. Such changes will be detected, and plants with such changes will be eliminated during transformation event-selection that precedes the commercialization of any new trait (Bradford et al., 2005; Sanvido et al., 2006), just as unintended or undesirable changes arising during conventional breeding are eliminated during variety selection. Therefore, in looking for unique properties of insect-resistant GM crops, we can focus exclusively on the insect resistance trait itself.

There are several unique properties of a pest management system based around GM crops expressing insect resistance traits compared with a pest management system based around conventional crops. First, the *Bt* proteins in commercially available crops are produced season-long throughout the crop plant and deployed as a prophylactic measure. Unlike with conventional insecticides, where variation in application equipment, timing, coverage, and environmental conditions can cause variable efficacy, *Bt* crop efficacy is far more predictable and consistent. Essentially all target pests within a *Bt* field are exposed to the proteins. There is also widespread and growing adoption of the technology such that the proportion of the cropping area using the technology may be greater than for a single insecticide. On the other hand, landscape-level exposure can be reduced compared with sprayed insecticides since only the insects feeding on crop tissues and the natural enemies of these herbivores are exposed. Off-plant exposure due to root exudates and movement of plant tissue (principally pollen, and decaying plant tissue after harvest) is clearly far more limited than off-plant exposure to broadcast insecticide applications.

The *Bt* δ-endotoxin proteins that are expressed in today's insect-protected GM crops are active against a very narrow spectrum of insects compared with most insecticides (e.g., Glare and O'Callaghan, 2000; Romeis et al., 2006), so where there is exposure, the potential for non-target effects is also much reduced. This applies both to pest and non-pest species. Cry1-type proteins only affect Lepidoptera; Cry2-type proteins only affect Lepidoptera and Diptera; Cry3-type proteins and Cry34/35 binary proteins only affect Coleoptera. Species outside of these insect orders are not directly affected. By contrast, the spectrum of activity of most conventional insecticides is considerably broader; indeed, grower expectations are that a single insecticide will control multiple pest species and they will not need multiple applications of different insecticides to achieve acceptable insect control.

The removal of a key target pest from the system that is possible with insect-resistant GM crops, and the resulting reduction in insecticide use, also permits increased flexibility in pest management practices for other pests. Secondary pests that were previously controlled by the broad-spectrum insecticides applied to control a key pest can increase in prominence as has been reported in *Bt* cotton (Naranjo et al., chapter 6). On the other hand, protection of beneficial arthropod populations and implementation of IPM practices may lead to suppression of non-target pest populations or augment the insecticidal trait activity against the target pests (Johnson et al., 1997; Romeis et al., chapter 4).

The heritability of the insecticidal traits is another important feature of GM crops compared with conventional pest management. Because production of the insecticidal protein is genetically based, the trait is passed on to the progeny. Those progeny can result from pollination of the crop within the field, pollination of conspecific crops in other fields, or pollination of wild relatives of the crop. Off-site movement of seed or grain could theoretically also lead to establishment of insect-resistant plants in the landscape. The potential for gene transfer to wild relatives or for the crop to become weedy have been proposed as possible negative effects of the GM plants (Dale et al., 2002).

Other potential effects theorized for *Bt* crops are not applicable at the landscape level or not qualitatively different from conventional management practices. These include off-crop exposure to the insecticide (pollen drift or other tissue movement equivalent to or less than spray drift), effects of root exudates or crop residues on soil biota (equivalent or less than other chemical inputs) or soil physico-chemistry (equivalent to any other crop) (e.g., Dale et al., 2002; Devare et al., 2007; Griffiths et al., 2007; Zwahlen et al., 2007).

10.3 Landscape-Level Effects on Herbivores

10.3.1 Suppression of Target Pest Species

The most direct and dramatic landscape-level effects of growing *Bt* crops would be expected for the primary consumers in the field that (a) are sensitive to the *Bt* protein, (b) consume the crop as their primary or sole food source, and (c) move across the landscape (Freeman and Smith, 1997). We can look to the spectrum of activity among herbivorous species to identify candidates for landscape-level effects. Such investigations typically are undertaken in the early stages of development of a *Bt* protein-based insecticide or trait to characterize the potential commercial uses. *Bt* cotton lines expressing one or two lepidopteran-active *Bt* proteins that have thus far been commercially cultivated cause very high levels of mortality of *Heliothis virescens* (tobacco budworm; Lepidoptera: Noctuidae), *Pectinophora gossypiella* (pink bollworm; Lepidoptera: Gelechiidae), and *Helicoverpa armigera* (the old world bollworm; Lepidoptera: Noctuidae) larvae, and somewhat lower levels of mortality of a range of other lepidopteran larvae, including *Spodoptera* spp. (armyworms; Lepidoptera: Noctuidae), *Trichoplusia ni* (cabbage looper; Lepidoptera: Noctuidae), *Pseudoplusia includens* (soybean looper; Lepidoptera: Noctuidae), *Helicoverpa zea* (bollworm; Lepidoptera: Noctuidae) (e.g., Greenberg et al., 2006; Siebert et al., 2007; Naranjo et al., chapter 6). *Bt* maize lines expressing lepidopteran-active *Bt* proteins that have been commercially cultivated cause very high levels of mortality of *Ostrinia nubilalis* (European corn borer; Lepidoptera: Crambidae), *Diatraea grandiosella* (southwestern corn borer; Lepidoptera: Crambidae), and *Diatraea saccharalis* (sugarcane borer; Lepidoptera: Crambidae) larvae and somewhat lower levels of larval mortality of a range of other lepidopteran pests, depending upon the *Bt* protein(s) being expressed, including *Agrotis ipsilon* (black cutworm; Lepidoptera: Noctuidae), *H. zea*, *Spodoptera frugiperda* (fall armyworm; Lepidoptera: Noctuidae), *Sesamia nonagrioides* (Mediterranean corn borer; Lepidoptera: Noctuidae), *Striacosta albicosta* (western bean cutworm; Lepidoptera: Noctuidae)

(Clark et al., 2000; Storer et al., 2001; Castro et al., 2004; Hellmich et al., chapter 5). *Bt* maize lines expressing coleopteran-active *Bt* proteins cause high levels of mortality of *Diabrotica* species (corn rootworms; Coleoptera: Chrysomelidae) (Storer et al., 2006; Oyediran et al., 2007; Hellmich et al., chapter 5). Of these insect species, landscape-level effects would be expected to be most dramatic for those that have a limited host range (perhaps just the crop of interest), at least for a major part of the season, and experience very high mortality on *Bt* crops that have been widely adopted for several years. Carrière et al. (2003) suggest that limited reproductive capacity and high mobility also tend to favor long-term population suppression.

P. gossypiella has a very narrow host range, being restricted to primarily plants of the Malvaceae family, and in some important cotton-growing regions, cultivated cotton is the only available host for much of the year (Carrière et al., 2003). *Bt* cotton has been deployed in Arizona since 1996, with greater than 70% adoption in recent years. All commercial lines of *Bt* cotton cause very high mortality of *P. gossypiella* larvae, which may penetrate the carpal wall of the cotton fruits, but fail to complete development (Liu et al., 2001). Furthermore, *P. gossypiella* is multivoltine with four to six generations on cotton each year; therefore cotton represents a relatively stable habitat within a growing season. Insecticidal sprays are not as effective as *Bt* cotton because once the larvae have penetrated the carpal wall as first instars, they are protected from exposure. Therefore, it is reasonable to expect that within a season, populations of *P. gossypiella* in areas dominated by *Bt* cotton would fail to grow at the rate experienced in cotton-growing areas without *Bt* cotton. Carrière et al. (2003) identified the population suppressive effects of *Bt* cotton use in Arizona as related to levels of *Bt* cotton adoption, after controlling for among-season variation in survival due to weather patterns and for the overall density of cotton fields. They showed that *Bt* cotton reached highest levels of penetration in areas where *P. gossypiella* populations were historically high prior to the introduction of *Bt* cotton, and that in these areas of highest *Bt* cotton use, the pest populations have become significantly reduced. Moth populations in the early spring in 1999 through 2001 were significantly negatively correlated with the proportion of *Bt* cotton used the previous year. Their data suggested a threshold adoption level of around 65% *Bt* cotton was necessary to see this effect; at these levels the net reproductive rate for a local population became less than 1.0.

P. gossypiella trap catch data from the Imperial Valley in California also suggest a population suppressive effect of *Bt* cotton (Chu et al., 2006). In the period 1989 to 1999, an area-wide management program used short-season cotton varieties to reduce overwintering populations of diapaused larvae. Annual pheromone trap catches of male pink bollworm moths suggested the program was effective until 1995 when there was a huge increase in the numbers caught. In 1999, short-season varieties were replaced with a program built around *Bt* cotton, which, up until 2003, appeared to be somewhat successful. Chu et al. (2006) concluded however, that cotton in the Mexicali Valley of northern Mexico provided an annual supply of pink bollworm moths to the Imperial Valley, and trap catches may be largely dependent on the area planted to cotton across the border rather than on local management practices.

Adamczyk and Hubbard (2006) described a similar dramatic decline in adult populations of *H. virescens* in Washington Co., Mississippi, during the first 9 years of commercial deployment of *Bt* cotton. While the evidence is circumstantial, it is likely that *Bt* cotton, which has been grown on >85% of all cotton acreage in the Mississippi Delta, played a major role in reducing the pest populations perhaps tenfold over this time period. Other changes in agronomic practices may have contributed to the decline in *H. virescens* populations during this time, including use of pre-emergent herbicides removing early season weed hosts (Adamczyk and Hubbard, 2006).

It is also important to consider the effects on *H. virescens* populations of herbicide tolerance traits in cotton. The majority of *Bt* cotton grown also expresses a trait conferring tolerance to glyphosate herbicides, which allows reduced tillage to manage weeds (Gianessi, 2005). While reduced tillage is believed to benefit soil quality, it also is less disturbing to overwintering habitats of soil organisms, which include pupae of lepidopteran pests (Schneider, 2003) and some important predators, such as fire ants (*Solenopsis invicta*) (e.g., Ruberson et al., 1997; Tillman et al., 2004). While reduced tillage could create conditions more conducive to *H. virescens* winter survival, Schneider (2003) found that cotton fields were not a major overwintering habitat for this pest in northeastern Mississippi, and concluded that changes in tillage practices would have little effect on area-wide populations.

Trends in blacklight trap monitoring over the past 35 years in Maryland provide evidence of regional suppression of targeted lepidopteran populations as a result of *Bt* maize use. To illustrate this, a subset of moth records recorded from >20 traps located in the mid and lower Eastern Shore were selected because this area in Maryland has the highest adoption rate of *Bt* maize (≈60%). Figure 10.1 compares the yearly total captures of *O. nubilalis* and *H. zea* moths with captures of *Feltia jaculifera* (dingy cutworm; Lepidoptera: Noctuidae) and *Xestia c-nigrum*) (spotted cutworm; Lepidoptera: Noctuidae). These latter species are general feeders on forages, vegetables, and forbs (not common in maize) and are not expected to be significantly affected by *Bt* maize. Thus, they serve as non-target control populations. Analysis of variance was used to test for differences in average yearly captures of moths between periods before and after introduction of *Bt* maize. It was hypothesized that captures of the targeted lepidopterans during the period of *Bt* maize use (1996–2007) would be significantly less, whereas the non-target cutworm species would show no overall change in moth activity. Linear regression was used to test the slope of response in moth captures as a function of year. In this analysis, it was predicted that captures of the targeted lepidopterans would exhibit a significant negative slope during 1996–2007, whereas the response slope of the cutworm species would not be different from zero.

Results of analyses support both predictions. *O. nubilalis* moth activity during 1996–2007 was 63% less than the long-term average captures prior to *Bt* maize use ($F_{(1,66)} = 17.3$, $P < 0.001$) and decreased linearly at a yearly rate of 44 fewer moths caught ($P < 0.001$). Similarly, *H. zea* activity declined significantly at a yearly rate of 34 fewer moths caught per trap ($P < 0.001$) and averaged 48% less during 1996–2007 ($F_{(1,66)} = 8.1$, $P = 0.006$). In contrast, neither cutworm species showed a significant difference or a decreasing trend in average yearly captures during the

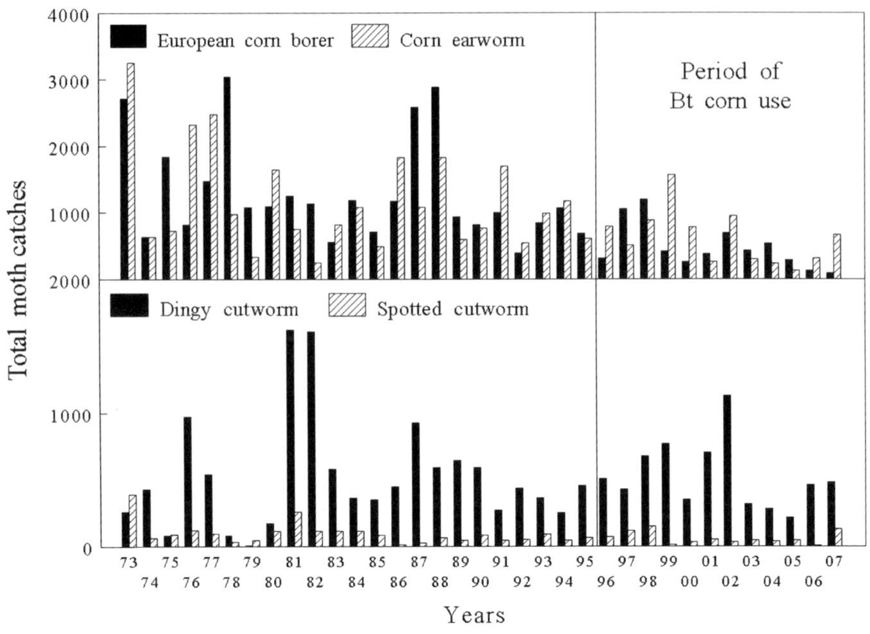

Fig. 10.1 Yearly moth captures of European corn borer (*Ostrinia nubilalis*), corn earworm (*Helicoverpa zea*), dingy cutworm (*Feltia jaculifera*), and spotted cutworm (*Xestia c-nigrum*) for the mid and lower Eastern Shore of Maryland. Data are derived from average daily captures of >20 blacklight traps operating from 1973 to 2007

period of *Bt* maize use. This implies that the decline in *O. nubilalis* and *H. zea* moth activity was unlikely due to weather trends or other environmental factors. Regional suppression of the resident populations of these major pest species has led to pest management benefits in other host crops, such as soybean and vegetables as reported by growers, extension agents and pesticide applicators.

Population suppression appears be less dramatic for *H. zea* than for *O. nubilalis*. Mortality of *H. zea* on single trait *Bt* crops is much lower than for European corn borer (Pilcher et al., 1997; Storer et al., 2001). The host range for *H. zea* is much broader than for *O. nubilalis*, and includes both maize and cotton, as well as a wide spectrum of other important crops and non-crops (Neunzig, 1969; Fitt, 1989; Gustafson et al., 2006). In addition, density-dependent mortality factors (cannibalism in maize ears and threshold-triggered insecticidal treatment of crops) are greater for *H. zea* than for *O. nubilalis* and would have less of an impact on the populations as they decline. The model of Storer et al. (2003) suggested that in eastern North Carolina, deployment of both *Bt* maize and *Bt* cotton in a mosaic would suppress populations by 50% to 65% once the adoption level of either technology exceeded 50%, which is consistent with these observations from Maryland. Long-distance migration across cropping regions as known from *Heliothis* spp. (Fitt, 1989; Pair et al., 1995; Gould et al., 2002) would also be expected to dampen population suppressive effects. (Note that whereas short-range dispersal within an agricultural ecosystem is thought to contribute to area-wide pest suppression by dampening the effects of local heterogeneity in distribution of *Bt* and non-*Bt* fields as discussed earlier [Carrière et al., 2003], long range migration from ecosystems with lower usage of *Bt* crops is likely to reduce area-wide population suppression.)

Adamczyk and Hubbard (2006) present similar evidence for population suppression of *H. zea* over recent years that may in part be attributable to the use of *Bt* crops. They posit that the combination of *Bt* maize and *Bt* cotton in Mississippi reduces the early summer population of adults. They further suggest that changes in other agricultural practices, such as changes in soybean (*Glycine max*) production practices that reduce its suitability as a host for *H. zea*, and more intensive pest management in *Bt* cotton for *H. zea* and other insect populations, including *Lygus lineolaris* (tarnished plant bug; Hemiptera: Miridae) may have contributed to this population reduction. A similar effect would be expected in Australia where *Bt* varieties of cotton resistant to *Helicoverpa* spp. are currently planted in greater than 90% of cotton land area.

O. nubilalis populations have also been examined for evidence of *Bt* maize-induced suppression in the Midwestern maize-growing region of the United States (Hutchison et al., 2007). Annual monitoring of population size, as measured by fall surveys of maize fields for over-wintering larvae, over 43 years revealed considerable annual variation in population size with a 7-year periodicity. Within this variation, the investigators were able to determine a significant population reduction since 1996 that correlated with increasing *Bt* maize use. Similar data from Illinois (University of Illinois Extension, 2007) indicate similar dynamics (Fig. 10.2) and strongly suggest that *Bt* maize could be contributing to population suppression since 1999 as the expected peak in numbers that occurs every 5–7 years was very much reduced in 2001–2002.

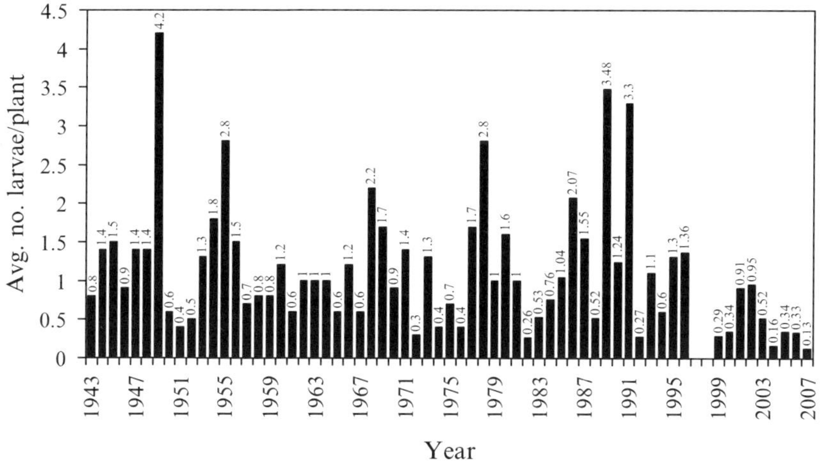

Fig. 10.2 Fall survey of *Ostrinia nubilalis* larvae in maize stalks in Illinois from 1943 to 2007. The data demonstrate periodicity in population sizes likely driven by density-dependent mortality, and suggest population suppression since the introduction of *Bt* maize in 1996 (Used by permission of Kevin Steffey, University of Illinois Extension)

It will be interesting to see over the coming years whether *Bt* maize for the control of *Diabrotica* species may contribute to changes in abundance of this important pest. The principal targets, *D. virgifera virgifera* and *D. barberi* (western and northern corn rootworm), are specialist feeders as larvae, restricted primarily to roots of maize, although other grass species can support development (Branson and Ortman, 1967, 1970; Clark and Hibbard, 2004). They are the major economically damaging pest of maize in North America, and can be difficult to control. They have evolved resistance both to chemical insecticides and to crop rotation (Hellmich et al., chapter 5). Therefore, it is anticipated that the adoption level of the technology will be very high in areas affected by the pest complex. Furthermore, their mobility as adults is believed to be somewhat limited allowing for the possibility of local areas of population suppression. Transgenic maize expressing coleopteran-active *Bt* proteins was first commercially available in 2003. The first commercial *Bt* trait giving protection from *Diabrotica* feeding, MON 863 (based on the Cry3Bb insecticidal protein), allowed 17% to 62% survival of *D. virgifera virgifera* to adulthood in the field compared with conventional counterparts (USEPA, 2002) which may limit its potential for population suppression. However, in 2006 the *Bt* event DAS-59122-7, based on the Cry34Ab1 and Cry35Ab1 binary proteins, was commercialized, and causes more than 99% mortality of *D. virgifera virgifera* larvae (Storer et al., 2006) and in 2007 new Cry3-based events (MON 88017 and MIR604) were released that may be similarly potent. The potential for area-wide population suppression in these species may be significantly mitigated, however, by the requirement to plant a refuge of maize that is not resistant to *Diabrotica* spp., and by reduced density-dependent mortality in corn rootworm pest populations in *Bt* fields (Onstad et al., 2001, 2006). This means that the actual level of corn rootworm population reduction in a *Bt* maize field can be significantly lower than the direct mortality caused by the proteins (Onstad et al., 2001; Storer et al., 2006).

Long-term trap catch data of target pests reveal huge variation in pest populations across time, driven by factors other than the presence of *Bt* crops, as illustrated in Figs. 10.1 and 10.2. Some of the variation may reflect endogenous population dynamics, perhaps affected by delayed density-dependent mortality factors, such as predators and disease (Hutchison et al., 2007). Other factors affecting this variation are exogenous and include weather patterns, local cropping patterns, distant cropping patterns, and other crop management practices. Long-term *O. nubilalis* trap data demonstrate dramatic cycles in corn borer numbers over a period of 60 years (e.g., Purdue Cooperative Extension Service, 2002; Hutchison et al., 2007; University of Illinois Extension, 2007), and one can speculate on a combination of any number of factors that could be causing this. The importance of *Bt* maize relative to other factors in long-term population dynamics is still to be determined. In Arizona, Carrière et al. (2003) showed that both weather-related factors and cropping patterns were important determinants of *P. gossypiella* trap catches, while Chu et al. (2006) showed that cotton acreage across a national border was probably more important in determining local *P. gossypiella* populations than were local management practices. They also showed that other pest management practices, in this case use of pheromone-based behavioral control coupled with short-season cotton varieties which can dramatically reduce the size of the winter population in diapause, can have at least as much effect on pest populations as does the use of high-efficacy *Bt* varieties.

The possibility of achieving pest population suppression using *Bt* crops can be exploited in specific cases to promote integrated pest management at the landscape level. The high mortality levels that can be achieved by whole-plant protection, coupled with biological, cultural, and behavioral management tools and targeted chemical use, means that insecticidal GM crops can enable the deployment of innovative IPM programs for the target and other pests. Such is the impact of *Bt* cotton on *P. gossypiella* populations, that the growing of *Bt* cotton without the non-*Bt* refuges typically used for resistance management (see Matten et al., chapter 2; Ferré et al., chapter 3), coupled with mass release of sterile males, has been implemented to eradicate the pest from parts of Arizona and California. The use of *Diabrotica*-protected *Bt* maize may be particularly appealing as part of an eradication program in Europe, where *D. v. virgifera* is a recently-introduced pest (Hummel, 2003). At a smaller geographic and temporal scale, potato growers have planted border rows of *Bt*-expressing varieties around conventional potato to reduce invasion of their fields by *Leptinotarsa decemlineata* (Colorado potato beetle; Coleoptera: Chrysomelidae) adults from field margins in the spring (Shelton and Badenes-Perez, 2006).

In contrast to the primary pest species, herbivorous species against which *Bt* proteins have low levels of activity, or which occur in multiple agricultural and non-agricultural habitats and for which the *Bt* crop is a minor habitat (e.g., grazers with incidental contact with the crop), would not be expected to exhibit population suppression as a result of *Bt* crop use. Secondary lepidopteran pests of maize, such as *A. ipsilon*, *S. frugiperda*, and *D. saccharalis* in North America, occur only sporadically on maize and have a wide range of alternate crop and non-crop hosts (Steffey et al., 1999). The data from the eastern shore of Maryland presented here support this finding for two cutworm species (Fig. 10.1). Similarly, other species that have incidental contact with *Bt* tissues, even if they were sensitive to the *Bt* proteins, would not to be expected to be affected at the landscape level.

10.3.2 Effects on Non-Target Pest Species

One can also speculate on a possible role of *Bt* crops in an increase in geographic importance of secondary pests. *S. albicosta* is native to the western USA. It was first reported as a pest of sweet and dent corn in Idaho (Douglas et al., 1957). Until 1980, *S. albicosta* distribution was concentrated in southern and western Nebraska and northeastern Colorado with only occasional detection in counties in Arizona, Idaho, Iowa, Kansas, New Mexico, South Dakota, Utah and Wyoming (Blickenstaff and Jolley, 1982). O'Rourke and Hutchison (2000) reported the first finding of sub-economic and sporadic populations in west-central and southern Minnesota during 1999 and 2000. Rice (2002) described *S. albicosta* as a rare pest of Iowa field maize before 2000 and reported detecting the species during 2001 and 2002 in Southwest to Northeast Iowa counties. Dorhout and Rice (2004) reported the first trap captures in Missouri and Illinois in 2004. In the most recent reports, the Western Bean Cutworm Monitoring Network has reported trap captures as far east and north as Ohio, Michigan, and Wisconsin, with large numbers caught in Iowa and northwestern Illinois (Iowa State University, 2007). *S. albicosta* is an ear-feeding caterpillar (Steffey et al., 1999), and probably competes with other ear-feeding caterpillars, including *O. nubilalis* and *H. zea*, where they occur together (Catangui and Berg, 2006). Furthermore, whereas *S. albicosta* is neither predatory nor cannibalistic (Seymour et al., 2004), *H. zea* is an aggressive competitor (Barber, 1936; Dial and Adler, 1990; Chilcutt, 2006) and will consume other caterpillars occurring in the same ear. While *Bt* maize is very effective at eliminating *O. nubilalis* larvae and suppressing *H. zea* larvae, Cry1Ab *Bt* maize appears to have little effect on *S. albicosta* (Catangui and Berg, 2006). In laboratory experiments, *S. albicosta* larvae have been shown to be equal competitors with *H. zea* when feeding on Cry1Ab-expressing maize tissue (Dorhout and Rice, 2006). (Although Cry1F-expressing maize more effectively controls *S. albicosta*, it has only been commercially available since 2003.) It is possible that reduced survival of *O. nubilalis* and *H. zea* larvae in Cry1Ab-expressing *Bt* maize ears has freed *S. albicosta* from population suppressive effects of competition and predation (Catangui and Berg, 2006).

There is stronger evidence that *Bt* cotton has released secondary pests from population suppression by insecticides applied to conventional cotton to control *H. virescens* and *H. zea*. Plant bugs and stink bugs (including *Euschistus servus*, *Acrosternum hilare*, *Nezara viridula* [all Hemiptera: Pentatomidae], and especially *L. lineolaris* [Hemiptera: Miridae]) were previously regarded as secondary pests of cotton, and pest management practices targeted at controlling the key pests – *Anthonomus grandis* (boll weevil; Coleoptera: Curculionidae) and the *Helicoverpa/Heliothis* complex – are thought to have also kept plant bug populations under control. However, with the combination of both *A. grandis* eradication and widespread *Bt* cotton deployment, traditional insecticide use in cotton has declined dramatically (Carpenter and Gianessi, 2001; Cattaneo et al., 2006; Fitt, chapter 11) and bugs, which are unaffected by the *Bt* proteins, have filled the vacated ecological niche (e.g., Roberts, 1999; Tillman, 2006). However, these effects are not consistent from study to study (e.g., Head et al., 2005; Naranjo, 2005).

Even where secondary pest species populations do expand to fill a vacated ecological niche, this effect should not be regarded as a negative landscape effect of the GM crop compared with the cultivation of conventional crops. The *Bt* crop still provides a productivity benefits by removing a key pest, even if conventional insecticide inputs are needed to control secondary pests. Generally, IPM approaches to secondary pest management can be more successful following elimination of the primary pest than they are in conventional crops.

10.3.3 Effects on Non-Target Lepidoptera

Non-target herbivores that do not directly feed on crops but occur within the agricultural landscape may have incidental exposure to the Bt proteins produced in insect-resistant GM crops through pollen deposition on their host plants. *Danaus plexippus* (monarch butterfly; Lepidoptera: Danaidae) and *Bt* maize is probably the best studied example of potential landscape effects on a sensitive non-target herbivorous population. Hellmich et al. (2001) showed that *D. plexippus* caterpillars are sensitive to Cry1Ab-protein when consumed at a high rate in the laboratory. However, in field circumstances, exposure to Cry1Ab protein is very limited, requiring spatial and temporal overlap between sensitive larvae and *Bt* maize pollen being deposited onto the foliage of milkweed growing in, or very close to, the maize field: only larvae that hatch on milkweed plants within or at the margins of maize fields at the time of anthesis are exposed to lethal levels of the *Bt* protein (Hellmich et al., 2001; Stanley-Horn et al., 2001). At the landscape level this temporal-spatial overlap is rare (Oberhauser et al., 2001; Pleasants et al., 2001) and population effects on *D. plexippus* are negligible (Sears et al., 2001; Dively et al., 2004).

The same risk assessment principal examining the overlap of hazard (species sensitivity) and exposure has been applied to other lepidopteran species. Wolt et al. (2005) concluded that Cry1F maize posed negligible risk to Japanese butterfly species that are distributed as larvae beyond the maize field or field margin. Peterson et al. (2006) similarly concluded that *Bt* maize pollen poses negligible risk to any endangered lepidopteran species in the USA. Losey et al. (2003) analyzed Lepidoptera associated with maize in the USA and found 229 species that feed on plants that have been reported to be associated with maize fields. However, they concluded that these species are not significantly threatened by pest management practices in maize, whether or not those practices involve planting of *Bt* hybrids. Schmitz et al. (2003) listed 96 lepidopteran species that could potentially be exposed to transgenic maize fields in Germany. However, nine of the species listed by Schmitz et al. (2003) feed on maize plants or are regarded as pests of maize or other crops (e.g., *A. ipsilon* and *Pieris rapae* [cabbage white butterfly; Lepidoptera: Pieridae]), and four are extinct. Only 47 of the species are anticipated to be present during anthesis. Of these, 43 are common species with broad habitat ranges, leaving only 4 species that may be affected maize at the population level by the use of *Bt*. The real importance of maize fields in the ecology of these species, whether the larvae may be exposed to maize pollen, and whether they are sensitive to the expressed *Bt* proteins, remain unknown. However, it is clear that the potential for any impact on non-target Lepidoptera is far smaller for *Bt* proteins from maize than for other agricultural practices such as weed management and use of insecticidal sprays.

In a similar finding to that for monarch butterfly larvae, Gathmann et al. (2006) found larval populations of two lepidopteran species known to be sensitive to *Bt* proteins, *Plutella xylostella* (diamondback moth; Lepidoptera: Plutellidae) and *P. rapae*, were unaffected in *Bt* maize when feeding on weed hosts growing within the maize plots, while they were reduced significantly by the use of insecticidal sprays. While these two lepidopteran species are pests in other systems and therefore not themselves of potential concern, they are representative of other potentially exposed and *Bt*-sensitive Lepidoptera that may be charismatic or threatened.

10.3.4 Potential for Target Pest Adaptation

A potential consequence of *Bt* crop deployment complementary to population suppression is population adaptation, or resistance evolution (Ferré et al., chapter 3). As insect genotypes that are susceptible to the *Bt* proteins become dramatically reduced, alternative genotypes conferring a higher fitness on *Bt* crops are expected to increase in relative frequency with use of the *Bt* crop. As such resistant genotypes begin to dominate a population, the population is predicted to rebound to the levels present before the introduction of *Bt* crops. If genes conferring resistance are initially rare and if heterozygotes (those insects bearing one copy of a resistance allele and one of a wild-type susceptible allele) do not have much of an advantage over susceptible insects across the landscape, resistance is expected to evolve slowly in the presence of a small refuge of non-*Bt* crops (Gould, 1998). Furthermore, fitness costs associated with resistance and incomplete resistance further limit the expected rate of adaptation (Carrière et al., 2006; Gould et al., 2006). On the other hand, if resistance is not initially rare, if heterozygotes have a significant advantage over susceptible insects, or if there are no spatial refugia from selection, resistance would be expected to evolve more rapidly (Gould, 1998).

Since resistance evolution is a landscape-scale phenomenon (Peck et al., 1999; Storer et al., 2003; Sisterson et al., 2005), under ideal circumstances, management at the landscape level could be most effective. In Arizona, a collaborative arrangement was established early to manage *Bt* cotton to reduce the risk of resistance (Carrière et al., 2001). The Arizona Bt Cotton Working Group consists of growers, university researchers, extension advisors, industry, and government agencies. Together, they developed refuge recommendations that were based on the best available science while accounting for the practical limitations faced by growers. The group also established an extensive and rigorous resistance-monitoring plan to identify incipient resistance in time to react so that it can be managed. The group devised a remediation plan to be implemented in the event that resistance is confirmed, involving actions by growers, industry, and the governmental agencies to reduce selection pressure and limit the spread of a resistant population (Matten et al., chapter 2). By taking a landscape view of the management and implementation of resistance management and engaging the full spectrum of stakeholders, Arizona has been able to implement a rigorous IRM strategy (Carrière et al., 2001) that has been effective despite the detection of resistance-conferring alleles in field populations (Tabashnik et al., 2005). This program works to a large extent due to a very strong grower organization in a discrete geopolitical area that coincides with spatial distribution of pest populations, with strong leadership from public sector scientists. Such circumstances are the exception and not the rule – pest populations do not respect geopolitical boundaries, growers may not have the resources to dedicate to long-term area-wide programs, or there may not be an appropriate broad stakeholder group with the expertise and inclusiveness to ensure success.

As new insect-resistant GM crops are introduced, it is becoming more important to understand the impact of the overall cropping landscape and pest life systems on the resistance evolution potential. Initial IRM plans were based on assumptions with regard to the landscape that require careful re-examination. For example, the plans were based on the worst-case assumption that there would be complete adoption of *Bt* crops and that only one type of *Bt* protein would be used. While adoption is rising as technology improves, maximum deployment only rarely occurs across ecologically relevant landscape scales. Furthermore, within the real landscapes, a diversity of *Bt* crops are grown. For example, for control of Lepidoptera, maize growers can choose among Cry1Ab-expressing and Cry1F-expressing hybrids and in cotton, Cry1Ac, Cry1F and Cry2Ab proteins are available (see Hellmich et al., chapter 5; Naranjo et al., chapter 6). This creates a landscape mosaic of different traits in a single crop or similar traits in different crops and, in understanding resistance potential, it becomes important to understand the potential effects of full or partial cross-resistance among the proteins. Increasingly, varieties of *Bt* crops are becoming available expressing combinations of *Bt* proteins with limited cross-resistance potential. These products are expected to reduce the risks of resistance evolving since insects with resistance to one protein would still be susceptible to the second and therefore not be able to pass their resistance on to the next generation. Simulation models indicate that such IRM pyramids have the potential to exponentially extend durability of insect-protected crops (Gould, 1986; Roush, 1994). However, when used alongside single-trait varieties containing one of the pyramided proteins, the durability gains can be greatly reduced (Zhao et al., 2005); although it is important to recognize that in these circumstances pyramids still have extended durability over single-trait versions. The desire for more robust durability prompted the US Environmental Protection Agency (EPA) Scientific Advisory Panel, when considering reducing or eliminating structured refuges for pyramided *Bt* cotton in the USA to recommend that single-gene Cry1Ac-expressing *Bt* cotton be removed from commercial availability (USEPA, 2006). This would result in a situation similar to that which occurred in Australia where single gene Bt cotton was phased out when Bollgard II was approved in 2002 (Zhao et al., 2005).

Complex spatial models (e.g., Storer et al., 2003) can be employed to help understand the implications of growing landscape complexities on resistance evolution. It rapidly becomes clear that the most effective landscape-level resistance management programs are difficult to implement on a product-by-product basis since the optimal program for one product may be sub-optimal for another. Furthermore, stringent policies, such as permitting only pyramids to be deployed (or heavily penalizing single-gene varieties), are likely to hinder development of new modes of action, with the result that unnecessary selective pressure are placed on the existing products and long-term sustainability at the landscape level is hampered.

Of course, the potential for pest adaptation is not unique to transgenic insecticidal crops. The Arthropod Pesticide Resistance Database maintained at Michigan State University records in excess of 7,500 cases of resistance encompassing 550 arthropod species exhibiting resistance, and 323 pesticides (http://www.pesticideresistance.org/). It is not unusual for isolated incidents of resistance to a pesticide to appear within 2–3 years of commercialization (e.g., Zhao et al., 2002). In fact, *Bt* crops offer a unique opportunity for risk mitigation through landscape manipulation by the creation of a spatial mosaic of *Bt* crops with non-*Bt* refugia (Carrière et al.,

2001). *Bt* crops offer further advantages from a resistance management standpoint. Dose cannot be manipulated by the user and is less prone to environmental variation such as the vagaries of weather and application methodology. For now, the limited number of different types of *Bt* crops means that polyphagous species are not exposed across large proportions of the populations (Gustafson et al., 2006). Fitness costs for resistant insects seem to be more extreme in the case of *Bt* resistance than is the case for some chemicals (Bird and Ackhurst, 2004; Tabashnik et al., 2005; Gahan et al., 2007), and the high levels of mortality caused by *Bt* crops appear to mean that alleles conferring less than complete resistance do not have a significant selective advantage on *Bt* plants (Tabashnik et al., 2005; Carrière et al., 2006; Huang et al., 2007).

At the landscape level, the consequence of pest adaptation to insect-resistant GM crops needs to be considered alongside the rate at which resistance could evolve. Superficially, pest adaptation would lead to a return to the pest management practices and economics prior to deployment of the GM crop; therefore nothing would be lost compared with what would have happened without technology. At the landscape level, the environmental sustainability benefits of even short-term widespread deployment of *Bt* proteins via GM crops would far outweigh any resistance-related loss of utility of a *Bt* protein in a microbial insecticide spray in specialized "organic" crop production systems. Indeed, the concern that resistance caused by *Bt* crops would lead to reduced utility of *Bt* pesticides to organic growers is limited to the special case of Cry1Ab/Cry1Ac resistance in pest populations that move between large-scale *Bt* crop areas and organic crops that are currently controlled by *Bt* sprays (and even in this situation, *Bt* sprays are likely to have on-going utility as they generally contain cocktails of several Cry proteins and various other insecticidal components). The other *Bt* proteins expressed in today's transgenic crops are not available as organic-labeled products. Similarly, the consequence of pest adaptation to future insect resistance traits should be regarded in the broader sustainability context.

10.4 Landscape-Level Effects on the Second Tier of Consumers: Predators and Parasitoids

Predators and parasitoids of pest insects are the agents of biological control; their preservation in the agricultural ecosystem can be important in maintaining pest populations at manageable levels and minimizing chemical intervention (see Romeis et al., chapter 4). Potential landscape-level effects of *Bt* crops on parasitoid and predatory arthropods are not qualitatively different from those of other insecticidal technologies; however, potential quantitative differences derive from the full season presence of the insecticidal protein and the near-elimination of a significant component of the primary herbivore community within *Bt* fields. Tempering this potential for more dramatic effects on predators and parasites than for chemicals is the lack of off-crop exposure and the narrow spectrum of activity of the *Bt* proteins.

10.4.1 Direct Effects

The same properties of *Bt* crops that pose unique challenges to herbivores also pose challenges to predators; i.e. potential exposure to the *Bt* proteins across an entire field for an entire season, and potential domination of the crop by *Bt* varieties. However, the level of exposure for secondary consumers is necessarily lower than for primary consumers in the food web. Predator and parasitoid exposure through consumption of prey containing the protein is low. In some cases, the *Bt* proteins are not ingested by the herbivore (e.g., aphids; Raps et al., 2001). If they are ingested, the *Bt* proteins can lose their activity through digestion by the herbivorous insects or pass more-or-less intact in the lumen of their guts (e.g., Raps et al., 2001; Brandt et al., 2004). In any case, the concentration of the *Bt* proteins within the prey will be lower than within the plant and exposure of predators to the proteins is therefore limited (Obrist et al., 2005, 2006; Vojtech et al., 2005; Torres et al., 2006; Romeis et al., chapter 4).

Given the narrow spectrum of activity of *Bt* Cry proteins, landscape-level effects of *Bt* crops would most likely occur in species belonging to the same taxonomic order as the target (i.e., are potentially sensitive to the proteins), especially those that exhibit some herbivory (highest potential exposure to the proteins). For lepidopteran-active *Bt* crops, there are no predatory or parasitoid non-target insects that meet these criteria (although charismatic or endangered herbivores deserve attention as discussed above). For coleopteran-active *Bt* crops, predatory beetles that might be expected to exhibit landscape-level effects include ladybird beetles (family Coccinellidae), ground beetles (family Carabidae), and rove beetles (family Staphylinidae).

Studies with predators and parasitoids have shown little or no direct effects of the *Bt* proteins expressed in plants (O'Callaghan et al., 2005; Romeis et al., 2006), although it is important to make this assessment for insect-resistant traits on a case-by-case basis as the spectrum of activity differs among proteins and levels of protein expression differ among transformation events.

10.4.2 Indirect Effects

Removal or reduction in fitness of the primary herbivore, the first step in the consumer element of the food chain, might be expected to have secondary on effects on the in-field ecosystem, especially for the primary predator and parasitoid communities (Schoenly et al., 2003; Romeis et al., chapter 4). This effect could potentially be more dramatic for insect-resistant GM crops than for insecticides due to greater efficacy and season-long presence of the insect control agent.

The few instances of reduced predator or parasitoid numbers observed in some studies comparing *Bt* plots with their unsprayed, non-*Bt* counterparts (Romeis et al., 2006; Marvier et al., 2007) can be attributed to reduction in prey quantity or quality. However, when such plots are compared to their non-*Bt* counterparts in which the target pests are managed with conventional insecticides, the *Bt* plots invariably contain equal or greater biodiversity (Romeis et al., 2006; Marvier et al., 2007). The improved efficacy of *Bt* crops compared with conventional insecticides is counterbalanced by the greater specificity of the *Bt* proteins.

10.4.3 Landscape-Level Effects

Concern has been raised that there may be long-term, landscape-level effects on non-target arthropods due to subtle sublethal effects ([e.g., development rates, fecundity, mating success, behavioral effects] [Schuler et al., 2004; Henry, 2006]). However, if direct effects of this type are small under laboratory conditions, it is unlikely that their consequences would be observable or ecologically relevant at the field level (Raybould, 2007a; Romeis et al., 2008). Further, if field impacts are slight, it is unlikely that such effects would have a measurable impact at the landscape level. As we have seen earlier in this chapter, even for target pest species, with highest exposure, highest sensitivity, and dramatic in-field effects, landscape-level effects are often subtle and hard to detect against the background of all the other ecological and operational factors that affect the populations.

If effects were to be seen at the field level that are either directly or indirectly attributable to the insect-resistant GM crop, possible impacts at the landscape level should be examined in a way that would take into account the ability of the species to recover, through dispersal and/or reproduction, within and across seasons. Effective beneficial insects need the same ecological adaptations to rapidly exploit temporally unstable ecosystems as do pest insects in an agricultural ecosystem.

For the currently commercialized *Bt* crops, the effects of *Bt* proteins on predators and parasitoids have been shown to be minimal or undetectable in field investigations. Across a large number of ecological studies looking for field-relevant direct or indirect effects on non-target insects, there are no reports of consistent adverse effects that are due to the *Bt* transgenes (Naranjo et al., 2005; Romeis et al., 2006; Marvier et al., 2007), even for insects that are in the same taxonomic order as the target pests (McManus et al., 2005 Dively et al., 2008). Head et al. (2005) in a large-scale multiyear study of non-target arthropod populations in commercial *Bt* cotton fields found no consistent differences from paired commercial non-*Bt* cotton fields, and where there were differences, non-target populations in the *Bt* fields were generally greater than in the non-*Bt* fields due to reduced insecticide use. Naranjo (2005) found that the ecological functioning of the non-target arthropods in cotton fields in Arizona was unaffected by the use of *Bt*-expressing varieties. Eizaguirre et al. (2006) studied non-target arthropod populations in commercial *Bt* maize fields for 3 years in Spain and similarly found no consistent effects of *Bt* maize on natural enemy populations. By contrast, a large number of field studies have been conducted to evaluate the impact of traditional agricultural practices on non-target organisms. The effects of soil tillage regimes, field margin management, crop variety choice, crop choice and crop rotation practices on beneficial arthropods are generally easily detected, even in small plot trials (e.g., Roach, 1980; Brust et al., 1985, 1986; Bradley et al., 1986; Hammond and Stinner, 1987; Terry et al., 1987; House and Alzugaray, 1989; Isenhour et al., 1989; Laub and Luna, 1992; Brust and King, 1994; Clark et al., 1997; Bruck and Lewis, 1998; Cottrell and Yeargan, 1998; Pfannenstiel and Yeargan, 1998a, b; Varchola and Dunn, 1999; Hawes et al., 2003; Holland et al., 2003).

Thus for the *Bt* crops currently deployed, which have a narrow spectrum of activity and limited exposure of non-target organisms, we conclude there is no evidence for adverse landscape-level effects on natural enemies resulting from widespread cultivation within the broader context of agricultural instability. Future insect-resistant crops should, however, be assessed on a case-by-case basis following a tiered risk assessment approach based upon careful problem formulation (Garcia-Alonso et al., 2006; Raybould, 2007b; Romeis et al., 2008). Indirect effects on specialist natural enemies due to loss of the primary herbivore have been observed in several studies. Landscape-level suppression of such natural enemies would be an expected effect of landscape-level suppression of their prey or host. However, the ecological consequence of such an effect can be regarded as negligible in the absence of the pest, except in the rare case where the crop conventionally serves as an important reservoir for the natural enemy.

10.5 Other Potential Effects of *Bt* Crops

10.5.1 Genetic Effects

One property of insect-resistant GM crops that is qualitatively different from conventional insecticides is the genetic control of the trait, and therefore the ability to pass on to subsequent generations the capability to produce the insecticide. Three potential landscape-level consequences of this have been proposed: transfer of the genes to nearby crops causing the trait to spread (intentionally or unintentionally) through agricultural production areas; "escape" of the insect resistance genes to non-cultivated related species (e.g., Letourneau et al., 2003; Baltazar et al., 2005); and increase in the potential of insect-resistant versions of crops to become established as weeds (Conner et al., 2003). These properties of insect resistance genes are shared with all other genes in crops that have been selected through breeding to provide agronomic benefits, including pest and disease resistance, vigor under cultivation, environmental tolerance, and yield. Transfer of GM genes cannot occur in isolation of the other crop genes: the entire crop genome is subject to transfer at the same time. The process of genetic modification does not allow the transgenes to move any more freely than any other gene: stability of the insertion into the genome is assessed during product development and demonstrated in the product characterization provided in regulatory packages. The landscape-level effects of insect resistance genes should therefore be in the context of the broader genetic background.

It is a normal consequence of crop husbandry that traits from one field can be spread to nearby sexually compatible plants through pollen flow, whether intentionally or not. Only in cases where crop genomes regularly became established in wild species could one envisage the insect-resistant genes also becoming established. Insect-resistant genes would only alter the probability of this occurring if the insect species that they provide protection against were important in regulating the population of the non-crop relatives (Letourneau et al., 2003). Even if such genes were to become established in a wild population, the ecological consequences of this would not be qualitatively different from the establishment of other (conventional) agronomic traits that favor improved yield. Indeed, it is equally possible to envisage positive ecological effects as negative, such as making weeds less suitable as reservoirs for pest organisms. We have seen that even in the case of very high levels of deployment of *Bt* crops, the ecological consequences are small relative to the effects of other ecological forces, and are generally restricted to the positive effect of reducing target pest populations and undetectable changes in communities of non-target organisms.

Finally, the chance that GM varieties of crops could become weeds themselves when the non-GM versions of the same crop do not is rather small. Crops generally do not become weeds in non-cultivated land since, in the absence of human intervention, they are poor competitors with native vegetation. Only if insect pressure were the driving force preventing a crop from becoming a weed could insect-resistant GM crops be established as weeds. In a study of the weediness potential of *Bt* cotton in northern Australia, Eastick and Hearnden (2006) found that *Bt* cotton did not differ from conventional cotton in seed germination or establishment and in only three of 13 sites did cotton plants persist at all across years. Persistence was not different for *Bt* and conventional cotton and neither type became invasive after 4 years. A similar study is being conducted by the Agricultural Biotechnology Stewardship Technical Committee in southern Texas, USA, to determine whether *Bt* maize is more likely than non-*Bt* maize to become established in conditions that approximate those in the center of origin for maize (L.S. Higgins, personal communication). The potential for a crop plant to escape husbandry and become an invasive weed is an important consideration in the introduction of any new crop. The likelihood that the addition of one or a few *Bt* genes would alter that potential is remote and is addressed in the regulatory process on a case-by-case basis before the GM plant is deregulated.

10.6 Conclusions – Insect-Resistant GM Crops and Sustainability

In assessing potential landscape effects of any new agricultural technology it is important to consider the context of the innate instability of the agricultural ecosystem, which has been artificially created to meet man's needs for food and fiber. Producers maximize yield by optimizing inputs and by reducing competition from other plants and pests. Agricultural ecosystems are also managed to be maintained over time frames that are much longer than the typical annual crop cycle to ensure the continuing ability to raise crops in a sustainable manner. These agricultural practices have dramatic effects on the species composition of the landscape. Soil tillage regimes, field margin management, crop variety choice, crop choice, planting date, planting density, irrigation, weed management, and crop rotation practices all dramatically affect the diversity and abundance of species within crop fields and nearby habitats. It is in this context of the highly specialized and disturbed agricultural ecosystem that the impact of GM crops should be evaluated.

Agricultural sustainability is improved if man's growing food, fuel, and fiber needs can be met with greater efficiency, and with maintenance or improvement of environmental quality within and beyond the agricultural setting. GM crop development and adoption are driven by the need for improvements in agricultural productivity, reducing both the area of land needed for production and the intensity of inputs. After more than a decade of widespread commercial use, the landscape effects of *Bt* crops have clearly favored agricultural sustainability. The insecticidal traits in these crops act in a highly specific manner against key target pests that compete with humans for food, fuel, and fiber. Area-wide suppression of the key target pests, the effect for which the expectation and evidence are strongest, promotes sustainability. It results in simplified crop management, reduced yield loss, and reduced inputs. These pest species dominate the cropping systems and are highly sensitive to the insecticidal proteins. However, we have seen that even changes to populations of these species are not ubiquitous and are often difficult to detect in the context of large natural variability of pest populations and on-going alterations in agricultural practices and the environment that are unrelated (or very loosely related) to the use of *Bt* crops. Replacement of target pests by secondary pests or adaptation to the crops by the target pests are other reasonably expected landscape effects of *Bt* crops. However, even these effects should be regarded as, at worst, neutral from a sustainability perspective, because such pests could still be controlled using pre-*Bt* management tools.

Other species, for which exposure is lower (at the landscape level) or which are less sensitive to the proteins, are not likely to exhibit landscape effects that are greater than the background effects of ecological disturbance resulting from the instability of agricultural ecosystems. Laboratory testing of the presently used *Bt* proteins has not identified significant direct effects against non-target beneficial or charismatic species, nor have sublethal or chronic effects been manifested in field studies. While removing a key herbivore from the ecosystem is expected to have secondary effects through the food web, such effects are rarely detectable and apparently very subtle. Thus, non-target populations do not appear to be significantly affected either directly or indirectly at current levels of *Bt* crop use, and this is in stark contrast to the effects commonly seen with traditional insecticides. It seems untenable to believe that subtle indirect effects within *Bt* fields would be magnified to produce meaningful landscape effects outside of *Bt* fields. Future insect resistance traits may not have a similarly narrow spectrum of activity, although the in-plant mode of production will inherently continue to limit non-target exposure compared with conventional insecticide sprays.

Therefore, while each trait needs to be assessed on its own merits, based on experience to date and on the general properties inherent to the technology, the landscape effects of insect protected GM crops are consistent with the goals of agricultural sustainability.

References

Adamczyk, J.J., Jr., and Hubbard, D., 2006. Changes in populations of *Heliothis virescens* (F.) (Lepidoptera: Noctuidae) and *Helicoverpa zea* (Boddie) (Lepidoptera: Noctuidae) in the Mississippi Delta from 1986 to 2005 as indicated by adult male pheromone traps. Journal of Cotton Science 10: 155–160.

Baltazar, B.M., de Jesus Sanchez-Gonzalez, J., de la Cruz-Larios, L., and Schoper, J.B., 2005. Pollination between maize and teosinte: an important determinant of gene flow in Mexico. Theoretical and Applied Genetics 110: 519–526.

Barber, G.W, 1936. The Cannibalistic Habits of the Corn Ear Worm. Technical Bulletin 499. United States Department of Agriculture, Washington, DC, USA.

Betz, F.S., Hammond, B.G., and Fuchs, R.L., 2000. Safety and advantages of *Bacillus thuringiensis*-protected plants to control insect pests. Regulatory Toxicology and Pharmacology 32: 156–173.

Bird, L.J., and Akhurst, R.J., 2004. Relative fitness of Cry1A-resistant and -susceptible *Helicoverpa armigera* (Lepidoptera: Noctuidae) on conventional and transgenic cotton. Journal of Economic Entomology 97: 1699–1709.

Blickenstaff, C.C., and Jolley, P.M., 1982. Host plants of western bean cutworm. Environmental Entomology 11: 421–425.

Bradford, K.J., Van Deynze, A., Gutterson, N., Parrott, W., and Strauss, S.H., 2005. Regulating transgenic crops sensibly: lessons from plant breeding, biotechnology and genomics. Nature Biotechnology 23: 439–444.

Bradley, J.R., Jr., Herzog, G.A., Roach, S.H., Stinner, R.E., and Terry, I., 1986. Cultural control in southeastern United States cropping systems. In: Theory and Tactics of *Heliothis* Population Management. I. Cultural and Biological Control, Southern Cooperative Series Bulletin 316, pp. 22–28.

Brandt, S.L., Coudron, T.A., Habibi, J., Brown, G.R., Ilagan, O.M., Wagner, R.M., Wright, M.K., Backus, E.A., and Huesing J.E., 2004. Interaction of two *Bacillus thuringiensis* δ-endotoxins with the digestive system of *Lygus hesperus*. Current Microbiology 48: 1–9.

Branson, T.F., and Ortman, E.E., 1967. Host range of larvae of the western corn rootworm. Journal of Economic Entomology 60: 201–203.

Branson, T.F., and Ortman, E.E., 1970. The host range of the larvae of the western corn rootworm: further studies. Journal of Economic Entomology 63: 800–803.

Bruck, D.J., and Lewis, L.C., 1998. Influence of adjacent cornfield habitat, trap location, and trap height on capture numbers of predators and parasitoid of the European corn borer (Lepidoptera: Pyralidae) in central Iowa. Environmental Entomology 27: 1557–1562.

Brust, G.E., and King, L.R., 1994. Effects of crop rotation and reduced chemical inputs on pests and predators in maize agroecosystems. Agriculture, Ecosystems and Environment 48: 77–89.

Brust, G.E., Stinner, B.R., and McCartney, D.A., 1985. Tillage and soil insecticide effects on predator-black cutworm (Lepidoptera: Noctuidae) interactions in corn agroecosystems. Journal of Economic Entomology 78: 1389–1392.

Brust, G.E., Stinner, B.R., and McCartney, D.A., 1986. Predatory activity and predation in corn agroecosystems. Environmental Entomology 15: 1017–1021.

Cannon, R.J.C., 2000. Bt transgenic crops: risks and benefits. Integrated Pest Management Reviews 5: 151–173.

Carlson, G.A., Sappie, G.P., and Hammig, M.D., 1989. Economic Returns to Boll Weevil Eradiation. USDA Agricultural Economics Report No. 621. USDA Economic Research Service, Washington, DC, USA.

Carpenter, J.E., and Gianessi, L.P., 2001. Agricultural Biotechnology: Updated Benefit Estimates. National Center for Food and Agricultural Policy, Washington, DC, USA.

Carrière, Y., Dennehy, T.J., Pedersen, B., Haller, S., Ellers-Kirk, C., Antilla, L., Liu, Y.-B., Willott, E., and Tabashnik, B.E., 2001. Large-scale management of insect resistance to transgenic cotton in Arizona: can transgenic insecticidal crops be sustained? Journal of Economic Entomology 94: 315–325.

Carrière, Y., Ellers-Kirk, C., Sisterson, M., Antilla, L., Whitlow, M., Dennehy, T.J., and Tabashnik, B.E., 2003. Long-term regional suppression of pink bollworm by *Bacillus thuringiensis* cotton. Proceedings of the National Academy of Sciences of the USA 100: 1519–1523.

Carrière, Y., Ellers-Kirk, C., Biggs, R.W., Nyboer, M.E, Unnithan, G.C., Dennehy, T.J., and Tabashnik, B.E., 2006. Cadherin-based resistance to *Bacillus thuringiensis* cotton in hybrid strains of pink bollworm: fitness costs and incomplete resistance. Journal of Economic Entomology 99: 1925–1935.

Castro, B.A., Leonard, B.R., and Riley, T.J., 2004. Management of feeding damage and survival of southwestern corn borer and sugarcane borer (Lepidoptera: Crambidae) with *Bacillus thuringiensis* transgenic field corn. Journal of Economic Entomology 97: 2106–2116.

Catangui, M.A., and Berg, R.K., 2006. Western bean cutworm, *Striacosta albicosta* (Smith) (Lepidoptera: Noctuidae), as a potential pest of transgenic Cry1Ab *Bacillus thuringiensis* corn hybrids in South Dakota. Environmental Entomology 35: 1439–1452.

Cattaneo, M.G., Yafuso, C., Schmidt, C., Huang, C.-Y., Rahman, M., Olson, C, Ellers-Kirk, C., Orr, B.J., Marsh S.E., Antilla, L., Dutilleul, P., and Carrière, Y., 2006. Farm-scale evaluation of the impacts of transgenic cotton on biodiversity, pesticide use, and yield. Proceedings of the National Academy of Sciences of the USA 103: 7571–7576.

Cellini, F., Chesson, A., Colquhoun, I., Constable, A., Davies, H.V., Engel, K.H., Gatehouse, A.M.R., Karenlampi, S., Kok, E.J., Leguay, J.J., Lehesranta, S., Noteborn, H., Pedersen, J., and Smith, M., 2004. Unintended effects and their detection in genetically modified crops. Food and Chemical Toxicology 42: 1089–1125.

Champion, G.T., May, M.J., Bennett, S., Brooks, D.R., Clark, S.J., Daniels, R.E., Firbank, L.G., Haughton, A.J., Hawes, C., Heard, M.S., Perry, J.N., Randle, Z., Rossall, M.J., Rothery, P., Skellern, M.P., Scott, R.J., Squire, G.R., and Thomas, M.R., 2003. Crop management and agronomic context of the Farm Scale Evaluations of genetically modified herbicide-tolerant crops. Philosophical Transactions of the Royal Society of London Series B Biological Sciences 358: 1801–1818.

Chilcutt, C.F., 2006. Cannibalism of *Helicoverpa zea* (Lepidoptera: Noctuidae) from *Bacillus thuringiensis* (Bt) transgenic corn versus non-Bt corn. Journal of Economic Entomology 99: 728–732.

Chu, C.-C., Natwick, E.T., Lopez, R.L., Dessert, J.R., and Henneberry, T. J., 2006. Pink bollworm moth (Lepidoptera: Gelechiidae) catches in the Imperial Valley, California from 1989 to 2003. Insect Science 13: 469–475.

Clark, M.S., Gage, S.H., and Spence, J.R., 1997. Habitats and management associated with common ground beetle (Coleoptera: Carabidae) in a Michigan agricultural landscape. Environmental Entomology 26: 519–527.

Clark, T.L., Foster, J.E., Kamble, S.T., and Heinrichs, E.A., 2000. Comparison of Bt (*Bacillus thuringiensis* Berliner) maize and conventional measures for control of the European corn borer (Lepidoptera: Crambidae). Journal of Entomological Science 35: 118–128.

Clark, T.L., and Hibbard, B.E., 2004. A comparison of non-maize hosts to support western corn rootworm (Coleoptera: Chrysomelidae) larval biology. Environmental Entomology 33: 681–689.

Conner, A.J., Glare, T.R., and Nap, J.P., 2003. The release of genetically modified crops into the environment. Part II: Overview of ecological risk assessment. The Plant Journal 33: 19–46.

Cottrell, T.E., and Yeargan, K.V., 1998. Influence of a native weed, *Acalypha ostraefolia* (Euphorbiaceae), on *Coleomegilla maculata* (Coleoptera: Coccinellidae) population density, predation, and cannibalism in sweet corn. Environmental Entomology 27: 1375–1385.

Dale, P.J., Clarke, B., Fontes, E.M., 2002. Potential for the environmental impact of transgenic crops. Nature Biotechnology 20: 567–574.

Devare, M., Londono-R, L.M., and Thies, J.E., 2007. Neither transgenic Bt maize (MON863) nor tefluthrin insecticide adversely affect soil microbial activity or biomass: a 3-year field analysis. Soil Biology and Biochemistry 39: 2038–2047.

Dial, C.I., and Adler, P.H., 1990. Larval behavior and cannibalism in *Heliothis zea* (Lepidoptera: Noctuidae). Annals of the Entomological Society of America 83: 258–263.

Dively, G.P., Rose, R, Sears, M.K., Hellmich, R.L., Stanley-Horn, D.E., Russo, J.M., Calvin, D.D., and Anderson, P.L., 2004. Effects on Monarch butterfly larvae (Lepidoptera: Danaidae) after continuous exposure to Cry1Ab-expressing corn during anthesis. Environmental Entomology 33: 1116–1125.

Dively G.P., Higgins, L.S., Storer, N.P., and Layton, R.J., 2008. Field-scale assessment of the effects of Cry34/35Ab1 Bt rootworm maize on non-target predaceous coleopterans and their prey. Environmental Entomology (submitted).

Dorhout, D.L., and Rice, M.E., 2004. First report of western bean cutworm, *Richia albicosta* (Noctuidae) in Illinois and Missouri. Online. Crop Management. doi:10.1094/CM-2004-1229-01-BR.

Dorhout, D. and Rice, M., 2006. Intraguild competition of lepidopteran larvae in maize: European corn borer, corn earworm, and western bean cutworm. Entomological Society of America Annual Meeting, December 10–13, 2006, Indianapolis, IN.

Douglas, J.R., Ingram, J.W., Gibson, K.E., and Peay, W.E., 1957. The western bean cutworm as a pest of corn in Idaho. Journal of Economic Entomology 50: 543–545.

Eastick, R.J., and Hearnden, M.N., 2006. Potential for weediness of Bt cotton in northern Australia. Weed Science 54: 1142–1151.

Eizaguirre, M., Albajes, R., López, C., Eras, J., Lumbierres, B., and Pons, X., 2006. Six years after the commercial introduction of Bt maize in Spain: field evaluation, impact and future prospects. Transgenic Research 15: 1–12.

Fitt, G.P., 1989. The Ecology of *Heliothis* species in relation to agroecosystems. Annual Review of Entomology 34: 17–52.

Freeman, B.L., and Smith, R.H., 1997. Insect scouting in Bt-transgenic cotton. Alabama Cooperative Extension System Publication ANR-1932.

Gahan, L.J., Gould, F., López, J.D., Jr., Micinski, S., and Heckel, D.G., 2007. A polymerase chain reaction screen of field populations of *Heliothis virescens* for a retrotransposon insertion conferring resistance to *Bacillus thuringiensis* toxin. Journal of Economic Entomology 100: 187–194.

Garcia-Alonso, M., Jacobs, E., Raybould, A., Nickson, T.E., Sowig, P., Willekens, H., Van Der Kouwe, P., Layton, R., Amijee, F., Fuentes, A.M., and Tencalla, F., 2006. A tiered system for assessing the risk of genetically modified plants to non-target organisms. Environmental Biosafety Research 5: 57–65.

Gathmann, A., Wirooks, L., Hothorn, L.A., Bartsch, D., and Schuphan, I., 2006. Impact of Bt maize pollen (MON810) on lepidopteran larvae living on accompanying weeds. Molecular Ecology 15: 2677–2685.

Gianessi, L.P., 2005. Economic and herbicide use impacts of glyphosate-resistant crops. Pest Management Science 61: 241–245.

Glare, T.R., and O'Callaghan, M., 2000. *Bacillus thuringiensis*: Biology, Ecology and Safety. Wiley, Chichester, UK.

Gould, F., 1986. Simulation models for predicting durability of insect-resistant germ plasm: a deterministic diploid, two-locus model. Environmental Entomology 15: 1–10.

Gould, F., 1998. Sustainability of transgenic insecticidal cultivars: integrating pest genetics and ecology. Annual Review of Entomology 43: 701–726.

Gould, F., Blair, N., Reid, M., Rennie, T.L., Lopez, J., and Micinski, S., 2002. *Bacillus thuringiensis*-toxin resistance management: stable isotope assessment of alternate host use by *Helicoverpa zea*. Proceedings of the National Academy of Sciences of the USA 99: 16581–16586.

Gould, F., Cohen, M.B., Bentur, J.S., Kennedy, G.G., and Van Duyn, J., 2006. Impact of small fitness costs on pest adaptation to crop varieties with multiple toxins: a heuristic model. Journal of Economic Entomology 99: 2091–2099.

Greenberg, S.M., Armstrong, J.S., Liu, T.-X., Coleman, R.J., and Adamczyk, J.J., 2006. Comparative efficacy of Bollgard II® and non-Bt cotton on the noctuid complex in the Lower Rio Grande Valley of Texas. Proceedings of the Beltwide Cotton Research Conferences, 2006, National Cotton Council of America, Memphis, TN, USA, Vol. 2, pp. 1073–1079.

Griffiths, B., Heckmann, L.-H., Caul, S., Thompson, J., Scrimgeour, C., and Krogh P.H., 2007. Varietal effects of eight paired lines of transgenic Bt maize and near-isogenic non-Bt maize on soil microbial and nematode community structure. Plant Biotechnology Journal 5: 60–68.

Gustafson, D.I., Head, G.P., and Caprio, M.A., 2006. Modeling the impact of alternative hosts on *Helicoverpa zea* adaptation to Bollgard cotton. Journal of Economic Entomology 99: 2116–2124.

Hammond, R.B., and Stinner, B.R., 1987. Soybean foliage insects in conservation tillage systems: effects of tillage, previous cropping history, and soil insecticide application. Environmental Entomology 16: 524–531.

Hawes, C., Haughton, A.J., Osborne, J.L., Roy, D.B., Clark, S.J., Perry, J.N., Rothery, P., D.A. Bohan, Brooks, D.R., Champion, G.T., Dewar, A.M., Heard, M.S., Woiwod, I.P., Daniels, R.E., Young, M.W., Parish, A.M., Scott, R.J., Firbank, L.G., and Squire, G.R., 2003. Responses of plants and invertebrate trophic groups to contrasting herbicide regimes in the Farm Scale Evaluations of genetically modified herbicide-tolerant crops. Philosophical Transactions of the Royal Society of London Series B, Biological Sciences 358: 1899–1915.

Head, G., Moar, W., Eubanks, M., Freeman, B., Ruberson, J., Hagerty, A., and Turnipseed, S., 2005. A multiyear, large-scale comparison of arthropod populations on commercially managed *Bt* and non-*Bt* cotton fields. Environmental Entomology 34: 1257–1266.

Hellmich, R.L., Siegfried, B.D., Sears, M.K., Stanley-Horn, D.E., Daniels, M.J., Mattila, H.R., Spencer, T., Bidne, K.G., and Lewis, L.C., 2001. Monarch larvae sensitivity to Bacillus thuringiensis- purified proteins and pollen. Proceedings of the National Academy of Sciences of the USA 98: 11925–11930.

Henry, C., 2006. Cumulative Long-Term Effects of Genetically Modified (GM) Crops on Human/ Animal Health and the Environment: Risk Assessment Methodologies. Final Report. Central Science Laboratory, York, UK.

Holland, J., Birkett, T., Begbie, M., Southway, S., and Thomas, C.F.G., 2003. The special dynamics of predatory arthropods and the importance of crop and adjacent margin habitats. IOBC/ WPRS Bulletin 26(4): 65–70.

House, G.J., and Alzugaray, M.D.R., 1989. Influence of cover cropping and no-tillage practices on community composition of soil arthropods in a North Carolina agroecosystem. Environmental Entomology 18: 302–307.

Huang, F., Leonard, B.R., and Andow, D.A., 2007. Sugarcane borer (Lepidoptera: Crambidae) resistance to transgenic *Bacillus thuringiensis* maize. Journal of Economic Entomology 100: 164–171.

Hummel, H.E., 2003. Introduction of *Diabrotica virgifera virgifera* into the Old World and its consequences: a recently acquired invasive alien pest species on *Zea mays* from North America. Communications in Agricultural and Applied Biological Sciences 68: 45–57.

Hutchison, W.D., Burkness, E.C., Moon, R., Leslie, T., Fleischer, S., and Abrahamson, M., 2007. Evidence for regional suppression of European corn borer populations in transgenic Bt maize in the Midwestern U.S.: analysis of long-term time series data from three states. Proceedings of the XVI International Plant Protection Congress 2007, pp 512–513.

Iowa State University, 2007. Western Bean Cutworm Monitoring Network, Iowa State University Department of Entomology. Online at http://www.ent.iastate.edu/trap/westernbeancutworm/ (accessed 7 December 2007).

Isenhour, D.J., Wiseman, B.R., and Layton, R.C., 1989. Enhanced predation by *Orius insidiosus* (Hemiptera: Anthocoridae) on larvae of *Heliothis zea* and *Spodoptera frugiperda* (Lepidoptera: Noctuidae) caused by prey feeding on resistant corn genotypes. Environmental Entomology 18: 418–422.

James, C., 2007. Global Status of Commercialized Biotech/GM Crops: 2007, ISAAA Brief No. 37, International Service for the Acquisition of Agri-Biotech Applications, Ithaca, NY, USA.

Johnson, M.T., Gould, F., and Kennedy, G.G., 1997. Effects of natural enemies on relative fitness of Heliothis virescens genotypes adapted and not adapted to resistant host plants. Entomologia Experimentalis et Applicata 82: 219–230.

Kennedy, G.G. and Storer, N.P., 2000. Life systems of polyphagous pests in temporally unstable cropping systems. Annual Review of Entomology 45: 467–493.

König, A., Cockburn, A., Crevel, R.W., Debruyne, E., Grafstroem, R., Hammerling, U., Kimber, I., Knudsen, I., Kuiper, H.A., Peijnenburg, A.A., Penninks, A.H., Poulsen, M., Schauzu, M., and Wal, J.M., 2004. Assessment of the safety of foods derived from genetically modified (GM) crops. Food Chemistry and Toxicology 42: 1047–1088.

Laub, C.A., and Luna, J.M., 1992. Winter cover crop suppression practices and natural enemies of armyworm (Lepidoptera: Noctuidae) in no-till corn. Environmental Entomology 21: 41–49.

Letourneau, D.K., Robinson, G.S., and Hagen, J.A., 2003. *Bt* crops: predicting effects of escaped transgenes on the fitness of wild plants and their herbivores. Environmental Biosafety Research 2: 219–246.

Liu, Y.B., Tabashnik, B.E., Dennehy, T.J., Patin, A.L., Sims, M.A., Meyer, S.K., and Carrière, Y., 2001. Effects of Bt cotton and CrylAc toxin on survival and development of pink bollworm (Lepidoptera: Gelechiidae). Journal of Economic Entomology 94: 1237–1242.

Losey, J.E., Hufbauer, R.A., and Hartzler, R.G., 2003. Enumerating lepidopteran species associated with maize as a first step in risk assessment in the USA. Environmental Biosafety Research 2: 247–261.

Marvier, M., McCreedy, C., Regetz, J., and Kareiva, P., 2007. A meta-analysis of effects of Bt cotton and maize on nontarget invertebrates. Science 316: 1475–1477

McManus, B.L., Fuller, B.W., Boetel, M.A., French, B.W., Ellsbury, M.M., and Head, G.P., 2005. Abundance of *Coleomegilla maculata* (Coleoptera: Coccinellidae) in corn rootworm-resistance Cry3Bb1 maize. Journal of Economic Entomology 98: 1992–1998.

Naranjo, S.E. 2005. Long-term assessment of the effects of transgenic *Bt* cotton on the function of the natural enemy community. Environmental Entomology 34: 1211–1223.

Naranjo, S.E., Head, G., and Dively, G.P., 2005. Field studies assessing arthropod nontarget effects in *Bt* transgenic crops: introduction. Environmental Entomology 34: 1178–1180.

Neunzig, H.H., 1969. The biology of the tobacco budworm and the corn earworm in North Carolina with particular reference to tobacco as a host. North Carolina Agricultural Experimental Station Technical Bulletin 196: 63.

O'Callaghan, M., Glare, T.R., Burgess, E.P.J., and Malone, L.A., 2005. Effects of plants genetically modified for insect resistance on non-target organisms. Annual Review of Entomology 50: 271–292.

O'Rourke, P.K., and Hutchison, W.D., 2000. First report of western bean cutworm, *Richia albicosta* (Smith) (Lepidoptera: Noctuidae), in Minnesota corn. Journal of Agricultural and Urban Entomology 17: 213–217.

Oberhauser, K.S., Prysby, M.D., Mattila, H.R., Stanley-Horn, D.E., Sears, M.K., Dively, G., Olson, E., Pleasants, J.M., Lam, W.K., and Hellmich, R.L., 2001. Temporal and spatial overlap between monarch larvae and corn pollen. Proceedings of the National Academy of Sciences of the USA 98: 11913–11918.

Obrist, L.B., Klein, H., Dutton, A., and Bigler, F., 2005. Effects of Bt maize on *Frankliniella tenuicornis* and exposure of thrips predators to prey-mediated Bt toxin. Entomologia Experimentalis et Applicata 115: 409–416.

Obrist, L.B., Dutton, A., Albajes, R., and Bigler, F., 2006. Exposure of arthropod predators to Cry1Ab toxin in Bt maize fields. Ecological Entomology 31: 143–154.

Obrycki, J.J., Losey, J.E., Taylor, O.R., and Jesse, L.C.H., 2001. Transgenic insecticidal corn: beyond insecticidal toxicity to ecological complexity. BioScience 51: 353–361.

Onstad, D.W., Guse, C.A., Spencer, J.L., Levine, E., and Gray, M.E., 2001. Modeling the dynamics of adaptation to transgenic corn by western corn rootworm (Coleoptera: Chrysomelidae). Journal of Economic Entomology 94: 529–540.

Onstad, D.W., Hibbard, B.E., Clark, T.L., Crowder, D.W., and Carter, K.G., 2006. Analysis of density-dependent survival of *Diabrotica* (Coleoptera: Chrysomelidae) in cornfields. Environmental Entomology 35: 1272–1278.

Oyediran, I.O., Higdon, M.L., Clark, T.L., and Hibbard, B.E., 2007. Interactions of alternate hosts, post-emergence grass control, and rootworm-resistant transgenic corn on western corn rootworm (Coleoptera: Chrysomelidae) damage and adult emergence. Journal of Economic Entomology 100: 557–565.

Pair, S.D., Raulston, J.R., Sparks, A.N., Westbrook, J.K., Wolf, W.W., and Goodenough, J.L., 1995. A prospectus impact of *Helicoverpa zea* (Bodie) (Lepidoptera: Noctuidae) production from corn in the lower Rio-Grande Valley on regional cropping systems. Southwestern Entomologist Supplement 18: 155–167.

Peck, S.L., Gould, F., and Ellner, S.P., 1999. Spread of resistance in spatially extended regions of transgenic cotton: implications for management of *Heliothis virescens*. Journal of Economic Entomology 92: 1–6.

Peterson, R.K., Meyer, S.J., Wolf, A.T., Wolt, J.D., and Davis, P.M., 2006. Genetically engineered plants, endangered species, and risk: a temporal and spatial exposure assessment for Karner blue butterfly larvae and Bt maize pollen. Risk Analysis 26: 845–858.

Pfannenstiel, R.S., and Yeargan, K.V., 1998a. Association of predaceous Hemiptera with selected crops. Environmental Entomology 27: 232–239.

Pfannenstiel, R.S., and Yeargan, K.V., 1998b. Ground beetles in three tillage plots in Ontario and observations on their importance as predators of the northern corn rootworm, *Diabrotica longicornis* (Coleoptera: Chrysomelidae). Proceedings of the Entomological Society of Ontario 110: 65–73.

Pilcher, C.D., Rice, M.E., Obrycki, J.J., and Lewis, L.C., 1997. Field and laboratory evaluations of transgenic *Bacillus thuringiensis* corn on secondary lepidopteran pests (Lepidoptera: Noctuidae). Journal of Economic Entomology 90: 669–678.

Pleasants, J.M., Hellmich, R.L., Dively, G.P., Sears, M.K., Stanley-Horn, D.E., Mattila, H.R., Foster, J.E., Clark, P., and Jones, G.D., 2001. Corn pollen deposition on milkweeds in and near cornfields. Proceedings of the National Academy of Sciences of the USA 98: 11919–11924.

Purdue Cooperative Extension Service, 2002. Pest and Crop. No. 26. October 18, 2002. http://www.entm.purdue.edu/Entomology/ext/targets/p&c/P&C2002/P&C26_2002.htm (accessed 7 December 2007).

Raps, A., Kehr, J., Gugerli, P., Moar, W.J., Bigler, F., and Hilbeck, A., 2001. Immunological analysis of phloem sap of *Bacillus thuringiensis* corn and of the nontarget herbivore *Rhopalosiphum padi* (Homoptera: Aphididae) for the presence of Cry1Ab. Molecular Ecology 10: 525–533.

Raybould, A., 2007a. Ecological versus ecotoxicological methods for assessing the environmental risks of transgenic crops. Plant Science 173: 589–602.

Raybould, A., 2007b. Problem formulation and hypothesis testing for environmental risk assessments of genetically modified crops. Environmental Biosafety Research 5: 119–125.

Rice, M.E., 2002. So Many Bugs, So Little Time: Reflections on the Minor Insect Pests of 2002. 2002 Proceedings of the 14th Annual Integrated Crop Management Conference. Iowa State University, University Extension, Ames, IA, USA, pp. 53–55.

Roach, S.H., 1980. Arthropod predators on cotton, corn, tobacco, and soybeans in South Carolina. Journal of the Georgia Entomological Society 15: 131–138.

Roberts, P., 1999. Observations of emerging pests in low spray environments. Proceedings of the Beltwide Cotton Research Conferences, 1999, National Cotton Council of America, Memphis, TN, USA, Vol. 2, p. 1034.

Romeis, J., Meissle, M., and Bigler, F., 2006. Transgenic crops expressing *Bacillus thuringiensis* toxins and biological control. Nature Biotechnology 24: 63–71.

Romeis, J., Bartsch, D., Bigler, F., Candolfi, M.P., Gielkens, M.M.C., Hartley, S.E., Hellmich, R.L., Huesing, J.E., Jepson, P.C., Layton, R., Quemada, H., Raybould, A., Rose, R.I., Schiemann, J., Sears, M.K., Shelton, A.M., Sweet, J., Vaituzis, Z., and Wolt, J.D., 2008. Assessment of risk of insect-resistant transgenic crops to nontarget arthropods. Nature Biotechnology 26: 203–208.

Roush, R.T., 1994. Managing pests and their resistance to *Bacillus thuringiensis*: can transgenic crops be better than sprays? Biocontrol Science and Technology 4: 501–506.

Ruberson, J.R., Phatak, S.C., and Lewis, W.J., 1997. Insect Populations in a Cover Crop/Strip Till System. In: Proceedings, Beltwide Cotton Conferences. National Cotton Council, Memphis, TN, USA, pp. 1121–1124.

Sanvido, O., Romeis, J., and Bigler, F., 2006. Ecological impacts of genetically modified crops: Experiences from ten years of experimental field research and commercial cultivation. Agroscope Reckenholz-Tänikon Research Station ART, ART Schriftenreihe 1, 84 p.

Schaber, B.D., Harper, A.M., and Entz, T., 1990. Effect of swathing alfalfa for hay on insect dispersal. Journal of Economic Entomology 83: 2427–2433.

Schmitz, G., Bartsch, D., and Pretscher, P., 2003. Selection of relevant non-target herbivores for monitoring the environmental effects of *Bt* maize pollen. Environmental Biosafety Research 2: 117–132.

Schneider, J.C., 2003. Overwintering of *Heliothis virescens* (F.) and *Helicoverpa zea* (Boddie) (Lepidoptera: Noctuidae) in cotton fields of northeast Mississippi. Journal of Economic Entomology 96: 1433–1447.

Schoenly, K.G., Cohn, B., Barrion, A.T., Zhang, W., Gaolach, B., and Viajante, V.D., 2003. Effects of *Bacillus thuringiensis* on non-target herbivore and natural enemy assemblages in tropical irrigated rice. Environmental Biosafety Research 2: 181–206.

Schuler, T.H., Denholm, I., Clark, S.J., Stewart, C.N., and Poppy G.M., 2004. Effects of Bt plants on the development and survival of the parasitoid *Cotesia plutellae* (Hymenoptera: Braconidae) in susceptible and Bt-resistant larvae of the diamondback moth, *Plutella xylostella* (Lepidoptera: Plutellidae). Journal of Insect Physiology 50: 435–443.

Sears, M.K., Hellmich, R.L., Stanley-Horn, D.E., Oberhauser, K.S., Pleasants, J.M., Mattila, H.R., Siegfried, B.D., and Dively, G.P., 2001. Impact of Bt corn pollen on monarch butterfly populations: a risk assessment. Proceedings of the National Academy of Sciences of the USA 98: 11937–11942.

Seymour, R.C., Hein, G.L., Wright, R.J., and Campbell, J.B., 2004. Western Bean Cutworm in Corn and Dry Beans. NebGuide G1359, University of Nebraska, Nebraska Cooperative Extension, Lincoln, NE, USA.

Shelton, A.M., and Badenes-Perez, F.R., 2006. Concept and applications of trap cropping in pest management. Annual Review of Entomology 51: 285–308.

Shelton, A.M., and North, R.C., 1986. Species composition and phenology of Thysanoptera within field crops adjacent to cabbage fields. Environmental Entomology 15: 513–519.

Siebert, M.W., Braxton, L.B., Huckaba, R.M., Walton, L.C., Haygood, R.A., Lassiter, R.B., Haile, F.J., and Thompson, G.D., 2007. Field performance of Dow AgroSciences' WideStrike Insect Protection against key lepidopteran pests in the mid-south and southeastern U.S. Proceedings of Beltwide Cotton Conferences 2007: 1114–1118

Sisterson, M.S., Carrière, Y., Dennehy, T.J., and Tabashnik, B.E., 2005. Evolution of resistance to transgenic crops: interaction between insect movement and field distribution. Journal of Economic Entomology 98: 1751–1762.

Smith, N., 2000. Seeds of Opportunity: An assessment of the benefits, safety, and oversight of plant genomics and agricultural biotechnology. Committee on Science Subcommittee on Basic Research, U. S. House of Representatives.

Southwood, T.R.E., 1962. Migration of terrestrial arthropods in relationship to habitat. Biological Reviews 37: 171–214.

Stanley-Horn, D.E., Dively, G.P., Hellmich, R.L., Mattila, H.R., Sears, M.K., Rose, R., Jesse, L.C., Losey, J.E., Obrycki, J.J., and Lewis, L., 2001. Assessing the impact of Cry1Ab-expressing corn pollen on monarch butterfly larvae in field studies. Proceedings of the National Academy of Sciences of the USA 98: 11931–11936.

Steffey, K.L., Rice, M.E., All, J., Andow, D.A., Gray, M.E., and Van Duyn, J.W., 1999. Handbook of Corn Insects. Entomological Society of America, Lanham, MD, USA.

Stinner, R.E., Barfield, C.S., Stimac, J.L., and Dohse, L., 1983. Dispersal and movement of insect pests. Annual Review Entomology 28: 319–335.

Storer, N.P., Van Duyn, J.W., and Kennedy, G.G., 2001. Life History Traits of *Helicoverpa zea* (Boddie) (Lepidoptera: Noctuidae) on non-Bt and Bt transgenic corn hybrids in eastern North Carolina. Journal of Economic Entomology 94: 1268–1279.

Storer, N.P., Peck, S.L., Gould, F., Van Duyn, J.W., and Kennedy, G.G., 2003. Spatial processes in the evolution of resistance in *Helicoverpa zea* (Lepidoptera: Noctuidae) to Bt transgenic corn and cotton: a biology-rich stochastic simulation model. Journal of Economic Entomology 96: 156–172.

Storer, N.P., Babcock, J.M., and Edwards, J.M., 2006. Field measures of western corn rootworm (Coleoptera: Chrysomelidae) mortality caused by Cry34/35Ab1 proteins expressed in maize event 59122 and implications for trait durability. Journal of Economic Entomology 99: 1381–1387.

Tabashnik, B.E., Dennehy, T.J., and Carrière, Y., 2005. Delayed resistance to transgenic cotton in pink bollworm. Proceedings of the National Academy of Sciences of the USA 43: 15389–15393.

Terry, I., Bradley, J.R., and Van Duyn, J.W., 1987. Population dynamics of *Heliothis zea* (Lepidoptera: Noctuidae) as influenced by selected soybean cultural practices. Environmental Entomology 16: 237–245.

Tillman, G., Schomberg, H., Phatak, S., Mullinix,B., Lachnicht, S., Timper, P., and Olson, D., 2004. Influence of cover crops on insect pests and predators in conservation tillage cotton. Journal of Economic Entomology 97: 1217–1232.

Tillman, P.G., 2006. Sorghum as a trap crop for *Nezara viridula* L. (Heteroptera: Pentatomidae) in cotton in the southern United States. Environmental Entomology 35: 771–783.

Torres, J.B., Ruberson, J.R., and Adang, M.J., 2006. Expression of *Bacillus thuringiensis* Cry1Ac protein in cotton plants, acquisition by pests and predators: a tritrophic analysis. Agricultural and Forest Entomology 8: 191–202.

USDA (United States Department of Agriculture), 2007. Adoption of Genetically-Engineered Crops in the U.S., Economic Research Service, Washington, DC, USA. http://www.ers.usda.gov/Data/BiotechCrops/alltables.xls (accessed 24 January 2008).

USEPA (United States Environmental Protection Agency), 2002. A Set of Scientific Issues Being Considered by the Environmental Protection Agency Regarding: Corn Rootworm Plant-incorporated Protectant Nontarget Insect and Insect Resistance Management Issues. Scientific Advisory Panel Meeting Minutes No. 2002-05A.

USEPA (United States Environmental Protection Agency), 2006. A Set of Scientific Issues Being Considered by the Environmental Protection Agency Regarding: Analysis of a Natural Refuge of Non-cotton Hosts for Monsanto's Bollgard II Cotton. Scientific Advisory Panel Minutes No. 2006-03.

University of Illinois Extension, 2007. Is the European Corn Borer an Endangered Species? The Bulletin No. 24 Article 3, November 9, 2007. http://ipm.uiuc.edu/bulletin/article.php?id=865 (accessed 21 January 2008).

Varchola, J.M., and Dunn, J.P., 1999. Changes in ground beetle (Coleoptera: Carabidae) assemblages in farming systems bordered by complex or simple roadside vegetation. Agriculture. Ecosystems and Environment 73: 41–49.

Vojtech, E., Meissle, M., and Poppy, G.M., 2005. Effects of Bt maize on the herbivore *Spodoptera littoralis* (Lepidoptera: Noctuidae) and parasitoid *Cotesia marginiventris* (Hymenoptera: Braconidae). Transgenic Research 14: 133–144.

Wallner, W.E., 1987. Factors affecting insect population dynamics: differences between outbreak and non-outbreak species. Annual Review of Entomology 32: 317–340.

Way, M.J., 1977. Pest and disease status in mixed stands vs. monocultures: the relevance of ecosystem stability. In: Origins of Pest, Parasite, Disease and Weed Problems, J.M. Cherrett and G.R. Sagar, eds., Blackwell, Oxford, UK, pp. 127–138.

Wolt, J.D., Conlan, C.A., and Majima, K., 2005. An ecological risk assessment of Cry1F maize pollen impact to pale grass blue butterfly. Environmental Biosafety Research 4: 243–251.

Zhao, J.Z., Li, Y.X., Collins, H.L., Gusukuma-Minuto, L., Mau, R.F., Thompson, G.D., and Shelton, A.M., 2002. Monitoring and characterization of diamondback moth (Lepidoptera: Plutellidae) resistance to spinosad. Journal of Economic Entomology 95: 430–436.

Zhao, J., Cao, J., Collins, H.C., Bates, S.L., Roush, R.T., Earle, E.D., and Shelton, A.M., 2005. Concurrent use of transgenic plants expressing a single and two Bt genes speeds insect adaptation to pyramided plants. Proceedings of the National Academy of Sciences of the USA 102: 8426–8430.

Zwahlen, C., Hilbeck, A., and Nentwig, W., 2007. Field decomposition of transgenic Bt maize residue and the impact on non-target soil invertebrates. Plant and Soil 300: 245–257.

Chapter 11
Have *Bt* Crops Led to Changes in Insecticide Use Patterns and Impacted IPM?

Gary P. Fitt*

Abstract GM crops have now been commercialised for over ten years and currently over 114 million hectares are grown in 23 countries (James, 2007). By incorporating a powerful pest management tactic within the plant these *Bt* crops overcome some, but not all of the problems with timing and variable rates of application of insecticides, which reduce efficacy and often result in higher than necessary concentrations being applied than is necessary. The aim of this chapter is to gather the current evidence for impacts of *Bt* crops, largely *Bt* cotton and *Bt* maize, on insecticide use and to reflect on their role in IPM. Analyses of *Bt* crop performance across a range from large-scale intensive production to smallholder production systems of varying levels of sophistication indicate significant reductions in insecticide input and in some systems, highly significant improvements in yield. However, economic performance is highly variable and seems dependent more on the market characteristics, support structures and culture of the systems in which *Bt* crops are deployed than on the *Bt* crops themselves. Given their specificity for key target pests and well demonstrated lack of impact on beneficial insects, *Bt* crops provide an important new platform for sustainable IPM systems, one that is compatible with a full range of other tactics. However, achieving that IPM outcome will often require ongoing education and extension support for farmers, particularly in smallholder systems, to ensure they can build confidence and gain sustainable benefit from a mix of new and established technologies in pest management.

11.1 Introduction

GM crops have now been grown commercially for over ten years. Currently over 114 million hectares are grown to GM crops in 23 countries (James, 2007). This first generation of GM crops has focussed exclusively on so-called input traits associated with pest, disease and weed management and thereby provide direct benefits to producers, but are perceived to provide little direct benefit to consumers.

CSIRO Entomology, Brisbane, Queensland, Australia

*To whom correspondence should be addressed. E-mail: gary.fitt@csiro.au

J. Romeis, A.M. Shelton, G.G. Kennedy (eds.), *Integration of Insect-Resistant Genetically Modified Crops within IPM Programs.*

The missing client here of course is the environment. In relation to *Bt* crops the potential to significantly reduce reliance on insecticides has also brought opportunities for substantial environmental benefits, which in most countries should be seen as a societal benefit as well. Of the 114 million hectares of GM crops about 37% express insecticidal traits (almost exclusively insecticidal proteins from *Bacillus thuringiensis – Bt*), and 40% are grown in developing countries, which is the fastest growing sector of the GM crop market.

Despite the claims of anti-GM proponents the commercialised *Bt* crops, largely *Bt* cotton and *Bt* maize, have not generated *Bt* resistance (Tabashnik et al., 2003, 2008) nor devastated populations of non-target organisms. The crops and pest management systems discussed extensively for maize (Hellmich et al., chapter 5), cotton (Naranjo et al., chapter 6), and potato (Grafius and Douches, chapter 7) provide excellent examples of *Bt* crop deployment where the motivation was to provide more sustainable management of key pests but with greatly reduced reliance on synthetic insecticides. By incorporating a powerful pest management tactic within the plant, these *Bt* crops overcome some but not all of the problems with timing and variable rates of application of insecticides, which reduce efficacy and often result in higher concentrations than necessary being applied than is necessary.

Broadly, *Bt* crops provide an important new tool for integrated pest management (IPM) systems, one that targets the key pests, is compatible with a full range of other tactics, particularly with the enhancement of beneficial insects, and so provides a platform for construction of sustainable IPM (Fitt, 2000; Fitt and Wilson, 2000; Way and van Emden, 2000; Kennedy, chapter 1; Romeis et al., chapter 4). While some purists will argue that the constitutive expression of *Bt* in crops is akin to prophylactic use of insecticides and thus inconsistent with IPM, this is an incorrect perspective. GM plants are little different to the deployment of traditional host plant resistance where antibiosis traits are introduced into crop plants and utilised as part of an IPM approach. While conventional host-plant resistance (HPR) traits most often provide only partial resistance to pests, whereas *Bt* genes aim for complete control of specific pests, the same opportunities exist for GM insecticidal plants to be components of IPM (Fitt, 2000; Way and van Emden, 2000). The aim of this chapter is to synthesize the current evidence for impacts of *Bt* crops on insecticide use and to reflect on their role in IPM. Elsewhere, Qaim et al. (chapter 12) provide a detailed analysis of the economic performance of GM crops, particularly in relation to their impact on insecticide use.

11.2 Impacts on Insecticide Use

11.2.1 Cotton

Cotton is among the most intensively sprayed of all field crops and there has been a real imperative to reduce the reliance of pest control on insecticides for economic, environmental and human health reasons. Of the 114 million hectares of GM crops grown in 2007 about 13% was GM cotton (James, 2007), 93% of which expressed *Bt* genes – mostly Cry1Ac alone, some with both Cry1Ac and Cry2Ab (the other 7% was for herbicide tolerance) (see also Table 6.1 in Naranjo et al., chapter 6). Collectively this was 35% of total world cotton area and Brookes and Barfoot (2006) estimate that during the first ten years of *Bt* cotton production (1996–2005) the volume of insecticide active (a.i.) ingredient applied was reduced by 94.5 million kilograms. This represents about a 19% reduction across the global cotton crop, of which *Bt* cotton represented about 38% in nine countries in 2006 (13.4 million hectares out of a total crop of 35 million hectares – James, 2006). However, across the 13.4 million hectares actually producing *Bt* cotton average reductions in insecticide use are about 50% (see later). The farm level economic value of using *Bt* cotton in 2005 was US$1.73 billion, while cumulatively since 1996, the farm income benefit has been over $7.51 billion.

Historically, almost half of the insecticide used in agriculture has been applied to cotton, with approximately 50% of that used against caterpillar pests. Heliothine caterpillars are thus the main target pests in all countries where *Bt* cotton has been deployed. In the USA, Mexico and Argentina the main target is *Heliothis virescens* (Lepidoptera: Noctuidae) a pest with high sensitivity to *Bt* toxins and a great propensity to develop resistance. The pink bollworm, *Pectinophora gossypiella* (Lepidoptera: Gelechiidae) is also a key target species, in both north and south America. For these species the *Bt* protein Cry1Ac, which is expressed in most *Bt* cotton varieties worldwide, is highly efficacious and from a resistance management viewpoint it is possible to achieve a high dose as part of a high dose/refuge strategy to manage resistance (Caprio, 1994; Tabashnik, 1994; Roush, 1998; Gould, 1998). Elsewhere in the world the main target is *Helicoverpa armigera* (Lepidoptera: Noctuidae) (Australia, China, India, South Africa, Indonesia) and the related species, *Helicoverpa zea* (Lepidoptera: Noctuidae), in the USA and Mexico and *Helicoverpa gelotopoeon* (Lepidoptera: Noctuidae) in Argentina. These species are up to tenfold more tolerant of the Cry1A and Cry2A toxins than is *H. virescens* and several other lepidopteran targets, and *Bt* cotton varieties with only a single Cry1Ac protein driven by the CAMV promoter are unlikely to provide a high dose and do not provide season long efficacy (Daly and Fitt, 1998). Where reliance is placed on single gene *Bt* cotton the importance of effective refuges becomes much more critical. Where pyramided two gene varieties (Cry1Ac/Cry2Ab) have been deployed the efficacy of the proteins is sufficient to effectively administer a high dose, and thus these varieties contribute significantly to resistance management. Several other Lepidopteran pests are also significant targets for *Bt* cottons in certain countries. These include cotton tipworm *Crocidosema plebejana*, (Tortricidae), cotton leaf perforator *Bucculatrix* spp. (Bucculatricidae) and pink-spotted bollworm (*Pectinophora scutigera*, Gelechiidae) in Australia, cotton leafworm (*Alabama argillacea*, Noctuidae) in South America, and red bollworm (*Diparopsis castanea*, Noctuidae) in southern Africa. In some cases these species may also represent a challenge for resistance management. Naranjo et al. (chapter 6) provide a comprehensive account of the deployment of *Bt* cotton globally and its present and future role in IPM.

Here I briefly summarise the available data on the impacts of Bt cotton on insecticide use and other consequences for IPM. Table 11.1 summarises available data on insecticide use for *Bt* cotton in several countries. It shows that significant reductions in insecticide usage have resulted in every country where *Bt* cotton has been grown.

These reductions in both the number of insecticide applications and in the quantity of active ingredient (a.i.) of insecticide, as a result of *Bt* cotton production, average about 50%. The level of reduction in insecticide use that is achievable is clearly dependent on the relative importance of Heliothines and other key lepidopteran pests in the pest complex (Fitt et al., 1994; Fitt, 2000; Hillocks, 2005), the intensity of pest infestations, the general level of insecticide application typical for the country prior to *Bt* cotton introduction and the overall effectiveness of the pest management regime. Substantial yield increases have been associated with *Bt* cotton in developing countries (e.g., India, China) where the effectiveness of pest management (sampling, insecticide application, insecticide efficacy) has historically been compromised, whereas in the USA and Australia yields gains if any have been relatively small. In these cases, pest management has previously been quite effective and *Bt* cotton has simply substituted for insecticides applied for the target pests.

Perhaps of more interest is the changing spectrum of toxicities of insecticides applied to *Bt* cotton crops, with evidence in some countries of even more significant reductions in the environmental impact quotient (EIQ) of insecticides now applied. For example Knox et al. (2006) calculated environmental impact quotients for conventional and Bollgard II cotton in Australia. Even when the Cry1Ac and Cry2Ab proteins were included in the calculation of environmental load, the authors demonstrated a 64% reduction in environmental impact associated with the use of Bollgard II cotton. Likewise Wossink and Denaux (2006) analysed the environmental impact and insecticide use efficiency of stacked insect resistance and herbicide tolerance traits in USA *Bt* cotton. They concluded that the adoption of stacked gene cotton led to a reduction in the environmental impact of insecticide use required for cotton production even though it did not necessarily benefit the producers through savings in pest control costs. There is little doubt that in some countries insecticides are drastically overused in cotton production and any technological advance could produce reductions. In many cases the adoption of IPM could achieve significant gains, and *Bt* cotton in conjunction with IPM could achieve even more.

11.2.1.1 Australia

In Australia there was a phased introduction of *Bt* technology with single gene (Cry1Ac) cotton (known as Ingard™) grown from 1996 until 2004, after which two gene varieties (Cry1Ac/Cry2Ab, known as Bollgard II™) completely replaced the single gene lines. Figure 11.1 shows that during the first 8 years of use of *Bt* cotton there was an average reduction of 44% in active ingredient (and a 59% reduction in insecticide applications – Fig. 11.2) applied for *Helicoverpa* spp. Following the introduction of Bollgard II varieties this reduction in a.i.

Table 11.1 Impacts of *Bt* cotton on insecticide use and yield of *Bt* cotton compared to conventional cotton in several countries worldwide

	Number of sprays			Average insecticide use (kg/ha)[a]			Yield (kg/ha)			
Country	*Bt* cotton	Non-*Bt* cotton	% change	*Bt* cotton	Non-*Bt* cotton	% change	*Bt* cotton	Non-*Bt* cotton	% change	References
Argentina	2.14	4.5	–52	1.85	4.15	–55	2,032	1,537	+32	Qaim and De Janvry (2005)
USA[b]	3.74	5.2	–28.1				689	666	+3.3	Williams (2003)
	3.75	6.7	–44.1				1,509	1,639	+8.6	Cattaneo et al. (2006)
Australia	6.23	11.7	–47	3.57	6.3	–43	4,762	4,790	0	Fitt (2003)
Ingard[c]	3.06	10.41	–71	0.48	3.23	–85				Pyke (2008)
Bollgard II[c]										
India	4.19	7.19	–42	0.43	1.41	–70	608	337	+80	Qaim and Zilberman (2003)
							1,238[e]	706[e]	+75	Bambawale et al. (2004)
								370[f]	+230	
	4.18	6.79	–38	2.07	4.17	–50	659	491	+34	Qaim et al. (2006)
China	8.1	19.8	–59	21.7[f]	65.6	–70	3,264	2,741	+19	Pray et al. (2002)
	6.6	19.8	–66	12.1	60.7	–80	3,290	3,186	+3	Huang et al. (2002)
				18	46	–61				Lu et al. (2002)
South Africa							464	348	+34	Yousouf et al. (2001)
small scale all dryland	3.8	11.2	–66.1							Ismael et al. (2002)
							471	261	+80	Kirsten et al. (2002)

(continued)

Table 11.1 (continued)

Country	Number of sprays			Average insecticide use (kg/ha)[a]			Yield (kg/ha)			References
	Bt cotton	Non-*Bt* cotton	% change	*Bt* cotton	Non-*Bt* cotton	% change	*Bt* cotton	Non-*Bt* cotton	% change	
				97[a]	129	–25	576	395	+46	Bennett et al. (2003)
	2.9	4.8	–40				760	670	+18	Hofs et al. (2006)
South Africa				226[a]	519	–56	4,046	3,413	+19	Kirsten et al. (2002)
large scale										
irrigated										
dryland										
							947	832	+21	Gouse et al. (2003)
Mexico	2.3	4.9	–54				1,645	1,480	+11	Traxler et al. (2003)

[a] Pesticide data for China and South Africa is total insecticide (active + inert ingredients) per hectare
[b] Combined data for five states (Arizona, Louisiana, Georgia, North Carolina, and South Carolina)
[c] Eight years data for single gene *Bt* cotton (Ingard) in Australia
[d] Four years data for two-gene *Bt* cotton (Bollgard II) in Australia
[e] *Bt* cotton and conventional cotton grown under an IPM system
[f] Conventional cotton grown without IPM

averaged 65–75% (with a 80–90% reduction in number of sprays, Brookes and Barfoot [2006], Fig. 11.2) on a per hectare basis. Some indication of the magnitude of this reduction is given by 1998/99 growing season when 1.75 million fewer liters of insecticide were applied on *Bt* cotton crops compared to conventional cotton (Fitt, 2003). While for the single gene *Bt* varieties reductions were mostly in early to midseason applications due the gradual loss of efficacy of these varieties (Fitt, 2000), Bollgard II varieties have essentially provided season long control of *Helicoverpa* spp. (Pyke, 2008) with only an occasional crop requiring one spray at the end of the season. Nonetheless, total insecticide use on Australian cotton fluctuates widely because the area of production changes among seasons due to water availability and seasonal conditions which directly influence pest abundance.

Figure 11.2 shows the average numbers of insecticide sprays applied for *Helicoverpa* spp. and all other pests combined for Ingard and Bollgard II varieties compared to conventional varieties in the same growing seasons. Note that despite the significant reductions in sprays for *Helicoverpa* spp. there has been little change in total number of sprays for all other pests (mirids, mites, aphids, thrips). This trend will be discussed below.

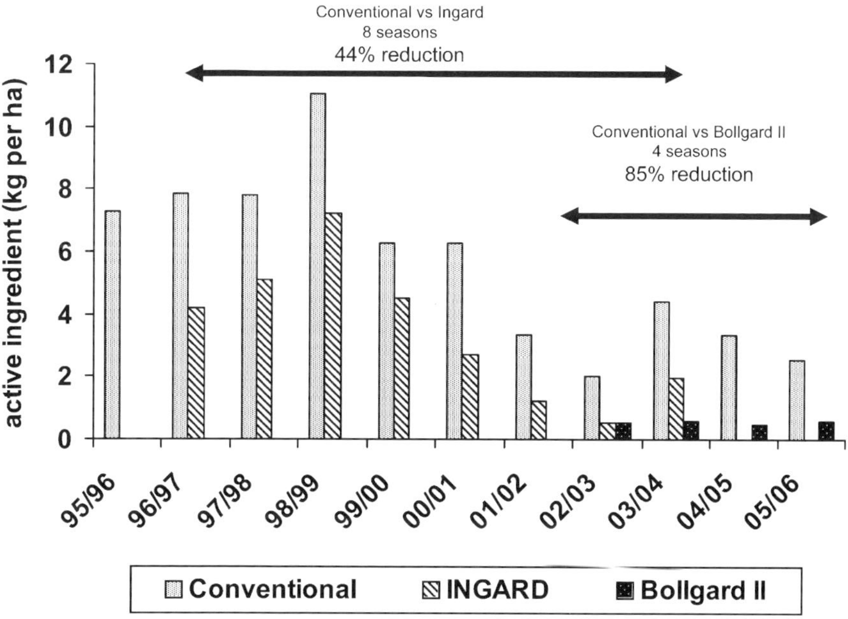

Fig. 11.1 Changes in active ingredient of insecticide applied to *Bt* (Ingard or BGII) cotton crops in Australia compared to conventional varieties grown in the same seasons

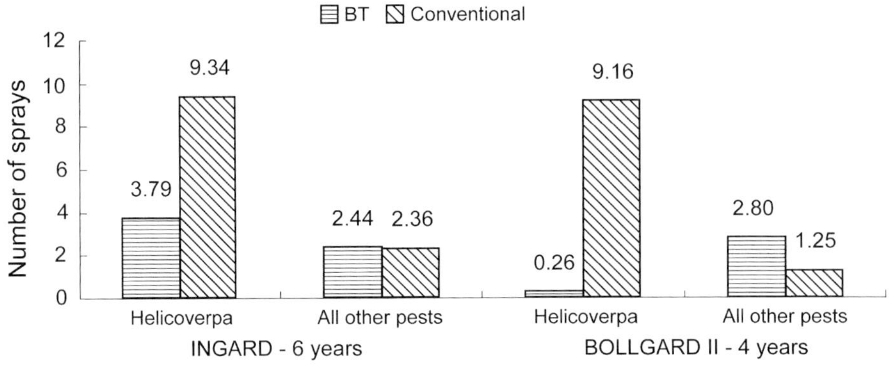

Fig. 11.2 Average numbers of sprays applied to *Bt* and conventional cotton crops during 6 years when Ingard (Cry 1Ac) cotton was grown and 4 years when Bollgard II was grown. Note that there has been little change in total number of sprays for all other pests (mirids, mites, aphids, thrips) despite a 60% to 95% reduction in number of applications targeting *Helicoverpa*

11.2.1.2 USA

In the United States, protocols for calculating reductions in insecticide usage as a result of planting *Bt* cotton were established by Fernandez-Cornejo and McBride (2000) and by Gianessi et al. (2002). Those studies indicated significant and in some cases dramatic reductions in insecticide use on Bollgard cotton with up to 35% fewer sprays and economic grower benefits of about US$123/ha. Williams (2003) summarised data across the US cotton belt which showed an average 28% reduction in insecticide applications for *Bt* cotton, but this varied considerably from state to state dependent on the relative importance of *H. virescens*. Cattaneo et al. (2006) indicate a 44% reduction in insecticide applications on *Bt* cotton in Arizona in a system where pink bollworm (*P. gossypiella*) is the major target. Most recently Brookes and Barfoot (2006) have applied the protocols developed by Gianessi et al. (2002) to provide updated figures across the USA cotton belt which indicate continued reductions in a.i. of 600,000–800,000 kg per annum (7–10% reduction in a.i.) for Bollgard cotton over the ten years up to 2005, together with a 25% to 35% reduction in EIQ. From 1996 to 2005 the cumulative saving in insecticide input to the environment was 5.14 million kilograms. It should be noted that the boll weevil eradication program and the pink bollworm eradication programs have also contributed to reduction in insecticide use in the US. Further reductions in the average number of insecticide applications per crop season are expected as Bollgard II cotton varieties, containing two *Bt* genes and with a broader spectrum and efficacy of pest insect control (Perlak et al., 2001), become more widely grown.

11.2.1.3 China

China is the world's largest cotton producer with 7.5 million small holder farmers now growing 3.5 million hectares of *Bt* cotton (66% of the total crop) spread across three major ecological regions: the Yangtze River Region, the Yellow River Region, and the Northwestern Region (Wu, 2007). As a result of the variability in pest complex and production systems across these regions (Wu and Guo, 2005) it is difficult to make generalisations about the impact of *Bt* cotton. Nonetheless Raney (2006) concludes that deployment of *Bt* cotton in China "represents the most successful case so far in terms of productivity, farmer incomes, equity and sustainability". Part of this success relates to the well developed agricultural research system, which has independently developed some transgenic constructs and has ensured a focus on locally adapted cotton varieties.

Several authors have documented reductions in insecticide applications and in active ingredient of between 60% to 70% (Huang et al., 2002; Pray et al., 2002; Hossain et al., 2004) among small-holders growing less than 0.5 ha of cotton (Table 11.1). Insecticide costs are likewise reduced by an average of 67% (Raney, 2006). These reductions vary across regions, but appear to have been reasonably consistent across years and have been correlated with real human health benefits from reductions in farmer poisonings (Pray et al., 2001; Lu et al., 2002). Pray and Huang (2003) demonstrated that small poor farmers gained the greatest economic benefit from *Bt* cotton. Further Hossain et al. (2004) demonstrate that adoption of *Bt* cotton can have significant benefits in reduction in famer poisonings through reduced exposure to insecticides. In their analysis of insecticide use changes over 9 years, Brookes and Barfoot (2006) show that by 2005 insecticide a.i. and in EIQ for cotton production across the whole of China had been reduced by 50% compared to if the area had remained as conventional cotton.

Nonetheless a number of authors question the general benefits of *Bt* cotton across Chinese production regions. Xu and Fok (2008) note that benefits have been less clear outside the Yellow River valley and some on-line reports suggest that secondary pests have now increased to the extent that Chinese cotton farmers are losing money. Men et al. (2005) provide an analysis over 3 years of *H. armigera* and sucking pest abundance in *Bt* and non-*Bt* cotton in Henan province in which they highlight the risk that sucking pests would be more of a problem on *Bt* cotton varieties. However, this study was clearly in an area where pests were not abundant as they applied a total of only eight to nine sprays in non-*Bt* cotton over 3 years and *H. armigera* was clearly not the key pest in this region in those years as they note that only two of the eight applications to non-*Bt* cotton were targeted at *H. armigera*.

These conclusions seem at odds with other results and the broad conclusions of Brookes and Barfoot (2006), although the significant emergence of black market *Bt* cotton seed in China (with less reliable performance) may well be distorting the potential performance and insecticide savings which were clearly evident in earlier years when seed availability was more restricted and controlled. Yang et al. (2005a) also demonstrate that IPM trained farmers achieve greater insecticide reductions with *Bt* cotton than those who had not previously practised IPM. They note that farmers "simply transfer the experience and knowledge developed on conventional cotton to *Bt* cotton" (Yang et al., 2005b). These uncertainties in China reflect a major challenge in implementing GM technology with smallholder famers with limited extension support. Yang et al. (2005a, b) clearly demonstrate the lack of fundamental understanding by Chinese farmers of the composition and dynamics of the pest and beneficial complex in their cotton crops. Although about 60% of farmers at least sampled their fields before applying insecticides the lack of awareness of the real abundance and damaging potential of different insects underpins the ongoing excessive use of insecticides in *Bt* crops, which must often be unjustified on economic or biological grounds.

A feature of *Bt* crops is that as a result of the removal of many disruptive insecticides, they are "living crops" with considerable activity of beneficial and secondary pest insects. Farmers with experience of IPM will also be more confident with seeing a "living crop". Despite all the caveats and while insecticide use may still be excessive in some areas there is little doubt that *Bt* cotton has the potential for widespread and significant savings in insecticide in China. Ongoing education which increases farmer capacity to manage inputs and develop confidence in the adoption of innovative technology like *Bt* cotton will be critical to maximizing the value achieved from *Bt* cotton in terms of environmental, economic and human health (Hossain et al., 2004; Yang et al., 2005a; Qaim et al., 2006; Pemsl et al., 2008).

11.2.1.4 India

India has the largest cotton production area in the world – more than 9 million hectares grown by more than five million farmers (James, 2007). Of this, more than 6 million hectares was *Bt* cotton (63%) by 2007. This vast production occurs across a wide latitudinal range in three broad regions (northern, central and southern) and in multiple states resulting in a diversity of pest complexes and pressures. Consequently it is difficult to generalise about impacts, but the scope for impact is great, since prior to the release of *Bt* cotton in India, the industry used more than $3.8 billion of insecticide annually.

Since *Bt* cotton was first grown in India in 2002 there has been much contention about its performance and suitability for that production system. Concerted efforts by protest groups, some NGOs and media have sought to stop *Bt* cotton cultivation. When first released, *Bt* genes were available in only a small number of hybrid varieties that were not well adapted to many of the production regions of India. Overall performance of *Bt* cotton hybrids was encouraging across 80% of the 45,052 ha where they were grown (Qaim, 2003; Qaim and Zilberman, 2003) with insecticide savings of US$40/ha (~$16/acre) and increased income of US$377/ha (~$154/acre). However, some genotypes among the limited range of hybrids available had problems with the unusual weather conditions (long dry-spells, unusually high temperatures during boll formation followed by heavy rain) and provided the basis for a plethora of negative and often irrational media coverage for GM cotton broadly in India. However, in subsequent years as more registered varieties became available the area rapidly increased (Raney, 2006), performance stabilised to some extent and *Bt* cotton now accounts for 80% of all hybrid cotton, which itself represents 70% of Indian production.

Qaim (2003) and Qaim and Zilberman (2003) document insecticide reductions in early trial years of about 80% against *H. armigera*, while total applications were reduced by 42%. Importantly, sprays for secondary pests were unchanged. They also document yield gains of up to 80%, which by 2007 has allowed India to become a net exporter of cotton. These reductions in insecticide use now appear to have stabilised at about 50%, but with variation across States from 19% reduction in Andra Pradesh to 73% reduction in Tamil Nadu (Raney, 2006). In a recent analysis, Qaim et al. (2006) use data for 2002/03 across four states in India to examine the insecticide impacts and economics of *Bt* cotton. Reductions in insecticide applications occurred across all states with an average reduction of 38% (and 50% reduction in a.i.), but a range from 21% reduction in Andhra Pradesh to 75% reduction in Tamil Nadu. Likewise yields increased on average by 34% but ranged from no change in Andhra Pradesh to a 72% increase in Karnataka. Overall Qaim et al. (2006, chapter 12) conclude that clear benefits have been realised but there is huge variability due to regional differences, varietal differences, impacts of black market varieties, unrealistic expectations, and lack of awareness or availability of information to farmers. Pemsl et al. (2004) note the overriding influence of potential yield on performance and perceptions of performance for *Bt* cotton, which is strongly influenced by the availability of adapted quality assured germplasm with *Bt* genes. They also note the inherent unreliability in some survey methods for post-hoc gathering of information from farmers on their insecticide use. Another variable in India is the range of illegal and mislabelled varieties available in the market (Qaim et al., 2006), which lead to confusion and lack of trust by farmers. By 2007 more than 130 *Bt* cotton varieties were commercially available in India (James, 2007).

As in other countries where single gene *Bt* cotton expressing Cry1Ac has been deployed against *H. armigera*, Indian *Bt* hybrids have displayed a distinct loss of efficacy as plants mature during the second half of the growing season (Kranthi et al., 2005). This decline had been well established in Australia as early as 1994 (Fitt et al., 1994) and was widely researched and reported (e.g., Fitt et al., 1998; Olsen and Daly, 2000; Olsen et al., 2005). In India the phenomenon added considerably to confusion in the market, despite attempts to clarify expectations of what *Bt* cotton could achieve, and was widely perceived among anti-GM protesters as part of a cover-up by seed companies and technology providers.

Adoption of *Bt* cotton in India has continued to increase at a rapid rate, so the majority of famers must perceive benefit, but "innovation adoption is a learning process, and farmers have to identify optimal input adjustments through experimentation and reliable external advice" (Qaim et al., 2006). This is exactly the issue for both IPM and *Bt* cotton as a component of IPM. Providing poorly educated farmers with the support to build knowledge around innovation will be essential in both India and China to ensure that inherent benefits in the technology are realised.

11.2.1.5 South Africa

Bt cotton was first introduced in South Africa in 1997/98, and was adopted by large-scale farmers growing irrigated and dryland cotton and by smallholder growing only dryland crops. Although 95% of South Africas cotton production is produced by 300 large scale farmers, much of the focus of researchers has been on the 3,000 smallholders in the Makhatini Flats of KwaZulu Natal province who have almost unanimously adopted *Bt* cotton and achieved significant insecticide reductions and yield benefits (Table 11.1). The experiences of this smallholder group is seen as a valuable case study for smallholder benefits of GM technology in other African countries, none of which have yet approved GM crops.

While there are clearly benefits from *Bt* technology in reduced spraying, these smallholders are still achieving very low yields due to other agronomic constraints and their economic returns also remain low. Hofs et al. (2006) note the reduction in insecticide use, but claim that cotton production has not increased among the smallholders in the Makhatini Flats and that benefits have not been as great as expected. For example in 1997/98, dryland yields were 600 kg seed cotton per hectare, with 0% GM cotton adoption. In 2004/05, with close to 100% GM cotton adoption, yields were once again 600 kg/ha. However, these yields are highly variable from year to year and influenced largely by variable rainfall and minimal management expertise of the farmers. Despite the pessimistic overview from Hofs et al. (2006), they document a 40% reduction in total sprays, including an 83% reduction in spraying for *H. armigera*, one of the key pests. Similarly Bennett et al. (2003) and Thirtle et al. (2003) document reduced insecticide use among smallholders over 3 years. They, and Hillocks (2005), emphasise the multitude of factors that contribute to benefit or dis-benefit of technologies like *Bt* cotton, but clearly document a difference in chemical costs of $50/ha between non-*Bt* growers ($132/ha) and *Bt* cotton growers ($83/ha). As in China, there is a real expectation of health benefits among smallholders growing *Bt* cotton in South Africa (Bennett et al., 2003).

Smallholder awareness of the importance of resistance management is also poor (Bennett et al., 2003) as is the overall level of crop management (Hillocks, 2005). This is clearly a case where *Bt* cotton had been viewed as a silver bullet technology whereas improvements in many aspects of overall crop management (planting, nutrition, weed control, pest sampling, timeliness of operations) are required to raise yield potential and realise the full benefits from *Bt* technology (Hillocks, 2005; Hofs et al., 2006). Gouse et al. (2003, 2004) provide an excellent comparative assessment of the benefits of *Bt* cotton for both large irrigated farmers and smallholder dryland farmers. They note that benefits were widely shared by all farm types, and in terms of economic benefits the return to smallholders (69% of total benefit) were proportionately greater than those for large-scale dryland farms (45% of benefit), but less so than large irrigated farms (75%). As with China there is a clear and ongoing need for farmer education and advisory support structures if IPM systems based on *Bt* cotton varieties are to flourish.

11.2.1.6 Argentina

Bt cotton was first released in Argentina in 1998, but its adoption has been constrained in part by the fact that it was covered by patent protection. This situation is in contrast to herbicide tolerant soybeans which had been widely adopted. In 2003 *Bt* cotton comprised only 8% of the total cotton area (Trigo and Cap, 2003). Qaim and De Janvry (2005) provide an assessment of the economic and environmental effects of *Bt* cotton use in Argentina after about 7 years of commercial use of Cry1Ac expressing cotton varieties. Based on farm surveys, they estimate that insecticide use has been reduced by 50%, almost of all of which is of highly toxic chemicals. In addition they found that *Bt* cotton adopters realised significantly improved yields.

11.2.2 Maize

In contrast to cotton, the initial motivation for use of *Bt* genes in field maize was based less on replacement of currently used insecticides and more on the potential to manage a previously intractable pest – the European corn borer *Ostrinia nubilalis* (Lepidoptera: Crambidae) (Gianessi et al., 2002; Hellmich et al., chapter 5). GM maize is now grown over 35 million hectares in 13 countries, including 8 countries in Europe, and represents 24% of the global maize area (James, 2007). The majority of this area (80%) was varieties expressing *Bt* genes for borer control.

Maize is the largest acreage crop grown in the US (80 million acres in 1998). Prior to the release of *Bt* maize, it is estimated that only 4% of the Corn Belt's acreage was treated with insecticides (primarily carbofuran) for European corn borer control (Gianessi et al., 2002), although this percentage varied across different states. Approximately 681,000 kg of active ingredient were used to control *O. nubilalis* each year, but about 80% of the crop was estimated to suffer yield losses of 5–10% from *O. nubilalis*. Consequently this pest was the target for the first *Bt* maize varieties released in 1996 that expressed Cry1Ab. While significant quantities of insecticide are applied for corn rootworms (*Diabrotica* spp., Coleoptera: Chrysomelidae), no viable GM approach for this pest was available until 2003 when Cry3Bb1 expressing maize was released (USEPA, 2003). Since soil insecticides applied for control of *Diabrotica* spp. represent the single largest use of insecticides in the USA, there is a great potential for *Bt*-transgenic varieties to reduce the application of insecticides to maize (Rice, 2004). Hellmich et al. (chapter 5) provide a comprehensive overview of the current and future prospects of GM maize in IPM including coverage of pest complexes and challenges.

Since the release of *Bt* maize hybrids in 1996 the average volume of insecticide use on maize in the USA had fallen by about 12.5% (about 0.6 million kilograms a.i. each year) by 2005 (Brookes and Barfoot, 2006), with cumulative savings of 6.4 million kilograms of active ingredient of insecticides.

Bt sweet corn expressing Cry1Ab was introduced commercially in the USA in 1998 into an industry that is highly sensitive to damage to corn ears from Lepidopteran pests. Highly effective control of European corn borer (*O. nubilalis*), corn earworm (*H. zea*) and fall armyworm (*Spodoptera frugiperda*, Lepidoptera: Noctuidae) provides the potential for 70–90% reductions in insecticide requirement (Musser and Shelton, 2003). Musser et al. (2006) show how *Bt* sweet corn combined with the action of predators can provide a truly integrated pest management system with only one foliar insecticide required.

In Africa, *Bt* maize targeting stalk boring Lepidoptera, has also been adopted and widely planted by smallholders in South Africa where Gouse et al. (2006) reported yield increases of 30–60% across different regions. Only a small proportion of farmers had previously sprayed their maize crops; so insecticide reductions are not significant.

Since 1998 *Bt* maize has also been grown in small quantities in Spain and is now firmly established, although it comprises only about 6% of the total crop (Demont and Tollens, 2004). In Spain and other European countries about 20% of the crop was sprayed for stalk borers (*O. nubilalis* and *Sesamia nonagrioides* [Lepidoptera: Noctuidae]) and so the opportunity to demonstrate real insecticide savings is much less than with cotton. Nonetheless this study also demonstrates significant yield and economic benefits to growers of *Bt* maize.

11.2.3 Potatoes

GM insect resistant potatoes were also grown commercially in the USA between 1996 and 2000. Potatoes suffer damage from a variety of pests, which attack both the aerial portions of the plant and the tubers. Of these, the most damaging and difficult to control is the Colarado potato beetle, *Leptinotarsa decemlineata* (Coleoptera: Chrysomelidae). Historically this pest evolved resistance to a number of insecticide groups and the introduction of potato varieties expressing Cry3A proteins provided a critical new management tactic. After introduction in 1995, *Bt* potatoes (which also had virus resistance) had the fastest rate of adoption of any potato variety in the USA (Thornton, 2003), reaching a peak of 50,000 acres in 1999. Significant insecticide savings were achieved and a comprehensive resistance management strategy had also been introduced. Thornton (2003) notes that average insecticide savings across 20 locations ranged from two to four applications (or $41 to $80 per acre). Gianessi et al. (2002) indicated that in the main potato growing states of Idaho, Oregon and Washington, insecticide use could be reduced by 1.45 million pounds per year, with a net economic impact of over $58 million, if insect and virus resistant GM potatoes were planted in those states.

Unfortunately this technology was withdrawn in 2001 largely as a result of activism by GM opponents and specifically in response to the decision of some leading potato processors and fast food outlets to stop using GM potatoes (Grafius and Douches, chapter 7). While this decision related to perceived public concerns, the GM potatoes had passed all regulatory and food safety hurdles and provided the producer and processor with a lower cost, higher yielding and more consistent product. It also delivered significant reductions in insecticide use (Carpenter and Gianessi, 2001).

11.3 Consequences for IPM

11.3.1 Secondary Pests and Resurgence

A common expectation with *Bt* crops is that secondary pests, which are generally not affected by the specific *Bt* toxin, may become more abundant after release from suppression by the insecticides originally targetted at the main pest (Fitt, 2000; Wu et al., 2002). This expectation is particularly relevant to cotton where the main targets of *Bt* plants are lepidopteran pests (usually noctuids such as *H. armigera*) while a suite of sucking pests (mirid bugs, stink bugs, aphids, mites) are not affected by the toxin. In some cotton systems there is evidence for resurgence of secondary pests in *Bt* cotton. Naranjo et al. (chapter 6) note resurgences of mirids and stink bugs in the eastern USA and various plant bugs in South Africa and India. Wu and Guo (2005) also note increases in mirids on *Bt* cotton in China. These secondary pests were often suppressed previously by insecticide sprays applied for the main Lepidopteran pest and with those sprays minimised, the habitat provided by the *Bt* crop becomes more favourable. While this is clearly true, there is as yet little evidence in any country that secondary or minor pests (not affected by *Bt*) have emerged as *major* problems requiring significant increases in insecticide to the extent that the reduction in insecticide requirement from the use of *Bt* cotton has been nullified.

In both Australia and China (Mainland) no differences were found in abundance of sucking pests on unsprayed crops of conventional and *Bt* cotton (Fitt et al., 1994; Wu et al., 2002; Whitehouse et al., 2005), although some sucking pests (mirids, aphids) are now the most important pest in *Bt* cotton. This is not to say that changes in abundance of minor pests have not been seen. In the USA, stink bugs have emerged as a pest requiring more focused management across much of the southern cotton belt (Greene et al., 2001; Williams, 2006). In many instances the well documented increases in abundance of beneficial insects in *Bt* cotton crops undoubtedly assist with ongoing suppression of sucking pests (Wilson et al., 2004; Romeis et al., chapter 4).

Wu and Guo (2005) note that the diversity of insects found in Chinese cotton fields and observe that *Bt* protein is directly toxic to only a narrow spectrum of lepidopteran species. They reported field experiments, which indicate that the removal of key early season insecticides reduces flaring of the cotton aphid, presumably because beneficial insects are able to provide effective control. However, the same was not true for mirid bugs. Mirid density was higher on nonsprayed *Bt* cotton than on sprayed non-*Bt* cotton. This is hardly surprising and although they suggest that mirids may become key pests in the *Bt* cotton system, it is unclear whether this has actually occurred or whether farmers are over-reacting to pest presence which could be tolerated.

Men et al. (2003) showed an increase in the diversity of arthropod communities and pest subcommunities in *Bt* cotton fields in China, but the diversity of natural enemy subcommunities was decreased. Similar results were obtained by Liu et al. (2002), who noted that the nonpest and secondary pest arthropods in *Bt* cotton fields were important in the food chain of natural enemies and helped to enhance the overall stability of arthropod communities. Wu and Guo (2005) conclude that *Bt* cotton can increase the stability of arthropod diversity in cotton ecosystems, which may provide overall benefits for pest management. Nonetheless there are some key pests, e.g. mirid bugs, for which predators and parasites seem to be generally ineffective and which do require careful management in *Bt* crops.

11.3.2 Enhanced Beneficial Insects and Other IPM Components

Due to their specificity, extensive research has shown little effect of *Bt* proteins expressed in crop plants on non-target species, including non-lepidopterous pests, beneficial insects, and other canopy dwelling and soil dwelling species (Whitehouse et al., 2005; Marvier et al., 2007; Romeis et al., 2006, chapter 4; Hellmich et al., chapter 5; Naranjo et al., chapter 6; Grafius and Douches, chapter 7; Cohen et al., chapter 8).

Survival of predators and parasitoids (beneficial insects) in *Bt* cotton crops is demonstrably higher than in conventional sprayed cotton (Fitt and Wilson, 2002; Wu and Guo, 2003; Cattaneo et al., 2006). The same is true in *Bt* sweet corn (Musser and Shelton, 2003; Rose and Dively, 2007). These enhanced populations of beneficials should in turn provide control for some secondary pests, particularly those which may have previously been induced pests in the sprayed crop (e.g. mites and aphids) (Naranjo et al., chapter 6). This potential will be further enhanced as more efficacious transgenic varieties are released.

Using selective chemicals only when it is essential will be an important component for IPM systems based on *Bt* crops. These options are discussed fully in Wilson et al. (2004) for cotton. Insecticidal transgenes in combination with other HPR characters introduced through classical plant breeding may also enhance the stability of IPM systems. Sachs et al. (1996) showed synergism between Cry IAb protein and high gossypol levels and some efforts are underway to combine these traits in commercial cotton cultivars. In both China (Mainland) and Australia *Bt* genes have been incorporated with other HPR traits in cotton.

However, one failing of *Bt* cotton introductions in some developing countries has been that the background germplasm often lacks HPR traits which have been accumulated in local germplasm to deal with the suite of pests present locally. Clear examples here are the introduction of *Bt* genes in glabrous varieties from the US into China, South Africa and India where they prove highly susceptible to sucking pests (Men et al., 2005; Matthews and Tunstall, 2006). Locally derived varieties in these countries are often hirsute and better able to tolerate jassids, mirds and other sucking pests. Clearly *Bt* genes should be introgressed into locally derived germplasm as part of ongoing breeding programs designed to enhance yield potential, *Bt* efficacy and traits for tolerance to other pests and diseases.

Plant compensation is another IPM component (Sadras and Fitt, 1997). Sadras (1998) showed that the addition of the Cry1Ac gene does not alter the compensatory capacity of cotton varieties in the Australian environment. Cultural techniques are highly compatible with *Bt* cottons. Soil cultivation of cotton crop residues to destroy any surviving *Helicoverpa* pupae through winter (Fitt and Daly, 1990) is a mandatory requirement of the resistance management strategy for Australian *Bt* cotton (Fitt et al., 2008) and crop residue destruction is also recommended for *Bt* sweet corn (D. Andow, personal communication).

11.3.3 Constraints on IPM Adoption – Confidence and Knowledge

Broadly IPM can be defined as "the careful consideration of all available pest control techniques and subsequent integration of appropriate measures that discourage the development of pest populations and keep insecticides and other interventions to levels that are economically justified and reduce or minimize risks to human health and the environment. IPM emphasizes the growth of a healthy crop with the least possible disruption to agro-ecosystems and encourages natural pest control mechanisms." (FAO, 2002). In essence *Bt* crops that target one or a few key production pests simply package pest management technology into the seed and provide a tool to be integrated with other IPM compatible approaches (Fitt, 2000; Matthews and Tunstall, 2006).

Bt crops should not be perceived as "magic bullets" for pest control, although in some cases, such as *Bt* cotton grown by smallholders with little capacity to implement complex IPM approaches, they often will be (Hillocks, 2005). While *Bt* crops clearly provide an opportunity to address significant environmental concerns about cotton production, their real value is as a foundation to build IPM systems which incorporate a broad range of biological and cultural tactics (e.g., Fitt, 2000; Bambawale et al., 2004; Wilson et al., 2004).

Achieving truly integrated pest management requires knowledge of all of the other factors (agronomic, varietal, pests, climate) operating in the cropping system and the skill to provide timely management inputs. As discussed earlier, *Bt* crops will often be "living crops" with considerable activity of beneficial and secondary pest insects, in contrast to the "biological deserts" more typical of systems dominated by broad-spectrum insecticides. Achieving gains in IPM adoption with or without *Bt* crops requires considerable investment in skill development and confidence of farmers and their advisors to allow them to implement integrated solutions and tolerate biological activity in their crops. This is just as important in developed

(Wilson et al., 2004; Fitt et al., 2008) as in developing countries (van den Berg and Jiggins, 2007). Unfortunately in many parts of the developing world farmer capability varies widely and, despite decades of support in terms of IPM training programs for farmers and extension personnel, many farmers do not have the education or necessary supporting structures to provide the level of management input needed. As a result opportunities to achieve IPM may falter (Yang et al., 2005a; Matthews and Tunstall, 2006). As Van den Berg and Jiggins (2007) note "Complex pest management information....does not readily diffuse among farmers but has to be acquired through experiential learning." Ongoing investments in farmer education and advisory support structures will be essential for IPM systems based on *Bt* crop varieties to flourish.

11.3.4 Landscape Scale Effects

Cotton and maize, the major field crops in which *Bt* genes have been commercialised to date often tend to dominate agricultural landscapes where they are grown, both in terms of the extent of production and their impact on the dynamics of key pests. With large proportions of those crops now expressing *Bt* genes (e.g. 70–90% of cotton in the USA and Australia is *Bt* cotton, 60–70% of maize in the USA and Argentina is *Bt* maize), there is real potential for landscape scale impacts on the abundance of key pests particularly for those that are quite specific to these crops.

Storer et al. (chapter 10) explores in detail a couple of now well documented examples of regional suppression of key pests by *Bt* cotton. Pink Bollworm (*P. gossypiella*), a cotton specialist is a prime example. Carrière et al. (2003) describe the suppression of this species in Arizona through the widespread adoption of *Bt* cotton. They were able to identify this effect when *Bt* cotton reached 65% of the cotton landscape. Similarly Adamczyk and Hubbard (2006) speculate that a decline in abundance of *H. virescens* relative to *H. zea* in southern cotton regions of the USA may be due to widespread plantings of *Bt* cotton. Their analysis of trap catches over a 20 year period showed a significant change in the ten years after *Bt* cotton release in 1996 compared to the previous ten years. Similar changes in the seasonal dynamics of *H. armigera* and *Helicoverpa punctigera* (Lepidoptera: Noctuidae) in association with *Bt* cotton have been observed in Australia (G. Baker, G. Fitt, C. Tann, unpublished data), although the change and link to cotton is less clear. Finally Storer et al. (chapter 10) also document the same suppressive effect of *Bt* maize on European corn borer (*O. nubilalis*) populations.

The potential for resistance evolution to *Bt* crops is another area where landscape scale effects are manifest. Extensive use of *Bt* crops will impose selection pressure across significant components of pest populations and hence the pre-emptive management strategies proposed to avoid resistance must be applied in a coordinated way across whole landscapes. In some systems one aspect of these strategies is a requirement for structured refuge crops (Roush, 1998; Andow et al., 2008; Ferré et al., chapter 3) which can be unsprayed crops that will add vegetational diversity to the agricultural landscape (Fitt et al., 2008). A range of environmental and biodiversity benefits result from these types of structured refuges.

Finally we could expect to see landscape scale effects on the dynamics of non-target organisms, particularly the predators and parasites which utilise *Bt* crops. These effects will in most cases be positive with increased population sizes of generalist predators due to the reduced input of disruptive insecticides applied to the whole system as discussed in Section 11.3.2. However for parasitoids which are specialised on the pests targetted by the *Bt* crops, and the even smaller suite of specialised predators, there may well be significant reductions in abundance correlated with the reduced abundance of their hosts (Fitt et al., 1994; Whitehouse et al., 2005; Romeis et al., chapter 4; Storer et al., chapter 10). On balance, however, these landscape scale effects on beneficial species are positive and if anything serve to enhance opportunities for IPM.

11.3.5 GM Crops as Platforms for Integrated Change

IPM systems of the future will, of necessity, be more complex than the insecticide based systems previously, and in some cases currently, in place. They will thus require greater effort on the part of crop managers whether they be professional consultants or farmers themselves. *Bt* crops with activity against one or more key pests offer great scope to dramatically reduce insecticide dependence and to allow the integration of a wide range of IPM compatible tactics, particularly to maximise the role of naturally occurring beneficial organisms (predators and parasites) which are likely to rebound significantly in cropping systems where insecticide use is reduced. Provided they are supported with well-researched resistance management strategies, transgenic *Bt* crops should provide a foundation for sustainable IPM systems. The real challenge for researchers is to achieve this integration of approaches that rivals the predictability of conventional insecticides and to conduct the practical, field based ecological research needed to underpin management strategies (Way and van Emden, 2000). Musser et al. (2006) outline how this might be approached in *Bt* sweet corn, while several authors have discussed the opportunities for true IPM founded on *Bt* cotton (Wilson et al., 2004).

A significant challenge for researchers and funding agencies alike is to recognise that work on a range of IPM components must continue alongside the increasing focus on investment in biotechnology and modelling. Transgenic insecticidal crops will not be sustainable technologies alone; they must be supported with other approaches, which will require continued research.

11.4 Future Opportunities for GM Crops in IPM

The current generation of insecticidal GM crops all express proteins from *Bacillus thuringiensis* with varying levels of specificity. New *Bt* crops are rapidly being deployed and there seems little doubt that at some point conflicts between crops will arise, particularly in compromising resistance management. For the future it is likely that a broader range of insecticidal transgenes will be commercialised and provide additional options for pest management. These may include VIP3A genes (Llewellyn et al., 2007), engineered toxins (Soberón et al., 2007) applications of RNAi technologies (Baum et al., 2007). However, we should not underestimate the costs involved in breeding, development and registration of crop varieties expressing a broad range of insecticidal proteins (either in different varieties or pyramided). Maintaining inventories of numerous varieties is also expensive. Overall these costs associated with GM crops are akin to the substantial discovery and development costs for conventional synthetic insecticides.

Other chapters in this volume outline the future opportunities for GM crops (maize, cotton, rice, horticultural crops) to play a significant role in IPM systems where they can provide a foundation for development of sustainable and truly integrated approaches.

To date the main focus of *Bt* crops has been in the broadacre systems, cotton and maize, with the only horticultural crop, potatoes, having been withdrawn. However a number of vegetables are heavily sprayed against insect attack, and insect-resistant GM varieties of several species are at varying stages of development. *Bt* eggplant will likely be commercialised in India in 2008/09, while *Bt* brassicas and additional *Bt* field crops (e.g., chickpeas) are also close to market. Shelton et al. (chapter 9) discuss some of these examples.

This proliferation of *Bt* crops will introduce new challenges in landscape management if the benefits are to be sustained. These challenges relate to the risk of target pests evolving resistance to *Bt* genes and the management requirements needed to minimise this risk (see Ferré et al., chapter 3). Strategies for resistance management of *Bt* crops have been exhaustively explored with population genetic models and innovative methods to modify the selection environment imposed by *Bt* crops on the pest (Tabashnik et al., 2003, 2004; Gould, 1998; Roush, 1998). Although debate about the appropriateness of certain strategies continues, the overwhelming body of evidence supports the use of a refuge strategy combined with the highest possible efficacy of the *Bt* plants (Tabashnik et al., 2003).

Resistance management strategies become more challenging where the same insecticidal genes are deployed against the same pest across multiple crops. For example, in the USA the release of *Bt* maize was restricted in southern states where *Bt* cotton had already been deployed and required larger refuges. Likewise in Australia, *Bt* cotton is the first GM field crop to be commercialised. Future releases of other *Bt* crops may well propose to use the same *Bt* genes (e.g., *Bt* sorghum, *Bt* chickpeas, *Bt* maize) and target the same pests. All these crops are significant hosts for *H. armigera*, but the economic extent of damage is small relative to that suffered by cotton. Managing the deployment of *Bt* genes in future grain crops, when the same genes may already be widely in use in crops like cotton requires careful consideration and probably a regulatory protocol which can balance the pros and cons of different applications of transgenes for pest management (Fitt, 1997).

11.5 Conclusions

Bt crops have been deployed widely for over ten years and now cover about 42 million hectares in both developed and developing economies. The published evidence analysed here indicates clearly that *Bt* crops, particularly *Bt* cotton, have achieved considerable reductions in insecticide input, and variable but significant increases in yield. With appropriate management support to avoid resistance and to integrate a range of pest management tactics, they can provide sustainable components of IPM systems where biological, agronomic and climatic factors are collectively managed to achieve production outcomes. Environmental and economic benefits from *Bt* crops are evident in extensive, developed agricultural industries and in smallholder systems. However, economic performance is highly variable and seems dependent more on the market characteristics, support structures and culture of the systems in which *Bt* crops are deployed than on the *Bt* crops themselves. There is clear evidence of the ongoing need for eduction and extension support, particularly in smallholder systems, and for focussed research to provide practical management strategies that maximise longevity and benefits of *Bt* crop technology.

References

Adamczyk, J.J., and Hubbard, D., 2006. Changes in populations of *Heliothis virescens* (F.) (Lepidoptera: Noctuidae) and *Helicoverpa zea* (Boddie) (Lepidoptera: Noctuidae) in the Mississippi Delta from 1986 to 2005 as indicated by adult male pheromone traps. Journal of Cotton Science 10: 155–160.

Andow, D.A., Fitt, G.P., Grafius, E.J., Jackson, R.E., Radcliffe, E.B., Ragsdale, D.W., and Rossiter, L., 2008. Pesticide and transgenic plant resistance management in the field. In: Global Pesticide Resistance in Arthropods, M.E. Whalon, D. Mota-Sanchez and R.M. Hollingworth, eds. CABI, Wallingford, UK, pp. 118–144.

Bambawale, O.M., Singh, A., Sharma, O.P., Bhosle, B.B., Lavekar, R.C., Dhandapani, A., Kanwar, V., Tanwar, R.K., Rathod, K.S., Patange, N.R., and Pawar, V.M., 2004. Performance of Bt cotton (MECH-162) under integrated pest management in farmers' participatory field trial in Nanded district, Central India. Current Science 86: 1628–1633.

Baum, J.A., Bogaert, T., Clinton, W., Heck, G.R., Feldmann, P., Ilagan, O., Johnson, S., Plaetinck, G., Munyikwa, T., Pleau, M., Vaughn, T., and Roberts, J., 2007. Control of coleopteran insect pests through RNA interference. Nature Biotechnology 25: 1322–1326.

Bennett, R., Buthelezi, T.J., Ismael, Y., and Morse, S., 2003. Bt cotton, pesticides, labour and health: A case study of smallholder farmers in the Makhathini Flats, Republic of South Africa. Outlook in Agriculture 32: 123–128.

Brookes, G., and Barfoot, P., 2006. GM Crops: The First Ten Years – Global Socio-Economic and Environmental Impacts. ISAAA Brief No. 36, International Service for the Acquisition of Agri Biotech Applications, Ithaca, NY, USA.

Caprio, M.A., 1994. *Bacillus thuringiensis* gene deployment and resistance management in single and multi-tactic environments. Biocontrol Science and Technology 4: 487–498.

Carpenter, J.E., and Gianessi, L.P., 2001. Agricultural Biotechnology: Updated Benefit Estimates. National Center for Food and Agricultural Policy (NCFAP), Washington, DC, USA. http://www.ncfap.org/reports/biotech/updatedbenefits.pdf (accessed 25 January 2008).

Carrière, Y., Ellers-Kirk, C., Sisterson, M., Antilla, L., Whitlow, M., Dennehy, T.J., and Tabashnik, B.E., 2003. Long-term regional suppression of pink bollworm by *Bacillus thuringiensis* cotton. Proceedings of the National Academy of Sciences of the USA 100: 1519–1523.

Cattaneo, M.G., Yafuso, C., Schmidt, C., Huang, C., Rahman, M., Olson, C., Ellers-Kirk, C., Orr, B.J., Marsh, S.E., Antilla, L., Dutilleul, P., and Carriere, Y., 2006. Farm-scale evaluation of the impacts of transgenic cotton on biodiversity, pesticide use, and yield. Proceedings of the National Academy of Sciences of the USA 103: 7571–7576.

Daly, J.C., and Fitt, G.P., 1998. Efficacy of *Bt* cotton plants in Australia - What is going on? In: New Frontiers in Cotton Research - Proceedings World Cotton Research Conference – 2, Athens, 6–12 September 1998, F. Gillham, ed., P. Petridis and Thessaloniki, Greece, pp. 675–678.

Demont, M., and Tollens, E., 2004. First impact of biotechnology in Europe: Bt maize adoption in Spain. Annals of Applied Biology 145: 197–207.

FAO, 2002. Code of Conduct on Pesticides' Definition. 2002: Integrated Pest Management (IPM).

Fernandez-Cornejo, J., and McBride, W., 2000. Genetically Engineered Crops for Pest Management in U.S. Agriculture. USDA, Agricultural Economics Report No. (AER786), 28 pp., May 2000. http://www.ers.usda.gov/publications/AER786/ (accessed 25 January 2008).

Fitt, G.P., 1997. Risks, deployment and integration of insect resistant crops expressing genes from *Bacillus thuringiensis*. In: Commercialisation of Transgenic Crops: Risk, Benefit and Trade Considerations, G.D. McLean, P.M. Waterhouse, G. Evans and M.J. Gibbs, eds., Cooperative Research Centre for Plant Science and Bureau of Resource Sciences, Canberra, Australia, pp. 273–284.

Fitt, G.P., 2000. An Australian approach to IPM in cotton: Integrating new technologies to minimise insecticide dependence. Crop Protection 19: 793–800.

Fitt, G.P., 2003. Deployment and impact of transgenic Bt cottons in Australia. In: The Economic and Environmental Impacts of Agbiotech: A Global Perspective, N.G. Kalaitzandonakes, ed., Kluwer/Plenum, New York, USA, pp. 141–164.

Fitt, G.P., and Daly, J.C., 1990. Abundance of overwintering pupae and the spring generation of *Helicoverpa* spp. (Lepidoptera: Noctuidae) in northern New South Wales, Australia: Consequences for pest management. Journal of Economic Entomology 83: 1827–1836.

Fitt, G.P., and Wilson, L.J., 2000. Genetic engineering in IPM: Bt cotton. In: Emerging Technologies in Integrated Pest Management: Concepts, Research and Implementation, G.G. Kennedy and T.B. Sutton, eds., APS Press, St. Paul, MN, USA, pp. 108–125.

Fitt, G.P., and Wilson, L.J., 2002. Non-target effects of Bt-cotton: A case study from Australia. In: Biotechnology of *Bacillus thuringiensis* and Its Environmental Impact: Proceedings of the 4th Pacific Rim Conference, R.J. Akhurst, C.E. Beard and P.A. Hughes, eds., CSIRO, Canberra, Australia, pp. 175–182.

Fitt, G.P., Mares, C.L., and Llewellyn, D.J., 1994. Field evaluation and potential ecological impact of transgenic cottons in Australia. Biocontrol Science and Technology 4: 535–548.

Fitt, G.P., Daly, J.C., Mares, C.L., and Olsen, K., 1998. Changing efficacy of transgenic Bt cotton - patterns and consequences. In: Pest Management - Future Challenges, Proceedings of the 6th Australian Applied Entomology Conference, Brisbane, M.P. Zalucki, R.A.I. Drew and G.G. White, eds., University of Qld Press, Brisbane, Australia, pp. 189–196.

Fitt, G.P., Wilson, L.J., Kelly, D., and Mensah, R., 2008. Advances with integrated pest management as a component of sustainable agriculture: The case of the Australian cotton industry. In: Integrated Pest Management: Innovation, Dissemination and Impact, R. Peshin, ed., Springer, New York, USA (in press).

Gianessi, L.P., Silvers, C.S., Sankula, S., and Carpenter, J.E., 2002. Plant Biotechnology: Current and Potential Impact for Improving Pest Management in U.S. Agriculture: An Analysis of 40 Case Studies. National Center for Food and Agricultural Policy (NCFAP), Washington, DC, USA. http://www.ncfap.org/40CaseStudies/MainReport.pdf (accessed 25 January 2008).

Gould, F., 1998. Sustainability of transgenic insecticidal cultivars: Integrating pest genetics and ecology. Annual Review of Entomology 43: 701–726.

Gouse, M., Kirsten, J.F., and Jenkins L., 2003. Bt cotton in South Africa: Adoption and the impact on farm incomes amongst small-scale and large scale farmers. Agrekon 42: 15–28.
Gouse, M., Pray, C., and Schimmelpfennig, D., 2004. The distribution of benefits from Bt cotton adoption in South Africa. AgBioForum 7(4): 187–194.
Gouse, M., Pray, C., Schimmelpfennig, D., and Kirsten, J., 2006. Three seasons of subsistence insect-resistant maize in South Africa: Have smallholders benefited? AgBioForum 9(1): 15–22.
Greene, J.K., Turnipseed, S.G., Sullivan, M.J., and May, O.L., 2001. Treatment thresholds for stink bugs in cotton. Journal of Economic Entomology 94: 403–409.
Hillocks, R.J., 2005. Is there a role for Bt cotton in IPM for smallholders in Africa? International Journal of Pest Management 51: 131–141.
Hofs, J., Fok, M., and Vaissayre, M., 2006. Impact of Bt cotton adoption on pesticide use by smallholders: A 2-year survey in Makhatini Flats (South Africa). Crop Protection 25: 984–988.
Hossain, F., Pray, C.E., Lu, Y., Huang, J., Fan, C., and Hu, R., 2004. Genetically modified cotton and farmers' health in China. International Journal of Occupational and Environmental Health 10: 296–303.
Huang, J., Rozelle, S., Pray, C., and Wang, Q., 2002. Plant biotechnology in China. Science 295: 674–677.
Ismael, Y., Bennett, R., and Morse, S., 2002. Do small-scale Bt cotton adopters in South Africa gain an economic advantage? 6th International ICABR Conference, 11–14 July 2002, Ravello, Italy, pp. 1–16.
James, C., 2007. Global Status of Commercialized Biotech/GM Crops: 2007. ISAAA Brief No. 37, International Service for the Acquisition of Agri-Biotech Applications, Ithaca, NY, USA.
Kirsten, J., Gouse, M., and Jenkins, L., 2002. Bt cotton in South Africa: Adoption and the impact on farm incomes amongst small-scale and large-scale farmers. Agricultural biotechnologies: New avenues for production, consumption and technology transfer. 6th International ICABR Conference, 11–14 July 2002, Ravello, Italy, pp. 1–11.
Knox, O.G., Constable, G.A., Pyke, B., and Gupta, V.V.S.R., 2006. Environmental impact of conventional and Bt insecticidal cotton expressing one and two Cry genes in Australia. Australian Journal of Agricultural Research 57: 501–509.
Kranthi, K.R., Naidu, S., Dhawad, C.S., Tatwawadi, A., Mate, K., Patil, E., Bharose, A.A., Behere, G.T., Wadaskar, R.M., and Kranthi, S., 2005. Temporal and intra-plant variability of Cry1Ac expression in Bt cotton and its influence on the survival of the cotton bollworm, *Helicoverpa armigera* (Hübner) (Noctuidae: Lepidoptera). Current Science 89: 291–298.
Liu, W., Wan, F., and Guo, J., 2002. Structure and seasonal dynamics of arthropods in transgenic Bt cotton field. Acta Ecologica Sinica 22: 729–735.
Llewellyn, D., Mares, C., and Fitt, G.P., 2007. Field performance and seasonal changes in the efficacy against *Helicoverpa armigera* (Hübner) of transgenic cotton (VipCot) expressing the insecticidal protein VIP3A. Agricultural and Forest Entomology 9: 93–101.
Lu, Y., Pray, C.E., and Hossain, F., 2002. An econometric analysis of the reduction in pesticide poisoning due to Bt cotton use in China. 6th International ICABR Conference, 11–14 July 2002, Ravello, Italy, pp. 1–21.
Marvier, M., McCreedy, C., Regetz, J., and Kareiva, P., 2007. A meta-analysis of effects of Bt cotton and maize on nontarget invertebrates. Science 316: 1475–1477.
Matthews, G.A, and Tunstall, J.P., 2006. Smallholder cotton production in sub-Saharan Africa: An assessment of the way forward. International Journal of Pest Management 52: 149–153.
Men, X., Ge, F., Liu, X., and Yardim, E.N, 2003. Diversity of arthropod communities in transgenic Bt cotton and non-transgenic cotton agroecosystems. Environmental Entomology 32: 270–275.
Men, X., Ge, F., Edwards, C.A., and Yardim, E.N., 2005. The influence of pesticide applications on *Helicoverpa armigera* Hubner and sucking pests in transgenic Bt cotton and non-transgenic cotton in China. Crop Protection 24: 319–324.
Musser, F.R., and Shelton, A.M., 2003. *Bt* sweet corn and selective insecticides: Impacts on pests and predators. Journal of Economic Entomology 96: 71–78.

Musser, F.R., Nyrop, J.P., and Shelton, A.M., 2006. Integrating biological and chemical contols in decision making: European corn borer (Lepidoptera: Crambidae) control in sweet corn as an example. Journal of Economic Entomology 55: 1538–1549.

Olsen, K.M., and Daly, J.C., 2000. Plant-toxin interactions in transgenic Bt cotton and their effect on mortality of *Helicoverpa armigera*. Journal of Economic Entomology 93: 1293–1299.

Olsen, K.M., Daly, J.C., Holt, H.E., and Finnegan, E.J., 2005. Season-long variation in expression of Cry1Ac gene and efficacy of *Bacillus thuringiensis* toxin in transgenic cotton against *Helicoverpa armigera* (Lepidoptera: Noctuidae). Journal of Economic Entomology 98: 1007–1017.

Pemsl, D., Waibel, H., and Orphal, J., 2004. A methodology to assess the profitability of Bt-cotton: Case study results from the state of Karnataka, India. Crop Protection 23: 1249–1257.

Pemsl, D., Gutierrez, A., and Waibel, H., 2008. The economics of biotechnology under ecosystem disruption. Ecological Economics 66: 177–183.

Perlak, F.J., Oppenhuizen, M., Gustafson, K., Voth, R., Sivasupramaniam, S., and Heering, D., 2001. Development and commercial use of Bollgard cotton in the USA: Early promises versus today's reality. The Plant Journal 27: 489–501.

Pray, C.E., and Huang, J., 2003. The impact of Bt-cotton in China. In: The Economic and Environmental Impacts of Agbiotech: A Global Perspective, N. Kalaitzandonakes, ed., Kluwer/Plenum, New York, USA, pp. 223–242.

Pray, C.E., Ma, D.M., Huang, J.K., and Qiao F.B., 2001. Impact of Bt cotton in China. World Development 29: 813–825.

Pray, C.E., Huang, J., Hu, R., and Rozelle, S., 2002. Five years of Bt cotton in China - the benefits continue. The Plant Journal 31: 423–430.

Pyke, B.A., 2008. The impact of high adoption of Bollgard®II cotton on pest management in Australia. In: Proceedings of the World Cotton Research Conference 4, Lubbock, TX, USA, August 9–14, 2007.

Qaim, M., 2003. Bt cotton in India: Field trial results and economic projections. World Development 31: 2115–2127.

Qaim, M., and de Janvry, A., 2005. Bt cotton and pesticide use in Argentina: Economic and environmental effects. Environment and Development Economics 10: 179–200.

Qaim, M., and Zilberman, D., 2003. Yield effects of genetically modified crops in developing countries. Science 299: 900–902.

Qaim, M., Subramanian, A., Naik, G., and Zilberman, D., 2006. Adoption of Bt cotton and impact variability: Insights from India. Review of Agricultural Economics 28: 48–58.

Raney, T., 2006. Economic impact of transgenic crops in developing countries. Current Opinion in Biotechnology 17: 1–5.

Rice, M.E., 2004. Transgenic rootworm corn: Assessing potential agronomic, economic, and environmental benefits. Plant Health Progress doi:10.1094/PHP-2004-0301-01 RV. http://www.plantmanagementnetwork.org/sub/php/review/2004/rootworm/Rice.pdf (accessed 1 December 2007).

Romeis, J., Meissle, M., and Bigler, F., 2006. Transgenic crops expressing *Bacillus thuringiensis* toxins and biological control. Nature Biotechnology 24: 63–71.

Rose, R., and Dively, G., 2007. Effects of insecticide-treated and lepidopteran-active *Bt* transgenic Sweet corn on the abundance and diversity of arthropods. Environmental Entomology 36: 1254–1268.

Roush, R.T., 1998. Two toxin strategies for management of insecticidal transgenic crops: Can pyramiding succeed where pesticide mixtures have not? Philosophical Transactions of the Royal Society London B 353: 1777–1786.

Sachs, E.S., Benedict, J.H., Taylor, J.F., Stelly, D.M., Davis, S.K., and Altman, D.W., 1996. Pyramiding CryIA(b) insecticidal protein and terpenoids in cotton to resist tobacco budworm (Lepidoptera: Noctuidae). Environmental Entomology 25: 1257–1266.

Sadras, V.O., 1998. Herbivory tolerance of cottons expressing insecticidal proteins from *Bacillus thuringiensis*: Response to damage caused by *Helicoverpa* spp. and to manual bud removal. Field Crops Research 56: 287–299.

Sadras, V.O., and Fitt, G.P., 1997. Resistance to insect herbivory of cotton lines: quantification of recovery capacity after damage. Field Crops Research 52: 129–136.

Soberón, M., Pardo-López, L., López, I., Gómez, I., Tabashnik, B.E., and Bravo, A., 2007. Engineering modified Bt toxins to counter insect resistance. Science 318: 1640–1642.

Tabashnik, B.E., 1994. Evolution of resistance to *Bacillus thuringiensis*. Annual Review of Entomology 39: 47–79.

Tabashnik, B.E., Carrière, Y., Dennehy, T.J., Morin, S., Sisterson, M., Roush, R.T., Shelton, A.M., and Zhao, J.Z., 2003. Insect resistance to transgenic Bt crops: Lessons from the laboratory and field. Journal of Economic Entomology 96: 1031–1038.

Tabashnik, B.E., Gould F., and Carrière, Y., 2004. Delaying evolution of insect resistance to transgenic crops by decreasing dominance and heritability. Journal of Evolutionary Biology 17: 904–912.

Thirtle, C., Beyers, L., Ismael, Y., and Piesse, J., 2003. Can GM-technologies help the poor? The impact of Bt cotton in Makhathini Flats, KwaZulu-Natal. World Development 31: 717–732.

Thornton, M., 2003. The Rise and Fall of NewLeaf Potatoes. Biotechnology: Science and Society at a Crossroad, National Agricultural Biotechnology Council (NABC) Report 15: 235–243. http://nabc.cals.cornell.edu/pubs/pubs_reports.html#nabc15 (accessed 22 January 2008).

Traxler, G., Godoy-Avila, S., Falck-Zepeda, J., and Espinosa-Arellano, J., 2003. Transgenic cotton in Mexico: Economic and environmental impacts. In: The Economic and Environmental Impacts of Agbiotech: A Global Perspective, N.G. Kalaitzandonakes, ed., Kluwer/Plenum, New York, pp. 183–202.

Trigo, E., and Cap, E., 2003. The impact of the introduction of transgenic crops in Argentinean agriculture. AgBioForum 6: 87–94.

USEPA (United States Environmental Protection Agency), 2003. *Bacillus thuringiensis* Cry3Bb1 protein and the genetic material necessary for its production (vector ZMIR13L) in event MON863 corn. http://www.epa.gov/pesticides/biopesticides/ingredients/fr_notices/frnotices_006430-006484.htm (accessed 25 January 2008).

Van Den Berg, H., and Jiggins J., 2007. Investing in farmers - the impacts of farmer field schools in relation to integrated pest management. World Development 35: 663–686.

Way, M.J., and van Emden, H.F., 2000. Integrated pest management in practice - pathways towards successful application. Crop Protection 19: 81–103.

Whitehouse, M.E.A., Wilson, L.J., and Fitt, G.P., 2005. A comparison of Arthropod communities in transgenic Bt and conventional cotton in Australia. Environmental Entomology 34: 1224–1241.

Williams, M.R., 2003. Cotton insect loss estimates – 2002. Proceedings Beltwide Cotton Conferences, National Cotton Council, Memphis, TN, USA, pp. 1217–1273.

Williams, M.R., 2006. Cotton insect losses 2005. In: Proceedings of the Beltwide Cotton Conferences, National Cotton Council, Memphis, TN, USA, pp. 1151–1204.

Wilson, L.J., Mensah, R.K., and Fitt, G.P., 2004. Implementing integrated pest management in Australian cotton. In: Insect Pest Management: Field and Protected Crops, A.R. Horowitz and I. Ishaaya, eds., Springer, Berlin/Heidelberg/New York, pp. 97–118.

Wossink, A., and Denaux, Z.S., 2006. Environmental and cost efficiency of pesticide use in transgenic and conventional cotton production. Agricultural Systems 90: 312–328.

Wu, K., 2007. Monitoring and management strategy for *Helicoverpa armigera* resistance to Bt cotton in China. Journal of Invertebrate Pathology 95: 220–223.

Wu, K.M., and Guo, Y.Y., 2003. Influences of *Bacillus thuringiensis* Berliner cotton planting on population dynamics of the cotton aphid, *Aphis gossypii* Glover, in northern China. Environmental Entomology 32: 312–318.

Wu, K.M., and Guo, Y.Y., 2005. The evolution of cotton pest management practices in China. Annual Review of Entomology 50: 31–52.

Wu, K., Li, W., Fena, H., and Guo, Y., 2002. Seasonal abundance of the mirids, *Lygus lucorum* and *Adelphocoris* spp. (Hemiptera: Miridae) on Bt cotton in northern China. Crop Protection 21: 997–1002.

Xu, N., and Fok, M., 2008. Multiple-factor adoption of GM Cotton in China: Influence of conventional technology development and rural change in Jiangsu Province. In: Proceedings of the World Cotton Research Conference 4, Lubbock, TX, USA, 9–14 August 2007.

Yang, P., Li, K., Shi, S., Xia J., Guo, R., Li, S., and Wang, L., 2005a. Impacts of transgenic Bt cotton and integrated pest management education on smallholder cotton farmers. International Journal of Pest Management 51: 231–244.

Yang, P., Iles, M., Yan, S., and Jolliffe, F., 2005b. Farmers' knowledge, perceptions and practices in transgenic Bt cotton in small producer systems in Northern China. Crop Protection 24: 229–239.

Yousouf, I., Bennett, R., and Morse, S., 2001. Farm level impact of Bt cotton in South Africa. Biotechnology and Development Monitor 48: 15–19.

Chapter 12
Economic and Social Considerations in the Adoption of *Bt* Crops

Matin Qaim[1,*], Carl E. Pray[2], and David Zilberman[3]

Abstract This chapter provides an overview of the economics of *Bt* crop adoption, with a special emphasis on *Bt* cotton. On average, the technology reduces insecticide applications and pest-related crop losses. In spite of higher seed prices, farmers realize substantial gains in cotton incomes. In India and China alone, *Bt* cotton produces annual welfare gains of several hundred million US dollars, with farmers being the main beneficiaries. In the USA, stronger IPR protection leads to larger benefit shares for biotechnology companies, but agricultural producers and consumers profit as well. More preliminary studies for maize, rice, and eggplant suggest that similar results can also be expected for other *Bt* crops. Furthermore, health benefits and aspects of consumer acceptance are analyzed. Finally, several institutional and regulatory issues are discussed, as these might have important ramifications for technology access, benefit distribution, seed market structures, and biodiversity.

12.1 Introduction

Bt (*Bacillus thuringiensis*) crops, which are resistant to different lepidopteran and coleopteran insect pests, were among the first genetically modified (GM) crops to be commercialized in the mid-1990s. The first country to grow *Bt* crops on a larger scale was the USA, where *Bt* maize, *Bt* cotton, and *Bt* potato were commercially approved in 1995. In Canada, *Bt* maize and *Bt* potatoes were approved in 1996, and in 1997 several other countries started to grow *Bt* cotton, including Australia, China, Mexico, and South Africa (James, 1997). In 2007, *Bt* crops were grown in 22 countries on a total of 42.1 million hectares, accounting for 37% of the global area under GM crops (James, 2007). The major countries growing *Bt* crops are

[1] Georg-August-University of Goettingen, Goettingen, Germany

[2] Rutgers University, New Brunswick, NJ, USA

[3] University of California, Berkeley, CA, USA

* To whom correspondence should be addressed. E-mail: mqaim@uni-goettingen.de

J. Romeis, A.M. Shelton, G.G. Kennedy (eds.), *Integration of Insect-Resistant Genetically Modified Crops within IPM Programs.*

Table 12.1 Major countries growing *Bt* crops (2007) (USDA, 2007; James, 2007, and communication with industry representatives)

Country	Crop	Estimated area (million ha)	Percent of total area under the particular crop
USA	*Bt* maize	18.6	49
	Bt cotton	3.1	72
India	*Bt* cotton	6.2	66
China	*Bt* cotton	3.8	69
Argentina	*Bt* maize	2.5	64
	Bt cotton	0.2	49
Canada	*Bt* maize	0.7	54
South Africa	*Bt* maize	1.2	43
	Bt cotton	<0.1	90
Australia	*Bt* cotton	0.1	90
Philippines	*Bt* maize	0.1	5
Spain	*Bt* maize	<0.1	21
Mexico	*Bt* cotton	<0.1	52

shown in Table 12.1. At present, *Bt* maize and *Bt* cotton are the only *Bt* crops with significant area shares.[1] Other *Bt* crops are at advanced stages of field testing, but have not yet been fully approved. Examples include *Bt* rice in China and Iran, and *Bt* vegetables in India and other countries of Asia (e.g., Huang et al., 2005; Krishna and Qaim, 2007; Cohen et al., chapter 8; Shelton et al., chapter 9).

Farmers only adopt new crop technologies when they can realize personal benefits in terms of productivity gains or other advantages. The fast process of technology diffusion over the last decade therefore suggests that there are significant benefits associated with *Bt* crops. Indeed, there are numerous studies showing that *Bt* crops allow sizeable insecticide savings and reductions in pest-related crop losses (e.g., Naseem and Pray, 2004; Qaim and Matuschke, 2005; Fernandez-Cornejo and Caswell, 2006). Nonetheless, there are still debates about the suitability of *Bt* crops for peasant farming systems, especially also regarding the economic and social implications (GRAIN, 2004). This chapter reviews the evolving literature on socioeconomic aspects of *Bt* crops from an international perspective. Since many of the controversies relate to issues in developing countries, examples from those countries will receive particular attention.

The chapter is organized as follows. The next section discusses agronomic and economic effects of *Bt* crops at the farm level as well as aggregate impacts on economic surplus. Special attention is given to the experience with *Bt* cotton in India, but also results from other countries are presented. In subsequent sections, the focus is broadened to include health effects of *Bt* crops (section 12.3), consumer acceptance/marketing issues (section 12.4), and institutional aspects such as biosafety

[1] *Bt* potatoes were taken from the North-American market in 2001, due to marketing problems (Kaniewski and Thomas, 2004; Grafius and Douches, chapter 7).

regulations, intellectual property rights (IPRs), and seed markets (section 12.5). The last section concludes.

12.2 Agronomic and Economic Effects of *Bt* Crops

12.2.1 Insecticide and Yield Effects

Bt crops produce Cry proteins derived from the soil bacterium *Bacillus thuringiensis* that are toxic to larvae of some lepidopteran and coleopteran insects. Therefore, *Bt* is a pest control agent that can be used as a substitute for traditional chemical insecticides. Following Lichtenberg and Zilberman (1986) and Zilberman et al. (2004), this can be expressed in a damage control framework:

$$Y = F(x)\,[1 - D\,(z, Bt; N)] \tag{1}$$

where Y is the effective crop yield, and $F(\cdot)$ is potential yield without insect damage, which depends on variable inputs, x. $D(\cdot)$ is the damage function determining the fraction of potential output being lost to insect pests; it can take values in the 0–1 interval. Crop losses depend on exogenous pest pressure, N, and they can be reduced through the application of chemical insecticides, z, and/or the use of *Bt* technology. If pest pressure is high and farmers use a lot of chemical insecticides in the conventional crop, *Bt* adoption should lead to substantial insecticide reductions.

However, *Bt* technology can also impact effective crop yields. While the *Bt* gene does not affect potential yield, $F(\cdot)$, it can lead to a reduction in crop losses, $D(\cdot)$, when there is previously uncontrolled pest damage, thus leading to a higher Y. Obviously, insecticide reduction and yield effects are closely related: farmers who use little amounts of insecticides in their conventional crop in spite of high pest pressure will realize a sizeable yield effect through *Bt* adoption, while the insecticide reduction effect will dominate in situations where farmers initially use higher amounts of chemical inputs. These linkages are visualized in Fig. 12.1 based on field trial data with *Bt* cotton in India. Figure 12.1 also demonstrates that *Bt* does not completely eliminate the need for insecticide sprays, as some crop damage still occurs when the technology is used. The reason is that the *Bt* toxin is very specific to certain pest species, while other insect pests, especially sucking pests, remain unaffected. Moreover, since *Bt* toxins usually do not cause 100% mortality, and toxin expression declines in aging cotton plants, insecticide sprays against *Bt* target pests are sometimes necessary when there is heavy infestation (Naranjo et al., chapter 6).

What do the agronomic impacts of *Bt* crops look like under practical farmer conditions? Since *Bt* cotton has been grown commercially in many countries and over several years, there are more impact studies available for this particular technology than for other GM crops. Table 12.2 confirms that both insecticide-reducing and yield-increasing effects can be observed internationally. Obviously, yield effects of *Bt* cotton are highest in Argentina and India. For Argentina, the explanation is simple:

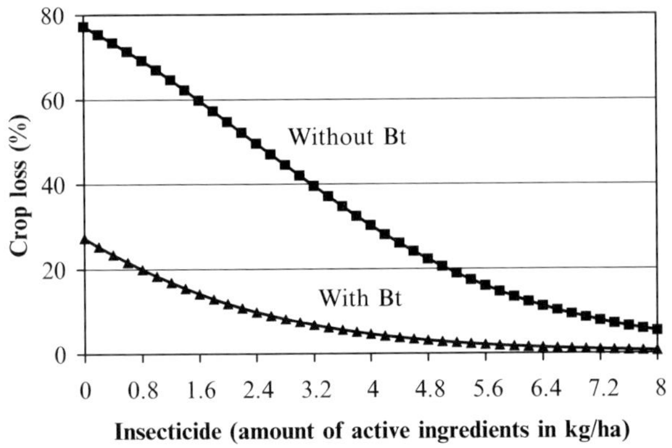

Fig. 12.1 Relationship between insecticide use and cotton crop losses with and without *Bt* in India (Qaim and Zilberman, 2003)

Table 12.2 Average agronomic effects of *Bt* crops

Country	Insecticide reduction (%)	Increase in effective yield (%)
	Bt cotton	
Argentina[a]	47	33
China[b]	65	24
India[c]	50	34
Mexico[d]	77	9
South Africa[e]	33	22
USA[f]	36	10
	Bt maize	
Argentina[g]	0	9
South Africa[h]	10	11
Spain[i]	63	5
USA[g, j]	8	5

[a] Qaim and de Janvry (2003, 2005); [b] Pray et al. (2002); [c] Qaim et al. (2006); [d] Traxler et al. (2003); [e] Thirtle et al. (2003); [f] Carpenter et al. (2002) and Falck-Zepeda et al. (2000); [g] Brookes and Barfoot (2005); [h] Gouse et al. (2006); [i] Gómez-Barbero and Rodríguez-Cerezo (2006); [j] Fernandez-Cornejo and Li (2005)

conventional cotton farmers under-use chemical insecticides – on average only about 2.5 kg are applied per season and hectare – so that insect pests are not effectively controlled (Qaim and de Janvry, 2005). In India, however, with an average of 10 kg per hectare, insecticide use in conventional cotton is much higher than in Argentina (Qaim et al., 2006). This suggests that there are also factors other than insecticide quantity influencing damage control in conventional cotton and thus the yield effects of *Bt* technology. Among others, these factors include insecticide quality, insecticide resistance, and the correct choice of products and timing of sprays.

For *Bt* maize, similar effects are observable as for *Bt* cotton, albeit generally at a lower magnitude (Table 12.2). Except for Spain, where the percentage reduction in insecticide use is large, the more important result of *Bt* maize is a slight increase in effective yields. In the USA, for instance, *Bt* maize is mainly used against the European corn borer (*Ostrinia nubilalis*; Lepidoptera: Crambidae), which is often not controlled by chemical means (Carpenter et al., 2002).[2] On average, *Bt* technology leads to 5% higher maize yields in the USA, although yield effects are bigger in years with high pest pressure (Fernandez-Cornejo and Caswell, 2006). In Argentina and South Africa, mean yield effects are somewhat higher, because there is more severe pest pressure than in temperate climates, and *Bt* technology is also used to control different stemborers (*Busseola fusca*; Lepidoptera: Noctuidae; and *Chilo partellus*; Lepidoptera: Pyralidae). The average yield gain of 11% in South Africa refers to large commercial farms. These farms have been growing yellow *Bt* maize hybrids for several years. Gouse et al. (2006) also analyzed on-farm trials that were carried out with 175 South African smallholder farmers and white *Bt* maize hybrids; they found average yield gains of 32% on *Bt* plots. In the Philippines, yield advantages of *Bt* maize were even reported to be in a magnitude of 40–60% (Gonzales, 2002).

Apart from *Bt* cotton and *Bt* maize, other *Bt* crops have not yet been adopted on a larger scale, although this might potentially change in the near future. *Bt* rice, which is resistant to rice stemborers (*Chilo suppressalis* and *Scirpophaga incertulas*; Lepidoptera: Pyralidae), has been tested in large-scale farmer managed trials in China (see Cohen et al., chapter 8 for a detailed discussion of *Bt* rice). The results indicate that *Bt* rice varieties only have a relatively small yield advantage over conventional varieties, while mean insecticide reductions are huge: on average, farmers use 21 kg of insecticides in conventional rice, an amount which could be reduced by 90% through *Bt* technology (Huang et al., 2005). *Bt* eggplant, which is resistant to the shoot and fruit borer (*Leucinodes orbonalis*; Lepidoptera: Pyralidae), is near to commercialization in India (Shelton et al., chapter 9). In on-station field trials that were carried out in different states of the country, the technology allowed insecticide reductions of 85%, while more than doubling effective yields (Krishna and Qaim, 2007). The exact results from such trials might not be replicable under all conditions, but they nevertheless suggest a large potential of *Bt* technology across different crop species.

While some of the available studies on the agronomic impacts of *Bt* crops are based on observations over several years, the long-term effects are not yet fully understood. In the first years of *Bt* crop deployment it was predicted that insect populations would soon develop *Bt* resistance, which would undermine the technology's effectiveness and lead to declining insecticide reductions, or even insecticide increases, over time. However, until now *Bt* resistance development has not

---

[2] More recently, a different Bt maize technology has been commercialized in the USA to also control the corn rootworm complex (*Diabrotica* spp.), against which significant amounts of chemical insecticides are used in conventional agriculture. However, representative studies on the impacts of this new Bt maize technology under farmer conditions are not yet available.

been observed under field conditions (but see Matten et al., chapter 2 for a recent putative example), which might partly be due to successful resistance management strategies, including the planting of non-*Bt* refuges (Bates et al., 2005; Ferré et al., chapter 3). But even in countries like China, where no deliberate resistance management strategy is implemented, *Bt* resistance has not yet been reported. Yet, there are also other factors that can lead to changes in *Bt* effects over time. In China, for instance, insecticide applications somewhat increased again after several years of *Bt* cotton use, in spite of the absence of *Bt* resistance. Wang et al. (2006) attributed this to secondary pests, which might have become more important through the *Bt*-induced reduction in broad-spectrum insecticides. Their analysis, however, was based on only one year of observations with increased insecticide applications and was disputed by others (e.g., Hu et al., 2006). Therefore, further studies should be carried out, analyzing whether secondary pests are really eroding the benefits of *Bt* cotton technology in China, or whether insecticide increases are only a temporary phenomenon driven by peculiar conditions in one particular year.

12.2.2 Seed Prices and Gross Margin Effects

Since most *Bt* crops available to date have been commercialized by the private sector, a technology fee is charged. In some countries, like the USA, the technology fee is displayed separately, while in many other countries, it is directly included in the seed price. In any case, the fee is associated with seed sales, so that seed costs for *Bt* adopting farmers increase. Table 12.3 shows average seed cost increases for *Bt* cotton and *Bt* maize in several countries. The technology fee (or seed price

Table 12.3 *Bt* seed cost increases and gross margin gains (US$/ha)

Country	Seed cost increase	Gross margin gain
	Bt cotton	
Argentina[a]	87	23
China[b]	32	470
India[c]	56	111
Mexico[d]	58	295
South Africa (small farms)[e]	23	52
South Africa (large farms)[e]	47	129
USA[f]	79	58
	Bt maize	
Argentina[g]	22	20
South Africa[g]	10	42
Spain[h]	40	116
USA[f, g]	28	10

[a] Qaim and de Janvry (2003); [b] Pray et al. (2002); [c] Qaim et al. (2006); [d] Traxler et al. (2003); [e] Gouse et al. (2004); [f] Naseem and Pray (2004); [g] Brookes and Barfoot (2005); [h] Gómez-Barbero and Rodríguez-Cerezo (2006)

markup), which companies can charge, depends on the value of the technology and the degree of market power in the national setting. Strong intellectual property rights (IPRs) and/or special seed sales contracts limit competition and reduce farmers' options to save seeds. This is observed for *Bt* cotton in the USA, Mexico, and Argentina. In India, *Bt* cotton seed prices were also relatively high during the first years of adoption, in spite of the fact that the technology is not patented there. The reason is that in India *Bt* is incorporated into cotton hybrids. Hence, there is a technical restriction for farmers to reproduce seeds, which also increases the seed companies' pricing scope.[3] In China, IPR protection is weak, and *Bt* is used in open-pollinated cotton varieties. Therefore, *Bt* seed costs are relatively low, and use of farm-saved seeds is widespread (Pray et al., 2001). In South Africa, Monsanto implements a system of price discrimination for *Bt* cotton seeds: small dryland farmers, which are dominant in the Makhathini Flats region, pay a significantly lower price than large-scale farmers (Gouse et al., 2004). For maize, *Bt* technology is only incorporated in hybrids in all countries, so that seed saving by farmers hardly occurs.

In spite of the technology fee, *Bt* adopting farmers benefit in terms of higher average gross margins (Table 12.3). That is, the economic advantages associated with insecticide savings and higher effective yields more than outweigh the technology fee charged on seeds. The absolute gains differ remarkably between countries and crops. Apart form agroecological differences and unequal technology fees, this is partly due to dissimilar agricultural policies. In the USA, China, and Mexico, the cotton sector is heavily subsidized, which encourages intensive production schemes and high overall yields. The situation is similar for maize in Spain. In Argentina, by contrast, farmers are not subsidized, but face world-market prices. Especially for cotton, world-market prices have been declining recently, which erodes the economic benefits resulting from technological yield gains.

India is now the country with the biggest *Bt* cotton area worldwide (James, 2007), and reports on Indian farmers' experiences with the technology have featured prominently in the global public GM crop debate (Sahai and Rahman, 2003; GRAIN, 2004). Therefore, a closer look at farm level economic impacts in India is particularly interesting. Table 12.4 shows cotton enterprise budgets with and without *Bt* technology for the 2002 growing season. The data were collected from randomly selected farms in four cotton-producing states. As expected, with *Bt* technology seed costs are higher and insecticide costs are lower on average. Notable differences also occur for labor costs. In Indian cotton systems, insecticides are mostly applied manually with knapsack sprayers. Hence, spraying is

[3] Partly as a result of high prices in formal markets, a black market for cheaper, unlabeled *Bt* cotton seeds emerged in India, which has reached sizeable proportions. More recently, following farmers' complaints, some state governments have issued decrees, specifying maximum retail prices for *Bt* seeds, which are significantly lower than the previous prices in formal markets. Since at these lower prices, demand is higher than supply, a new form of black market has emerged, where input dealers are now selling labeled *Bt* seeds unofficially at rates far exceeding the official maximum retail prices.

Table 12.4 Crop enterprise budgets for cotton in India with and without *Bt* (2002) (Qaim et al., 2006)

	With *Bt*	Without *Bt*
Number of insecticide sprays	4.2	6.8
Insecticide use (kg/ha)	5.1	10.3
Yield of raw cotton (kg/ha)	1,628	1,213
Variable production cost (US$/ha)		
Seed	81.0	25.2
Insecticides	64.8	109.5
Fertilizer and manure	96.9	85.4
Hired labor for field operations	88.6	72.4
Hired labor for harvesting	61.7	43.6
Other cost	41.5	35.7
Total variable cost (US$/ha)	434.5	371.9
Revenue (US$/ha)	707.1	533.2
Gross margin (US$/ha)	272.5	161.3

labor-intensive, and a reduction in the number of sprays is associated with labor savings. Yet, in the Indian context, these savings through *Bt* technology are offset by more labor being used for other operations, especially harvesting, so that overall more labor is used in *Bt* than in conventional cotton.[4] Deducting production costs from sales revenues results in the gross margin per hectare, which can be interpreted as net cotton income, because fixed cost components are negligible. Table 12.4 shows that gross margins for *Bt* cotton adopters were $111 per hectare higher than for conventional cotton growers in 2002. Similar data were also collected in 2004 with an average difference in gross margins of $142 per hectare.

Distributional impacts by farm size have also been analyzed. Most of the cotton-growing farms in India are small; the average farm size of *Bt* cotton adopters is around 5 ha. Table 12.5 shows that the farm level benefits for small producers are very similar to those of their larger colleagues. In China and South Africa, small farms were even shown to benefit more from *Bt* cotton than large farms (Huang et al., 2002; Morse et al., 2004). While more research on the impacts of *Bt* cotton on poverty and income distribution is needed, the evidence available so far disproves the widespread notion that *Bt* technology as such is biased against smallholder farmers.

An important aspect to consider, however, is the timing of liquidity requirements for poor farm households. Buying relatively expensive seeds presupposes the availability of financial resources at the beginning of the growing season. Therefore, credit

[4] In India, cotton harvesting is primarily a female activity, so that *Bt* cotton technology especially improves employment opportunities for women. It should be noted, though, that the net labor effect of *Bt* technology is situation specific. In China, for instance, the reduction in the number of sprays is bigger and the yield increase is smaller than in India, so that overall *Bt* is labor-saving there (Pray et al., 2002).

Table 12.5 Farm level effects of *Bt* cotton for small and large farms in India (2002) (Own survey data. The survey covered a total of 434 plot observations randomly selected in four different states of central and southern India, namely Maharashtra, Karnataka, Andhra Pradesh, and Tamil Nadu)

	Small farms (<5 ha) n = 299	Large farms (>5 ha) n = 135
Insecticide reduction (%)	50.6	50.6
Increase in effective yield (%)	33.4	38.2
Gross margin increase (US$/ha)	102.3	134.8

constraints can limit *Bt* technology adoption, as was shown by Thirtle et al. (2003) and Qaim and de Janvry (2003). This access problem should be addressed through appropriate rural policies. While it generally applies to most new agricultural technologies, it might be more severe in the case of GM crops developed by the private sector, especially when the level of IPR protection increases (Basu and Qaim, 2007).

12.2.3 Variability of Effects

While the results reported on agronomic and economic effects in previous sub-sections clearly underline the overall advantages of *Bt* technology, they mask the fact that there can be significant impact variability. The suitability of insect-resistant *Bt* crops depends on local pest infestation levels, which can vary regionally and seasonally. In China, for instance, infestation levels of lepidopteran pests are highest in the northern and eastern parts of the country, so that the benefits of *Bt* cotton are most pronounced there. This is reflected in much higher adoption rates, as compared to western China (Pray et al., 2002). In the USA, due to diverging pest infestation levels, *Bt* cotton adoption rates are lower in California than in other cotton-growing states (USDA, 2006).

Table 12.6 displays regional variability of *Bt* cotton impacts in India. While *Bt* adopters in Maharashtra, Karnataka, and Tamil Nadu realized significant net benefits in 2002, their colleagues in Andhra Pradesh suffered a loss in average incomes. Strikingly, most of the studies carried out by biotechnology critics during the early stages of *Bt* cotton diffusion in India placed heavy emphasis on observations from Andhra Pradesh (Sahai and Rahman, 2003; GRAIN, 2004). Overall, cotton in Andhra Pradesh is sprayed more often than in other states of India. Therefore, crop losses in conventional cotton are lower, and the expected *Bt* yield effect is small, especially in years with only moderate pest pressure. This small positive yield effect due to *Bt* technology itself was counteracted by a negative germplasm effect. In 2002, many farmers in Andhra Pradesh were affected by severe drought conditions, to which the hybrids carrying the *Bt* gene were not optimally adapted. Although the *Bt* gene itself does not alter the cotton plant's performance under water stress, the underlying germplasm was not particularly well suited for extreme drought situations. The number of *Bt* hybrids approved in India increased from 3 in 2002 to 131 in 2007. Many of the new hybrids are also suitable for conditions in Andhra Pradesh, so that average farm level benefits there increased substantially. Nonetheless, the example demonstrates that *Bt* technology can only be successful when combined with locally adapted germplasm.

Also within a region, *Bt* crop impacts can vary, as has been shown by Bennett et al. (2006). Apart from agroecological factors, this can be due to differences in conventional pest control strategies and other farm and household characteristics. In the early stages of diffusion, farmers usually experiment with a new technology, and they re-consider their adoption decision based on personal experiences made (Qaim, 2005). The adoption dynamics for *Bt* cotton in India are shown in Table 12.7 for a sample of 375 typical farms in central and southern states of the country. Although adoption levels within the sample increased substantially over the first years of technological diffusion, the process is not unidirectional. After the first season in 2002, almost half of the adopters abandoned *Bt* technology, because they were

Table 12.6 Differences of *Bt* cotton effects in India by state (2002) (Qaim et al., 2006)

	Maharashtra	Karnataka	Tamil Nadu	Andhra Pradesh
Insecticide reduction (%)	46	62	78	34
Increase in effective yield (%)	32	73	43	–3
Gross margin increase (US$/ha)	92	270	247	–69

Table 12.7 Adoption and disadoption of *Bt* cotton in a sample of 375 farms in India (Qaim, 2005)

	2002	2003	2004	2005
Number of adopters	113	108	165	251
Number of disadopters after the season	51	26	18	n.a.
Out of the disadopters, number of farmers, who re-adopted in any of the following seasons	38	14	n.a.	n.a.

Note: n.a. means not available

not fully satisfied. Some also did not know how to use the technology properly and continued to spray substantially against lepidopteran pests. Also in subsequent seasons, some disadoption was observed, albeit the percentage of dropouts has been decreasing over time. Interestingly, a remarkable share of the disadopters re-adopted *Bt* technology after a break of one or two years. These patterns demonstrate that *Bt* crop adoption and disadoption are not irreversible decisions for farmers; they are part of a normal learning process.

12.2.4 Economic Surplus Effects

So far, we have only looked at the farm level effects of *Bt* crops, neglecting wider market impacts. Market impacts are important when analyzing the aggregate welfare outcomes of new technologies, which economists usually refer to as economic surplus effects. Whenever new crop technologies are adopted on a larger scale, the productivity increase will cause the crop's supply curve to shift downwards. This is because the marginal cost of production decreases. When the price is determined by market forces and the demand curve for the crop has the usual negative slope, the supply curve shift will lead to a lower equilibrium price. This is shown in Fig. 12.2, where D represents the demand curve, S_0 and S_1 are supply curves before and after the introduction of the new technology, and p_0 and p_1 are initial and new equilibrium prices. Consumers clearly benefit from the price decrease; the gain in consumer surplus can be calculated as area $(a + b + c)$. For farmers, the price decrease leads to a loss, which however is usually lower than their gain through the marginal cost reduction. The change in producer surplus can be calculated as area $(e + f)$ minus area a. In addition to these consumer and producer surplus effects, the technology-developing company will capture an innovation rent through the technology fee charged on seed sales. Different authors have used such a partial equilibrium

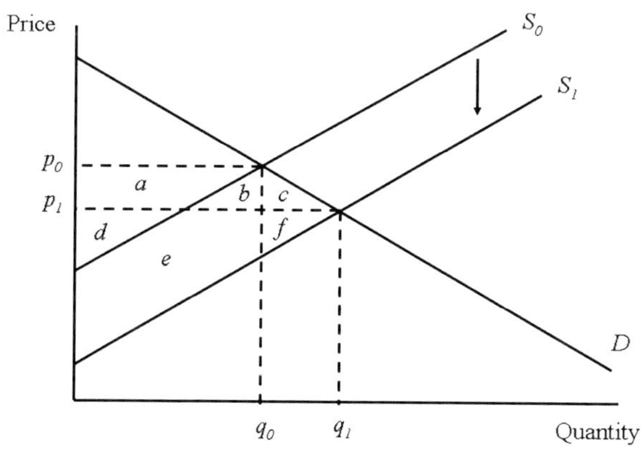

Fig. 12.2 Market impacts of new technology

framework to evaluate the welfare and distribution effects of *Bt* crops in different countries.

Price et al. (2003) estimated, that in the late 1990s, *Bt* cotton generated a total annual economic surplus gain of around $164 million in the USA, of which 37% was captured by farmers, 18% by consumers, and 45% by the innovating companies. Similar results were also reported by Falck-Zepeda et al. (2000). Since *Bt* cotton adoption in the USA has further increased since then, absolute surplus gains are higher today, but relative surplus distribution is still similar (Fernandez-Cornejo and Caswell, 2006). For *Bt* cotton in China, Pray et al. (2001) estimated economic surplus gains of about $140 million in 1999, with only 1.5% going to the innovating companies and the rest being captured by farmers. IPR protection in China is weak, and use of farm-saved *Bt* cotton seeds widespread. Moreover, private sector *Bt* cotton varieties in China face competition from publicly developed *Bt* varieties. Under these conditions, it is difficult for companies to capture innovation rents, so that farmers are the main beneficiaries. Cotton consumers did not benefit in 1999, because the government controlled output markets thus preventing a decrease in the equilibrium price. Prices were somewhat liberalized more recently, so that Chinese consumers now also benefit from *Bt* cotton technology. In India, *Bt* cotton surplus gains were projected at $315 million for 2005 (Qaim, 2003). Cotton prices in India are not fully liberalized, so that consumer benefits were not considered. Farmers capture two-thirds of the overall surplus gains; the rest accrues to biotech and seed companies. As *Bt* cotton in India is commercialized in hybrids, use of farm-saved seeds is low. Thus the private sector innovation rent is higher than in China.

For other *Bt* crops, much less research has been carried out on aggregate welfare effects. Wu (2002) estimated that *Bt* maize generated a total surplus gain of $334 million in the USA in 2001. Nearly 53% of this was due to increases in farm producer surplus, followed by industry profits (31%). The consumer share was relatively small. For *Bt* maize in Spain, Demont and Tollens (2004) calculated welfare gains of about $2 million in 2003, of which 60% went to farmers and 40% to seed companies.[5] The relatively low absolute gain is due to the fact that *Bt* maize in 2003 only covered an area of around 25,000 ha in Spain; the *Bt* maize area has tripled since then. In an ex ante projection, Krishna and Qaim (2008a) estimated that *Bt* eggplant technology could generate annual surplus gains of over $100 million in India in the future, with eggplant consumers capturing over 50% of the benefits. The distribution of the remaining surplus gains between farmers and companies will much depend on prices to be charged for *Bt* eggplant seeds.

These examples demonstrate that *Bt* crops can generate sizeable economic surplus gains and that the distribution effects depend on the particular situation. In developing countries, farmers and sometimes also consumers seem to be the main beneficiaries of *Bt* crops up till now. In developed countries, the private sector benefit shares are partly somewhat larger, because of more effective IPR protection. But even here farmers mostly realize significant positive welfare effects.

---

[5] Based on more detailed farm survey data, Gómez-Barbero and Rodríguez-Cerezo (2006) calculated an even higher benefit share of 77% for Spanish *Bt* maize farmers.

12.3 Health Effects of *Bt* Crops

Apart from the agronomic and economic effects of *Bt* crops, these crops may also have significant environmental and health effects. Biotechnology critics often put potential environmental risks, such as undesirable impacts on non-target organisms, into the fore. Also, food safety concerns are being raised. Shelton et al. (2002) have reviewed such risks of *Bt* crops, concluding that they are manageable and generally lower than the risks of alternative pest-control technologies. Further aspects of risk assessment and management are discussed in other chapters of this book. Here, the focus is on potential health benefits of *Bt* crops.

Direct health benefits accrue to farmers and farm laborers due to less insecticide exposure during spraying operations. The health hazards for farmers and farm workers applying pesticides have been analyzed in different countries (e.g., Sunding and Zivin, 2000; Maumbe and Swinton, 2003). Often, the problems are greater in developing than in developed countries, because environmental and health regulations are laxer, pesticides are mostly applied manually, and farmers are less educated and less informed about negative side effects. But also for consumers, *Bt* crops can bring about potential health benefits through lower pesticide residues in food and water. Furthermore, *Bt* crops may reduce contamination of foods with mycotoxins. While not all of these effects have been analyzed conclusively up till now, a brief overview of selected studies is given in the following.

12.3.1 Effects on Farmer Health

Several authors showed that *Bt* cotton leads to reductions especially in highly toxic insecticides, belonging to toxicity levels I and II of the World Health Organization (WHO) classification (Qaim and Zilberman, 2003; Qaim and de Janvry, 2005). More in-depth studies about the actual health implications for farmers have been carried out in China, where insecticide use in cotton is particularly heavy. Between 1999–2001, the Center for Chinese Agricultural Policy and Rutgers University had conducted a series of surveys of about 400 small farmers in the northern cotton growing-region – Hebei, Shandong, and Henan Provinces – and in the Yangtze Region – Hubei and Jiangsu (Pray et al., 2002; Hossain et al., 2004). Randomly selected farmers were interviewed on cotton production aspects, including details of insecticide applications. In 2000, *Bt* cotton adopters had sprayed about 20 kg of formulated insecticides per hectare, while conventional cotton growers had used 46 kg. Table 12.8 further categorizes the insecticides used by chemical type. The most widely used insecticides were organophosphates and pyrethroids, and for these the *Bt*-related reductions were particularly pronounced. Furthermore, farmers were asked to give details on the frequency and type of insecticide poisonings experienced during sprays or immediately afterwards. Typical health symptoms include eye and skin irritations, headache, nausea, and breathing problems, among others. Table 12.9 demonstrates that many of the poisoning cases reported are related to organophosphates and pyrethroids, so that *Bt* technology has likely

Table 12.8 Average insecticide quantities used by cotton farmers in China (2000) (Hossain et al., 2004)

	Average quantity (kg/ha)		
	With *Bt*	Without *Bt*	Reduction (%)
Organophosphates	8.8	21.0	58
Pyrethroids	5.2	13.0	60
Organosulphates	2.8	6.0	53
Organochlorines	1.6	3.9	58
Amino-formicdacid esters	0.3	0.4	25
Other insecticides	0.8	2.1	64
Total	19.5	46.4	58

Table 12.9 Type and toxicity levels of insecticides causing poisonings in a sample of 400 Chinese cotton farmers (1995–2000) (Hossain et al., 2004)

	WHO toxicity level	Poisoning cases
Organophosphates		
Chlordimeform	I	94
Parathion-methyl	I	65
Acephate	I	19
Carbofuran (furadan)	I	9
Phorate	I	9
Parathion	III	8
Monocrotophos	I	5
Pyrethroids		
Cypermethrin	II	12
Killingthrin 39	III	6

reduced the negative health effects. Indeed, while one-third of the conventional cotton growers had reported cases of poisoning in the 2000 season, the share among the *Bt* adopters was only 9%.

The study by Hossain et al. (2004) also provides the first econometric evidence of the link between the adoption of *Bt* crops and improvements in human health. The authors modeled the linkage as a health-production function, with farmers' reports of poisonings as the dependent variable and insecticide use, farmers' characteristics, and other controlling factors as independent variables. It was hypothesized that the main impact of *Bt* technology on poisoning would be through its impact on insecticide quantities. Therefore, at first the net impact of *Bt* cotton adoption on insecticide quantities was estimated, resulting in a negative and highly significant coefficient. Then, the health-production function was estimated, using instrumented insecticide quantities from the first-stage regression. As expected, pesticide poisonings decrease with a reduction in insecticide quantities. In addition, a *Bt* cotton variable was directly included in the health-production function, to capture possible impacts on poisoning other than those related to insecticide quantities (e.g., differential toxicity levels). This additional variable also showed a small negative impact on the probability of poisoning.

Bennett et al. (2003) have analyzed the health effects of *Bt* cotton in South Africa. Although they used a smaller sample and less sophisticated methodologies, they came to the same conclusion: with an increase in the uptake of *Bt* cotton among smallholder farmers, the rates of accidental insecticide poisoning have been declining. Since cotton is the number one insecticide-consuming crop worldwide, the magnitude of the effects might not be directly transferable to other crops. Nonetheless, there is evidence that the general linkages also apply in other situations where chemical insecticides are used intensively.

Using data from on-farm field trials with *Bt* rice in China, Huang et al. (2005) could show that the technology might have important positive effects for farmers' health. Krishna and Qaim (2008a) carried out a survey of 360 eggplant farmers in different states of India and found that, on average, eggplants are sprayed 30 times per season with insecticides.[6] Twenty-five percent of all farmers had experienced cases of pesticide poisoning during the 12 months prior to the survey. While these cases might also be related to sprays in other crops, an econometric model was used to show that insecticide applications in eggplant accounted for almost half of these poisonings, suggesting that *Bt* technology can improve the situation considerably. Using realistic assumptions on technology related reductions in the number of sprays under farmers' conditions and a detailed breakdown of the cost-of-illness associated with pesticide poisonings, Krishna and Qaim (2008a) estimated that the health benefits of *Bt* eggplant technology for Indian farmers could be worth around $4 million per year. It should be noted that this is rather a conservative estimate, because it only captures the immediate effects of pesticide poisonings, neglecting possible long-term adverse outcomes for human health. In conclusion, *Bt* crops can be associated with considerable health advantages for farmers and farm laborers, especially in the smallholder production systems of developing countries.

12.3.2 Effects on Consumer Health

For consumers, reductions in insecticide applications might potentially be associated with lower contamination of water and food. Especially when insecticides are overused, runoff into surface waters or leaching into the groundwater can occur. In rural areas, such water is often used directly for human consumption without further treatment, which can lead to significant intake and accumulation of toxic chemicals. But also after water treatment and processing, pesticide residues are often not completely removed, so that urban consumers can be affected, too. Sample analyses in India have shown that even bottled water and soft drinks sold by international brands regularly exceed rich-country maximum residue levels (Umali-Deininger and Sur, 2006).

---

[6] In West Bengal, a major eggplant producing state of India, farmers even apply insecticides 66 times on average. Some farmers spray more than 80 times during a single season.

With respect to food, pesticide residues are of concern especially in fruits and vegetables, because these are often sprayed intensively and consumed raw or with little processing. The resulting health risks explain part of the increasing demand for organically produced horticultural crops in many developed countries (Florax et al., 2005). Also in developing countries, consumer awareness of food safety issues is increasing. Because of laxer environmental and health standards, residue problems are more widespread in developing countries, including for pesticides which are banned in developed-country agriculture. A recent survey of urban households in India showed that over 60% associate pesticide residues in common vegetables with medium or high risks for human health; accordingly, the average consumer would be willing to pay a 57% price premium for residue-free vegetables (Krishna and Qaim, 2008b). As mentioned above, *Bt* eggplant and other *Bt* vegetables are near the end of the research and development (R&D) pipeline (Shelton et al., chapter 9). They could reduce consumer-related risks considerably.

Other potential health benefits of *Bt* crops for consumers relate to lower mycotoxin contamination in foods. Food-borne mycotoxins are secondary metabolites of fungi. Especially in maize, it has been shown that insect damage is one factor that contributes to mycotoxin contamination, because damage by insects encourages fungal colonization, and insects themselves are a vector by which fungal spores move from plant to plant (Munkvold and Hellmich, 1999). There are different types of mycotoxins. Two of the most important ones in maize are aflatoxins and fumonisins. Aflatoxins, mostly produced by the fungus *Aspergillus flavus*, were among the first mycotoxins discovered. They are the most potent chemical liver carcinogen known. Aflatoxins have a synergistic effect with the carcinogenic hepatitis viruses, and they can also cause stunting in children and immune system disorders (Turner et al., 2003). The effects on animals can be equally devastating. Fumonisins are produced by *Fusarium verticillioides* and *Fusarium proliferatum* on maize and other crops. High levels of fumonisins have been associated with esophageal cancer and neural tube defects in various parts of Africa, Central America, and in Asia (Marasas, 2001). Animal effects include equine leukoencephalomalacia, porcine pulmonary edema, and liver and kidney cancer in rodents.

In a variety of field studies, *Bt* maize was shown to contain significantly lower levels of certain mycotoxins. The technology has been particularly successful in lowering fumonisin contamination in a number of countries (Munkvold and Hellmich, 1999; de la Campa et al., 2005). Given that *Bt* target insects are not as important in predisposing plants to infection by *A. flavus* as they are for *F. verticillioides* and *F. proliferatum*, *Bt* technology has been less successful in reducing aflatoxins as compared to fumonisins. So far, *Bt* maize has a mixed record of reducing aflatoxins in the USA, and related evidence from other countries is scant (Wu, 2006).

Whether *Bt* maize so far has really reduced cancer or other health disorders is uncertain. In the USA and other developed countries, maize was already carefully graded before *Bt* commercialization, so that only small amounts of mycotoxins seem to find their way into the food supply. The same holds true for countries like Argentina that export significant quantities of maize to Europe. Therefore, in these countries mycotoxins are associated with economic costs for testing and grading, while actual health costs are relatively small. This is different in certain parts of the developing world, where maize is grown and consumed by small-scale farmers as a major subsistence food. Examples include South Africa, the Philippines, and some areas of Mexico, where *Bt* maize is grown. Yet, up till now the impact of *Bt* technology on subsistence farmers is small. First, *Bt* technology is commercialized only in hybrids, whereas subsistence farmers often use open-pollinated varieties. Second, most of the available *Bt* maize is yellow maize, which is primarily fed to animals in many developing countries. Only in South Africa, some white *Bt* maize hybrids have been commercialized, and adoption levels among smallholder farmers are gradually increasing.

If small farmers start to adopt *Bt* maize more extensively, it is possible but far from certain that the technology would actually reduce the amount of fumonisins consumed. A study of farmers in KwaZulu Natal (South Africa) by scholars from the South African Medical Research Council, the University of Pretoria, and Rutgers University in the USA is underway to measure how much fumonisin is found in *Bt* hybrid maize, conventional hybrid maize, and local varieties of smallholder farmers. Maize was sampled in two villages and two years, and the amount of fumonisin was measured. *Bt* hybrids had less fumonisin than conventional hybrids in three of four cases, *Bt* maize had less than local varieties in two of four cases, and local varieties had less than conventional hybrids in three of four cases. Further studies are planned to find out whether *Bt* maize can reduce fumonisin on a regular basis and whether this can lead to actual health improvements under local dietary patterns.

In any case, the discussion has shown that *Bt* crops can have important positive health effects, especially in developing countries. Health advantages for farmers and farm laborers through lower pesticide exposure have already been proven. On the consumer side, potential health benefits have been identified, but the evidence available is still more tentative. More research is needed to better understand these effects and integrate them into economic impact analysis.

12.4 Consumer Acceptance/Marketing Issues

In spite of the tangible benefits that GM crops may have, critics remain suspicious about the wider impacts of agricultural biotechnology. Critics argue that GM food crops could be associated with health risks for consumers, claiming that newly introduced proteins could act as allergens or toxins and alter the metabolism of the crop. In addition, there are environmental, social, and ethical concerns, which heavily influence the public debate (Welsh and Ervin, 2006). Understanding public attitudes and how they emerge is important, in order to design appropriate policies and assess the future evolution of GM crops and other new technologies. There is a growing body of literature on consumer acceptance and public attitudes towards GM crops, especially GM foods that have been commercialized or that are likely to be commercialized soon. Lusk et al. (2005) provide a meta-analysis of 25 studies from different countries. Across all studies, consumers on average placed 19% lower value for GM relative to non-GM foods. However, a number of factors significantly affected this value estimate. In particular, European consumers placed a lower value on GM foods than consumers from North America. Moreover, for fresh fruits and vegetables the GM discount was bigger than for processed food products, and it was lower when the GM foods were associated with direct benefits to consumers. While these results are relatively unsurprising, it should be mentioned that almost all of the underlying studies were carried out in developed countries. Comparatively little is known on consumer perceptions in developing countries.

Curtis et al. (2004) reviewed the few available studies on GM crop acceptance in developing countries. They conclude that perceptions are generally more positive than in developed countries, which could be explained by more urgent needs in terms of food availability and nutritional content. This general tendency is confirmed by a recent study in India, where Krishna and Qaim (2008b) interviewed 645 urban households on their attitudes towards *Bt* vegetables. While awareness of GM food crops is still relatively low in India, the majority of consumers were quite open to the innovation. Sixty-eight percent of the respondents supported the introduction of *Bt* vegetables, while only 17% were mildly or strongly opposing the new technology. The mean willingness to pay (WTP) for *Bt* vegetables was estimated at 1.5% above market prices of currently purchased non-*Bt* vegetables. Indeed, Fig. 12.3 shows that 55% of the respondents would purchase *Bt* vegetables at current market prices. More than 80% would purchase at a price discount of 10%. Evidently,

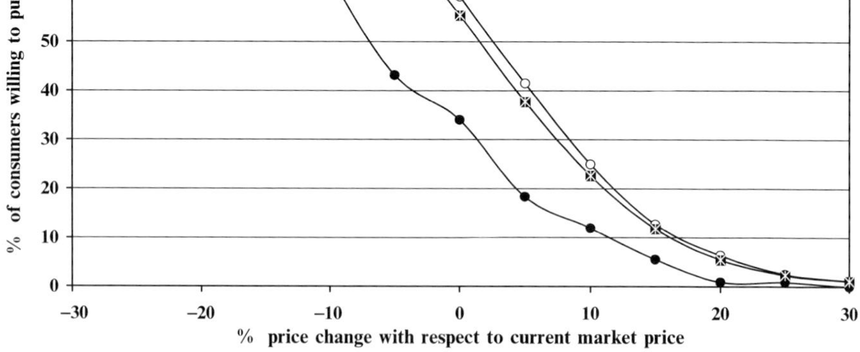

Fig. 12.3 Estimated percentage of consumers willing to purchase *Bt* vegetables at different price levels in urban India (survey is based on 645 urban households) (Krishna and Qaim, 2008b)

respondents who generally oppose the introduction of the technology would require larger price discounts. Still, over 30% of them would be willing to purchase *Bt* vegetables also without any discounts, indicating that the opposition is not very pronounced. However, it was also found that trust in public food safety authorities and exposure to mass media has a significant influence on consumer attitudes. Therefore, acceptance levels can potentially change rapidly with different types of information or misinformation becoming available. This needs to be taken into account when designing strategies to introduce and disseminate *Bt* food crops in India and other developing countries.

One possible explanation for the positive attitude of Indian consumers towards *Bt* vegetables is the reduction in pesticide residues, because this was explicitly discussed during the interviews. As mentioned above, urban consumers in India are well aware of the high residue levels occurring in vegetables and associated risks for human health. Strikingly, however, the WTP for *Bt* vegetables was found to be negatively related to the WTP for residue-free vegetables, with a highly significant correlation coefficient (Krishna and Qaim, 2008b). That is, people who are most concerned about pesticide residues are least willing to accept *Bt* vegetables. Hence, consumer decision making appears to be largely driven by general risk attitudes. Those who do not pay a lot of attention to the risks of pesticide residues are also not much concerned about potential GM food risks. Risk-averse consumers, in contrast, are concerned about both types of risks, and for them a reduction in one risk cannot easily compensate for an increase in the other. While they would clearly value a reduction in pesticide residues, the disliked GM attribute ways heavier and seems to overshadow the benefits. In this respect, it does not matter whether GM foods are really associated with human health risks from a scientific point of view; a subjective feeling suffices for risk-averse consumers to lower acceptance considerably. This phenomenon is often exploited by anti-biotech pressure groups.

More research about consumer attitudes towards *Bt* crops under different conditions is certainly needed. A better understanding of how consumers value different types of risks and benefits can help improve the innovation process.

12.5 Institutional Issues

12.5.1 Economics of Biosafety Regulation

Most people and countries in the world agree that there is a role for governments in ensuring that novel foods are safe for human consumption and that novel agricultural inputs do not cause major negative impacts on the environment and on long term agricultural production possibilities. Most countries, with the notable exception of the USA, consider crops produced by genetic engineering to be novel foods, no matter what the characteristics of their final product are. The USA does not automatically regulate GM food unless it alters the nutritional profile or possibly adds an allergen to the food, is a pesticide, or has the potential to harm other crops or the environment. Most other countries have developed or are developing new laws and regulatory institutions to regulate GM crops. These regulations have become a major barrier to the spread of *Bt* and other transgenes around the world. Large biotechnology companies will not sell their products to farmers in countries where there is no biosafety regulatory system, because of potential liability problems and the public-relations nightmare it would create when anti-biotech groups publicized this unregulated introduction. Many smaller developing countries have not been able to legislate and operate a biosafety regulatory system so far, and so they are shut off from technology from large biotechnology firms.

In countries where there is a biosafety system in place, most of regulators' efforts are put into reducing the possibility that products which might harm people or the environment are commercialized (also called reducing type I errors). Often, regulators are extremely cautious, requiring many regulatory trials over a long period of time. If this system is effective in identifying potential problems and then enforcing bans or restrictions on those technologies that have a potential to create problems, it could have several important benefits for society: first, less health problems such as sickness or death from allergens and anti-nutritional factors; second, less environmental problems such as reductions in biodiversity; and third, less agricultural problems from allowing technology that could create complications in the future. There are also benefits from effective regulations for the companies that are developing new products. Effective regulation gives consumers more confidence that the new products are safe, and this will increase the demand for the products that are approved by the government (Pray et al., 2006a).

However, these benefits of biosafety regulations come at a cost. The greatest cost to society in some situations is the foregone benefit to farmers and consumers who are not getting access to an advantageous product that is safe for consumption and the environment (preventing the use of safe products is referred to as type II errors). In one of the only attempts that have been made to measure this cost with GM crops, Pray et al. (2005) found that a two-year delay in the approval of *Bt* cotton in India led to aggregated losses to farmers of over $100 million.

Strict regulations and cautious regulators also impose costs on the companies that are trying to obtain regulatory clearance. Regulators are particularly cautious when the GM technology involves a major food crop or when the country is the center of biodiversity for the species. Table 12.10 indicates companies' expected costs of meeting biosafety regulatory requirements in India. The expected costs of new GM food crops are higher than the costs of bringing a new non-food GM crop like cotton to the market. New genes and transformation events, that have not been tested elsewhere, are the most expensive, especially when there are potential crop exports, because then the firm needs to meet not only the Indian biosafety requirements but also the requirements of the main export markets. These costs and the uncertainty about when, if ever, a product will be approved have caused some large companies to stop testing products in a country entirely, and they certainly act as a market entry barrier for smaller firms (Pray et al., 2005).

Table 12.10 Private companies' estimates of the full costs of meeting biosafety regulations in India in the future (Pray et al., 2005)

Type of crop (example)	Event approved in US, Europe, Canada, Australia, or Japan	Event approved in India	Estimated costs of meeting regulations ($)
Non-food crop (e.g., cotton)	Yes	No	500,000–1,000,000
Food crop (e.g., maize)	Yes	No	500,000–1,500,000
Non-food crop (e.g., jute)	No	No	1,000,000–1,500,000
Food crop (e.g., rice)	No	No	1,500,000–2,000,000
Food crop – possible exports (e.g., vegetables)	No (additional approval in export markets required)	No	4,000,000

To obtain the benefits from effective regulation, governments must be able to enforce the regulations that they establish. So far, this has proven difficult, particularly when the product they are trying to control is economically very profitable to farmers and has no obvious health or environmental risks associated with it. Regulators have been able to prevent large biotechnology companies from selling unapproved GM seeds, but it is very difficult for governments to control farmer-to-farmer spread of unapproved technology or small local companies, trading products across country boundaries (Herring, 2007). Brazilian farmers grew millions of hectares of Roundup Ready soybeans, which came in from Argentina, before they were legally allowed to plant them. At one point, at least 20% of the *Bt* cotton in China contained an unapproved gene, and in India, two-thirds of the total *Bt* cotton area were "illegal" in 2004 (Pray et al., 2006b). These examples are typical type II errors by regulators – not allowing safe technologies – but they reduce people's confidence in the system. The primary way that governments have been able control the spread of these illegal varieties has been to speed up the approval of other superior GM varieties, which are hoped to gradually push out the older unapproved ones. This strategy has been used effectively in China against the illegal *Bt* gene, and India is currently trying the same with unapproved *Bt* cotton hybrids (Pray et al., 2006b).

As in most developing countries biosafety systems are still evolving, it is likely that approval processes and enforcement of legislations will be handled more efficiently in the future, when more experience has been gained. However, the costs of effective biosafety systems, including the testing costs for innovating private or public sector organizations, will remain relatively high and might constitute an obstacle for small developing countries to get access to suitable GM technologies. Ways to reduce these costs have to be sought for, while ensuring that safety levels are not jeopardized. One possible option is closer regional cooperation between neighboring countries with similar agroecological conditions.

12.5.2 Intellectual Property Rights, Agrobiodiversity, and Local Seed Markets

IPRs confer a temporary monopoly to the inventor of a new technology. Thus, competitors can be excluded from its use, or they have to pay a license fee. The fee is added to the market price of seeds, and the extra revenue is captured by the IPR holder as an innovation rent, which is meant to compensate for the R&D investments made. Especially with respect to developing countries, there are concerns that excessive fees would lead to very high market prices of GM seeds, so that all the benefits from using these seeds would be captured by the innovating companies. This could potentially lead to an exploitation of smallholder farmers, with undesirable social consequences. As was shown above, IPRs in most developing countries are relatively weak, so that farmers are currently the main beneficiaries of *Bt* crops and other GM technologies. A strengthening of IPR protection and enforcement would increase GM seed prices, which would change the distribution of benefits. This, however, would unlikely lead to an exploitation of farmers. Rather, excessive GM seed prices would constitute a technology access problem, because farmers would decide not to adopt (Qaim and de Janvry, 2003; Qaim, 2005; Basu and Qaim, 2007). Therefore, the appropriate level of IPR protection has to be country and situation specific, balancing farmers' interests on the one hand, and incentives to invest in R&D on the other.

These general linkages hold for all types of seed technologies. Yet there are also related aspects which are more specific to GM crops. Modern biotechnology permits a separation between the act of developing a specific crop trait and the breeding of locally adapted germplasm. Thus, unlike previous high-yielding varieties (HYVs), the outcome of GM research is not a particular new variety, but a transformation event, or a GM trait, which can be used for backcrossing into numerous locally adapted varieties. Thus, GM technologies can be made available for various agroecological environments. Farmers located in marginal zones were largely bypassed by previous crop innovations, because the cost of developing particular varieties for these areas was relatively high compared to the expected productivity gains. This could change with modern biotechnology, since backcrossing available GM traits into locally adapted varieties is relatively straightforward. The incorporation of GM traits into many local varieties could also reduce the loss of agrobiodiversity observed in many countries during the Green Revolution. Instead of replacing many local varieties with only a few HYVs, GM versions of these local varieties could be made available at relatively low cost. Whether this will happen in reality is largely a question of institutional arrangements (Qaim et al., 2005). Table 12.11 shows the estimated number of varieties available for *Bt* maize and *Bt* cotton in selected countries.

Unsurprisingly, the number of *Bt* varieties is positively correlated with the area under the technology. A minimum market size per variety appears to be necessary to justify the additional cost of backcrossing. Furthermore, IPRs play an important role. When a GM technology is patented, breeders or seed companies need to get a

Table 12.11 Estimated number of *Bt* varieties available in selected countries (2007) (James, 2007; USDA, 2007; Qaim et al., 2005; and communication with industry representatives)

Country	Technology	Area under technology (ha)	Total number of *Bt* varieties/ hybrids	Based on locally adapted germplasm	Based on imported germplasm
USA	*Bt* maize	18.6 million	750	750	0
	Bt cotton	3.1 million	19	19	0
India	*Bt* cotton	6.2 million	131	131	0
China	*Bt* cotton	3.8 million	150	144	6
Argentina	*Bt* maize	2.5 million	25	19	6
	Bt cotton	195,000	3	0	3
Mexico	*Bt* cotton	60,000	2	0	2
South Africa	*Bt* maize	1.2 million	11	4	7
	Bt cotton	10,000	3	0	3

license before they can use it in their own germplasm. Non-exclusive licenses can result in a large number of GM varieties, as the example of *Bt* maize in the USA demonstrates. An exclusive license with only one seed company, however, can potentially result in a loss of agrobiodiversity, especially when the GM trait is so powerful that farmers adopt it even when it is not incorporated into varieties optimally suited to their conditions. This can also be associated with a re-structuring in seed markets. For instance, Delta and Pineland's exclusive license for Monsanto's *Bt* cotton in the USA, Mexico, and South Africa led to a notable increase in the company's seed market share in these countries (Carpenter et al., 2002; Traxler et al., 2003; Gouse et al., 2004). Similarly, in Argentina, Monsanto and Delta and Pineland started a joint venture with only one local seed company, Ciagro (Qaim and de Janvry, 2003).

When a GM technology is not IPR protected in a country, local breeders can use it without a license. A case in point is China, where *Bt* cotton varieties are produced and marketed by several local breeding stations and seed companies (Pray et al., 2006b). In India, *Bt* cotton is not patented, but between 2002 and 2006 every single *Bt* hybrid had to be approved by the Genetic Engineering Approval Committee (GEAC), also when it was based on an already sanctioned transformation event. In the first years after *Bt* cotton commercialization in India, only three *Bt* hybrids had been approved, which were grown on large areas in different regions (Qaim et al., 2006). These three hybrids were released by Mahyco, Monsanto's local joint venture partner, whose share in cotton seed markets increased dramatically through incorporation of the *Bt* gene. In the meantime, Mahyco and Monsanto have also licensed *Bt* cotton technology to other Indian seed companies, and the GEAC has adopted an event-based approval process; that is, approval for new *Bt* hybrids is much faster when the applicant can prove that it is based on the same transformation event. Accordingly, the number of *Bt* cotton hybrids sold by different seed companies in India increased substantially to a total of 131 in 2007, and is likely to further grow in the future.

An interesting question in the Indian context is why seed companies need a license from Monsanto anyway, because *Bt* cotton technology is not patented there. The answer is related to the biosafety approval process rather than to IPRs. Seed companies can get quick approval for their *Bt* hybrids when they prove that these are based on the Mahyco-Monsanto event, and the way to prove this easily is through an official license. Without a license, other seed companies could still use the technology, but they might have to run through the complete lengthy and costly biosafety process. Unlike commonly believed, black market *Bt* cotton seeds in India are illegal because of a violation of biosafety laws, not IPR laws. In any case, *Bt* technology in India has had significant impacts on cotton seed markets, with the share of private sector hybrids increasing at the expense of public varieties and hybrids (Murugkar et al., 2006).

Most of the countries where GM crops have been commercialized so far have a relatively strong breeding sector, so that foreign biotechnology companies could work together with locally active seed companies or public organizations. GM crop adoption in poorer developing countries is still limited. Lack of local breeding capacities can be a serious obstacle for technology transfer. The outcome could be that either new seed technologies will not become available at all in these countries or that foreign technologies are used without local adaptation. Both outcomes are highly undesirable on economic, social, and environmental grounds. Public policy support, including from the international community, should be targeted at reducing related institutional constraints.

12.6 Conclusions

This chapter has analyzed various issues in the adoption of *Bt* crops, including agronomic and socioeconomic effects, potential health benefits, consumer acceptance, and institutional implications. Empirical evidence from developed and developing countries has been reviewed, with a special emphasis on *Bt* cotton, as this is currently the most widely grown *Bt* crop in different regions of the world. On average, *Bt* cotton significantly reduces insecticide applications and pest-related crop losses, leading to higher effective yields. Yield effects tend to be bigger in developing countries, especially in the tropics, where pest infestation levels are often more severe than in temperate climates, and where farmers do not always control pests effectively through pesticides due to various constraints. These clear benefits of *Bt* technology for farmers come at the cost of higher seed prices. The magnitude of the technology fee charged by private companies on GM seeds depends on the strength of IPR protection and enforcement in a country. Overall, the extra cost is lower than the benefits, so that farmers realize substantial gains in gross margins and cotton incomes. However, given seasonal and regional variability in impacts, there are also cases where individual farmers did not benefit in a particular year. Disappointed farmers tend to stop using the technology in the next year, but the rapid overall increase in adoption clearly indicates that the majority is satisfied with *Bt*. Economic surplus studies show that the aggregate welfare effects for society are large and increasing with adoption levels. In India alone, *Bt* cotton produces annual welfare gains of over $300 million, with farmers capturing the largest share. Also in China, smallholder farmers are the main beneficiaries of *Bt* cotton. In the USA and other developed countries, stronger IPR protection leads to a larger share of the overall gains being captured by biotechnology companies, but agricultural producers and consumers still benefit to a considerable degree. More preliminary studies for other *Bt* crops – including *Bt* maize, *Bt* eggplant, and *Bt* rice – suggest that these results are not confined to cotton alone.

In addition to the direct economic effects, *Bt* crops might bring about environmental and health benefits. Lower insecticide application rates can reduce farmers' exposure to toxic chemicals during sprays, as well as environmental contamination and pesticide residues in foods. Such benefits will be bigger in developing than in developed countries, because environmental and health standards are lower there, and pesticides are often applied manually by farmers with relatively little educational background. For China and South Africa, reductions in pesticide poisonings through *Bt* have already been shown, and similar effects are also expected in other countries and crops. Furthermore, *Bt* technology might reduce mycotoxin contamination in foods and thus lead to health improvements, especially in subsistence agriculture and local markets where food safety regulations are not in place or not effective. More research on such possible externalities of *Bt* crops is needed, in order to assess the impacts more comprehensively.

Consumer acceptance and marketing problems have slowed down the commercialization of *Bt* food crops in developed countries. In general, European consumers are more concerned about possible risks than North American consumers and perhaps more likely to be influenced by organizations critical of GM crops. Although less is known on public acceptance in developing countries, the few available studies indicate that consumer attitudes towards GM foods are more positive there, which bodes well for the future of *Bt* food crops from an international perspective. Yet, the general level of consumer awareness about GM foods is relatively low in developing countries, so that attitudes are still evolving. Hence, the introduction of *Bt* food crops should be accompanied by efforts to spread objective and easily accessible information.

Finally, the costs and benefits of biosafety and IPR regulations have been discussed. Such institutional issues can have important ramifications for technology access, benefit distribution, seed market structures, and agrobiodiversity. Proper regulation is necessary for sustainable technology management, but over-regulation can also result in significant efficiency losses. Especially small developing countries are at a disadvantage, because they usually lack the human and financial resources to establish and strengthen appropriate institutions. International support will be required to enable those countries to take part in the biotechnology evolution. Moreover, further research is necessary to identify options of how to reduce regulatory costs without unintended side-effects.

References

Basu, A.K., and Qaim, M., 2007. On the adoption of genetically modified seeds in developing countries and the optimal types of government intervention. American Journal of Agricultural Economics 89: 784–804.

Bates, S.L., Zhao, J.Z., Roush, R.T., and Shelton, A.M., 2005. Insect resistance management in GM crops: Past, present and future. Nature Biotechnology 23: 57–62.

Bennett, R., Morse, S., and Ismael, Y., 2003. Bt cotton, pesticides, labour and health: A case study of smallholder farmers in the Makhathini Flats, Republic of South Africa. Outlook on Agriculture 32: 123–128.

Bennett, R., Kambhampati, U., Morse, S., and Ismael, Y., 2006. Farm-level economic performance of genetically modified cotton in Maharashtra, India. Review of Agricultural Economics 28: 59–71.

Brookes, G., and Barfoot, P., 2005. GM Crops: The Global Socioeconomic and Environmental Impact – The First Nine Years. PG Economics Ltd., Dorchester, The Netherlands.

Carpenter, J., Felsot, A., Goode, T., Hammig, M., Onstad, D., and Sankula, S., 2002. Comparative Environmental Impacts of Biotechnology-derived and Traditional Soybean, Corn, and Cotton Crops. Council for Agricultural Science and Technology, Ames, IA, USA.

Curtis, K.R., McCluskey, J.J., and Wahl, T.I., 2004. Consumer acceptance of genetically modified food products in the developing world. AgBioForum 7: 70–75.

de la Campa, R., Hooker, D.C., Miller, J.D., Schaafsma, A.W., and Hammond, B.G., 2005. Modeling effects of environment, insect damage, and Bt genotypes on fumonisin accumulation in maize in Argentina and the Philippines. Mycopathologia 159: 539–552.

Demont, M., and Tollens, E., 2004. First impact of biotechnology in the EU: Bt maize adoption in Spain. Annals of Applied Biology 145: 197–207.

Falck-Zepeda, J.B., Traxler, G., and Nelson, R.G., 2000. Surplus distribution from the introduction of a biotechnology innovation. American Journal of Agricultural Economics 82: 360–369.

Fernandez-Cornejo, J., and Caswell, M., 2006. The First Decade of Genetically Engineered Crops in the United States. Economic Information Bulletin No. 11, United States Department of Agriculture, Washington, DC, USA.

Fernandez-Cornejo, J., and Li, J., 2005. The impacts of adopting genetically engineered crops in the USA: The case of Bt corn. Selected paper presented at the American Agricultural Economics Association Annual Meetings, Providence, RI, USA.

Florax, R.J.G.M., Travisi, C.M., and Nijkamp, P., 2005. A meta-analysis of the willingness to pay for reductions in pesticide risk exposure. European Review of Agricultural Economics 32: 441–467.

Gómez-Barbero, M., and Rodríguez-Cerezo, E., 2006. GM Crops in EU Agriculture; Case Study for the BIO4EU Project. European Commission, Institute for Prospective Technological Studies, Seville, Spain.

Gonzales, L., 2002. Likely transcendental effects of agri-biotechnology: The case of Bt hybrid corn in the Philippines. Paper presented at the Symposium on Bt Technology, University of the Philippines, Los Baños.

Gouse, M., Pray, C., and Schimmelpfennig, D., 2004. The distribution of benefits from Bt cotton adoption in South Africa. AgBioForum 7: 187–194.

Gouse, M., Pray, C., Schimmelpfennig, D., and Kirsten, J., 2006. Three seasons of subsistence insect-resistant maize in South Africa: Have smallholders benefited? AgBioForum 9: 15–22.

GRAIN (Genetic Resource Action International), 2004. New Studies Contradict FAO Report and Show That Genetically Engineered Bt Cotton Fails to Benefit Farmers. Genetic Resource Action International, Barcelona, Spain. http://www.grain.org/research/btcotton.cfm?id = 129 (accessed 9 November 2007).

Herring, R., 2007. Stealth seeds: Biopiracy, biosafety, biopolitics. Journal of Development Studies 43: 130–157.

Hossain, F., Pray, C.E., Lu, Y., Huang, J., and Hu, R., 2004. Genetically modified cotton and farmers' health in China. International Journal of Occupational and Environmental Health 10: 296–303.

Hu, R., Huang, J., Lin, H., and Rozelle, S., 2006. *Bt* cotton in China: Are secondary insect infestations offsetting the benefits in farmer fields? Paper presented at the 10th International Consortium of Agricultural Biotechnology Research Conference, June 29–July 2, Ravello, Italy.

Huang, J., Hu, R., Rozelle, S., Qiao, F., and Pray, C.E., 2002. Transgenic varieties and productivity of smallholder cotton farmers in China. Australian Journal of Agricultural and Resource Economics 46: 367–387.

Huang, J., Hu, R., Rozelle, S., and Pray, C., 2005. Insect-resistant GM rice in farmers' fields: Assessing productivity and health effects in China. Science 308: 688–690.

James, C., 1997. Global Status of Transgenic Crops in 1997. ISAAA Brief No. 5, International Service for the Acquisition of Agri-biotech Applications, Ithaca, NY, USA.

James, C., 2007. Global Status of Commercialized Biotech/GM Crops: 2007. ISAAA Brief No. 37, International Service for the Acquisition of Agri-biotech Applications, Ithaca, NY, USA.

Kaniewski, W.K., and Thomas, P.E., 2004. The potato story. AgBioForum 7: 41–46.

Krishna, V.V., and Qaim, M., 2007. Estimating the adoption of Bt eggplant in India: Who benefits from public-private partnership? Food Policy 32: 523–543.

Krishna, V.V., and Qaim, M., 2008a. Potential impacts of Bt eggplant on economic surplus and farmers' health in India. Agricultural Economics 38: 167–180.

Krishna, V.V., and Qaim, M., 2008b. Consumer attitudes towards GM food and pesticide residues in India. Review of Agricultural Economics 30: 233–251.

Lichtenberg, E., and Zilberman, D., 1986. The economics of damage control: Why specification matters. American Journal of Agricultural Economics 68: 261–273.

Lusk, J.L., Jamal, M., Kurlander, L., Roucan, M., and Taulman, L., 2005. A meta-analysis of genetically modified food valuation studies. Journal of Agricultural and Resource Economics 30: 28–44.

Marasas, W.F.O., 2001. Discovery and occurrence of the fumonisins: A historical perspective. Environmental Health Perspectives 109 (Supplement 2): 239–243.

Maumbe, B.M., and Swinton, S.M., 2003. Hidden health costs of pesticide use in Zimbabwe's smallholder cotton growers. Social Science and Medicine 57: 1559–1571.

Morse, S., Bennett, R., and Ismael, Y., 2004. Why Bt cotton pays for small-scale producers in South Africa. Nature Biotechnology 22: 379–380.

Munkvold, G.P., and Hellmich, R.L., 1999. Comparison of fumonisin concentrations in kernels of transgenic Bt maize hybrids and nontransgenic hybrids. Plant Disease 83: 130–138.

Murugkar, M., Ramaswami, B., and Shelar, M., 2006. Liberalization, Biotechnology, and the Private Seed Sector: The Case of India's Cotton Seed Market. Mimeo, Indian Statistical Institute, New Delhi, India.

Naseem, A., and Pray, C., 2004. Economic impact analysis of genetically modified crops. In: Handbook of Plant Biotechnology, P. Christou and H. Klee, eds., Wiley, Chichester, UK, pp. 959–991.

Pray, C.E., Ma, D., Huang, J., and Qiao, F., 2001. Impact of Bt cotton in China. World Development 29: 813–825.

Pray, C.E., Huang, J., Hu, R., and Rozelle, S., 2002. Five years of Bt cotton in China - the benefits continue. The Plant Journal 31: 423–430.

Pray, C.E., Bengali, P., and Ramaswami, B., 2005. The cost of biosafety regulations: The Indian experience. Quarterly Journal of International Agriculture 44: 267–289.

Pray, C.E., Huang, J., Hu, R., Wang, Q., Ramaswami, B., and Bengali, P., 2006a. Benefits and costs of biosafety regulation in India and China. In: Regulating Agricultural Biotechnology: Economics and Policy, R.E. Just, J.M. Alston and D. Zilberman, eds., Springer, New York, USA.

Pray, C.E., Ramaswami, B., Huang, J., Bengali, P., Hu, R., and Zhang, H., 2006b. Costs and enforcement of biosafety regulation in India and China. International Journal of Technology and Globalization 2: 137–157.

Price, G.K., Lin, W., Falck-Zepeda, J.B., and Fernandez-Cornejo, J., 2003. The Size and Distribution of Market Benefits from Adopting Agricultural Biotechnology. Technical Bulletin No. 1906, United States Department of Agriculture, Washington, DC, USA.

Qaim, M., 2003. Bt cotton in India: Field trial results and economic projections. World Development 31: 2115–2127.

Qaim, M., 2005. Agricultural biotechnology adoption in developing countries. American Journal of Agricultural Economics 87: 1317–1324.

Qaim, M., and de Janvry, A., 2003. Genetically modified crops, corporate pricing strategies, and farmers' adoption: The case of Bt cotton in Argentina. American Journal of Agricultural Economics 85: 814–828.

Qaim, M., and de Janvry, A., 2005. Bt cotton and pesticide use in Argentina: Economic and environmental effects. Environment and Development Economics 10: 179–200.

Qaim, M., and Matuschke, I., 2005. Impacts of genetically modified crops in developing countries: A survey. Quarterly Journal of International Agriculture 44: 207–227.

Qaim, M., and Zilberman, D., 2003. Yield effects of genetically modified crops in developing countries. Science 299: 900–902.

Qaim, M., Yarkin, C., and Zilberman, D., 2005. Impact of biotechnology on crop genetic diversity. In: Agricultural Biodiversity and Biotechnology in Economic Development, J. Cooper, L.M. Lipper and D. Zilberman, eds., Springer, New York, USA, pp. 283–307.

Qaim, M., Subramanian, A., Naik, G., and Zilberman, D., 2006. Adoption of Bt cotton and impact variability: Insights from India. Review of Agricultural Economics 28: 48–58.

Sahai, S., and Rahman, S., 2003. Performance of Bt Cotton in India: Data from the First Commercial Crop. Gene Campaign, New Delhi, India.

Shelton, A.M., Zhao, J.-Z., and Roush, R.T., 2002. Economic, ecological, food safety, and social consequences of the deployment of Bt transgenic plants. Annual Review of Entomology 47: 845–881.

Sunding, D., and Zivin, J., 2000. Insect population dynamics, pesticide use and farm-worker health. American Journal of Agricultural Economics 82: 527–540.

Thirtle, C., Beyers, L., Ismael, Y., and Piesse, J., 2003. Can GM-technologies help the poor? The impact of Bt cotton in Makhathini Flats, KwaZulu-Natal. World Development 31: 717–732.

Traxler, G., Godoy-Avila, S., Falck-Zepeda, J., and Espinoza-Arellano, J., 2003. Transgenic cotton in Mexico: Economic and environmental impacts. In: Economic and Environmental Impacts of First Generation Biotechnologies, N. Kalaitzandonakes, ed., Kluwer, New York, USA, pp. 183–202.

Turner, P.C., Moore, S.E., Hall, A.J., Prentice, A.M., and Wild, C.P., 2003. Modification of immune function through exposure to dietary aflatoxin in Gambian children. Environmental Health Perspectives 111: 217–220.

Umali-Deininger, D., and Sur, M., 2006. Food safety in a globalizing world: Opportunities and challenges for India. Plenary paper presented at the 26th International Association of Agricultural Economists (IAAE) Conference, August 12–18, Gold Coast, Australia.

USDA (United States Department of Agriculture), 2007. Adoption of Genetically Engineered Crops in the US. United States Department of Agriculture, Economic Research Service, Washington, DC, USA.

Wang, S., Just, D., and Pinstrup-Andersen, P., 2006. Tarnishing silver bullets: Bt technology adoption, bounded rationality and the outbreak of secondary pest infestations in China. Selected paper presented at the American Agricultural Economics Association Annual Meetings, Long Beach, CA, USA.

Welsh, R., and Ervin, D., 2006. Precaution as an approach to technology development: The case of transgenic crops. Science, Technology and Human Values 31: 153–172.

Wu, F., 2002. Bt or not Bt? Tools for regulatory decisions concerning genetically modified corn. Ph.D. dissertation, Engineering and Public Policy, Carnegie Mellon University, Pittsburgh, PA, USA.

Wu, F., 2006. Mycotoxin reduction in Bt corn: Potential economic, health, and regulatory impacts. ISB News Report, September, pp. 8–11.

Zilberman, D., Ameden, H., Graff, G., and Qaim, M., 2004. Agricultural biotechnology: Productivity, biodiversity, and intellectual property rights. Journal of Agricultural & Food Industrial Organization (online), 2 (2), article 3. Available at: http//www.bepress.com/jafio/vol2/iss2/art3.

Chapter 13
Beyond *Bt*: Alternative Strategies for Insect-Resistant Genetically Modified Crops

Louise A. Malone[1,*], Angharad M.R. Gatehouse[2], and Barbara I.P. Barratt[3]

Abstract *Bacillus thuringiensis* (*Bt*) plants dominate today's commercial market for insect-resistant transgenic crops. However, not all pests are susceptible to *Bt* Cry toxins and there are concerns that even susceptible species may evolve to become resistant to these crops. The search for alternatives is well under way, with significant progress already made towards producing transgenic crops expressing insecticidal compounds from plants, such as protease inhibitors, lectins and alpha-amylase inhibitors. New types of proteins from *B. thuringiensis*, such as the vegetative insecticidal proteins, are also being exploited. At an earlier stage of development but attracting much research interest are other insecticidal compounds, such as chitinases, defensins, enhancins, biotin-binding proteins, proteases and toxins, sourced from bacteria, viruses, plants and arthropods. Fusion proteins, combining the features of different insecticidal proteins, have significant potential for extending the range of insect species which could be controlled via transgenic plants. In the future, metabolic engineering of plants could allow us to alter with great precision the ways in which plants and insects interact. The compatibility of these novel insect control strategies with biological control and integrated pest management is discussed.

13.1 Introduction

Since the first insect-resistant transgenic crops were planted commercially in 1996 there has been a relentless and remarkable increase in their adoption, such that 22 countries now grow 20.1 million hectares of *Bt* (*Bacillus thuringiensis*) crops and

[1] The Horticulture and Food Research Institute of New Zealand Limited, Auckland, New Zealand

[2] University of Newcastle, UK

[3] AgResearch Invermay, PB 50034, Mosgiel, New Zealand

* To whom correspondence should be addressed. E-mail: lmalone@hortresearch.co.nz

J. Romeis, A.M. Shelton, G.G. Kennedy (eds.), *Integration of Insect-Resistant Genetically Modified Crops within IPM Programs.*

a further 22 million hectares of *Bt*/HT (herbicide tolerant) stacked trait GM crops (James, 2007). Cotton and corn are the predominant plant species and, with the exception of the serine protease inhibitor, cowpea trypsin inhibitor (CpTI), expressed in combination with *Bt* in cotton in China, Cry toxins are the only commercialised insect resistance traits for GM crops.

Early commercial varieties of insect-resistant transgenic crops expressed single Cry proteins with specific activity against lepidopteran pests, e.g., Monsanto's Bollgard® cotton expressing Cry1Ac and Syngenta's Attribute® maize expressing Cry1Ab. Subsequently other lepidopteran-active *Bt* toxins, such as Cry1F and Cry2Ab2, have been added to the list of commercialised traits, and often these are presented as pyramided genes in a single variety (e.g., Dow Agrosciences' Widestrike® cotton (Cry1F + Cry1Ac) and Monsanto's Bollgard II® (Cry1Ac + Cry2Ab2)). A chimeric *Bt* gene encoding Cry1A.105, which is composed of parts of Cry1Ab, Cry1Ac and Cry1F, is also being commercialised in maize (USEPA, 2006a). This pyramiding of *Bt* genes has several advantages. Effectiveness in controlling different species within a suite of lepidopteran pests may be enhanced, since different Cry proteins vary in their effectiveness against individual species, even within the same order. Pyramiding may reduce the likelihood of resistance evolving in the pest insects, especially if the pyramided *Bt* toxins use different gut receptors (Tabashnik et al., 2002; Bates et al., 2005), although resistance management practices involving refuges will still need to be employed (Manyangarirwa et al., 2006). Finally, the availability of a range of *Bt* genes with essentially similar types of activity has allowed different companies to produce competing products without patent infringement.

In addition to the lepidopteran-active Cry1 and Cry2 toxins, Cry3 toxins with activity against coleopteran pests are also being used in commercial transgenic crops, particularly maize, which can be significantly damaged by chrysomelid rootworms (e.g., Monsanto's Yieldgard Rootworm® maize expressing Cry3Bb1, Dow Agrosciences' Herculex RW® maize expressing Cry34Ab1 and Cry35Ab1, stacked with a HT gene, and Syngenta's Agrisure RW® maize expressing a modified version of Cry3A). More recently released transgenic maize varieties have these genes stacked with lepidopteran-active *cry* genes to give simultaneous protection against caterpillars and rootworms, and often have HT genes included as well (e.g., Monsanto's Yieldgard Plus®, Dow Agroscience's Herculex XTRA®, and Syngenta's Agrisure CB/RW®).

Bt transgenic varieties of other high-acreage crops with significant lepidopteran and/or coleopteran pests, such as rice, potatoes, tomatoes, eggplants and brassicas, have also been developed and are very likely to be widely adopted in the next few years. In the future, we are likely to see the development of new, modified *Bt* transgenes using domain swapping techniques to improve efficacy (Naimov et al., 2003; Singh et al., 2004).

Producing new varieties of *Bt* transgenic cotton and maize continues to be an area of significant commercial activity for several reasons: returns are maximised by concentrating on developing new variations on what is now a familiar technology for users such as cotton and maize growers, and the costs of meeting regulatory requirements for human and environmental health are minimised by using traits for which there is already a significant body of biosafety data.

However, there are some significant limitations with *Bt* biotechnology that are driving a search for alternative transgenic strategies. The first of these is the potential for insects to evolve resistance to Cry toxins expressed by transgenic *Bt* crops. This has been amply demonstrated in laboratory experiments in which resistant insects were selected for by exposing successive generations to purified Cry toxins (Tabashnik et al., 2003; Ferré et al., chapter 3). Resistance was widely expected to arise rapidly once *Bt* crops were planted in the field (Gould et al., 1997) but even after 11 years of planting, field resistance has not yet been convincingly demonstrated, probably because of the success of resistance management strategies involving the planting of non-*Bt* refuges which have been mandated in many countries (Tabashnik et al., 2003). Biotechnology companies have invested significantly in developing and implementing resistant management strategies and pyramiding *Bt* genes in crops to delay the spread of Cry toxin resistance genes among pest populations. Future development of transgenic perennial crops expressing insect resistance proteins over long periods of time presents another challenge to the durability of pest control strategies based solely on *Bt* toxins. The use of non-*Bt* insect resistance traits with totally different modes of action, such as protease inhibitors or lectins, solely or in combination with *Bt*, has long been advocated as a means of delaying selection for resistant pests (Boulter et al., 1990).

The second driver for finding alternatives to *Bt* is the very specific activities of different Cry toxins. Although this is a strength of the technology in the sense that non-target insects may be less affected, it can also be a limitation to its effectiveness. A large number of different Cry proteins has now been identified and although generalisations can be made about groups of toxins (e.g., those of the Cry1 family are all active against Lepidoptera), individual toxins within each group can vary in their ability to control different pest species (van Frankenhuyzen and Nystrom, 2002). One of the major commercial advantages of transgenic crops is that spraying of traditional synthetic insecticides is significantly reduced. This has created a commercial imperative to "build into the seed" as complete a pest control technology as possible. The characterisation and isolation of Cry proteins effective in controlling Coleoptera has gone some way towards achieving this goal, and transgenic crops with stacked toxins protected from both caterpillar and beetle attack are some of the newest products on the market. Cry toxins with activity against Diptera are also well-characterised and could also be used in transgenic crops. However, there are as yet no *Bt* toxins ready for market which could be used for control of sucking pests such as mirids, thrips, bugs, hoppers or aphids. Mirids and stinkbugs are significant secondary pests of *Bt* cotton in some countries, and aphids and hoppers are important pests and disease vectors of cereals.

A third (non-biological) factor driving the development of alternatives to *Bt* is the commercial requirement for new players in this market to find novel insect resistance genes to avoid infringing existing patents.

Thus, for some time now there has been a search for alternatives to *Bt* Cry toxins for incorporation into transgenic plants. Of these, the most advanced in their development are the protease inhibitors (PIs) and the vegetative insecticidal proteins (VIPs), isolated from *Bt*. Crops expressing CpTI have been in commercial production in China since 2000 (Song and Wang, 2001) and VIP-expressing cotton is apparently close to market in the USA. There has also been considerable research and development work on alpha-amylase inhibitors, lectins and biotin-binding proteins. Another recent development has been the creation of fusion proteins, which combine different domains from various types of proteins to produce novel proteins with enhanced insecticidal properties (Christou et al., 2006). Many of these alternatives to *Bt* crops have a broader range of activity than the currently commercialised Cry toxins, and the need to stack several Cry genes to achieve toxicity to an adequate range of pest species will be avoided. On the other hand, few of the alternatives offer the speed of kill that *Bt* Cry toxins do. Creating fusions and raising expression levels may address this, and such plants may follow a commercialisation pathway similar to that of *Bt* transgenic crops. However, others may be used in integrated pest management (IPM) systems of the future in a way that more resembles today's use of plant varieties bred conventionally to have enhanced resistance to pests or diseases, rather than as a complete substitute for synthetic insecticides. Recently researched alternatives to *Bt* Cry toxins are described below, their potential for impacts on natural enemies discussed, and their likely compatibility with IPM programs, in particular their potential impacts on biological control, considered.

The preservation of natural enemy species and the biological control function they provide is a central requirement of most IPM systems. Novel insecticidal proteins expressed in transgenic plants may have direct or indirect impacts on natural enemy species. Direct effects may occur if the natural enemy species is exposed to (and susceptible to) the protein itself. A natural enemy may be directly exposed to an insecticidal protein if it is expressed in the pollen or nectar of the transgenic plant which the natural enemy uses to supplement its diet, or if the prey/host insect has active residues of the insecticidal protein in its body, perhaps located in the gut lumen or even transported into the hemolymph. Indirect effects may occur when the effects of the protein on the target insect are such that it becomes a less (or more) attractive or nutritious item of prey. Negative effects on natural enemy populations arising as a consequence of the prey insect being controlled by the transgenic plant are an inevitable consequence of pest control, and are not considered here, as the issues involved are not peculiar to transgenic plants. They are no different from those encountered with any insecticidal technology (Romeis et al., chapter 4). IPM uses a variety of insect control technologies in concert to achieve reductions in pest populations such that damage to the crop is kept below economic thresholds with minimal use of synthetic insecticides. Where economic damage thresholds are low, this may equate to complete elimination of the pest from the crop. Where thresholds are higher, complete eradication of the pest species may not be necessary. In both cases, if biological control is an important component of IPM, then some pests will need to survive within the agricultural landscape in order to support self-sustaining populations of natural enemies. Because of this, a major concern with using GM insect-resistant plants in IPM is the possibility of effects on natural enemies that go beyond those already accepted when synthetic insecticides are used.

Other impacts on IPM systems are possible. For example, some of the plants discussed here are slower to kill pests than either *Bt* plants or synthetic insecticides, and different management systems may well be required to incorporate them effectively into an IPM system. These effects are likely to vary from crop to crop, however, and any discussion at this stage of their development would be highly speculative. Thus the focus of this chapter is the compatibility of non-*Bt*, pest-resistant transgenic plants with biological control and the preservation of natural enemies.

13.2 Protease Inhibitors

13.2.1 Protease Inhibitors as Insecticidal Proteins

The contribution of PIs produced by plants for protection against pest attack has been extensively researched since early observations of insecticidal properties of soybeans. The first insect-resistant transgenic plants expressing a PI were produced in 1987 (Hilder et al., 1987). Many other crop plant transformations with PIs have followed (see reviews such as Jouanin et al., 1998; Hilder and Boulter, 1999; Carlini and Grossi-de-Sa, 2002; Lawrence and Koundal, 2002; Ferry et al., 2006), as has the discovery that insects can sometimes respond to the ingestion of particular PIs by producing new resistant forms of digestive proteases (Bown et al., 1997; Jongsma and Bolter, 1997). Overproduction of proteases and production of inactive proteases also seems to occur (Christeller et al., 2005). There is now an understanding that the production of some PIs by plants and the responses of insects to them form just part of a complex, dynamic system of signalling between plants and herbivores (Ferry et al., 2006).

13.2.2 Serine Protease Inhibitors

The effects on insect herbivores of plant-derived inhibitors of serine proteases such as trypsin, chymotrypsin and elastase have been particularly well-studied (Gatehouse et al., 2000). A defence role for serine PIs in plants has been suggested since they commonly occur in seeds, which need long-term protection from pest attack, and also their synthesis can be induced by plant wounding, as occurs with insect feeding (Gatehouse et al., 2000). Furthermore, it has been noted that the major protein-processing functions in plants seem to involve cysteine, as opposed to serine proteases, suggesting that the abundant serine PIs are not so much involved in endogenous protease regulation and must have a different role (Green and Ryan, 1972).

Serine PIs inhibit the activity of specific serine proteases in the guts of susceptible insect species and may cause mortality in both Lepidoptera and Coleoptera. This antimetabolic effect is apparently the direct result of a loss of digestive function via PI-protease binding and inactivation, and perhaps also the consequence of a loss of amino acids needed for synthesis of new proteases (Gatehouse et al., 2000).

The first plant PI to be expressed in an insect-resistant transgenic plant was the serine PI CpTI (Hilder et al., 1987) and plants expressing this PI are the most advanced in terms of commercial development today. Table 13.1 lists some recently-studied serine PIs and gives their current status as potential or actual transgenic insect-resistant crops. Some have been stacked with other insect-resistance proteins (see section 13.9 and Table 13.8 below). Earlier (pre-2001) reports of trials with insect resistant PI-expressing transgenic plants are summarised elsewhere (Hilder and Boulter, 1999; Carlini and Grossi-de-Sa, 2002).

13.2.3 Serine Protease Inhibitors and Natural Enemies

The potential for compatibility of serine-PI-expressing transgenic plants with IPM systems has been investigated throughout their development and there have been many studies of their impacts on natural enemy species (Table 13.2).

One concern with serine PIs has been that direct effects may occur if the natural enemy's gut protease profile suggests a reliance on serine proteases and there is sufficient opportunity for direct exposure to the protein. *In vitro* inhibition of gut proteases by serine PIs has been demonstrated in the carabid beetles, *Pterostichus madidus* and *Ctenognathus novaezelandiae*, although tri-trophic experiments with transgenic plants and these predators showed only minor transient negative impacts, suggesting either insufficient exposure to the PIs or the action of compensatory mechanisms in their guts (Glare et al., 2004; Ferry et al., 2005; Burgess et al., 2008). Larvae and adults of the hymenopteran parasitoids *Aphelinus abdominalis* and *Aphidius ervi* use predominantly serine proteases for digestion and their fitness is impaired when they parasitise aphids fed on diet containing soybean Bowman-Birk inhibitor (Azzouz et al., 2005a, b).

Indirect negative effects, via inferior quality prey, have been demonstrated in the predatory stink bug, *Podisus maculiventris* preying on *Lacanobia oleracea* injected with CpTI, but not when CpTI-expressing transgenic plants were used as the prey's food (Bell et al., 2003). CpTI expression levels in these transgenic plants were 1% of total soluble protein and thus within the range where insecticidal effects on herbivorous larvae have been observed (Gatehouse et al., 1997; Bell et al., 2001a). To eliminate the possibility of tri-trophic effects due simply to the prey being stunted by its exposure to the plants, the prey larvae used were large (3rd instar) and were fed with the transgenic leaves for only 6–8 days, so that their growth would not be significantly altered by the treatment. This experiment therefore examined the possibilities that the predators might be affected by prey being nutritionally sub-optimal

Table 13.1 Some recently researched serine protease inhibitors conferring insect resistance on transgenic crop plants

Proteins	Target pests	Target crops	Status	References
Cowpea trypsin inhibitor (modified) (CpTI)	*Scripophaga incertulas* (Lepidoptera: Pyralidae)	Rice	Field trials (China)	Huang et al., 2005
	Otiorhynchus sulcatus (Coleoptera: Curculionidae)	Strawberry	Field trials (United Kingdom)	Graham et al., 2002
	Lacanobia oleracea (Lepidoptera: Noctuidae)	Potato	Greenhouse trials (United Kingdom)	Bell et al., 2001a, 2003
	Pieris rapae (Lepidoptera: Pieridae)	Cauliflower	Experimental plants produced	Lingling et al., 2005
	Sitotroga cerealella (Lepidoptera: Gelechiidae)	Wheat	Experimental plants produced	Bi et al., 2006
Nicotiana alata protease inhibitor (NaPI)	*Helicoverpa armigera, Helicoverpa punctigera* and others (Lepidoptera: Noctuidae)	Cotton	Application for field trial made (Australia)	OGTR, 2004
	Epiphyas postvittana (Lepidoptera: Tortricidae)	Apple	Greenhouse trials (Australia)	Maheswaran et al., 2007
Potato protease inhibitor 1 (Pot 1)	*H. armigera*, *H. punctigera* and others	Cotton	Application for field trial made (Australia)	OGTR, 2004
Barley trypsin inhibitor (CMe)	*Sitophilas oryzae* (Coleoptera: Curculionidae)	Rice	Experimental plants produced	Alfonso-Rubi et al., 2003
Soybean trypsin inhibitor (SBTI) + snowdrop lectin (GNA)	*Nilaparvata lugens* (Hemiptera: Delphacidae), *Cnaphalocrocis medinalis* (Lepidoptera: Pyralidae)	Rice	Experimental plants produced	Li et al., 2005
Arrowhead proteinase inhibitor (API-B) + Cry 1Ac	*H. armigera*	Cotton	Field trials; safety certificate for production trials issued	Guo et al., 2003; Wu et al., 2005
Mustard trypsin inhibitor (MTI-2)	*Plutella xylostella* (Lepidoptera: Plutellidae)	Oilseed rape	Experimental plants produced	De Leo et al., 2001
Maize proteinase inhibitor (MPI)	*Chilo suppressalis* (Lepidoptera: Crambidae)	Rice	Experimental plants produced	Vila et al., 2005
Potato proteinase inhibitor (PINII-2x)	*C. suppressalis*	Rice	Experimental plants produced	Bu et al., 2006
Locust proteinase inhibitor peptides (SGCI, SGTI)	*Leptinotarsa decemlineata* (Coleoptera: Chrysomelidae)	Potato	Experimental plants produced	Kondrak et al., 2005

Table 13.2 Effects of serine protease inhibitors on insect predators and parasitoids

Proteins	Host/prey insect	Transgenic plant or diet; concentration or expression level	Potential effects on hosts/prey	Effects on hosts/prey used in this trial	Natural enemy	Stage tested	Effects on natural enemy of protein-exposed hosts/prey	References
Cowpea trypsin inhibitor (CpTI)	*Lacanobia oleracea* (Lepidoptera: Noctuidae)	Injected prey; 5 or 10 µg per late 4th instar larva	N/A[1]	None detected as prey's head capsule crushed immediately after injection	*Podisus maculiventris* (Hemiptera: Pentatomidae)	Newly-hatched nymphs	Female adult bugs weighed less than controls if fed as nymphs on prey injected with 10 µg CpTI	Bell et al., 2003
		Potato; 1% of total soluble protein	Prey biomass reduced to 45% cf. controls if fed as neonate larvae	None detected as prey deliberately fed at 3rd instar and used in experiment 6–8 days later (4th instar)	*P. maculiventris*	2nd instar nymphs	No effects on development time (2nd instar to adult) or adult bug weights	Gatehouse et al., 1997; Bell et al., 2003
		Diet; 2% of soluble protein	Prey larvae significantly smaller, slower growing if fed from 1st instar	Minimal as prey deliberately fed at 3rd and used in experiment at 4th instar	*Eulophus pennicornis* (Hymenoptera: Eulophidae)	48 h old adults	No effect on rate of parasitism, parasite emergence or development time, or fecundity of emergent adults	Bell et al., 2001a

		Potato; up to 1% soluble protein	Prey larvae weighed 50% less than controls at 4th instar after feeding from 3rd instar onwards	Minimal as prey deliberately fed at 3rd or 4th instar and used in experiment at 4th or 5th instar	*E. pennicornis*	48 h old adults	Fewer prey parasitised than controls, only if the prey had been exposed to the plants from 3rd instar (those exposed from 4th instar were not affected); fecundity of F_1 wasps unaffected	Bell et al., 2001a
	N/A; adult parasitoids fed directly	Honey solution (50% aqueous) containing 0.01%, 0.1%, or 1.0% w:v	N/A	N/A	*E. pennicornis*	Newly-emerged adults fed continuously	Mean adult longevity, egg loads, numbers of hosts attacked, numbers of progeny produced, their pupation and subsequent adult emergence success not affected	Bell et al., 2004

(continued)

Table 13.2 (contined)

Proteins	Host/prey insect	Transgenic plant or diet; concentration or expression level	Potential effects on hosts/prey	Effects on hosts/prey used in this trial	Natural enemy	Stage tested	Effects on natural enemy of protein-exposed hosts/prey	References
	Otiorhynchus sulcatus (Coleoptera: Curculionidae)	Strawberry ("high" or "moderate" expression level as determined by trypsin activity assay)	Feeding damage to strawberry roots reduced in glasshouse trial	Numbers of pupae reduced; plant growth improved c.f. controls	*Trechus quadristriatus*, *Calanthus melanocephalus* and other unidentified spp. (Coleoptera: Carabidae), unidentified spp. of spiders (Araneae), Collembola, Diptera, Dermaptera, slugs (Mollusca), millipedes and centipedes (Myriapoda)	Field study	Field abundance not affected	Graham et al., 1997, 2002
Soybean Bowman-Birk inhibitor (SbBBI)	*Macrosiphum euphorbiae* (Hemiptera: Aphididae)	Diet; 20, 100, $500\,\mu g.mL^{-1}$	No effect on nymphal survival; adults fed 100 or $500\,\mu g.mL^{-1}$ had reduced fecundity	Used only hosts fed with $100\,\mu g.mL^{-1}$	*Aphelinus abdominalis* (Hymenoptera: Aphelinidae)	Females exposed to 5-day-old aphids	Emergence success reduced; adult resistance to starvation increased	Azzouz et al., 2005b

		Diet; 100, 500 µg.mL^{-1}	As above	As above	*Aphidius ervi* (Hymenopetra: Braconidae)	As above	Female size and sex ratio altered	Azzouz et al., 2005a
	N/A; adult parasitoids fed directly	Sugar solution containing 100 µg.mL^{-1}	N/A	N/A	*A. ervi*	1-day-old females fed daily	No effect on longevity; digestive protease activity partially inhibited	Azzouz et al., 2005a
		As above	N/A	N/A	*A. abdominalis*	As above	Longevity not affected	Azzouz et al., 2005b
Mustard trypsin inhibitor (MTI-2)	*Plutella xylostella* (Lepidoptera: Plutellidae)	Oilseed rape (expression levels not given but all inhibited >30% of trypsin activity)	Larval survival reduced after 5 days; survivors had lower mean weights in final instar (if fed MTI-2 plants from egg-hatch)	Only prey that had reached final instar used (all had received transgenic plant food for 5 days)	*Pterostichus madidus* (Coleoptera: Carabidae)	Mated adults fed unlimited prey, or only 50 mg prey per week	No effects on adult survival; female weight gain reduced; no effect on fecundity	Ferry et al., 2005
Aprotinin (bovine pancreatic or bovine spleen trypsin inhibitor) (BPTI/BSTI)	*Helicoverpa armigera* (Lepidoptera: Noctuidae)	Diet; 1% w:w of total protein content	Transgenic tobacco expressing 0.5% BSTI reduced larval growth and survival	Prey larvae exposed to diet for 7–8 days (3rd instar)	*Harpalus affinis* (Coleoptera: Carabidae)	Adults fed on BPTI-fed prey for 24 h only; no-choice	Beetles consumed less prey	Jorgensen and Lovei, 1999; Christeller et al., 2002
		Diet, 0.5% w:w of total fresh weight	As above	Prey bodies (4th instar) contained 22.6 µg.g^{-1} aprotinin	*Nebria brevicollis* (Coleoptera: Carabidae)	Adults	Transient minor changes in beetle weights	Burgess et al., 2002a; Christeller et al., 2002

(continued)

Table 13.2 (continued)

Proteins	Host/prey insect	Transgenic plant or diet; concentration or expression level	Potential effects on hosts/prey	Effects on hosts/prey used in this trial	Natural enemy	Stage tested	Effects on natural enemy of protein-exposed hosts/prey	References
	Spodoptera litura (Lepidoptera: Noctuidae)	Tobacco; 0.34–0.46% w:w of total protein	As above	Only 3rd instar larvae used; same amounts of weighed prey used to counteract effects on prey size	*Ctenognathus novaezelandiae* (Coleoptera: Carabidae)	Larvae fed for 91 days; adults for 280 days	No effects on adult or larval survival, food consumption by larvae or adult fecundity; less female weight gain	E.P.J. Burgess, personal communication; Burgess et al., 2008; Glare et al., 2004

[1] Not applicable.

(although of "normal" size) and/or via direct exposure by eating large larvae with guts filled with transgenic leaf material. Neither produced a significant effect in this case.

Studies thus far indicate that although the potential exists for negative impacts of some serine PI-expressing transgenic plants on some natural enemies, such effects are not likely to be major. If the PI is effective enough to be used as a control technology, many of the prey will be killed at an early stage and thus be unavailable to most natural enemies (as with all successful control technologies). However, the studies described here are concerned with the possibility that natural enemies may be exposed to prey that have survived ingesting the PI, perhaps because they were not exposed until a late instar, or only exposed for a short time. No insect control technology is perfect and there will still be some pest insects in the crop even after economic control thresholds have been reached. These insects could well be important in maintaining natural enemy populations. The tri-trophic studies presented here show that prey exposed to but not killed by PI-expressing plants are unlikely to significantly affect their natural enemies, probably because direct exposure of the predator or parasitoid to any active PI remaining in the prey's body is likely to be at a low level and impacts on prey quality do not seem to alter natural enemy fitness significantly in most cases.

13.2.4 Case Study: CpTI Cotton

Cotton is an economically important fibre crop, and in 2005 approximately 25 million metric tons, grown in over 100 countries, was produced, with four countries (China, the USA, India and Pakistan) accounting for two thirds of the world output (ICAC, 2004). Recent figures estimate the annual value at US$35 billion per year. Due to the very high levels of loss as a result of insect damage, and the high levels of insecticides required, the development of insect-resistant transgenic cotton was an obvious goal for the industry. *Bt* (Cry1Ac) expressing cotton has now been commercially available for over 10 years, and following its introduction there have been major environmental gains. Since 1996 there has been an estimated 24% reduction in the environmental impact, and a 19% decrease in the volume of insecticides applied (Brookes and Barfoot, 2006). Furthermore, there have been substantial economic gains (through a combination of higher yields and lower costs) (see Naranjo et al., chapter 6; Qaim et al., chapter 12). In 2005, cotton farm income levels in countries adopting transgenic cotton were $1.9 billion higher, and since 1996 the sector has benefited from an additional $8.44 billion. The 2005 income gains are equivalent to adding 13.3% to the value of the cotton crop in these countries, or 7.3% to the value of total global cotton production. Also in that year, it was estimated that cotton accounted for 11% of the transgenic crops grown (approximately 10 million hectares). In the USA, South Africa, Australia, India and China, *Bt* and *Bt*/HT transgenic cotton crops, as a percentage of their total cotton plantings in 2007, were estimated to be 72%, 85%, 90%, 66% and 69%, respectively (James, 2007).

Whilst the adoption of *Bt*-expressing cotton, for control of the budworm-bollworm complex (*Helicoverpa* spp., *Heliothis virescens*, *Pectinophora gossypiella*), saw a significant reduction in insecticide usage, this was accompanied, in some regions, by the emergence of other insects as pests of cotton. For example, in Australia aphids, mirids, thrips and jassids are now recognized as emerging pests of cotton (see Naranjo et al., chapter 6). Many laboratories are thus actively identifying other useful transgenes to express in cotton, not just with the view to increasing the spectrum of insect pests targeted, but also to increase durability of *Bt* cotton in the field. Whilst one strategy to ensure durability of control against the lepidopteran pests has been to pyramid different *Bt* genes (e.g., Cry1Ac + Cry2Ab), other approaches involve the co-expression of genes whose products act on different targets within the insect (see section 13.3.4 and Naranjo et al., chapter 6).

In China cotton is attacked by a number of different insect pests, but the major ones are the cotton bollworm (*Helicoverpa armigera*), cotton pink bollworm (*P. gossypiella*) and the cotton aphid (*Aphis gossypii*), all of which can cause serious yield loss if left uncontrolled (Wu and Guo, 2005). To address damage by the two major lepidopteran pests, and to address issues concerning durability, scientists at the Institute of Genetics of the Chinese Academy of Sciences (CAS) and the Biotechnology Research Centre of the Chinese Academy of Agricultural Sciences (CAAS) successfully transformed the plant expression vector pGBI121S4ABC, harboring both a synthesized *Bt* Cry1Ac gene and the modified CpTI gene, into the elite cotton cultivars SGK 321, CCRI 19, 3517 and 541, using the pollen tube pathway method (Guo et al., 1999); some of the resulting transformants, expressing both transgenes, were highly toxic to larvae (mortality up to 96%). The major reason for co-expressing *Bt* and CpTI in cotton was to reduce the likelihood of the insects becoming resistant to this cultivar and to extend its effective life. In addition to its high levels of resistance to lepidopteran pests, SGK 321 also exhibited desirable yield characteristics and fiber quality. Prior to commercial release, cultivars were evaluated for performance in small- and then large-scale field trials (Zhang et al., 2000). Since larval feeding behavior will influence insecticidal efficacy of the transgenic crops and the evolution of resistance within the pest population, differences in feeding behavior of *H. armigera* on non-transgenic cotton, *Bt* cotton and CpTI + *Bt* cotton were compared. The results demonstrated that larvae had the ability/tendency to avoid *Bt* and/or CpTI + *Bt* cotton, but that the differences in behavior between the two transgenic lines were not significant (Zhang et al., 2004). CpTI + *Bt* cotton cultivars were commercially released in China in 2000 (Song and Wang, 2001), and in 2005 accounted for approximately 15% of the cotton grown (He et al., 2008).

13.2.5 Cysteine Protease Inhibitors

Plants are also good sources of cysteine PIs, which are thought to regulate endogenous plant processes involving cysteine proteases, as well as providing protection against insect herbivores that use these proteases for their digestion (Arai and Abe, 2000).

Cysteine proteases appear to be particularly important digestive enzymes for beetle species and so the development of cysteine-PI-expressing transgenic plants has been proposed as an effective way of targeting coleopteran pests. Table 13.3 lists some recent (post-2000) reports of insect-resistant transgenic plants expressing cysteine PIs.

Oryzacystatin 1 (OC1) is a well-studied cysteine PI from rice seeds which has been successfully introduced into several different crop plant species where it protects against attack by beetles and, in some cases, aphids (Table 13.3). OC1-oilseed rape has been proposed for control of the suite of coleopteran pests of oilseed rape in Europe and OC1-potato for Colorado potato beetle control, but there is as yet no sign that such crops are close to commercialisation. A cystatin from barley has also provided protection against Colorado potato beetle, *Leptinotarsa decemlineata* (Álvarez-Alfageme et al., 2007).

Beetle pests of trees are also likely targets for transgenic cysteine-PI-plants and poplars expressing OC1 or *Arabidopsis thaliana* cysteine protease inhibitor have both been shown to exert a controlling effect on cerambycid beetle species (Leple et al., 1995; Delledonne et al., 2001).

Cysteine PIs may also be isolated from non-plant sources and recently equistatin, from the sea anemone *Actinia equina*, has been expressed in potatoes for control of Western flower thrips (Outchkourov et al., 2004a), although a synthetic multidomain cysteine PI incorporating domains from equistatin, kininogen, stefin A, cystatin C, and potato cystatin gave better control (Outchkourov et al., 2004b).

13.2.6 Cysteine Protease Inhibitors and Natural Enemies

As with the serine PIs, there has been extensive investigation of the potential non-target impacts of cysteine PIs (Table 13.4). Because cysteine PIs were found to be particularly useful for control of coleopteran pests, there was concern that natural enemies such as predatory beetles might also be deleteriously affected. However, experiments so far have shown that predatory beetles respond to prey fed with cysteine PIs by either up-regulation of native forms or synthesis of novel forms of their digestive proteases, and that they are not negatively affected by such prey (Bouchard et al., 2003a; Ferry et al., 2003; Mulligan et al., 2006). The predatory bug, *P. maculiventris*, was unaffected by potato expressing barley cystatin (Álvarez-Alfageme et al., 2007). Results with hymenopteran parasitoids have varied, with negative effects shown for *A. abdominalis* and *A. ervi* (Azzouz et al., 2005a, b), positive effects on *Aphidius nigripes* (Ashouri et al., 2001a), and varied results, although no change in pest control effectiveness, with *Diaeretiella rapae* (Schuler et al., 2001). A field study with OC1-potatoes, designed to be resistant to potato cyst nematodes, showed that numbers of non-target aphids (and their natural enemies) were significantly reduced on non-transgenic plants treated with nematicide but not on untreated non-transgenic or transgenic potatoes (Cowgill et al., 2004). Parasitoid communities were more diverse on the transgenic OC1-potatoes than on the non-transgenic plants not treated with nematicide.

Table 13.3 Some recently researched cysteine protease inhibitors conferring insect resistance on transgenic crop plants

Proteins	Target pests	Target crops	Status	References
Oryzacystatin (OC1)	*Myzus persicae* (Hemiptera: Aphididae)	Oilseed rape	Experimental plants produced	Rahbé et al., 2003
	Leptinotarsa decemlineata (Coleoptera: Chrysomelidae)	Potato	Experimental plants produced	Lecardonnel et al., 1999
	M. persicae, *Macrosiphum euphorbiae* (Hemiptera: Aphididae)	Eggplant	Experimental plants produced	Ribeiro et al., 2006
	Species not specified (Coleoptera, Hemiptera) (insect damage to plants in field reduced)	Cabbage	Experimental plants produced	Lei et al., 2006
	Chrysomela tremulae (Coleoptera: Chrysomelidae)	Poplar	Experimental plants produced	Leple et al., 1995
Arabidopsis thaliana cysteine protease inhibitor	*Chrysomela populi* (Coleoptera: Chrysomelidae)	White poplar	Experimental plants produced	Delledonne et al., 2001
Barley cystatin (HvCPI-1)	*L. decemlineata*	Potato	Experimental plants produced	Álvarez-Alfageme et al., 2007
Equistatin, sea anemone cysteine protease inhibitor	*Frankliniella occidentalis* (Thysanoptera: Thripidae)	Potato	Experimental plants produced, US patent issued 2000	Outchkourov et al., 2004a

Table 13.4 Effects of cysteine protease inhibitors on insect predators and parasitoids[1]

Proteins	Host/prey insect	Transgenic plant or diet; concentration or expression level	Potential effects on hosts/prey	Effects on hosts/prey used in this trial	Natural enemy	Stage tested	Effects on natural enemy of protein-exposed hosts/prey	References
Oryzacystatin (OC1)	*Deroceras reticulatum* (Mollusca)	Oilseed rape; 0.01–0.06% w:w of total soluble protein	Negligible effects over 30 days on survival and weight gain of juvenile slugs	Not stated; slugs fed for a minimum of 2 days on OC1 plants	*Pterostichus melanarius* (Coleoptera: Carabidae)	Adults; limited or unlimited feeding for 30 days	No effects on mortality, weight gain, or food consumption	Mulligan et al., 2006
	Plutella xylostella (Lepidoptera: Plutellidae)	Oilseed rape; 0.03–0.06% w:w of total soluble protein	No effects on larval growth, development or survival	No effects; OC-1 accumulates in larval tissues up to 3 ng per gut	*Harmonia axyridis* (Coleoptera: Coccinellidae)	Entire life cycle	No effects on larval development, adult or larval survival, adult fecundity, egg viability	Ferry et al., 2003
	Leptinotarsa decemlineata (Coleoptera: Chrysomelidae)	Potato; 0.8% w:w of total soluble protein	Larvae develop faster; adults consume foliage faster, no effects on weight gain or eggs laid	No effects	*Perillus bioculatus* (Hemiptera: Pentatomidae)	Larvae; exposed to prey eggs and larvae for 5 days	No effects on development, growth or survival	Clouthier and Michaud 2000; Bouchard et al., 2003b

(continued)

Table 13.4 (continued)

Proteins	Host/prey insect	Transgenic plant or diet; concentration or expression level	Potential effects on hosts/prey	Effects on hosts/prey used in this trial	Natural enemy	Stage tested	Effects on natural enemy of protein-exposed hosts/prey	References
	Macrosiphum euphorbiae (Hemiptera: Aphididae)	Diet; 20, 100, 250, or 500 $\mu g.mL^{-1}$	Nymphal survival reduced; no adults emerged (all doses)	Only used hosts fed with 100 $\mu g.mL^{-1}$	*Aphelinus abdominalis* (Hymenoptera: Aphelinidae)	Female parasitoids exposed to 5-day-old aphids that had been exposed to OC1	Larval development time extended; mummification rate and nymphal development time not affected; mummy weights reduced; adult emergence reduced	Azzouz et al., 2005b
		Diet; 100, 500 $\mu g.mL^{-1}$	As above	As above	*Aphidius ervi* (Hymenoptera: Braconidae)	As above	Reduced percent mummified; no adults emerged	Azzouz et al., 2005a
		Potato; 1% w:w of total soluble protein	Larvae develop faster, adults heavier, more progeny	No effects; hosts fed on potatoes from neonate to 3rd instar only	*Aphidius nigripes* (Hymenoptera: Braconidae)	Entire life cycle	No effects on larval development time or total survival; newly-emerged adults weighed more and laid more eggs per day in OC1-potato-fed aphids than controls	Ashouri et al., 2001a, b

	N/A; adult parasitoids fed directly	Sugar solution containing 100 μg.mL^{-1}	N/A	N/A	*A. ervi*	One-day-old females fed daily	No effect on longevity; digestive protease activity partially inhibited	Azzouz et al., 2005a
	Myzus persicae (Hemiptera: Aphididae)	Oilseed rape; 0.1–0.5% of total soluble protein	No effects	No effects; 2nd instar nymphs used	*Diaeretiella rapae* (Hymenoptera: Braconidae)	Parasitoids released into cages with plants and many aphids; destructively sampled after 14 days	No consistent effects on adult emergence and sex ratio; no effects on control of aphids	Schuler et al., 2001
OC1 IΔD86 (for nematode control)	*M. euphorbiae*, *M. persicae*	Potato; 264.5 or 335.1 ng. g fresh weight^{-1}	No effects	No effects on aphid population densities	(Diptera: Syrphidae), (Hemiptera: Anthocoridae), *A. ervi*, *Alloxysta victrix*, *Phaenoglyphis villosa* (Hymenoptera: Figitidae), *Asaphes vulgaris* (Hymenoptera: Pteromalidae), *Dendrocerus carpenteri*, *Dendrocerus serricornis* (Hymenoptera: Megaspilidae)	Relative abundance of predators, parasitoids and hyperparasitoids (field study)	No effects on percent parasitism, adult emergence; parasitoid communities more diverse on transgenic plants	Cowgill et al., 2004

(continued)

Table 13.4 (continued)

Proteins	Host/prey insect	Transgenic plant or diet; concentration or expression level	Potential effects on hosts/prey	Effects on hosts/prey used in this trial	Natural enemy	Stage tested	Effects on natural enemy of protein-exposed hosts/prey	References
Barley cystatin (HvCPI-1 C68 → G)	*L. decemlineata*	Potato; (expression levels not given but plants used significantly reduced papain activity)	Survival not affected; larval weight gain over 4 days reduced	Larvae smaller	*Podisus maculiventris* (Hemiptera: Pentatomidae)	Nymphs fed on potato-fed prey until 5th instar	No effects on nymph weights	Álvarez-Alfageme et al., 2007
	Spodoptera littoralis (Lepidoptera: Noctuidae)	Potato; (expression levels not given but plants used significantly reduced papain activity)	Survival not affected; larval weight gain over 4 days increased	Larvae larger	*P. maculiventris*	Nymphs fed on potato-fed prey until 5th instar	No effects on nymph weights	Álvarez-Alfageme et al., 2007

[1] Not applicable.

Results thus far suggest that cysteine PI-expressing transgenic plants will not have major negative impacts on natural enemies, and are likely to be compatible with IPM. As with all successful pest control technologies, the inevitable reduction in pest populations will have negative impacts on numbers of predators and parasitoids. However, unlike pesticides with broad toxicity, direct mortality of natural enemies via exposure to the cysteine PIs (either directly or via ingestion of prey containing the PI) seems unlikely, from the evidence available thus far. Two recent studies have compared the impacts of cysteine PI-plants with those arising from the use of conventional pesticides and shown that the transgenic technology compares very favourably in terms of compatibility with natural enemies (Cowgill et al., 2004; Mulligan et al., 2006).

13.3 Lectins

13.3.1 Lectins from Plants

The insecticidal properties of some lectins for pest species from several orders have been known for some time now, and the first transgenic lectin-expressing plants demonstrating insecticidal activity were produced in 1990 (Boulter et al., 1990). Over recent years lectins have attracted significant research interest since several have been shown to be toxic to many species of pest insects, whilst exhibiting no/low mammalian toxicity. Generally the insecticidal effects of lectins manifest as significant reductions in feeding damage to plants via mortality of some of the pest population and reductions in the feeding, size and growth of the rest (Fitches et al., 1997; Gatehouse et al., 1997). Intergenerational pest control can also result from reductions in fecundity (Gatehouse et al., 1996). Of particular interest are those lectins effective against phloem-feeding insects such as aphids and hoppers, since they are not susceptible to *Bt* Cry proteins or most PIs. Although their mode of action is not yet fully understood, some lectins have been shown to bind to insect gut brush-border membranes and others to the peritrophic membrane (Chrispeels and Raikhel, 1991; Peumans and Van Damme, 1995; Powell et al., 1998). However, whilst binding appears to be a prerequisite for toxicity, not all lectins that bind to these membranes exert subsequent insecticidal effects. Although some of the initial studies to investigate the potential of these molecules for insect control, such as wheatgerm agglutinin (WGA), had the potential disadvantage of having some mammalian toxicity, later candidates such as the snowdrop lectin, *Galanthus nivalis* agglutinin (GNA), were shown not to exhibit mammalian toxicity and have gone on to reach commercialisation stage (see case study below). The insecticidal potential of lectins has been reviewed by Legaspi et al. (2004). Table 13.5 lists some recent publications on lectins introduced into transgenic plants and shown to have negative effects on insects in feeding assays. GNA plants are the most advanced in terms of commercialisation, and there is now significant research effort going into finding similar lectin genes from various other species of bulb-forming plants (garlic, lilies, etc.) (Table 13.5).

Table 13.5 Some recently researched lectins conferring insect resistance on transgenic crop plants

Proteins	Target pests	Target crops	Status	Reference
Galanthus nivalis agglutinin or snowdrop lectin (GNA)	*Nilaparvata lugens*, *Laodelphax striatellus*, *Sogatella furcifera* (Hemiptera: Delphacidae), *Nephotettix virescens* (Hemiptera: Cicadellidae)	Rice	Field trials (India, China)	Sun et al., 2002; Wu et al., 2002; Nagadhara et al., 2003, 2004
	Aphis gossypii (Hemiptera: Aphididae)	Cotton	Field trials (China)	www.im.ac.cn
	Lacanobia oleraceae (Lepidoptera: Noctuidae)	Potato	Greenhouse trials	Bell et al., 2001b
	Eoreuma loftini (Lepidoptera: Pyralidae), *Diatraea saccharalis* (Lepidoptera: Crambidae)	Sugarcane	Experimental plants produced	Bernal and Sétamou, 2003
	Ostrinia furnacalis (Lepidoptera: Crambidae), *Rhopalosiphum maidis* (Hemiptera: Aphididae)	Maize/Corn	Experimental plants produced	Wang et al., 2005b
	Rhopalosiphum padi (Hemiptera: Aphididae)	Wheat	Experimental plants produced	Shah et al., 2005
	Myzus persicae (Hemiptera: Aphididae)	Tobacco	Experimental plants produced	Yuan et al., 2001
	Cnaphalocrocis medinalis (Lepidoptera: Pyralidae), *N. lugens*	Rice	Experimental plants produced	Li et al., 2005
Concanavalin A (ConA)	*L. oleraceae*, *M. persicae*	Tomato	Experimental plants produced	Gatehouse et al., 1999
Pea lectin	*Meligethes aeneus* (Coleoptera: Nitidulidae)	Oilseed rape	Experimental plants produced	Melander et al., 2003
Leaf and bulb lectins from garlic; ASAL (leaf), ASAII (bulb)	*Alabama argillacea* (Lepidoptera: Noctuidae)	Tobacco	Experimental plants produced	Sadeghi et al., 2008

Garlic leaf lectin (ASAL)	*Lipaphis erysimi* (Hemiptera: Aphididae)	Indian mustard	Experimental plants produced	Dutta et al., 2005a, b
	N. lugens, *N. virescens*	Rice	Experimental plants produced	Saha et al., 2006
Pinella ternata (a herb) agglutinin (PTA)	*M. persicae*	Tobacco	Experimental plants produced	Yao et al., 2003
Arisaema heterophyllum (dancing crane cobra lily) (AHA)	*M. persicae*	Tobacco	Experimental plants produced	Yao et al., 2004
Amaranthus caudatus (love lies bleeding) agglutinin (ACA)	*M. persicae*	Tobacco	Experimental plants produced	Guo et al., 2004
	A. gossypii	Cotton	Experimental plants produced	Wu et al., 2006
Zephyranthes candida (fairy lily) agglutinin gene (ZCA)	*M. persicae*	Tobacco	Experimental plants produced	Pang et al., 2004

13.3.2 Lectins and Natural Enemies

Transgenic plants expressing GNA have demonstrated what has been termed "partial resistance" to the target pests, reducing pest damage without the near-complete mortality wrought by the use of synthetic insecticides. As such, the use of these plants has always been envisaged as part of an IPM system and because of this their potential interactions with natural enemies have been well-studied (Table 13.6). There have been a few examples of GNA having negative impacts on predators and parasitoids, but these tend to be reductions in longevity or reproduction rather than acute mortality, as is often also the case with their impacts on the target pests (Table 13.6). Both direct and indirect impacts of lectins on natural enemies have been investigated. By feeding natural enemies directly with lectins in artificial diet, potential exposure via direct feeding on pollen or nectar of transgenic plants or honeydew excreted by sap-sucking insects has been investigated. Tritrophic systems (plant/pest/natural enemy) have also been used. Although these experiments seem more realistic, they can have the disadvantage of confounding direct impacts of ingesting lectins contained within the prey's body and indirect impacts on the natural enemy of consuming prey that is "sub-optimal" because it has been exposed to the lectin. This difficulty can be overcome by using a non-susceptible (non-pest) prey species that nevertheless contains significant quantities of the lectin in the system; this will demonstrate direct effects only (e.g., Sétamou et al., 2002a, c; Wakefield et al., 2006).

Tomov et al. (2003) described a scenario where the use of the GNA might complement biological control. Larvae of the pyralid pest of sugarcane, *Eoreuma loftini*, that had received a high dose of the lectin were less attractive to their parasitoid, but were also more stunted and thus less capable of causing crop damage. Larvae that escaped these effects of the lectin grew faster, but also retained their appeal for the parasitoid and so could be controlled in that way. Interestingly, most of the negative effects so far reported for GNA occurred when the lectin was provided directly to the natural enemy at high doses via artificial diet (Romeis et al., 2003; Bell et al., 2004; Hogervorst et al., 2006), and thus usually represents a "worst case scenario". It is also important when carrying out all risk assessment studies with insect-resistant transgenic crops to ascertain whether the impact on natural enemy populations is just a consequence of the inevitable loss of their prey when control methods succeed, or whether it is due to toxicity *per se*, whether it is direct (via ingestion of the insecticidal protein) or indirect (via ingestion of sub-lethally affected prey). Another important consideration, again applicable not only to lectins but to all the transgenic plants listed here is, how do these impacts compare with those of currently-used pest control methods? The development and field testing of GNA crop plants (see case studies below) will provide valuable data to help answer these questions.

Table 13.6 Effects of lectins on insect predators and parasitoids

Proteins	Host/prey insect	Transgenic plant or diet; concentration or expression level	Potential effects on hosts/prey	Effects on hosts/prey used in this trial	Natural enemy	Stage tested	Effects on natural enemy of protein-exposed hosts/prey	References
Galanthus nivalis agglutinin or snow-drop lectin (GNA)	N/A[1]; predators fed directly	Sugar solution; 1% w:v	N/A	N/A	*Chrysoperla carnea* (Neuroptera: Chrysopidae)	Larvae	Longevity of final instars, but not first instars, reduced; not repellent	Hogervorst et al., 2006
					Adalia bipunctata (Coleoptera: Coccinellidae)	Larvae	Longevity of both final and first instars reduced; not repellent	Hogervorst et al., 2006
					Coccinella septempunctata (Coleoptera: Coccinellidae)	Larvae	Longevity of both final and first instars reduced; not repellent	Hogervorst et al., 2006
	N/A; parasitoids fed directly	Sugar solution; 0.01%, 0.1%, or 1% w:v	N/A	N/A	*Trichogramma brassicae* (Hymenoptera: Trichogrammatidae)	Adults; 1-day-old mated females	Antifeedant; high dose reduced longevity	Romeis et al., 2003
					Aphidius colemani (Hymenoptera: Braconidae)	Adults; 1-day-old mated females	Higher doses reduced longevity	Romeis et al., 2003
					Cotesia glomerata (Hymenoptera: Braconidae)	Adults; 1-day-old mated females	Higher doses reduced longevity	Romeis et al., 2003

(continued)

Table 13.6 (continued)

Proteins	Host/prey insect	Transgenic plant or diet; concentration or expression level	Potential effects on hosts/prey	Effects on hosts/prey used in this trial	Natural enemy	Stage tested	Effects on natural enemy of protein-exposed hosts/prey	References
		Honey solution: 0.01%, 0.1%, 1% w:v	N/A	N/A	*Eulophus pennicornis* (Hymenoptera: Eulophidae)	Newly-emerged female adults	0.1% and 1% doses reduced adult longevity; 1% dose reduced fecundity	Bell et al., 2004
	Myzus persicae (Hemiptera: Aphididae)	Artificial diet; 0.1% w:v	Plants expressing 0.5–1% of total soluble protein reduce fecundity of *M. persicae*	Adults fed GNA diet for 4 days were smaller than controls	*Adalia bipunctata* (Coleoptera: Coccinellidae)	Larvae	No effects on survival; marginal effects on larval development	Gatehouse et al., 1996; Down et al., 2000
		Potato; 0.02–0.1% of total leaf protein		Aphid numbers reduced by 50%; nymphs exposed for 14 days	*A. bipunctata*	Adults (age not given)	Reduced longevity and reproduction of adults	Birch et al., 1999
		Potato; 0.1–0.2% of total soluble protein		Data not given	*A. bipunctata*	Larvae	No effects on larval development or survival; reduced viability of eggs laid by emergent adults	Down et al., 2003
		Diet; 0.1% w:v		No effects	*A. bipunctata*	Adults	No effects on longevity; increased fertility	Down et al., 2003

	Diet; 0.02%, 0.045%, 0.1% w:v		Adults fed 0.1% GNA diet for 4 days were smaller than controls	*Aphidius ervi* (Hymenoptera: Braconidae)	Adults	Dose-dependant effects on development time, mummy formation, adult emergence, adult weights	Couty et al., 2001b
	Potato; 0.01–0.12% of total soluble protein in leaves			*A. ervi*	Adults	No effects on parasitism or emergence success, adult weights	Couty et al., 2001b
Metopolophium dirhodum (Hemiptera: Aphididae)	Wheat; 0.05–0.08% of total soluble protein	Data not given	No effects	*Pandora neoaphidis* (Zygomycota: Entomophthorales) (pathogenic fungus)	Adults	No effects on infection; no effect on aphid fecundity	Shah et al., 2005
Macrosiphum euphorbiae (Hemiptera: Aphididae)	Diet: 0.1% w:v	Aphids fed for 6 days on GNA-diet are 24% smaller than controls	Used aphids fed GNA for 7–9 days and thus the same size as 6-day-fed control aphids, as well as smaller aphids fed GNA for 6 days	*Aphelinus abdominalis* (Hymenoptera: Aphelinidae)	Female adults exposed to mates for 48 h, then 3 days exposure to aphids for oviposition	Reduced size and longevity of adults provided with smaller 6-day GNA aphids, but not the "normal-sized" 7–9-day GNA aphids; fecundity lower in both groups of GNA-exposed females; emergence of progeny and their sex ratio not affected	Couty et al., 2001a

(continued)

Table 13.6 (continued)

Proteins	Host/prey insect	Transgenic plant or diet; concentration or expression level	Potential effects on hosts/prey	Effects on hosts/prey used in this trial	Natural enemy	Stage tested	Effects on natural enemy of protein-exposed hosts/prey	References
	Diatraea saccharalis (Lepidoptera: Crambidae)	Diet; containing transgenic GNA-sugarcane, 0.49% of total protein content	This diet has no effects on survival, development, larval or pupal weights, adult fecundity or egg viability	No effects on 18-day-old larvae used here	*Cotesia flavipes* (Hymenoptera: Braconidae)	Adults	Small negative effects on parasitism rates, cocoon formation, adult emergence, adult lifespan	Sétamou et al., 2002a, c
		Sugarcane; 0.9% of total protein		No effects on 10-day-old larvae used here	*C. flavipes*	Adults	No effects on host location (olfactometer tests) or parasitism (choice and no-choice tests; field cage tests)	Sétamou et al., 2002b
	Eoreuma loftini (Lepidoptera: Pyralidae)	Diet; containing transgenic GNA-sugarcane, 0.47% of total protein content	Larval feeding reduced	Larvae fed for 26–30 days on GNA diet had reduced activity; no change in frass volatiles	*Parallorhogus pyralophagus* (Hymenoptera: Braconidae)	Mated female adults, 2–4 days old; 24 h exposure	Parasitism rates (drilling and oviposition) reduced	Tomov et al., 2003

Lacanobia oleracea (Lepidoptera: Noctuidae)	Injected prey larvae; 5–10 μg GNA per day given to each predator	N/A	No effects; 4th instar larvae injected with GNA and presented to predator	*Podisus maculiventris* (Hemiptera: Pentatomidae)	Nymphs 2nd instar	Reduced growth	Bell et al., 2003
	Potato: 0.4% of total soluble protein	Potato (<2% expression) has 50% l ess leaf damage, 50% less insect biomass; larval survival reduced by 20%	No effects: 3rd instar larvae fed GNA potato for 6–8 days	*P. maculiventris*	Nymphs 2nd instar to adulthood	Reduced fecundity	Gatehouse et al., 1997; Bell et al., 2003
	Potato: 0.08% of total soluble protein	Larval biomass, food consumption reduced; development slowed	Larvae fed from beginning of 3rd or 4th instar, used in experiment at 5th	*E. pennicornis*	Pairs of adults exposed to 5th instar GNA-fed larvae until 24 h after parasitism or until host larva dead or pupated	No significant effects on: percent of hosts parasitised, percent hosts in which wasp larvae developed, numbers of wasp larvae per host, numbers of adult wasp progeny emerging per host	Fitches et al., 1997; Bell et al., 1999

(continued)

Table 13.6 (continued)

Proteins	Host/prey insect	Transgenic plant or diet; concentration or expression level	Potential effects on hosts/prey	Effects on hosts/prey used in this trial	Natural enemy	Stage tested	Effects on natural enemy of protein-exposed hosts/prey	References
		Tomato: 2% of total soluble protein	Larvae fed on GNA-tomato plants from hatching develop faster, and more survive to adulthood than controls; GNA detected in body tissues	Larvae fed until 4th instar and then used	*Meteorus gyrator* (Hymenoptera: Braconidae)	Female adults (<7 days old) each exposed to 5 host larvae at 4th instar	Development and survival of wasp progeny not affected; GNA not detected in adult progeny	Wakefield et al., 2006
		Diet: 2% of protein	As above	As above	*M. gyrator*	As above; 4th instar host larvae	As above	Wakefield et al., 2006
		Potato: 0.1% of total soluble protein	Larval biomass, food consumption reduced; development slowed	As above	*M. gyrator*	As above, but hosts at 4th or 5th instar	As above	Fitches et al., 1997; Wakefield et al., 2006
Concanavalin A	N/A; parasitoids fed directly	Honey solution: 0.01%, 0.1%, 1% w:v	N/A	N/A	*E. pennicornis*	Newly-emerged female adults	0.1% and 1% doses reduced adult longevity; 1% dose reduced fecundity	Bell et al., 2004

[1] Not applicable

13.3.3 Case Study: GNA Rice

During the 1980s the Rockefeller Foundation implemented a long-term programme on rice biotechnology, one of whose primary objectives was the enhancing of resistance to insect pests. Damage caused by the rice brown planthopper (*Nilaparvata lugens*, BPH) is recognised as one of the key constraints on rice production in South East Asia. It not only causes symptoms known as "hopperburn", but, probably more importantly, acts as a vector for viral diseases, including grassy stunt and ragged stunt virus.

The use of recombinant DNA technology was recognised as a viable approach to enhancing resistance to rice pests. Early studies focussed on expression of genes encoding Cry proteins for control of stem borers (Bennett et al., 1997). Given that those Cry proteins available were ineffective against sap-sucking insects (Hemiptera), alternative insecticidal proteins effective against this insect order had to be identified. The mannose specific lectin from snowdrop (*Galanthus nivalis*, GNA) was shown to be toxic towards a number of agricultural pests from several different orders, including hemipteran pests (Down et al., 1996). It was particularly effective against BPH (Powell et al., 1998), where it was shown to bind to midgut epithelial cells, with subsequent systemic effects; later studies identified ferritin as the major binding protein for GNA, suggesting that these molecules play a role in metal homeostatis (Du et al., 2000). Following these initial studies, transgenic rice plants were generated where transgene expression was driven by a phloem specific promoter (from the rice sucrose synthase *RSs1* gene) and by a constitutive promoter (from the maize ubiquitin *ubi1* gene). Irrespective of the promoter used, these plants were shown to be significantly resistant to BPH in terms of insect development and survival, with subsequent fecundity being significantly reduced (Rao et al., 1998). Subsequently, many similar studies were carried out in different laboratories using different elite rice lines and extending the spectrum of hemipteran pests targeted. These studies have thus verified the efficacy of GNA to significantly enhance resistance of rice to such pests (Foissac et al., 2000; Tinjuangjun et al., 2000; Loc et al., 2002; Nagadhara et al., 2004; Li et al., 2005).

To date, only very limited amounts of transgenic rice of any type have been grown commercially. James (2005) reported that "several hundred farmers" grew 4,000 hectares of *Bt* rice in Iran in 2005, but local sources have reported that at present commercial production of transgenic rice is not permitted in Iran, although field trials continue (Cohen et al., chapter 8). China, which is currently the top rice producer globally (in 2003, producing >28% of the global market) and has conducted large field trials with *Bt* rice since 1998 (Cohen et al., chapter 8), has no immediate plans for commercialisation of GM rice, although decisions are expected in the near future. However, despite this, extensive field trials of GNA expressing rice lines are currently being carried out (G.Y. Ye, personal communication). These trials involve the growing of GNA rice lines in different regions and in different years to evaluate overall agronomic performance under different environmental and climatic conditions.

13.3.4 Case Study: GNA Cotton

Whilst cotton cultivars expressing *Bt* and CpTI + *Bt* have been commercially deployed and are effective at controlling lepidopteran pests (see Naranjo et al., chapter 6), such cultivars remain susceptible to hemipteran pests. Some reports, indeed, suggest that the growing of *Bt* cotton has exacerbated the situation since the accompanying reduction in insecticide usage has resulted, in some regions, in the emergence of other insects as important pests of cotton. This is particularly true in northern China where the cotton aphid (*A. gossypii*) is now the second most important pest of cotton, particularly of the seedling stage (Zhang et al., 2000).

The nature of the damage caused by cotton aphid varies both seasonally and with the growth stage of the plant; as expected, significant damage appears more likely when environmental conditions such as dry weather are already stressing cotton growth. Light aphid populations on mid-season cotton often do not generate any obvious damage symptoms, whereas heavy aphid populations cause symptoms similar to those observed on seedling cotton, i.e. cupped, crinkled leaves, honeydew accumulations, sooty mold and, in extreme cases, limited defoliation. At this stage of the growing season heavy infestations can also decrease the size of bolls, stunt plant growth, and may increase square and boll shedding. In addition to direct reductions in yield, heavy aphid infestations can also cause spoilage of the crop since they can cause a condition known as "sticky cotton" whereby the honeydew accumulates on cotton lint, lowers the grade of the cotton, and can cause production problems during fiber processing and yarn manufacturing.

To address some of the problems associated with aphid infestations on cotton, and subsequent loss of yield/and or quality, scientists at the Cotton Research Institute (CRI) of CAAS, in collaboration with scientists at Fudan University, have expressed a gene encoding snowdrop lectin (*gna*) in cotton via *Agrobacterium*-mediated transformation. Similarly, scientists at the Biotechnology Research Centre (BRC), together with scientists at CAS, have also produced GNA expressing cotton. Laboratory-based insect bioassays demonstrated that these transgenic lines were significantly more resistant to aphid attack compared with their respective parental lines. Following these initial studies, GNA has now been co-expressed with other insect-resistance genes in at least two of the four major cotton producing countries of the world, namely China and Pakistan. For example, scientists at the Institute of Microbiology, CAS, have recently carried out field trials in Xinjiang and Hunan of transgenic color cotton and Xiang-mian cotton expressing both Cry1Ac and GNA (www.im.ac.cn). Similar studies have also been undertaken in Pakistan with the same set of transgenes; in this case the rationale was primarily to increase field durability of the *Bt*-transgenic cotton (Syed, 2002).

GNA is not the only lectin that has been shown to be toxic to cotton aphids, since a lectin from *Amaranthus caudatus* (ACA) has also been shown to be effective (Wu et al., 2006). Genes encoding both these insecticidal lectins were co-expressed in cotton at the Shanxi Cotton Institute, again via *Agrobacterium*-mediated transformation. In addition to cotton, these transgenes were also co-expressed in tobacco plants to evaluate the efficacy of pyramiding these two gene constructs for aphid resistance. Fifty of the resulting transgenic tobacco plants expressing both GNA and ACA were tested for aphid-resistance and the results showed that the average decrease in population build-up was approx. 84%, with 58% of those plants exhibiting a decrease of 90%. As anticipated, the levels of aphid resistance achieved were higher in plants expressing both transgenes, compared to those transformed with only one of the lectin genes.

In addition to a transgenic approach to the control of cotton aphid, complementary studies are currently investigating the potential for the use of a wheat-cotton intercropping system. In one such study, four wheat varieties that were either resistant or susceptible to wheat aphid (*Sitobium avenae*) were evaluated with respect to their role in conserving arthropod natural enemies and suppressing cotton aphids. The results demonstrated that this approach preserved and augmented natural enemies to a greater extent than a cotton monoculture and that as a consequence more predators were available to suppress cotton aphids during the cotton seedling stage (Ma et al., 2006). It is important when developing effective pest control strategies not to rely on a single method, but to consider a more integrated approach. It would be interesting to see how the GNA expressing cultivars performed in such cropping systems; all the evidence to date suggests that recombinant DNA technology, including expression of GNA, is compatible with biological control. Such an integrated approach will certainly help delay the inevitable evolution of resistance of the pest Lepidoptera to the currently deployed Cry proteins.

13.4 Other Insect Resistance Proteins from Plants

Other plant-derived proteins, thought to be involved in plant defence against insect attack and with potential for use in insect-resistant transgenic crops, include the alpha-amylase inhibitors, defensins, chitinases, and lipid acyl hydrolases.

Alpha-amylase inhibitors are commonly found in many seeds, where they are thought to act as plant defence chemicals, and many have been shown to interfere with an insect's ability to digest starch. For a recent review of alpha-amylase inhibitors for insect control see Franco et al. (2002). A number of proteinaceous alpha-amylase inhibitors from legumes have been expressed in transgenic plants with the aim of controlling weevil pests of seeds, as these insects rely heavily on starch as a food source (Table 13.7). Of these, the most advanced in terms of technical development are peas, chickpeas and adzuki beans expressing *Phaseolus vulgaris* (bean) alpha-amylase inhibitor-1 (α-AI1) (Shade et al., 1994; Ishimoto et al., 1996; Sarmah et al., 2004). Transgenic pea plants expressing 0.8–1.0% α-AI1 in the seeds caused complete mortality of 1st and 2nd instar larvae in laboratory tests with *Bruchus pisorum* (pea weevil), *Callosobruchus maculatus* (cowpea weevil) and *Callosobruchus chinensis* (adzuki bean weevil) (Shade et al., 1994; Schroeder et al., 1995). Expression of the gene was stable to the T_5 generation (Schroeder et al., 1995) and effective *B. pisorum* control was also obtained in field trials with α-AI1-expressing transgenic peas (Morton et al., 2000). Transgenic adzuki beans expressing

0.9% (of dry weight) α-AI1 were completely resistant to *C. chinensis*, *C. maculatus* and *Callosobruchus analis* (Graham bean weevil) attack, but not to that caused by *Zabrotes subfasciatus* (Mexican bean weevil) (Ishimoto et al., 1996). A second inhibitor from beans, α-AI2, was very effective at killing *Z. subfasciatus* in the laboratory (Ishimoto et al., 1996), but provided only partial protection against *B. pisorum* on transgenic peas in the field (Morton et al., 2000), suggesting that these transgenes could be pyramided to provide greater pest species coverage. Field tests in Australia showed that transgenic peas expressing bean α-AI1 provided effective pest control, but these were abandoned in November 2005 when it was shown that the expressed protein had a different structure and different allergenicity from the native protein (Prescott et al., 2005). When mice were fed with the transgenic peas they produced antibodies to α-AI1 and demonstrated mild lung inflammation when challenged with pea α-AI1; mice fed with purified α-AI1 did not produce this response (Prescott et al., 2005). Furthermore, the ileal dry matter of pigs fed with transgenic α-AI1-expressing peas was significantly reduced compared to that of pigs fed non-transgenic diets, indicating differences in starch digestibility (Collins et al., 2006). A similar reduction in starch digestibility was observed in chickens when their diets included uncooked α-AI1 pea seed meal (Li et al., 2006).

Defensins are a family of peptides isolated from many plant species with activity against a range of plant pathogenic microbes (for a recent review see Lay and Anderson, 2005). Some have also been shown to have insecticidal activity (Lay et al., 2003; Liu et al., 2006b). Some floral defensins from solanaceous plants have been patented in Australia and presumably there are plans to exploit the insecticidal and antimicrobial activities of these commercially (Table 13.7).

Since chitin is an important structural component of both fungi and insects, chitinases isolated from plants, microbes and animals have attracted interest for engineering into plants for control of fungal diseases or insects. Chitinase genes of plant origin are being used experimentally in transgenic plants for control of fungal diseases, but recent research on chitinases for insect control has focused more on insect-derived chitinases (see section 13.8 below).

The first demonstration of insecticidal activity of a lipid acyl hydrolase was in 1995, when Strickland et al. (1995) showed that patatin, an abundant lipid acyl hydrolase from potato tubers, inhibited the growth of southern and western corn rootworms. Subsequently, several patents have been issued covering the insecticidal potential of these compounds (e.g., by Monsanto [Alibhai and Rydal, 2003]). However, currently available public records do not show evidence of commercial development of transgenic patatin-expressing insecticidal plants or any other lipid acyl hydrolases.

Thus, of these transgenic insect-resistant plants expressing plant defence proteins, those expressing alpha-amylase inhibitors are the most advanced in their development. Their compatibility with biological control has not yet been investigated, although their potential for activity against Coleoptera suggests that an investigation of potential impacts on predatory beetles might be advisable if commercial deployment is contemplated.

Table 13.7 Recent research on transgenic plants expressing novel insect resistance proteins from plants

Proteins	Target pests	Target crops	Status	References
Phaseolus vulgaris (bean) alpha amylase inhibitor (α-AI1)	*Bruchus pisorum*, *Callosobruchus maculatus*, *Callosobruchus chinensis* (Coleoptera: Bruchidae)	Peas	Field trials (Australia)	Shade et al., 1994; Schroeder et al., 1995; Morton et al., 2000
	C. maculatus, *C. chinensis*, *Callosobruchus analis*	Adzuki beans	Experimental plants produced	Ishimoto et al., 1996
	C. maculatus, *C. chinensis*	Chickpeas	Greenhouse trials	Sarmah et al., 2004; Ignacimuthu and Prakash, 2006
Floral defensin from *Nicotiana alata* (tobacco) (NaD1)	*Helicoverpa armigera*, *Helicoverpa punctigera* (Lepidoptera: Noctuidae)	Not stated	US patent issued 2002	Lay et al., 2003
Patatin	*Diabrotica undecimpunctata*, *Diabrotica virgifera* (Coleoptera: Chrysomelidae)	Not stated	US patent issued 2003	Alibhai and Rydal, 2003

13.5 Non-Cry Toxins from *Bacillus thuringiensis* (*Bt*)

13.5.1 VIPs and Other Insecticidal Proteins from Bacillus thuringiensis

In addition to the well-known crystal (Cry) proteins, which are endotoxins produced in the spores, *B. thuringiensis* also secretes a number of insecticidal exotoxins. The best-known and closest to commercialisation is the vegetative insecticidal protein (VIP) 3A(a), which has been engineered into cotton to protect against lepidopteran pests (Estruch et al., 1996; ICAC, 2003). Unlike the Cry proteins, VIPs do not need to be solubilised in the insect gut before they can act. They bind to receptors in the insect gut, but not the same receptors as Cry proteins (Lee et al., 2003, 2006), and Cry1Ac-resistant strains of *Heliothis virescens* have been shown to be as susceptible to VIP3A as Cry1Ac-susceptible strains of this pest (Jackson et al.,

2007). Thus VIPs have the potential to delay the development of resistance if used in crops where Cry proteins have also been used. They also exhibit activity against different spectra of host species from some of the currently available Cry toxins and therefore complement current *Bt*-expressing crops (ICAC, 2003).

Since its discovery in 1994, VIP3A(a) has been transformed into the 'Coker' variety of cotton, under the control of various promoters, and field-tested in the United States (2003/03) (ICAC, 2003). Some of these cotton varieties are currently being trialled in Australia by Deltapine Australia Pty Ltd (OGTR, 2005). Despite press releases in 2003 suggesting VIP cotton would be available to growers in the United States in 2004/05 (Syngenta, 2003), this did not eventuate. By 2007 Monsanto had agreed to buy Delta and Pine Land Company, the cotton breeding company that had licensed the VIP gene from Syngenta (A.M. Shelton, personal communication). At that time, Syngenta's public website noted that VIP cotton had "reached the final stage of selection of the technology options for later commercialization" and had continued to progress towards launch (Syngenta, 2007). VIP3A has also been incorporated into field maize (Pacha event), and stacked with Cry1Ac to produce transgenic hybrid field maize (Dively, 2005). In 2006, Syngenta applied to the USEPA for an experimental use permit for transgenic maize expressing VIP3A + Cry1Ab + modified Cry3A for lepidopteran and coleopteran control (USEPA, 2006b), with an estimated launch in 2009 (Syngenta, 2007).

In addition to the VIPs, other families of exotoxins have been identified in the supernatants of *Bacillus* cultures. Mycogen has a 2003 patent on three such families, the SUP proteins, which have activity against Lepidoptera, the MIS proteins which have toxicity to Coleoptera and the WAR proteins (Narva et al., 2003). Mycogen is now owned by Dow AgroSciences and presumably these inventions will be commercialised at some stage.

13.5.2 Compatibility with Natural Enemies

Cotton is attacked by a wide range of pest species wherever it is grown. Lepidoptera of the "heliothine complex" in particular are significant pests in most countries, along with armyworms and loopers. Sucking pests such as *Lygus* bugs and mirids are often secondary pests, causing significant damage after Lepidoptera have been controlled. Cotton crops are also the habitat for many natural enemy species. Predators include lacewings, big-eyed bugs, lady beetles, nabid bugs, thrips and mites, and hymenopteran parasitoids also abound. IPM programmes have been developed and are used on cotton crops in the USA, Australia, India, China and other countries, where they have significantly reduced the numbers of insecticide sprays needed to grow cotton (e.g., Fitt, 2000; Brookes and Barfoot, 2006; USDA, undated; Fitt, chapter 11). *Bt* Cry toxin-expressing transgenic cotton plants are completely compatible with these programmes as these Cry toxins do not affect natural enemy species.

Similarly, the *Bt* exotoxins being proposed for use in transgenic plants appear to have well-defined and reasonably restricted host specificities. For example, VIP3A(a) is very active against black cutworm (*Agrotis ipsilon*), fall armyworm (*Spodoptera frugiperda*) and beet armyworm (*S. exigua*), but not very effective against European corn borer (*Ostrinia nubilalis*) (ICAC, 2003). Deltapine Australia Pty Ltd's application to field-test VIP cotton in Australia includes results from bio-assays with a range of insects and, although the predatory lacewing *Chrysoperla carnea* was the only recognised natural enemy listed, it was not susceptible to the toxin and neither were representatives of the Hymenoptera, Diptera or Coleoptera tested (OGTR, 2005). This suggests that predatory beetles and wasp or fly parasitoids that may be important in IPM programs will not be harmed by VIP-expressing transgenic plants. An Australian field trial at two locations over one season (2003–2004) compared invertebrate communities on unsprayed VIP-cotton (Coker312Vip3A, event 102) and on unsprayed conventional cotton (Sicala 40) and revealed no significant differences in species richness or diversity that could be attributed to the modification, except for the expected decrease in the target pests *Helicoverpa* (Whitehouse et al., 2007). In another detailed field study (2000–2002), transgenic hybrid maize expressing both VIP3A and Cry1Ab did not significantly affect non-target arthropod communities, including saprovores, herbivores, predators or parasitoids, when compared to isogenic maize control plots. Significant changes (positive and negative) did occur in some taxa in the transgenic maize plots but in total the effects observed in these plots were less than the community disturbances caused by insecticide applications on some of the control plots (Dively, 2005). Similarly, no significant changes in the abundance of predators and an egg parasitoid were observed in a comparison of hybrid *Bt* VIP maize and non-transgenic maize in a field trial in Brazil (Fernandes et al., 2007). Results with transgenic maize and cotton suggest that VIP3A toxins will be as compatible with the use of natural enemies as the Cry toxins. Furthermore, there is a long history of safe IPM using *Bt* biopesticide sprays, which are comprised of mixtures of *Bt* spores, crystals, vegetative cells, and their secreted proteins, including exotoxins.

13.6 Other Toxins from Microbes

Although the soil-dwelling bacterium *B. thuringiensis* remains the most productive source of insecticidal proteins, other microbes, particularly the insect pathogens and their relatives, have also been found to yield some promising candidates for expression in transgenic plants.

Brevibacillus (*Bacillus*) *laterosporus* is a spore-forming bacterium with strains which produce insecticidal toxins with potential to control a wide range of pests, including beetles, nematodes, molluscs, mosquitoes and blackflies (de Oliveira et al., 2004). Although insecticidal secreted proteins (ISPs) from this bacterium have been patented for use in transgenic plants to control corn rootworms (*Diabrotica* spp.), Colorado potato beetle (*L. decemlineata*) and cotton boll weevil (*Anthonomus grandis*) (Boets et al., 2006), commercialisation plans have not been publicly announced. The extremely wide spectrum of activity of these proteins suggests that extensive biosafety testing will be required to ensure the safety of natural enemy species and compatibility with IPM if they are deployed in transgenic crops.

"Toxin complex" or *tc* genes are widespread among Gram-negative bacteria (Waterfield et al., 2001). The best known of these are the toxins produced by *Photorhabdus luminescens*, a bacterium which is symbiotic with entomopathogenic *Heterorhabditis* spp. nematodes. The bacteria and nematodes enter the host insect's hemolymph, where the bacteria secrete their toxins and kill the insect. Thus the toxicity of these proteins when administered directly *per os* to insects, has been a somewhat surprising if not useful finding (Waterfield et al., 2005). Toxin A from *P. luminescens* has been successfully expressed in transgenic *A. thaliana* which were toxic to tobacco hornworm and southern corn rootworm larvae, suggesting potential for these toxins to control both Lepidoptera and Coleoptera (Liu et al., 2003). Similar toxins identified from another nematode-associated bacterium, *Xenorhabdus nematophilus*, are effective in killing several lepidopteran species and the dipteran blowfly, *Lucilia cuprina* (Brown et al., 2006). A *Xenorhabdus* toxin has also been successfully expressed in *Escherichia coli* (Lee et al., 2004). Dow AgroSciences holds a patent on toxins from both *Photorhabdus and Xenorhabdus* bacteria with claims for activity against Coleoptera, Lepidoptera and Homoptera (Hemiptera) (Petell et al., 2004).

Photorhabdus insect related (Pir) proteins are another group of toxins from *P. luminescens* with activity against a range of insects (Duchaud et al., 2003). For example, Pir B was recently found to be very effective against the diamondback moth when the purified protein was applied to leaf discs and fed to the insects (Blackburn et al., 2006). Pir B has some amino acid sequence homology with leptinotarsin, a potent neurotoxin (to insects and vertebrates) isolated from the Colorado potato beetle, and it has been suggested that they may have similar modes of action. If this is the case, then Pir proteins will need to be thoroughly tested to ensure that they will not harm non-target organisms, if they are to be used in agricultural production systems.

Entomopathogenic viruses have also proven to be a rich source of genes with insecticidal properties in their own right or with the ability to enhance the activity of other compounds or pathogens (Liu et al., 2006a). Several homologous insecticidal proteins with chitin-binding properties have been isolated from insect viruses, e.g., enhancins and GP37 proteins from baculoviruses, and fusolins and spindle proteins from entomopoxviruses (Dall et al., 2001). Some enhancins also have metalloprotease activity (Lepore et al., 1996). (Interestingly, similar metalloproteases have recently been isolated from bacteria of the *Bacillus cereus* group, including a strain of *B. thuringiensis* [Hajaij-Ellouze et al., 2006].) Enhancins digest components of the peritrophic membrane and thus facilitate the passage of the virus into the host's gut. They can also have direct impacts on insects when ingested in the absence of any pathogens. For example, an enhancin from the granulovirus of *Trichoplusia ni* has been expressed in transgenic tobacco where it slowed the development of *Pseudaletia separata* and *Spodoptera exigua* larvae

(Hayakawa et al., 2004). The use of transgenic plants expressing proteins from insect viruses in order to enhance infection of pests with natural populations of baculoviruses in the field has also been proposed (e.g., rice expressing fusolin from an entomopox virus to enhance baculovirus infection of *Pseudoletia unipuncta* [Hukuhara et al., 1999]). There have been no tests with such proteins and beneficial non-target invertebrates, but their effects appear to be quite specific and reasonably subtle, suggesting a high probability of environmental safety for natural enemies. Indeed, where the transgenic plant must work in concert with a pre-existing insect pathogen, then it is, of necessity, part of an IPM system.

In conclusion, none of these microbe-derived insecticidal proteins, which could be used in transgenic plants, is sufficiently far advanced in its development for tests with natural enemies to have been published. Most have broad ranges of activity against insects of different orders, suggesting that such tests will be necessary to determine their compatibility with IPM.

13.7 Biotin-Binding Proteins

13.7.1 Biotin-Binding Proteins

Biotin-binding proteins (BBPs), such as avidin from chicken egg white and streptavidin from the bacterium *Streptomyces avidiniii*, have been shown to be significantly toxic to insect pests from many different orders (e.g., Morgan et al., 1993; Markwick et al., 2001; Malone et al., 2002a). Both avidin and streptavidin have been successfully expressed in transgenic plants which have been shown to be resistant to insect attack (Kramer et al., 2000; Burgess et al., 2002b; Markwick et al., 2003; Yoza et al., 2005). Expression of BBPs in transgenic plants is not straightforward, since plants require biotin and expression of BBPs directly in the cytoplasm, as would occur with most conventional transgene constructs, results in plant mortality. Targeting BBP expression to sub-cellular organelles can circumvent this difficulty and viable, insect-resistant transgenic plants expressing BBPs in the leaves, stems and other tissues have been produced using signal sequences from potato protease inhibitor to direct expression to the vacuoles (Burgess et al., 2002b; Murray et al., 2002; Markwick et al., 2003) or a barley alpha-amylase signal sequence directing it specifically to the extracellular compartments of seeds (Kramer et al., 2000; Yoza et al., 2005).

The insecticidal mode of action of BBPs is not completely understood, but it appears that they bind to dietary biotin, making it unavailable to the insects, which then die from a deficiency of this vitamin. Thus the balance between levels of biotin and BBPs in the insects' food is critical to the success of this control strategy. Beetle larvae have been "rescued" from the effects of avidin by the provision of extra biotin in their food (Morgan et al., 1993; Kramer et al., 2000). Honey bee larvae are unaffected by doses of avidin that would be expected to kill lepidopteran larvae, and the high concentration of biotin in the jelly and pollen food of bee larvae, compared with the lower concentrations found in leaves (Christeller and Phung, 1998), has been suggested as the reason for this lack of effect (Malone et al., 2002b). Although the biotin requirements of different insects have not been quantified, biotin is known to be essential for insect growth and moulting, as it is a cofactor of major carboxylases involved in glucogenesis, lipogenesis, fatty acid and amino acid catabolism (Wood and Barden, 1977; Knowles, 1989). Larvae fed with BBPs often die during moulting (Markwick et al., 2001, 2003; Burgess et al., 2002b; Malone et al., 2002a), whilst adult insects of the same species may be unharmed by them (e.g., clover root weevil [Malone et al., 2002a]).

Because they target a nutritional need for many insects, the BBPs are effective against a wide range of species. Unlike many other insect-resistance proteins, they do not require the insect to possess a particular gut receptor for activity, but depend more upon the target insect having a high requirement for dietary biotin relative to that available to it from its normal diet. Avidin and *Bt* Cry proteins complement each other and greater insecticidal effects may be obtained by combining the two (Burgess et al., 2002b; Zhu et al., 2005; Cooper et al., 2006).

To date, BBPs have been successfully expressed in tobacco (Burgess et al., 2002b), maize (Kramer et al., 2000), apple (Markwick et al., 2003), potato (Meiyalaghan et al., 2005), rice (Yoza et al., 2005), eucalyptus (E.P.J. Burgess, personal communication) and sugarcane plants (CRCSIIB, 2006). Commercial partners are currently being sought for development of the vacuolar-targeted BBP technology (E.T. Stark, personal communication).

13.7.2 Compatibility with Natural Enemies

Avidin-expressing transgenic plants have not yet been commercialised for pest control and so exhaustive environmental biosafety tests have not yet been undertaken. However, there has been some investigation into their potential impacts on non-target invertebrates, such as honeybees, predators, and their persistence in the environment (Malone et al., 2002b; Glare et al., 2004; Christeller et al., 2005; Christeller et al., 2006; Lawo and Romeis, 2008).

Theoretically any insect requiring biotin could be susceptible to a BBP. The likelihood of a significant biological effect will depend on the magnitude of that requirement and the insect's access to dietary biotin, relative to its exposure to the BBP. Lepidopteran larvae fed with transgenic avidin-expressing plants were found to contain active avidin in their bodies and to excrete it with their frass, but its concentration in both places was less than that in the transgenic leaves (Christeller et al., 2005). This "dilution effect" could explain why tri-trophic experiments in which predatory carabid beetle larvae and adults (*C. novaezelandiae*) were unaffected by exclusive diets of avidin-tobacco-fed prey (*S. litura* larvae) (Glare et al., 2004; Burgess et al., 2008).

In a study with three stored products pests exposed to avidin and a parasitoid, it was shown that applying transgenic avidin-maize powder to non-GM maize kernels controlled *Tribolium castaneum* and *Cryptolestes ferrugineus*, two beetle pests which feed on the outside of kernels, and introducing a hymenopteran parasitoid, *Thecolax elegans*, controlled the third pest *Sitophilus zeamais* (which lives within the kernels and was not exposed to the avidin powder) (Flinn et al., 2006). Overall, better control of the pest complex was achieved when the two treatments were combined than when used individually; the avidin maize powder did not interfere with the performance of the parasitoid.

Tri-trophic impacts of avidin-expressing plants on parasitoids of herbivores have not yet been investigated. The concentrations of avidin detected in the bodies of *S. litura* larvae feeding on transgenic avidin-plants suggest that the avidin activity resides within food remnants in the insect's gut lumen and does not penetrate the hemocoel (Christeller et al., 2005). This would suggest that parasitoids are unlikely to be exposed to avidin by this route. However, further research is needed to confirm this.

13.8 Toxins from Arthropods

Insects themselves and other arthropods have been investigated as potential sources of insecticidal transgenes for some time now. There are two rationales behind such studies. One suggests that insects possess genes encoding compounds which regulate essential metabolic processes and that some of these could be manipulated in such a way as to cause insect mortality, e.g., insect chitinases and proteases. The second rationale is to use gene-based defences (or weapons) employed by predatory invertebrates as a means to control insects, e.g., spider and scorpion venoms.

Chitinases are attractive candidates for use in insect control, since chitin occurs naturally only in arthropods, fungi, nematodes and some algae, and so side-effects on vertebrates are extremely unlikely. Recently, a chitinase gene from the tobacco hornworm, *Manduca sexta*, has been used to produce transgenic papaya plants resistant to attack from carmine spider mite (McCafferty et al., 2006). Transgenic cotton plants expressing *M. sexta* chitinase have also been produced, but their effectiveness in controlling insect pests has not yet been reported (Hao et al., 2005). There are as yet no published studies of potential impacts of chitinase-expressing plants on natural enemy species. Theoretically, any arthropod could be sensitive to a chitinase, but it would need to be sufficiently exposed to it during an appropriate life stage for there to be an effect. Studies to determine and quantify the presence of the chitinase in parts of the plant eaten by natural enemies (e.g., pollen) and in the bodies of their herbivore prey will be important for assessing the potential risks of this technology.

A cathepsin L-like protease from the flesh fly, *Sarcophaga peregrina*, has been found to digest insect basement membranes, which are located on the hemocoel side of the gut epithelium (Liu et al., 2006a). The gene encoding this protease has been isolated and introduced into a transgenic baculovirus that had enhanced activity against *H. virescens* larvae (Harrison and Bonning, 2001). Cathepsins are papain-like cysteine proteases, and insecticidal properties have also been demonstrated in papain itself and a number of other plant cysteine proteases (Pechan et al., 2000; Konno et al., 2004; Malone et al., 2005). The mode of action of these proteases is not yet fully understood and their potential impacts on biological control agents have not yet been considered.

The insecticidal properties of scorpion toxins have been utilised primarily to enhance the action of baculoviruses and to reduce their time to kill. For example, *Autographa californica* nuclear polyhedrosis virus was genetically modified with a venom gene from the Algerian scorpion *Androctonus australis* to produce a more effective virus against *H. virescens* (Hoover et al., 1995). Subsequently, this toxin gene was introduced into *Helicoverpa zea* nuclear polyhedrosis virus to provide good control of both *H. zea* and *H. virescens* in field trials (Treacy et al., 2000). More recently, scorpion toxins have been engineered into crop plants, where they have been shown to be effective in killing pest insects if combined with a second gene to facilitate transport of the toxin to the insect's hemolymph. Successful examples include transgenic oilseed rape expressing *M. sexta* chitinase and a toxin from the scorpion *Buthus martensii*, which caused mortality of diamondback moth larvae (Wang et al., 2005a), and a fusion protein, comprised of snowdrop lectin (GNA) and a lepidopteran-specific toxin from the South Indian red scorpion *Mesobuthus anulus*, which was toxic to *L. oleracea* larvae and *N. lugens* when fed with the purified protein (expressed in *Pichia pastoris*) (Trung et al., 2006).

Spiders also produce insecticidal toxins and a similar strategy to that described above has been used to transport the toxin to the insect's hemolymph. Fitches et al. (2004) expressed a fusion protein containing GNA and a neurotoxin from the spider *Segestria florentina* (SFI1) in *P. pastoris*, and noted mortality of early instar larvae and growth reductions in older larvae of the tomato moth fed with the purified protein. A component of the venom of the Australian funnel web spider *Hadronyche versuta* (Hvt) has recently been expressed in transgenic tobacco plants which were effectively protected from *H. armigera* and *S. littoralis* larval attack (Khan et al., 2006).

The impacts of insect control strategies employing scorpion or spider toxins on biological control have not yet been investigated, because of the early stage of their development. Observations that the host specificity of some of these toxins can be broadened when they are "chaperoned" through the gut epithelium by a lectin or a baculovirus suggests that pre-release biosafety tests with natural enemies will be particularly important with such technologies.

13.9 Stacked/Pyramided Traits and Fusion Proteins

The concept of "stacking" (different types of traits) or "pyramiding" (same types of traits) insecticidal genes in transgenic plants, as a means to improve efficacy and delay the evolution of resistance, has been proposed for some time (Boulter et al., 1990). Plants with pyramided *Bt* Cry genes are now available commercially (e.g., Herculex XTRA® maize, Bollgard II® cotton), as is cotton expressing CpTI

Table 13.8 Some recently researched insect resistant transgenic plants expressing fusion proteins or stacked traits (plants expressing only pyramided *Bt* Cry genes not included)

Proteins	Target pests	Target crops	Status	References
Synthetic *Bt* toxin Cry 1A + modified cowpea trypsin inhibitor (Bt + CpTI)	*Helicoverpa armigera* (Lepidoptera: Noctuidae), *Pectinophora gossypiella* (Lepidoptera: Gelechiidae)	Cotton	Commercialised (China)	Liu et al., 2005
Bt toxin Cry1Ac + cowpea trypsin inhibitor (Cry1Ac + CpTI)	*Chilo suppressalis* (Lepidoptera: Pyralidae)	Rice	Field trials (China)	Han et al., 2006
Bt toxin Cry1Ac + *Galanthus nivalis* agglutinin or snow-drop lectin (Cry1Ac + GNA)	*C. suppressalis*, *Nilaparvata lugens* (Hemiptera: Delphacidae)	Rice	Experimental plants produced	Loc et al., 2002
Bt toxins Cry1Ac + Cry2A + *Galanthus nivalis* agglutinin or snowdrop lectin (Cry1Ac + Cry2A + GNA)	*N. lugens*, *Cnaphalocrocis medinalis*, *Scirpophaga incertulas* (Lepidoptera: Pyralidae)	Rice	Experimental plants produced	Maqbool et al., 2001
Bt Cry1Ac toxin + galactose-binding domain of ricin B-chain (Cry1Ac-RB)	*C. suppressalis*, *Spodoptera littoralis* (Lepidoptera: Noctuidae), *Cicadulina mbila* (Hemiptera: Cicadellidae)	Rice, maize	Experimental plants produced	Mehlo et al., 2005
Synthetic multidomain cysteine PIs incorporating domains from kininogen, stefin A, cystatin C, and potato cystatin, with or without equistatin (KACP EIMKACP)	*Frankliniella occidentalis* (Thysanoptera: Thripidae)	Potato	Greenhouse trials	Outchkourov et al., 2004b
Potato carboxypeptidase inhibitor + potato serine protease inhibitor 2 (PCI + PI-II)	*H. armigera*	Tomato	Experimental plants produced	Abdeen et al., 2005

and Cry1Ac (Liu et al., 2005; Han et al., 2006); field tests with rice expressing CpTI + Cry1Ac are under way (Han et al., 2006). Better understanding of the structures and functions of insecticidal proteins has led to the technique of "domain-swapping" to make novel transgene constructs and the ability to synthesize hybrid toxins (Table 13.8). In the case of *Bt* Cry toxins, this technique has been used to broaden toxicity and delay the onset of resistance in pest populations (e.g., Naimov et al., 2003; Singh et al., 2004). Hybrid proteinase inhibitors have also been proposed as a means for overcoming insect adaptation to and compensation for ingestion of single PIs (Outchkourov et al., 2004b; Abdeen et al., 2005; Brunelle et al., 2005) and a PI/alpha-amylase inhibitor combination has been suggested for improved toxicity (Amirhusin et al., 2004). Lectins are being used as "carriers" for other insecticidal proteins which need to cross the midgut and enter the hemocoel for maximum toxicity (Fitches et al., 2004; Trung et al., 2006) or as a means of "tethering" PIs to gut epithelium-interacting lectins (Zhu-Salzman et al., 2003). Novel combinations of insecticidal proteins or their domains have the potential to expand the pool of candidate transgenic pest-resistant plants greatly, and it is highly likely that new commercial products targeting previously intractable pests will result from this approach.

Unless synergistic effects have been demonstrated or are expected, the compatibility with natural enemies of transgenic plants expressing two insecticidal proteins simultaneously should not be markedly different from those expected with plants expressing each protein singly. In the case of *Bt* plants, the United States Environmental Protection Agency does not require a completely new set of biosafety test results when performing environmental risk assessments of plants expressing two *Bt* Cry toxins that have previously been approved for single use, provided the dual-trait plants have been produced via conventional crossing of two transgenic varieties and there is evidence that the two expressed proteins are similar and functionally equivalent to those expressed in the single-trait varieties (USEPA, 2007). Pollen from transgenic cotton plants expressing both Cry1Ac and CpTI did not affect parasitoid wasps in laboratory bioassays (Geng et al., 2006), and the potential effects of these plants on natural enemies are not expected to by any greater than those that might occur with Cry1Ac plants or CpTI plants.

However, fusion proteins which significantly alter the specificity of an insecticidal protein will require extensive testing to ensure that natural enemies will not be directly harmed should they be exposed to the plant's toxins either via ingestion of prey with plant matter in the gut (as with predatory beetles) or through eating pollen from the plants (as with ladybird beetles for example). For instance, the fusion of the *Bt* toxin Cry1Ac with the galactose-binding domain of ricin B toxin extended the range of this usually lepidopteran-specific toxin to include a hemipteran pest (Mehlo et al., 2005). This strategy circumvents the need for the *Bt* toxin to bind with a single receptor type in the insect's gut and could significantly extend the range of insects that may be negatively affected, although no-effect results from a test with the cereal aphid showed that the new protein was not universally toxic to insects (Mehlo et al., 2005).

13.10 Secondary Metabolites

All the transgenic plants listed above express insecticidal proteins, and thus this approach to pest control is ultimately limited by the availability of such proteins in the natural world or our ability to synthesise variations on them. "Metabolic engineering" to alter the concentrations of secondary metabolites in plants will extend the range of available insect-resistance compounds considerably (for a recent review see Aharoni et al., 2005). The ability to change the volatiles emitted by plants opens up the possibility of altering insect behavior via plant biotechnology. Transgenic tobacco plants with altered levels of cembratriene-ol deterred and were toxic to aphids (Wang et al., 2001), *Arabidopsis* plants producing linalool induced altered aphid behaviour (Aharoni et al., 2003), and transgenic chrysanthemums producing linalool repelled *Frankliniella occidentalis* (Western flower thrips) (Aharoni et al., 2005). Since olfactory cues from both prey and the plants they feed on are important in parasitoid host-finding behaviour, the interplay between plants with altered profiles of volatiles, pest herbivores and beneficial parasitoids will need to be carefully analysed to ensure the compatibility of such plants with IPM.

Microbial sources of insecticidal metabolites are also being explored. For example, the gene cluster responsible for biosynthesis of peramine, an alkaloid produced by a fungus endophytic on grasses and known to be repellent and toxic to various insects (Rowan et al., 1990), has been isolated and proposed for engineering into ryegrass (Bryan et al., 2006). Diet studies have shown that peramine can induce behavioral changes in pest weevils that make them less attractive to a hymenopteran parasitoid (Gerard, 2000), suggesting that a control strategy employing peramine might not be compatible with biological control.

Case-by-case non-target risk assessments will be required to ensure that the products of metabolic engineering will not adversely affect IPM programs.

13.11 Conclusions

Experience thus far with *Bt* crops expressing Cry proteins has indicated that insect-resistant transgenic plants are very compatible with biological control (O'Callaghan et al., 2005; Romeis et al., 2006) and can work well as part of an IPM system (e.g., Hellmich et al., chapter 5; Naranjo et al., chapter 6; Grafius and Douches, chapter 7; Cohen et al., chapter 8). As might be expected from past experience with *Bt* biopesticide sprays, *Bt* incorporated into plants provides an effective means for controlling target pests without harming non-target natural enemies. The specificity of the various *Bt* Cry toxins means that non-target effects are minimal and predictable, and natural enemies are not likely to be endangered by direct ingestion of the toxins. *Bt*'s efficacy and speed of kill are comparable to synthetic insecticides, making them a readily acceptable alternative. The major shortcomings with *Bt* crops are the lack of efficacy of some of the toxins against some pest species and the potential for resistance to arise. By pyramiding different *Bt* Cry toxins, the effective range of target pests can be extended, although there are probably limits to this in that some orders of insects, e.g. the Hemiptera, may be entirely resistant to *Bt*'s effects.

The non-*Bt* alternatives presented here could be employed in different ways to contribute to IPM. Firstly, those with efficacy approaching that of *Bt* Cry toxins, such as the VIPs and other *Bt* exotoxins, some of the alpha-amylase inhibitors, the biotin-binding proteins, and some of the fusion proteins, could be deployed in much the same way as the Cry toxins are at present. They could be used on their own in a transgenic crop or combined with other toxins to extend the range of pests targeted and the effective life of technology. These traits have modes of action that are completely different from that of the Cry toxins, suggesting a significant reduction in the probability that resistance will arise. For example, biotin-binding proteins bind to dietary biotin and deprive the insect of this vitamin; they have no molecular interaction with the insect itself, and so it is very difficult to envisage how an insect could become resistant to them.

Other traits, such as the PIs, lectins, defensins and chitinases, are perhaps more likely to be used in a manner similar to the current use of conventionally-bred insect-resistant plant varieties, i.e. not as a substitute for synthetic insecticides but as an adjunct to other insect control measures. So far, the concept of using plant-based resistances in crop breeding programmes has been exploited more for plant disease control than for insect control, but a great many traits are known to help protect plants from insects (Sadasivam and Thayumanavan, 2003). The slowness of conventional plant breeding (and the ready availability of synthetic insecticides) has limited the utilisation of such traits thus far, but advances in plant genomics and biotechnology are likely to accelerate their incorporation into elite plant cultivars, and there could also be greater public acceptance of these "cisgenic" crops. Natural plant defences could also be combined with transgenic traits, as has been investigated for Colorado potato beetle control (Cooper et al., 2004, 2006).

Given that the major goal of IPM is to maintain adequate, sustainable insect control while reducing inputs of synthetic insecticides, the novel insect-resistant crops described here offer obvious additions to our arsenal of pest control technologies. At this stage of their development, major concerns about their compatibility with IPM arise only in relation to potential impacts on natural enemies. Notwithstanding particular, crop-specific changes in cultivation practices that may arise, there is no reason to suppose that using these plants would negatively affect IPM tactics such as timing crop growth to avoid pest damage, intercropping, preserving field margins or refuges to provide habitats and resources for natural enemies, choosing neighboring crops to minimize pest invasions, carrying out hygienic practices to eliminate pests, or using attract-and-kill or semiochemical-based technologies.

Of greater concern is the potential for toxicity or significant sub-lethal impacts on natural enemies, above and beyond those effects currently tolerated as a consequence of the inevitable reductions in pest populations required for effective pest control. Such effects may occur if the expressed protein has broad toxicity to many insect species (most described in this chapter are less specific than the *Bt* Cry toxins) and if there is a plausible route by which the natural enemy may be exposed to sufficiently high levels of the protein for this to happen. In all cases, accurate assessment of potential exposure levels in the field will be of paramount importance in designing meaningful pre-release biosafety tests. This has not been so crucial with *Bt* Cry toxins, where the extreme specificity of the toxins has often made the outcomes of natural enemy biosafety tests with very high concentrations of Cry toxins predictably benign.

Results obtained with PI-expressing transgenic plants so far suggest that exposure to these will not cause significant mortality of natural enemies, but some reductions in predator weights, parasitoid emergence or fecundity might occur in some cases. Similarly, negligible effects have been observed in studies with biotin-binding proteins and predators, even though potentially one might expect these proteins to affect any moulting insect. Both proteins have potentially broad host ranges and ensuring that tests accurately reflect field exposure levels will be very important.

Although not dramatically toxic to natural enemies, lectins have demonstrated significant effects on longevity and fecundity of parasitoids and predators when ingested directly via artificial diets. Quantitative field data on the ingestion of nectar and pollen by these natural enemies and accurate measurements of proteins expression levels in different parts of these transgenic plants will be vitally important in determining the significance of these results in terms of impacts on natural enemy populations in the field. Reductions in growth, longevity, and fecundity of natural enemies have also been observed in tri-trophic studies with lectins. The significance of these findings in terms of impacts on biological control outcomes in the field needs to be established.

Many of the novel insecticidal proteins described in this chapter are still in the early stages of development. Current research on these potential technologies focuses on establishing efficacy against target pests, rather than assessing possible non-target impacts. Once again, a vital early step in the biosafety testing process will be the determination of exposure pathways and likely exposure concentrations.

There is now a growing body of scientific literature on non-target impacts of novel insect-resistance traits. While this contributes to our ability to assess potential compatibility with IPM, there is still a need for generic tools to help us relate results obtained in the laboratory with actual environmental impacts in the field. At present the ecological relevance of the assessment endpoints used in some studies is questionable, and more research is needed to better place these results in the context of use of new technologies on farms.

References

Abdeen, A., Virgos, A., Olivella, E., Villanueva, J., Aviles, X., Gabarra, R., and Prat, S., 2005. Multiple insect resistance in transgenic tomato plants over-expressing two families of plant proteinase inhibitors. Plant Molecular Biology 57: 189–202.

Aharoni, A., Giri, A.P., Deuerlein, S., Griepink, F., de Kogel, W.J., Verstappen, F.W.A., Verhoeven, H.A., Jongsma, M.A., Schwab, W., and Bouwmeester, H.J., 2003. Terpenoid metabolism in wild-type and transgenic *Arabidopsis* plants. Plant Cell 15: 2866–2884.

Aharoni, A., Jongsma, M.A., and Bouwmeester, H.J., 2005. Volatile science? Metabolic engineering of terpenoids in plants. Trends in Plant Science 10: 594–602.

Alfonso-Rubi, J., Ortego, F., Castanera, P., Carbonero, P., and Diaz, I., 2003. Transgenic expression of trypsin inhibitor CMe from barley in *indica* and *japonica* rice, confers resistance to the rice weevil *Sitophilas oryzae*. Transgenic Research 12: 23–31.

Alibhai, M.F., and Rydal, T.J., 2003. Insect inhibitory lipid acyl hydrolases. United States Patent 6657046.

Álvarez-Alfageme, F., Martinez, M., Pascual-Ruiz, S., Castanera, P., Diaz, I., and Ortego, F., 2007. Effects of potato plants expressing a barley cystatin on the predatory bug *Podisus maculiventris* via herbivorous prey feeding on the plant. Transgenic Research 16: 1–13.

Amirhusin, B., Shade, R.E., Koiwa, H., Hasegawa, P.M., Bressan, R.A., Murdock, L.L., and Zhu-Salzman, K., 2004. Soyacystatin N inhibits proteolysis of wheat alpha-amylase inhibitor and potentiates toxicity against cowpea weevil. Journal of Economic Entomology 97: 2095–2100.

Arai, S., and Abe, K., 2000. Cystatin-based control of insects, with special reference to oryzacystatin. In: Recombinant Protease Inhibitors in Plants, D. Michaud, ed., Eurekah.com, Georgetown, TX, USA, pp. 27–42.

Ashouri, A., Michaud, D., and Cloutier, C., 2001a. Recombinant and classically selected factors of potato plant resistance to the Colorado potato beetle, *Leptinotarsa decemlineata*, variously affect the potato aphid parasitoid *Aphidius nigripes*. BioControl 46: 401–418.

Ashouri, A., Michaud, D., and Cloutier, C., 2001b. Unexpected effects of different potato resistance factors to the Colorado potato beetle (Coleoptera: Chrysomelidae) on the potato aphid (Homoptera: Aphididae). Environmental Entomology 30: 524–532.

Azzouz, H., Campan, E.D.M., Cherqui, A., Saguez, J., Couty, A., Jouanin, L., Giordanengo, P., and Kaiser, L., 2005a. Potential effects of plant protease inhibitors, oryzacystatin I and soybean Bowman-Birk inhibitor, on the aphid parasitoid *Aphidius ervi* Haliday (Hymenoptera, Braconidae). Journal of Insect Physiology 51: 941–951.

Azzouz, H., Cherqui, A., Campan, E.D.M., Rahbé, Y., Duport, G., Jouanin, L., Kaiser, L., and Giordanengo, P., 2005b. Effects of plant protease inhibitors, oryzacystatin I and soybean Bowman-Birk inhibitor, on the aphid *Macrosiphum euphorbiae* (Homoptera, Aphididae) and its parasitoid *Aphelinus abdominalis* (Hymenoptera, Aphelinidae). Journal of Insect Physiology 51: 75–86.

Bates, S.L., Zhao, J.-Z., Roush, R.T., and Shelton, A.M., 2005. Insect resistance management in GM crops: past, present and future. Nature Biotechnology 23: 57–62.

Bell, H.A., Fitches, E.C., Down, R.E., Marris, G.C., Edwards, J.P., Gatehouse, J.A., and Gatehouse, A.M.R., 1999. The effect of snowdrop lectin (GNA) delivered via artificial diet and transgenic plants on *Eulophus pennicornis* (Hymenoptera: Eulophidae), a parasitoid of the tomato moth *Lacanobia oleracea* (Lepidoptera: Noctuidae). Journal of Insect Physiology 45: 983–991.

Bell, H.A., Fitches, E.C., Down, R.E., Ford, L., Marris, G.C., Edwards, J.P., Gatehouse, J.A., and Gatehouse, A.M.R., 2001a. Effect of dietary cowpea trypsin inhibitor (CpTI) on the growth and development of the tomato moth *Lacanobia oleracea* (Lepidoptera: Noctuidae) and on the success of the gregarious ectoparasitoid *Eulophus pennicornis* (Hymenoptera: Eulophidae). Pest Management Science 57: 57–65.

Bell, H.A., Fitches, E.C., Marris, G.C., Bell, J., Edwards, J.P., Gatehouse, J.A., and Gatehouse, A.M.R., 2001b. Transgenic GNA expressing potato plants augment the beneficial biocontrol of *Lacanobia oleracea* (Lepidoptera: Noctuidae) by the parasitoid *Eulophus pennicornis* (Hymenoptera: Eulophidae). Transgenic Research 10: 35–42.

Bell, H.A., Down, R.E., Fitches, E.C., Edwards, J.P., and Gatehouse, A.M.R., 2003. Impact of genetically modified potato expressing plant-derived insect resistance genes on the predatory bug *Podisus maculiventris* (Heteroptera: Pentatomidae). Biocontrol Science and Technology 13: 729–741.

Bell, H.A., Kirkbride-Smith, A.E., Marris, G.C., Edwards, J.P., and Gatehouse, A.M.R., 2004. Oral toxicity and impact on fecundity of three insecticidal proteins on the gregarious ectoparasitoid *Eulophus pennicornis* (Hymenoptera: Eulophidae). Agricultural and Forest Entomology 6: 215–222.

Bennett, J., Cohen, M.B., Katiyar, S.K., Ghareyazie, B., and Khush, G.S., 1997. Enhancing insect resistance in rice through biotechnology. In: Advances in Insect Control, N. Carozzi and M. Koziel, eds., Taylor & Francis, London, UK, pp. 75–93.

Bernal, J.S., and Sétamou, M., 2003. Fortuitous antixenosis in transgenic sugarcane: antibiosis-expressing cultivar is refractory to ovipositing herbivore pests. Environmental Entomology 32: 886–894.

Bi, R.-M., Jia, H.-Y., and Feng, D.-S., 2006. Production and analysis of transgenic wheat (*Triticum aestivum* L.) with improved insect resistance by the introduction of cowpea trypsin inhibitor gene. Euphytica 151: 351–360.

Birch, N.E., Geoghegan, I.E., Majerus, M.E.N., McNicol, J.W., Hackett, C.A., Gatehouse, A. M.R., and Gatehouse, J.A., 1999. Tri-trophic interactions involving pest aphids, predatory 2-spot ladybirds and transgenic potatoes expressing snowdrop lectin for aphid resistance. Molecular Breeding 5: 75–83.

Blackburn, M.B., Farrar, R.R., Novak, N.G., and Lawrence, S.D., 2006. Remarkable susceptibility of the diamondback moth (*Plutella xylostella*) to ingestion of Pir toxins from *Photorhabdus luminescens*. Entomologia Experimentalis et Applicata 121: 31–37.

Boets, A., Arnaut, G., Van Rie, J., and Damme, N., 2006. Transgenic plants expressing insecticidal proteins and methods of producing the same. United States Patent 7091399.

Bouchard, E., Cloutier, C., and Michaud, D., 2003a. Oryzacystatin I expressed in transgenic potato induces digestive compensation in an insect natural predator via its herbivorous prey feeding on the plant. Molecular Ecology 12: 2439–2446.

Bouchard, E., Cloutier, C., and Michaud, D., 2003b. Molecular interactions between an insect predator and its herbivore prey on transgenic potato expressing a cysteine protease inhibitor from rice. Molecular Ecology 12: 2429–2437.

Boulter, D., Edwards, G.A., Gatehouse, A.M.R., Gatehouse, J.A., and Hilder, V.A., 1990. Additive protective effects of different plant-derived insect resistance genes in transgenic tobacco plants. Crop Protection 9: 351–354.

Bown, D.P., Wilkinson, H.S., and Gatehouse, J.A., 1997. Differentially regulated inhibitor-sensitive and insensitive protease genes from the phytophagous insect pest, *Helicoverpa armigera*, are members of complex multigene families. Insect Biochemistry and Molecular Biology 27: 625–638.

Brookes, G., and Barfoot, P., 2006. Global impact of biotech crops: socio-economic and environmental effects in the first ten years of commercial use. AgBioForum 9: 139–151.

Brown, S.E., Cao, A.T., Dobson, P., Hines, E.R., Akhurst, R.J., and East, P.D., 2006. Txp40, a ubiquitous insecticidal toxin protein from *Xenorhabdus* and *Photorhabdus* bacteria. Applied and Environmental Microbiology 72: 1653–1662.

Brunelle, F., Girard, C., Cloutier, C., and Michaud, D., 2005. A hybrid, broad-spectrum inhibitor of Colorado potato beetle aspartate and cysteine digestive proteinases. Archives of Insect Biochemistry and Physiology 60: 20–31.

Bryan, G.T., Johnson, G.T., Parker, E.J., Scott, D.B., Tanaka, A., and Voisey, C.R., 2006. Peramine biosynthesis. PCT/NZ2006/000148.

Bu, Q.Y., Wu, L., Yang, S.H., and Wan, J.M., 2006. Cloning of a potato proteinase inhibitor gene PINII-2x from diploid potato (*Solanum phurejia* L.) and transgenic investigation of its potential to confer insect resistance in rice. Journal of Integrative Plant Biology 48: 732–739.

Burgess, E.P.J., Lovei, G.L., Malone, L.A., Nielson, I.W., Gatehouse, H.S., and Christeller, J.T., 2002a. Prey-mediated effects of the protease inhibitor aprotinin on the predatory carabid beetle *Nebria brevicollis*. Journal of Insect Physiology 48: 1093–1101.

Burgess, E.P.J., Malone, L.A., Christeller, J.T., Lester, M.T., Murray, C., Philip, B.A., Phung, M.M., and Tregidga, E.L., 2002b. Avidin expressed in transgenic tobacco leaves confers resistance to two noctuid pests, *Helicoverpa armigera* and *Spodoptera litura*. Transgenic Research 11: 185–198.

Burgess, E.P.J., Philip, B.A., Christeller, J.T., Page, N.E.M., Marshall, R.K., and Wohlers, M.W., 2008. Tri-trophic effects of transgenic insect-resistant tobacco expressing a protease inhibitor or biotin-binding protein on adults of the predatory carabid beetle *Ctenognathus novaezelandiae*. Journal of Insect Physiology 54: 518–528.

Carlini, C.R., and Grossi-de-Sa, M.F., 2002. Plant toxic proteins with insecticidal properties. A review on their potentialities as bioinsecticides. Toxicon 40: 1515–1539.

Chrispeels, M.J., and Raikhel, N.V., 1991. Lectins, lectin genes, and their role in plant defense. Plant Cell 3: 1–9.

Christeller, J.T., and Phung, M.M., 1998. Changes in biotin levels in the leaves of two apple cultivars during the season. New Zealand Journal of Crop and Horticultural Science 26: 39–43.

Christeller, J.T., Burgess, E.P.J., Mett, V., Gatehouse, H.S., Markwick, N.P., Murray, C., Malone, L.A., Wright, M.A., Philip, B.A., Watt, D., Gatehouse, L.N., Lovei, G.L., Shannon, A.L., Phung, M.M., Watson, L.M., and Laing, W.A., 2002. The expression of a mammalian proteinase inhibitor, bovine spleen trypsin inhibitor in tobacco and its effects on *Helicoverpa armigera* larvae. Transgenic Research 11: 161–173.

Christeller, J.T., Malone, L.A., Todd, J.H., Marshall, R.M., Burgess, E.P.J., and Philip, B.A., 2005. Distribution and residual activity of two insecticidal proteins, avidin and aprotinin, expressed in transgenic tobacco plants, in the bodies and frass of *Spodoptera litura* larvae following feeding. Journal of Insect Physiology 51: 1117–1126.

Christeller, J.T., Markwick, N.P., Poulton, J., and O'Callaghan, M., 2006. Binding of an insecticidal transgene product to soil: biological activity of soil-bound avidin and the effects of time and microbial activity. Soil Biology & Biochemistry 38: 2043–2052.

Christou, P., Capell, T., Kohli, A., Gatehouse, J.A., and Gatehouse, A.M.R., 2006. Recent developments and future prospects in insect pest control in transgenic crops. Trends in Plant Science 11: 302–308.

Clouthier, C., and Michaud, D., 2000. Expression of protease inhibitors in potato. In: Recombinant Protease Inhibitors in Plants, D. Michaud, ed., Eurekah.com, Georgetown, TX, USA, pp. 148–166.

Collins, C.L., Eason, P.J., Dunshea, F.R., Higgins, T.J.V., and King, R.H., 2006. Starch but not protein digestibility is altered in pigs fed transgenic peas containing α-amylase inhibitor. Journal of the Science of Food and Agriculture 86: 1894–1899.

Cooper, S.G., Douches, D.S., and Grafius, E.J., 2004. Combining genetic engineering and traditional breeding to provide elevated resistance in potatoes to Colorado potato beetle. Entomologia Experimentalis et Applicata 112: 37–46.

Cooper, S.G., Douches, D.S., and Grafius, E.J., 2006. Insecticidal activity of avidin combined with genetically engineered and traditional host plant resistance against Colorado potato beetle (Coleoptera: Chrysomelidae) larvae. Journal of Economic Entomology 99: 527–536.

Couty, A., and Poppy, G., 2001. Does host-feeding on GNA-intoxicated aphids by *Aphelinus abdominalis* affect their longevity and/or fecundity? Entomologia Experimentalis et Applicata 100: 331–337.

Couty, A., Clark, S.J., and Poppy, G.M., 2001a. Are fecundity and longevity of female *Aphelinus abdominalis* affected by development in GNA-dosed *Macrosiphum euphorbiae*? Physiological Entomology 26: 287–293.

Couty, A., Down, R.E., Gatehouse, A.M.R., Kaiser, L., Pham-Delegue, M.H., and Poppy, G.M., 2001b. Effects of artificial diet containing GNA and GNA-expressing potatoes on the development of the aphid parasitoid *Aphidius ervi* Haliday (Hymenoptera: Aphidiidae). Journal of Insect Physiology 47: 1357–1366.

Cowgill, S.E., Danks, C., and Atkinson, H.J., 2004. Multitrophic interactions involving genetically modified potatoes, nontarget aphids, natural enemies and hyperparasitoids. Molecular Ecology 13: 639–647.

CRCSIIB, 2006. Cooperative Research Centre for Sugar Industry Innovation through Biotechnology 2005/06 Annual Report, Brisbane, Australia, 52 pp.

Dall, D., Luque, T., and O'Reilly, D., 2001. Insect-virus relationships: sifting by informatics. BioEssays 23: 184–193.

De Leo, F., Bonade-Bottino, M., Ceci, L.R., Gallerani, R., and Jouanin, L., 2001. Effects of a mustard trypsin inhibitor expressed in different plants on three lepidopteran pests. Insect Biochemistry and Molecular Biology 31: 593–602.

de Oliveira, E.J., Rabinovitch, L., Monnerat, R.G., Passos, L.K.J., and Zahner, V., 2004. Molecular characterization of *Brevibacillus laterosporus* and its potential use in biological control. Applied and Environmental Microbiology 70: 6657–6664.

Delledonne, M., Allegro, G., Belenghi, B., Balestrazzi, A., Picco, F., Levine, A., Zelasco, S., Calligari, P., and Confalonieri, M., 2001. Transformation of white poplar (*Populus alba* L.) with a novel *Arabidopsis thaliana* cysteine proteinase inhibitor and analysis of insect pest resistance. Molecular Breeding 7: 35–42.

Dively, G.P., 2005. Impact of transgenic VIP3A x Cry1Ab lepidopteran-resistant field corn on the nontarget arthropod community. Environmental Entomology 34: 1267–1291.

Down, R.E., Gatehouse, A.M.R., Hamilton, W.D.O., and Gatehouse, J.A., 1996. Snowdrop lectin inhibits development and decreases fecundity of the glasshouse potato aphid (*Aulacorthum solani*) when administered *in vitro* and via transgenic plants both in laboratory and glasshouse trials. Journal of Insect Physiology 42: 1035–1045.

Down, R.E., Ford, L., Woodhouse, S.D., Raemaekers, R.J.M., Leitch, B., Gatehouse, J.A., and Gatehouse, A.M.R., 2000. Snowdrop lectin (GNA) has no acute toxic effects on a beneficial insect predator, the 2-spot ladybird (*Adalia bipunctata* L.). Journal of Insect Physiology 46: 379–391.

Down, R.E., Ford, L., Woodhouse, S.D., Davison, G.M., Majerus, M.E.N., Gatehouse, J.A., and Gatehouse, A.M.R., 2003. Tritrophic interactions between transgenic potato expressing snowdrop lectin (GNA), an aphid pest (peach-potato aphid; *Myzus persicae* (Sulz.) and a beneficial predator (2-spot ladybird; *Adalia bipunctata* L.). Transgenic Research 12: 229–241.

Du, J., Foissac, X., Carss, A., Gatehouse, A.M.R., and Gatehouse, J.A., 2000. Ferritin acts as the most abundant binding protein for snowdrop lectin in the mid-guts of brown plant hopper (*Nilaparvata lugens*). Insect Biochemistry and Molecular Biology 30: 297–305.

Duchaud, E., Rusniok, C., Frangeul, L., Buchrieser, C., Givaudan, A., Taourit, S., Bocs, S., Boursaux-Eude, C., Chandler, M., Charles, J.F., Dassa, E., Derose, R., Derzelle, S., Freyssinet, G., Gaudriault, S., Medigue, C., Lanois, A., Powell, K., Siguier, P., Vincent, R., Wingate, V., Zouine, M., Glaser, P., Boemare, N., Danchin, A., and Kunst, F., 2003. The genome sequence of the entomopathogenic bacterium *Photorhabdus luminescens*. Nature Biotechnology 21: 1307–1313.

Dutta, I., Majumder, P., Saha, P., Ray, K., and Das, S., 2005a. Constitutive and phloem specific expression of *Allium sativum* leaf agglutinin (ASAL) to engineer aphid (*Lipaphis erysimi*) resistance in transgenic Indian mustard (*Brassica juncea*). Plant Science 169: 996–1007.

Dutta, I., Saha, P., Majumder, P., Sarkar, A., Chakraborti, D., Banerjee, S., and Das, S., 2005b. The efficacy of a novel insecticidal protein, *Allium sativum* leaf lectin (ASAL), against homopteran insects monitored in transgenic tobacco. Plant Biotechnology Journal 3: 601–611.

Estruch, J.J., Warren, G.W., Mullins, M.A., Nye, G.J., Craig, J.A., and Koziel, M.G., 1996. Vip3A, a novel *Bacillus thuringiensis* vegetative insecticidal protein with a wide spectrum of activities against lepidopteran insects. Proceedings of the National Academy of Sciences of the USA 93: 5389–5394.

Fernandes, O.A., Faria, M., Martinelli, S., Schmidt, F., Carvalho, V.F., and Moro, G., 2007. Short-term assessment of Bt maize on non-target arthropods in Brazil. Scientia Agricola 64: 249–255.

Ferry, N., Raemaekers, R.J.M., Majerus, M.E.N., Jouanin, L., Port, G., Gatehouse, J.A., and Gatehouse, A.M.R., 2003. Impact of oilseed rape expressing the insecticidal cysteine protease inhibitor oryzacystatin on the beneficial predator *Harmonia axyridis* (multicoloured Asian ladybeetle). Molecular Ecology 12: 493–504.

Ferry, N., Jouanin, L., Ceci, L.R., Mulligan, A., Emami, K., Gatehouse, J.A., and Gatehouse, A.M.R., 2005. Impact of oilseed rape expressing the insecticidal serine protease inhibitor, mustard trypsin inhibitor-2 on the beneficial predator *Pterostichus madidus*. Molecular Ecology 14: 337–349.

Ferry, N., Edwards, M., Gatehouse, J., Capell, T., Christou, P., and Gatehouse, A., 2006. Transgenic plants for insect pest control: a forward looking scientific perspective. Transgenic Research 15: 13–19.

Fitches, E., Gatehouse, A.M.R., and Gatehouse, J.A., 1997. Effects of snowdrop lectin (GNA) delivered via artificial diet and transgenic plants on the development of tomato moth (*Lacanobia oleracea*) larvae in laboratory and glasshouse trials. Journal of Insect Physiology 43: 727–739.

Fitches, E., Edwards, M.G., Mee, C., Grishin, E., Gatehouse, A.M.R., Edwards, J.P., and Gatehouse, J.A., 2004. Fusion proteins containing insect-specific toxins as pest control agents: snowdrop lectin delivers fused insecticidal spider venom toxin to insect haemolymph following oral ingestion. Journal of Insect Physiology 50: 61–71.

Fitt, G.P., 2000. An Australian approach to IPM in cotton: integrating new technologies to minimise insecticide dependence. Crop Protection 19: 793–800.

Flinn, P.W., Kramer, K.J., Throne, J.E., and Morgan, T.D., 2006. Protection of stored maize from insect pests using a two-component biological control method consisting of a hymenopteran parasitoid, *Theocolax elegans*, and transgenic avidin maize powder. Journal of Stored Products Research 42: 218–225.

Foissac, X., Loc, N.T., Christou, P., Gatehouse, A.M.R., and Gatehouse, J.A., 2000. Resistance to green leafhopper (*Nephotettix virescens*) and brown planthopper (*Nilaparvata lugens*) in transgenic rice expressing snowdrop lectin (*Galanthus nivalis* agglutinin; GNA). Journal of Insect Physiology 46: 573–583.

Franco, O.L., Rigden, D.J., Melo, F.R., and Grossi-de-Sa, M.F., 2002. Plant alpha-amylase inhibitors and their interaction with insect alpha-amylases - Structure, function and potential for crop protection. European Journal of Biochemistry 269: 397–412.

Gatehouse, A.M.R., Down, R.E., Powell, K.S., Sauvion, N., Rahbé, Y., Newell, C.A., Merryweather, A., and Gatehouse, J.A., 1996. Effects of GNA-expressing transgenic potato plants on peach-potato aphid, *Myzus persicae*. Entomologia Experimentalis et Applicata 79: 295–307.

Gatehouse, A.M.R., Davison, G.M., Newell, C.A., Merryweather, A., Hamilton, W.D.O., Burgess, E.P.J., Gilbert, R.J.C., and Gatehouse, J.A., 1997. Transgenic potato plants with enhanced resistance to the tomato moth, *Lacanobia oleracea*: growth room trials. Molecular Breeding 3: 49–63.

Gatehouse, A.M.R., Davison, G.M., Stewart, J.N., Galehouse, L.N., Kumar, A., Geoghegan, I.E., Birch, A.N.E., and Gatehouse, J.A., 1999. Concanavalin A inhibits development of tomato moth (*Lacanobia oleracea*) and peach-potato aphid (*Myzus persicae*) when expressed in transgenic potato plants. Molecular Breeding 5: 153–165.

Gatehouse, J.A., Gatehouse, A.M.R., and Bown, D.P., 2000. Control of phytophagous insect pests using serine proteinase inhibitors. In: Recombinant Protease Inhibitors in Plants, D. Michaud, ed., Eurekah.com, Georgetown, TX, USA, pp. 9–26.

Geng, J.H., Shen, Z.R., Song, K., and Zheng, L., 2006. Effect of pollen of regular cotton and transgenic Bt plus CpTI cotton on the survival and reproduction of the parasitoid wasp *Trichogramma chilonis* (Hymenoptera: Trichogrammatidae) in the laboratory. Environmental Entomology 35: 1661–1668.

Gerard, P.J., 2000. Ryegrass endophyte infection affects Argentine stem weevil adult behaviour and susceptibility to parasitism. New Zealand Plant Protection 53: 406–409.

Glare, T.R., O'Callaghan, M., Malone, L.A., Burgess, E.P.J., and Philip, B.A., 2004. Measuring environmental impacts of genetically modified crops in New Zealand. In: GM Crops – Ecological Dimensions, H.F. van Emden and A.J. Gray, eds., The Association of Applied Biologists, Warwick, UK, pp. 91–99.

Gould, F., Anderson, A., Jones, A., Sumerford, D., Heckel, D.G., Lopez, J., Micinski, S., Leonard, R., and Laster, M., 1997. Initial frequency of alleles for resistance to *Bacillus thuringiensis* toxins in field populations of *Heliothis virescens*. Proceedings of the National Academy of Sciences of the USA 94: 3519–3523.

Graham, J., Gordon, S.C., and McNicol, J.W., 1997. The effects of the CpTi gene in strawberry against attack by vine weevil (*Otiorhynchus sulcatus* F., Coleoptera: Curculionidae). Annals of Applied Biology 131: 133–139.

Graham, J., Gordon, S.C., Smith, K., McNicol, R.J., and McNicol, J.W., 2002. The effect of the cowpea trypspin inhibitor in strawberry on damage by vine weevil under field conditions. Journal of Horticultural Science & Biotechnology 77: 33–40.

Green, T.R., and Ryan, C.A., 1972. Wound-induced proteinase inhibitor in plant leaves: a possible defense mechanism against insects. Science 175: 776–777.

Guo, H.N., Jia, Y.T., Zhou, Y.G., Zhang, Z.S., Ouyang, Q., Jiang, Y., and Tian, Y.C., 2004. Effects of transgenic tobacco plants expressing ACA gene from *Amaranthus caudatus* on the population development of *Myzus persicae*. Acta Botanica Sinica 46: 1100–1105.

Guo, H.N., Wu, J.-H., Chen, X.-Y., Luo, X.-L., Lu, R., Shi, Y.-J., Qin, H.-M., Xiao, J.-L., and Tian, Y.-C., 2003. Cotton plants transformed with the activated chimeric Cry1Ac and API-B genes. Acta Botanica Sinica 45: 108–113.

Guo, S.D., Cui, H.Z., Xia, L.Q., and Wu, D., 1999. Effects of CpTI-Bt transgenic cotton and Bt transgenic cotton on survival, growth and nutrition utilization of *Helicoverpa armigera* (Hübner). Scientia Agricultura Sinica 32: 1–7.

Hajaij-Ellouze, M., Fedhila, S., Lereclus, D., and Nielsen-LeRoux, C., 2006. The enhancin-like metalloprotease from the *Bacillus cereus* group is regulated by the pleiotropic transcriptional activator PlcR but is not essential for larvicidal activity. FEMS Microbiology Letters 260: 9–16.

Han, L.Z., Wu, K.M., Peng, Y.F., Wang, F., and Guo, Y.Y., 2006. Evaluation of transgenic rice expressing Cry1Ac and CpTI against *Chilo suppressalis* and intrapopulation variation in susceptibility to Cry1Ac. Environmental Entomology 35: 1453–1459.

Hao, C., Chai, B., Wang, W., Sun, Y., and Liang, A., 2005. Polyclonal antibody against *Manduca sexta* chitinase and detection of chitinase expressed in transgenic cotton. Biotechnology Letters 27: 97–102.

Harrison, R.L., and Bonning, B.C., 2001. Use of proteases to improve the insecticidal activity of baculoviruses. Biological Control 20: 199–209.

Hayakawa, T., Hashimoto, Y., Mori, M., Kaido, M., Shimojo, E., Furusawa, I., and Granados, R.R., 2004. Transgenic tobacco transformed with the *Trichopusia ni* granulovirus enhancin gene affects insect development. Biocontrol Science and Technology 14: 211–214.

He, H., Wang, Z., and Zhang, Y., 2008. Monitoring Bt resistance in the field; China as a case study. In: Environmental Impact of Genetically Modified/Novel Crops, N. Ferry and A.M.R. Gatehouse, eds., CAB International, Wallingford, UK, in press.

Hilder, V.A., and Boulter, D., 1999. Genetic engineering of crop plants for insect resistance - a critical review. Crop Protection 18: 177–191.

Hilder, V.A., Gatehouse, A.M.R., Sheerman, S., Barker, R.F., and Boulter, D., 1987. A novel mechanism of insect resistance engineered into tobacco. Nature 330: 160–163.

Hogervorst, P.A.M., Ferry, N., Gatehouse, A.M.R., Wäckers, F.L., and Romeis, J., 2006. Direct effects of snowdrop lectin (GNA) on larvae of three aphid predators and fate of GNA after ingestion. Journal of Insect Physiology 52: 614–624.

Hoover, K., Schultz, C.M., Lane, S.S., Bonning, B.C., Duffey, S.S., McCutchen, B.F., and Hammock, B.D., 1995. Reduction in damage to cotton plants by a recombinant baculovirus that knocks moribund larvae of *Heliothis virescens* off the plant. Biological Control 5: 419–426.

Huang, J.K., Hu, R.F., Rozelle, S., and Pray, C., 2005. Insect-resistant GM rice in farmers' fields: assessing productivity and health effects in China. Science 308(5722): 688–690.

Hukuhara, T., Hayakawa, T., and Wijonarko, A., 1999. Increased baculovirus susceptibility of armyworm larvae feeding on transgenic rice plants expressing an entomopoxvirus gene. Nature Biotechnology 17: 1122–1124.

ICAC, 2003. VIP cotton: a new type of transgenic cotton. The ICAC Recorder 21(2): 3–6.

ICAC, 2004. Cotton: Review of the World Situation. Volume 58, Number 2. November–December 2004. International Cotton Advisory Committee, Washington, DC, USA. http://www.icac.

org/cotton_info/publications/samples/reviews/erev_november_04.pdf (accessed 15 January 2008).

Ignacimuthu, S., and Prakash, S., 2006. *Agrobacterium*-mediated transformation of chickpea with alpha-amylase inhibitor gene for insect resistance. Journal of Biosciences 31: 339–345.

Ishimoto, M., Sato, T., Chrispeels, M.J., and Kitamura, K., 1996. Bruchid resistance of transgenic azuki bean expressing seed alpha-amylase inhibitor of common bean. Entomologia Experimentalis et Applicata 79: 309–315.

Jackson, R.E., Marcus, M.A., Gould, F., Bradley, J.R., Jr., and Van Duyn, J.W., 2007. Cross-resistance responses of Cry1Ac-selected *Heliothis virescens* (Lepidoptera: Noctuidae) to the *Bacillus thuringiensis* protein Vip3A. Journal of Economic Entomology 100: 180–186.

James, C., 2005. Global Status of Commercialized Transgenic Crops: 2005. ISAAA Brief No. 34, International Service for the Acquisition of Agri-Biotech Applications, Ithaca, NY, USA.

James, C., 2007. Global Status of Commercialized Biotech/GM Crops: 2007. ISAAA Brief No. 37, International Service for the Acquisition of Agri-Biotech Applications, Ithaca, NY, USA.

Jongsma, M.A., and Bolter, C., 1997. The adaptation of insects to plant protease inhibitors. Journal of Insect Physiology 43: 885–895.

Jorgensen, H.B., and Lovei, G.L., 1999. Tri-trophic effect on predator feeding: consumption by the carabid *Harpalus affinis* of *Heliothis armigera* caterpillars fed on proteinase inhibitor-containing diet. Entomologia Experimentalis et Applicata 93: 113–116.

Jouanin, L., Bonade-Bottino, M., Girard, C., Morrot, G., and Giband, M., 1998. Transgenic plants for insect resistance. Plant Science 131: 1–11.

Khan, S.A., Zafar, Y., Briddon, R.W., Malik, K.A., and Mukhtar, Z., 2006. Spider venom toxin protects plants from insect attack. Transgenic Research 15: 349–357.

Knowles, J.R., 1989. The mechanism of biotin-dependent enzymes. Annual Review of Biochemistry 58: 195–221.

Kondrak, M., Kutas, J., Szenthe, B., Patthy, A., Banfalvi, Z., Nadasy, M., Graf, L., and Asboth, B., 2005. Inhibition of Colorado potato beetle larvae by a locust proteinase inhibitor peptide expressed in potato. Biotechnology Letters 27: 829–834.

Konno, K., Hirayama, C., Nakamura, M., Tateishi, K., Tamura, Y., Hattori, M., and Kohno, K., 2004. Papain protects papaya trees from herbivorous insects: role of cysteine proteases in latex. Plant Journal 37: 370–378.

Kramer, K.J., Morgan, T.D., Thorne, J.E., Dowell, F.E., Bailey, M., and Howard, J.A., 2000. Transgenic avidin maize is resistant to storage insect pests. Nature Biotechnology 18: 670–674.

Lawo, N.C., and Romeis, J., 2008. Assessing the utilization of a carbohydrate food source and the impact of insecticidal proteins on larvae of the green lacewing, *Chrysoperla carnea*. Biological Control 44: 389–398.

Lawrence, P.K., and Koundal, K.P., 2002. Plant protease inhibitors in control of phytophagous insects. Electronic Journal of Biotechnology [online]. 15 April 2002, 5(1). Available from: http://www.ejbiotechnology.info/content/vol5/issue1/full/3/index.html

Lay, F.T., and Anderson, M.A., 2005. Defensins – components of the innate immune system in plants. Current Protein & Peptide Science 6: 85–101.

Lay, F.T., Schirra, H.J., Scanlon, M.J., Anderson, M.A., and Craik, D.J., 2003. The three-dimensional solution structure of NaD1, a new floral defensin from *Nicotiana alata* and its application to a homology model of the crop defense protein alfAFP. Journal of Molecular Biology 325: 175–188.

Lecardonnel, A., Chauvin, L., Jouanin, L., Beaujean, A., Prevost, G., and Sangwan-Norreel, B., 1999. Effects of rice cystatin I expression in transgenic potato on Colorado potato beetle larvae. Plant Science 140: 71–79.

Lee, M.K., Walters, F.S., Hart, H., Palekar, N., and Chen, J.S., 2003. Mode of action of the *Bacillus thuringiensis* vegetative insecticidal protein Vip3A differs from that of Cry1Ab delta-endotoxin. Applied and Environmental Microbiology 69: 4648–4657.

Lee, P.J., Ahn, J.Y., Kim, Y.H., Kim, S.W., Kim, J.Y., Park, J.S., and Lee, J., 2004. Cloning and heterologous expression of a novel insecticidal gene (tccC1) from X*enorhabdus nematophilus* strain. Biochemical and Biophysical Research Communications 319: 1110–1116.

Lee, M.K., Miles, P., and Chen, J.S., 2006. Brush border membrane binding properties of *Bacillus thuringiensis* Vip3A toxin to *Heliothis virescens* and *Helicoverpa zea* midguts. Biochemical and Biophysical Research Communications 339: 1043–1047.

Legaspi, J.C., Legaspi, B.C., Jr., and Sétamou, M., 2004. Insect-resistant transgenic crops expressing plant lectins. In: Transgenic Crop Protection: Concepts and Strategies, O. Koul and D.S. Dhaliwal, eds., Science Publishers, Inc., Enfield, USA, pp. 85–116.

Lei, J.J., Yang, W.J., Yuan, S.H., Ying, F.Y., and Qiong, L.C., 2006. Study on transformation of cysteine proteinase inhibitor gene into cabbage (*Brassica oleracea* var. *capitata* L.). In: Proceedings of the Joint Meeting of the Fourteenth Crucifer Genetics Workshop and Fourth I.S.H.S. Symposium on Brassicas, Y.P. Lim, ed., International Society Horticultural Science, Leuven 1, pp. 231–238.

Leple, J.C., Bonade-Bottino, M., Augustin, S., Pilate, G., le Tan, V.D., Delplanque, A., Cornu, D., and Jouanin, L., 1995. Toxicity to *Chrysomela tremulae* (Coleoptera: Chrysomelidae) of transgenic poplars expressing a cysteine proteinase inhibitor. Molecular Breeding 1: 319–328.

Lepore, L.S., Roelvink, P.R., and Granados, R.R., 1996. Enhancin, the granulosis virus protein that facilitates nucleopolyhedrovirus (NPV) infections, is a metalloprotease. Journal of Invertebrate Pathology 68: 131–140.

Li, G.Y., Xu, X.P., Xing, H.T., Zhu, H.C., and Fan, Q., 2005. Insect resistance to *Nilaparvata lugens* and *Cnaphalocrocis medinalis* in transgenic indica rice and the inheritance of *gna* plus *sbti* transgenes. Pest Management Science 61: 390–396.

Li, X., Higgins, T.J.V., and Bryden, W.L., 2006. Biological response of broiler chickens fed peas (*Pisum sativum* L.) expressing the bean (*Phaseolus vulgaris* L.) α-amylase inhibitor transgene. Journal of the Science of Food and Agriculture 86: 1900–1907.

Lingling, L.V., Jianjun, L., Ming, S., Liyun, L., and Bihao, C., 2005. Study on transformation of cowpea trypsin inhibitor gene into cauliflower (*Brassica oleracea* L. var. *botrytis*). African Journal of Biotechnology 4: 45–49.

Liu, D., Burton, S., Glancy, T., Li, Z.S., Hampton, R., Meade, T., and Merlo, D.J., 2003. Insect resistance conferred by 283-kDa *Photorhabdus luminescens* protein TcdA in *Arabidopsis thaliana*. Nature Biotechnology 21: 1222–1228.

Liu, X.X., Zhang, Q.W., Zhao, J.Z., Li, H.C., Xu, B.L., and Ma, X.M., 2005. Effects of Bt transgenic cotton lines on the cotton bollworm parasitoid *Microplitis mediator* in the laboratory. Biological Control 35: 134–141.

Liu, S.J., Li, H.R., Sivakumar, S., and Bonning, B.C., 2006a. Virus-derived genes for insect-resistant transgenic plants. In: Insect Viruses: Biotechnological Applications, B.C. Bonning, ed., Elsevier, San Diego, CA, USA, pp. 427–457.

Liu, Y.J., Cheng, C.S., Lai, S.M., Hsu, M.P., Chen, C.S., and Lyu, P.C., 2006b. Solution structure of the plant defensin VrD1 from mung bean and its possible role in insecticidal activity against bruchids. Proteins-Structure Function and Bioinformatics 63: 777–786.

Loc, N.T., Tinjuangjun, P., Gatehouse, A.M.R., Christou, P., and Gatehouse, J.A., 2002. Linear transgene constructs lacking vector backbone sequences generate transgenic rice plants which accumulate higher levels of proteins conferring insect resistance. Molecular Breeding 9: 231–244.

Ma, X.M., Liu, X.X., Zhang, Q.W., Zhao, J.Z., Cai, Q.N., Ma, Y.A., and Chen, D.M., 2006. Assessment of cotton aphids, *Aphis gossypii*, and their natural enemies on aphid-resistant and aphid-susceptible wheat varieties in a wheat-cotton relay intercropping system. Entomologia Experimentalis et Applicata 121: 235–241.

Maheswaran, G., Pridmore, L., Franz, P., and Anderson, M.A., 2007. A proteinase inhibitor from *Nicotiana alata* inhibits the normal development of light-brown apple moth, *Epiphyas postvittana* in transgenic apple plants. Plant Cell Reports 26: 773–782.

Malone, L.A., Burgess, E.P.J., Mercer, C.F., Christeller, J.T., Murray, C., Phung, M.M., Philip, B.A., Tregidga, E.L., and Todd, J.H., 2002a. Effects of biotin-binding proteins on eight species of pasture invertebrates. New Zealand Plant Protection 55: 411–420.

Malone, L.A., Tregidga, E.L., Todd, J.H., Burgess, E.P.J., Philip, B.A., Markwick, N.P., Poulton, J., Christeller, J.T., Lester, M.T., and Gatehouse, H.S., 2002b. Effects of ingestion of a biotin-binding protein on adult and larval honey bees. Apidologie 33: 447–458.

Malone, L.A., Todd, J.H., Burgess, E.P.J., Philip, B.A., and Christeller, J.T., 2005. Effects of kiwifruit (*Actinidia deliciosa*) cysteine protease on growth and survival of *Spodoptera litura* larvae (Lepidoptera: Noctuidae) fed with control or transgenic avidin-expressing tobacco. New Zealand Journal of Crop and Horticultural Science 33: 99–105.

Manyangarirwa, W., Turnbull, M., McCutcheon, G.S., and Smith, J., 2006. Gene pyramiding as a Bt resistance management strategy: how sustainable is this strategy? African Journal of Biotechnology 5: 781–785.

Maqbool, S.B., Riazuddin, S., Loc, N.T., Gatehouse, A.M.R., Gatehouse, J.A., and Christou, P., 2001. Expression of multiple insecticidal genes confers broad resistance against a range of different rice pests. Molecular Breeding 7: 85–93.

Markwick, N.P., Christeller, J.T., Docherty, L.C., and Lilley, C.M., 2001. Insecticidal activity of avidin and streptavidin against four species of pest Lepidoptera. Entomologia Experimentalis et Applicata 98: 59–66.

Markwick, N.P., Docherty, L.C., Phung, M.M., Lester, M.T., Murray, C., Yao, J.L., Mitra, D.S., Cohen, D., Beuning, L.L., Kutty-Amma, S., and Christeller, J.T., 2003. Transgenic tobacco and apple plants expressing biotin-binding proteins are resistant to two cosmopolitan insect pests, potato tuber moth and lightbrown apple moth, respectively. Transgenic Research 12: 671–681.

McCafferty, H.R.K., Moore, P.H., and Zhu, Y.J., 2006. Improved *Carica papaya* tolerance to carmine spider mite by the expression of *Manduca sexta* chitinase transgene. Transgenic Research 15: 337–347.

Mehlo, L., Gahakwa, D., Nghia, P.T., Loc, N.T., Capell, T., Gatehouse, J.A., Gatehouse, A.M.R., and Christou, P., 2005. An alternative strategy for sustainable pest resistance in genetically enhanced crops. Proceedings of the National Academy of Sciences of the USA 102: 7812–7816.

Meiyalaghan, S., Takla, M.F.G., Jaimess, O.O., Shang, Y.J., Davidson, M.M., Cooper, P.A., Barrell, P.J., Jacobs, J.M.E., Wratten, S.D., and Conner, A.J., 2005. Evaluation of transgenic approaches for controlling tuber moth in potatoes. Communications in Agricultural and Applied Biological Sciences (Faculteit Landbouwkundige en Toegepaste Biologische Wetenschappen, Ghent, Belgium) 70: 641–650.

Melander, M., Ahman, I., Kamnert, I., and Stromdahl, A.C., 2003. Pea lectin expressed transgenically in oilseed rape reduces growth rate of pollen beetle larvae. Transgenic Research 12: 555–567.

Morgan, T.D., Oppert, B., Czapla, T.H., and Kramer, K.J., 1993. Avidin and streptavidin as insecticidal and growth-inhibiting dietary proteins. Entomologia Experimentalis et Applicata 69: 97–108.

Morton, R.L., Schroeder, H.E., Bateman, K.S., Chrispeels, M.J., Armstrong, E., and Higgins, T. J.V., 2000. Bean alpha-amylase inhibitor 1 in transgenic peas (*Pisum sativum*) provides complete protection from pea weevil (*Bruchus pisorum*) under field conditions. Proceedings of the National Academy of Sciences of the USA 97: 3820–3825.

Mulligan, E.A., Ferry, N., Jouanin, L., Walters, K.F.A., Port, G.R., and Gatehouse, A.M.R., 2006. Comparing the impact of conventional pesticide and use of a transgenic pest-resistant crop on the beneficial carabid beetle *Pterostichus melanarius*. Pest Management Science 62: 999–1012.

Murray, C., Sutherland, P.W., Phung, M.M., Lester, M.T., Marshall, R.K., and Christeller, J.T., 2002. Expression of biotin-binding proteins, avidin and streptavidin, in plant tissues using plant vacuolar targeting sequences. Transgenic Research 11: 199–214.

Nagadhara, D., Ramesh, S., Pasalu, I.C., Rao, Y.K., Krishnaiah, N.V., Sarma, N.P., Bown, D.P., Gatehouse, J.A., Reddy, V.D., and Rao, K.V., 2003. Transgenic indica rice resistant to sap-sucking insects. Plant Biotechnology Journal 1: 231–240.

Nagadhara, D., Ramesh, S., Pasalu, I.C., Rao, Y.K., Sarma, N.P., Reddy, V.D., and Rao, K.V., 2004. Transgenic rice plants expressing the snowdrop lectin gene (gna) exhibit high level resistance to the whitebacked planthopper (*Sogatella furcifera*). Theoretical and Applied Genetics 109: 1399–1405.

Naimov, S., Dukiandjiev, S., and de Maagd, R.A., 2003. A hybrid *Bacillus thuringiensis* delta-endotoxin gives resistance against a coleopteran and a lepidopteran pest in transgenic potato. Plant Biotechnology Journal 1: 51–57.

Narva, K.E., Schnepf, H.E., Feitelson, J.S., Stockhoff, B.A., Schmeits, J., Loewer, D., Dullum, C.J., Muller-Cohn, J., Stamp, L., Morrill, G., and Finstad-Lee, S., 2003. Plants and cells transformed with a nucleic acid from *Bacillus thuringiensis* strain KB59A4–6 encoding a novel SUP toxin. United States Patent 6603063.

O'Callaghan, M., Glare, T.R., Burgess, E.P.J., and Malone, L.A., 2005. Effects of plants genetically modified for insect resistance on nontarget organisms. Annual Review of Entomology 50: 271–292.

OGTR (Office of the Gene Technology Regulator), 2004. Risk Assessment and Risk Management Plan. Application for licence for dealings involving an intentional release of GMOs into the environment. DIR 048/2003. Title: Field trial to assess transgenic insecticidal cotton expressing natural plant genes. Applicant: Hexima Limited. July 2004. Office of the Gene Technology Regulator, Australian Government, Woden, Australia. http://www.ogtr.gov.au/pdf/ir/dir048finalrarmp.pdf (accessed 15 January 2008).

OGTR (Office of the Gene Technology Regulator), 2005. Risk Assessment and Risk Management Plan for DIR 058/2005. Limited and controlled release of insect resistant (VIP) GM cotton. Applicant: Deltapine Australia Pty Ltd. October 2005. Office of the Gene Technology Regulator, Australian Government, Woden, Australia. http://www.ogtr.gov.au/pdf/ir/dir058finalrarmp1.pdf (accessed 15 January 2008).

Outchkourov, N.S., de Kogel, W.J., Schuurman-de Bruin, A., Abrahamson, M., and Jongsma, M. A., 2004a. Specific cysteine protease inhibitors act as deterrents of western flower thrips, *Frankliniella occidentalis* (Pergande), in transgenic potato. Plant Biotechnology Journal 2: 439–448.

Outchkourov, N.S., de Kogel, W.J., Wiegers, G.L., Abrahamson, M., and Jongsma, M.A., 2004b. Engineered multidomain cysteine protease inhibitors yield resistance against western flower thrips (*Frankliniella occidentalis*) in greenhouse trials. Plant Biotechnology Journal 2: 449–458.

Pang, Y., Shen, G., Qi, H., Tan, E., Sun, X., and Tang, K., 2004. Transgenic tobacco expressing *Zephyranthes candida* agglutinin showing enhanced resistance to aphids. Engineering in Life Sciences 4: 155–159.

Pechan, T., Ye, L.J., Chang, Y.M., Mitra, A., Lin, L., Davis, F.M., Williams, W.P., and Luthe, D.S., 2000. A unique 33-kD cysteine proteinase accumulates in response to larval feeding in maize genotypes resistant to fall armyworm and other lepidoptera. Plant Cell 12: 1031–1040.

Petell, J.K., Merlo, D.J., Herman, R.A., Roberts, J.L., Guo, L., Schafer, B.W., Sukhapinda, K., and Merlo, A.O., 2004. Transgenic plants expressing photorhabdus toxin. United States Patent 6717035.

Peumans, W.J., and Van Damme, E.J.M., 1995. Lectins as plant defense proteins. Plant Physiology 109: 347–352.

Powell, K.S., Spence, J., Bharathi, M., Gatehouse, J.A., and Gatehouse, A.M.R., 1998. Immunohistochemical and developmental studies to elucidate the mechanism of action of the snowdrop lectin on the rice brown planthopper, *Nilaparvata lugens* (Stal). Journal of Insect Physiology 44: 529–539.

Prescott, V.E., Campbell, P.M., Moore, A., Mattes, J., Rothenberg, M.E., Foster, P.S., Higgins, T.J.V., and Hogan, S.P., 2005. Transgenic expression of bean α-amylase inhibitor in peas results in altered structure and immunogenicity. Journal of Agricultural and Food Chemistry 53: 9023–9030.

Rahbé, Y., Deraison, C., Bonade-Bottino, M., Girard, C., Nardon, C., and Jouanin, L., 2003. Effects of the cysteine protease inhibitor oryzacystatin (OC-I) on different aphids and reduced performance of *Myzus persicae* on OC-I expressing transgenic oilseed rape. Plant Science 164: 441–450.

Rao, K.V., Rathore, K.S., Hodges, T.K., Fu, X., E., S., Sudhakar, D., Williams, S., Christou, P., Bharathi, M., Bown, D.P., Powell, K.S., Spence, J., Gatehouse, A.M.R., and Gatehouse, J.A., 1998. Expression of snowdrop lectin (GNA) in transgenic rice plants confers resistance to rice brown planthopper. Plant Journal 15: 469–477.

Ribeiro, A.P.O., Pereira, E.J.G., Galvan, T.L., Picanco, M.C., Picoli, E.A.T., da Silva, D.J.H., Fari, M.G., and Otoni, W.C., 2006. Effect of eggplant transformed with oryzacystatin gene on *Myzus persicae* and *Macrosiphum euphorbiae*. Journal of Applied Entomology 130: 84–90.

Romeis, J., Babendreier, D., and Wäckers, F.L., 2003. Consumption of snowdrop lectin (*Galanthus nivalis* agglutinin) causes direct effects on adult parasitic wasps. Oecologia 134: 528–536.

Romeis, J., Meissle M., and Bigler, F., 2006. Transgenic crops expressing *Bacillus thuringiensis* toxins and biological control. Nature Biotechnology 24: 63–71.

Rowan, D.D., Dymock, J.J., and Brimble, M.A., 1990. Effect of fungal metabolite peramine and analogs on feeding and development of Argentine stem weevil (*Listronotus bonariensis*). Journal of Chemical Ecology 16: 1683–1695.

Sadasivam, S., and Thayumanavan, B., 2003. Molecular Host Plant Resistance to Pests. Marcel Dekker, New York, USA.

Sadeghi, A., Smagghe, G., Broeders, S., Hernalsteens, J.-P., De Greve, H., Peumans, W.J., and Van Damme, E.J.M., 2008. Ectopically expressed leaf and bulb lectins from garlic (*Allium sativum* L.) protect transgenic tobacco plants against cotton leafworm (*Spodoptera littoralis*). Transgenic Research 17: 9–18.

Saha, P., Majumder, P., Dutta, I., Ray, T., Roy, S.C., and Das, S., 2006. Transgenic rice expressing *Allium sativum* leaf lectin with enhanced resistance against sap-sucking insect pests. Planta 223: 1329–1343.

Sarmah, B.K., Moore, A., Tate, W., Molvig, L., Morton, R.L., Rees, D.P., Chiaiese, P., Chrispeels, M.J., Tabe, L.M., and Higgins, T.J.V., 2004. Transgenic chickpea seeds expressing high levels of a bean alpha-amylase inhibitor. Molecular Breeding 14: 73–82.

Schroeder, H.E., Gollasch, S., Moore, A., Tabe, L.M., Craig, S., Hardie, D.C., Chrispeels, M.J., Spencer, D., and Higgins, T.J.V., 1995. Bean alpha-amylase inhibitor confers resistance to the pea weevil (*Bruchus pisorum*) in transgenic peas (*Pisum sativum* L). Plant Physiology 107: 1233–1239.

Schuler, T.H., Denholm, I., Jouanin, L., Clark, S.J., Clark, A.J., and Poppy, G.M., 2001. Population-scale laboratory studies of the effect of transgenic plants on nontarget insects. Molecular Ecology 10: 1845–1853.

Sétamou, M., Bernal, J.S., Legaspi, J.C., and Mirkov, T.E., 2002a. Effects of snowdrop lectin (*Galanthus nivalis* agglutinin) expressed in transgenic sugarcane on fitness of *Cotesia flavipes* (Hymenoptera: Braconidae), a parasitoid of the nontarget pest *Diatraea saccharalis* (Lepidoptera: Crambidae). Annals of the Entomological Society of America 95: 75–83.

Sétamou, M., Bernal, J.S., Legaspi, J.C., and Mirkov, T.E., 2002b. Parasitism and location of sugarcane borer (Lepidoptera: Pyralidae) by *Cotesia flavipes* (Hymenoptera: Braconidae) on transgenic and conventional sugarcane. Environmental Entomology 31: 1219–1225.

Sétamou, M., Bernal, J.S., Legaspi, J.C., Mirkov, T.E., and Legaspi, B.C., Jr., 2002c. Evaluation of lectin-expressing transgenic sugarcane against stalkborers (Lepidoptera: Pyralidae): effects on life history parameters and damage. Journal of Economic Entomology 95: 469–477.

Shade, R.E., Schroeder, H.E., Pueyo, J.J., Tabe, L.M., Murdock, L.L., Higgins, T.J.V., and Chrispeels, M.J., 1994. Transgenic pea-seeds expressing the alpha-amylase inhibitor of the common bean are resistant to bruchid beetles. BioTechnology 12: 793–796.

Shah, P.A., Gatehouse, A.M.R., Clark, S.J., and Pell, J.K., 2005. Wheat containing snowdrop lectin (GNA) does not affect infection of the cereal aphid *Metopolophium dirhodum* by the fungal natural enemy *Pandora neoaphidis*. Transgenic Research 14: 473–476.

Singh, P.K., Kumar, M., Chaturvedi, C.P., Yadav, D., and Tuli, R., 2004. Development of a hybrid delta-endotoxin and its expression in tobacco and cotton for control of a polyphagous pest *Spodoptera litura*. Transgenic Research 13: 397–410.

Song, X.X., and Wang, S.M., 2001. Status and evaluation on the expression of cotton varieties in the production in China in the past 20 years. Cotton Science 13: 315–320.

Strickland, J.A., Orr, G.L., and Walsh, T.A., 1995. Inhibition of *Diabrotica* larval growth by patatin, the lipid acyl hydrolase from potato tubers. Plant Physiology 109: 667–674.

Sun, X., Wu, A., and Tang, K., 2002. Transgenic rice lines with enhanced resistance to the small brown planthopper. Crop Protection 21: 511–514.

Syed, S.H., 2002. Genetic transformation of cotton with *Galanthus nivalis* agglutinin (GNA) gene and Cry1Ac + GNA. Ph.D. thesis, University of Punjab, Lahore, Pakistan.

Syngenta, 2003. Syngenta Plans to Introduce a New Choice for Transgenic Control of Worms in Cotton. http://www.syngentacropprotection-us.com/media/article.asp?article_id = 303 (accessed 15 January 2008).

Syngenta, 2007. About Syngenta: Biotechnology. http://www.syngenta.com/en/about_syngenta/biotech_pipeline.aspx (accessed 15 January 2008).

Tabashnik, B.E., Roush, R.T., Earle, E.D., and Shelton, A.M., 2002. Resistance to Bt toxins. Science 287(5450): 41.

Tabashnik, B.E., Carriere, Y., Dennehy, T.J., Morin, S., Sisterson, M.S., Roush, R.T., Shelton, A.M., and Zhao, J.Z., 2003. Insect resistance to transgenic Bt crops: lessons from the laboratory and field. Journal of Economic Entomology 96: 1031–1038.

Tinjuangjun, P., Loc, N.T., Gatehouse, A.M.R., Gatehouse, J.A., and Christou, P., 2000. Enhanced insect resistance in Thai rice varieties generated by particle bombardment. Molecular Breeding 6: 391–399.

Tomov, B.W., Bernal, J.S., and Vinson, S.B., 2003. Impacts of transgenic sugarcane expressing GNA lectin on parasitism of Mexican rice borer by *Parallorhogas pyralophagus* (Marsh) (Hymenoptera: Braconidae). Environmental Entomology 32: 866–872.

Treacy, M.F., Rensner, P.E., and All, J.N., 2000. Comparative insecticidal properties of two nucleopolyhedrovirus vectors encoding a similar toxin gene chimer. Journal of Economic Entomology 93: 1096–1104.

Trung, N.P., Fitches, E., and Gatehouse, J.A., 2006. A fusion protein containing a lepidopteran-specific toxin from the South Indian red scorpion (*Mesobuthus tamulus*) and snowdrop lectin shows oral toxicity to target insects. Biomed Central Biotechnology [online]. 16 March 2006, 6(18). Available from: http://www.biomedcentral.com/1472–6750/6/18

USDA (United States Department of Agriculture), undated. National Information System for the Regional IPM Centers.United States Department of Agriculture. http://www.ipmcenters.org/newsletters.cfm?USDARegion = National%20Site&site (accessed 15 January 2008).

USEPA (United States Environmental Protection Agency), 2006a. Experimental use permit: receipt of application. Federal Register: May 26, 2006 (Volume 71, Number 102). United States Environmental Protection Agency. http://www.epa.gov/fedrgstr/EPA-PEST/2006/May/Day-26/p8042.htm (accessed 15 January 2008).

USEPA (United States Environmental Protection Agency), 2006b. *Bacillus thuringiensis* Vip3A Protein and the Genetic Material Necessary for its Production in Corn; Notice of Filing of a Pesticide. Petition to Establish a Temporary Tolerance. Federal Register: November 1, 2006 (Volume 71, Number 211). United States Environmental Protection Agency. http://www.epa.gov/EPA-PEST/2006/November/Day-01/p18425.htm (accessed 21 January 2008).

USEPA (United States Environmental Protection Agency), 2007. Biopesticides Registration Action Document *Bacillus thuringiensis* Cry3Bb1 corn. United States Environmental Protection Agency. http://www.epa.gov/pesticides/biopesticides/ingredients/tech_docs/cry3bb1/2_c_cry3bb1_environl.pdf (accessed 15 January 2008).

van Frankenhuyzen, K., and Nystrom, C., 2002. The *Bacillus thuringiensis* Toxin Specificity Database. http://www.glfc.cfs.nrcan.gc.ca/bacillus (accessed 15 January 2008).

Vila, L., Quilis, J., Meynard, D., Breitler, J.C., Marfa, V., Murillo, I., Vassal, J.M., Messeguer, J., Guiderdoni, E., and San Segundo, B., 2005. Expression of the maize proteinase inhibitor (mpi) gene in rice plants enhances resistance against the striped stem borer (*Chilo suppressalis*): effects on larval growth and insect gut proteinases. Plant Biotechnology Journal 3: 187–202.

Wakefield, M.E., Bell, H.A., Fitches, E.C., Edwards, J.P., and Gatehouse, A.M.R., 2006. Effects of *Galanthus nivalis* agglutinin (GNA) expressed in tomato leaves on larvae of the tomato moth *Lacanobia oleracea* (Lepidoptera: Noctuidae) and the effect of GNA on the development of the endoparasitoid *Meteorus gyrator* (Hymenoptera: Braconidae). Bulletin of Entomological Research 96: 43–52.

Wang, E., Wang, R., DeParasis, J., Loughrin, J.H., Gan, S., and Wagner, G.J., 2001. Suppression of a P450 hydroxlase gene in plant trichome glands enhances natural-product-based aphid resistance. Nature Biotechnology 19: 371–374.

Wang, J.X., Chen, Z.L., Du, J.Z., Sun, Y., and Liang, A.H., 2005a. Novel insect resistance in *Brassica napus* developed by transformation of chitinase and scorpion toxin genes. Plant Cell Reports 24: 549–555.

Wang, Z., Zhang, K., Sun, X., Tang, K., and Zhang, J., 2005b. Enhancement of resistance to aphids by introducing the snowdrop lectin gene gna into maize plants. Journal of Biosciences 30: 627–638.

Waterfield, N.R., Bowen, D.J., Fetherston, J.D., Perry, R.D., and ffrench-Constant, R.H., 2001. The tc genes of *Photorhabdus*: a growing family. Trends in Microbiology 9: 185–191.

Waterfield, N., Hares, M., Yang, G., Dowling, A., and ffrench-Constant, R., 2005. Potentiation and cellular phenotypes of the insecticidal toxin complexes of *Photorhabdus* bacteria. Cellular Microbiology 7: 373–382.

Whitehouse, M.E.A., Wilson, L.J., and Constable, G.A., 2007. Target and non-target effects on the invertebrate community of Vip cotton, a new insecticidal transgenic. Australian Journal of Agricultural Research 58: 273–285.

Wood, H.G., and Barden, R.E., 1977. Biotin enzymes. Annual Review of Biochemistry 46: 385–413.

Wu, K.M., and Guo, Y.Y., 2005. The evolution of cotton pest management practices in China. Annual Review of Entomology 50: 31–52.

Wu, A., Sun, X., Pang, Y., and Tang, K., 2002. Homozygous transgenic rice lines expressing GNA with enhanced resistance to the rice sap-sucking pest *Laodelphax striatellus*. Plant Breeding 121: 93–95.

Wu, J., Zhang, Z., Nie, Y., and Luo, X., 2005. High efficiency transformation of *Gossypium hirsutum* embryonic calli mediated by *Agrobacterium tumefaciens* and regeneration of insect-reistant plants. Plant Breeding 124: 142.

Wu, J., Luo, X., Guo, H., Xiao, J., and Tian, Y., 2006. Transgenic cotton, expressing *Amaranthus caudatus* agglutinin, confers enhanced resistance to aphids. Plant Breeding 125: 390–394.

Yao, J., Pang, Y., Qi, H., Wan, B., Zhao, X., Kong, W., Sun, X., and Tang, K., 2003. Transgenic tobacco expressing *Pinellia ternata* agglutinin confers enhanced resistance to aphids. Transgenic Research 12: 715–722.

Yao, J., Zhao, X., Qi, H., Wan, B., Chen, F., Sun, X., Yu, S., and Tang, K., 2004. Transgenic tobacco expressing an *Arisaema heterophyllum* agglutinin gene displays enhanced resistance to aphids. Canadian Journal of Plant Science 84: 785–790.

Yoza, K., Imamura, T., Kramer, K.J., Morgan, T.D., Nakamura, S., Akiyama, K., Kawasaki, S., Takaiwa, F., and Ohtsubo, K., 2005. Avidin expressed in transgenic rice confers resistance to the stored-product insect pests *Tribolium confusum* and *Sitotroga cerealella*. Bioscience, Biotechnology and Biochemistry 69: 966–971.

Yuan, Z.Q., Zhao, C.Y., Zhou, Y., and Tian, Y.C., 2001. Aphid-resistant transgenic tobacco plants expressing modified gna gene. Acta Botanica Sinica 43: 592–597.

Zhang, B.-H., Liu, F., Yao, C.-B., and Wang, K.-B., 2000. Recent progress in cotton biotechnology and genetic engineering in China. Current Science 79: 37–44.

Zhang, J.-H., Wang, C.-Z., Qin, J.-D., and Guo, S.-D., 2004. Feeding behaviour of *Helicoverpa armigera* on insect-resistant transgenic cotton and non-transgenic cotton. Journal Applied Entomology 128: 218–225.

Zhu-Salzman, K., Ahn, J.E., Salzman, R.A., Koiwa, H., Shade, R.E., and Balfe, S., 2003. Fusion of a soybean cysteine protease inhibitor and a legume lectin enhances anti-insect activity synergistically. Agricultural and Forest Entomology 5: 317–323.

Zhu, Y.C., Adamczyk, J.J., and West, S., 2005. Avidin, a potential biopesticide and synergist to *Bacillus thuringiensis* toxins against field crop insects. Journal of Economic Entomology 98: 1566–1571.

Chapter 14
IPM and Insect-Protected Transgenic Plants: Thoughts for the Future

Anthony M. Shelton[1,*], Jörg Romeis[2], and George G. Kennedy[3]

Abstract Political, economic, social and biological forces have altered agricultural practices in the last several decades. One result has been the development of transgenic maize and cotton varieties expressing insecticidal proteins from the bacterium, *Bacillus thuringiensis* (*Bt*), which have become important components in IPM programs globally. In 2007, *Bt* maize and *Bt* cotton were grown in 13 and 9 countries, respectively. *Bt* plants have provided simultaneous opportunities and challenges to managing insect pest complexes, but overall their use has resulted in strong economic and environmental benefits. Their adoption rate has been dramatic. In only the 12th year after their introduction, they were grown on 42.1 million hectares in 2007. Thus, *Bt* plants have turned what was once a minor foliar insecticide (*Bt*) into a major control strategy. Other *Bt* commodities are expected to be registered in the near future. Additionally, new insecticide molecules are being developed for expression in plants and plant genes are being altered to affect biochemical pathways that elicit insect resistance. However, adoption of *Bt* plants should be viewed within the larger context of food systems, cultures, human values, politics and the roles and responsibilities of science in the modern world. Such a context helps explain the variable adoption rates of *Bt* plants on a global basis and helps provide insights for the future deployment of insect-resistant plants.

14.1 Introduction

In the last decade of the 20th century and continuing today, agricultural scientists have found themselves at the center of a worldwide debate about food systems, cultures, human values, politics and the roles and responsibilities of science in the

[1] Department of Entomology, Cornell University/NYSAES, Geneva, NY 14456, USA

[2] Agroscope Reckenholz-Tänikon Research Station ART, 8046 Zurich, Switzerland

[3] Department of Entomology, North Carolina State University, Raleigh, NC, USA

* To whom correspondence should be addressed. E-mail: ams5@cornell.edu

J. Romeis, A.M. Shelton, G.G. Kennedy (eds.), *Integration of Insect-Resistant Genetically Modified Crops within IPM Programs*.

modern world. This is not the first time such controversies have occurred, as will be discussed later in this chapter, but it has profound consequences for our present and future food systems. Agricultural scientists generally tend to be problem solvers addressing questions of how to improve the quantity and quality of foods and fibers. Animal and plant breeding methods have helped solve many agricultural problems, and generally been supported by public funds without much controversy. For example, there was little, if any, social controversy when Norman Borlaug and his colleagues initiated the Green Revolution by introducing high-yielding wheat varieties. This new technology is credited with saving hundreds of millions of lives and for his efforts Borlaug became the only person in the 20th century to be awarded a Nobel Peace Prize for work in agriculture (Hesser, 2006). However, Borlaug's mission – to cause the environment to produce significantly more food – later was seen, at least by "some securely affluent commentators", as perhaps better left undone (Easterbrook, 1997). Those opposed to his work argue that food sustains human population growth, which they see as "antithetical to the natural world". Others argue that high-yield agriculture slows population growth rather than accelerates it, by starting the progression from the high-birth-rate, high-death-rate societies of feudal cultures toward the low-birth-rate, low-death-rate societies of Western nations (Easterbrook, 1997).

Likewise, pesticides and fertilizers, once seen as an aid to modern agricultural practices, are often viewed by the general public as problems rather than solutions to improving the quality and availability of foods. Much of this may have been derived by the interpretation (or misinterpretation) of Rachel Carson's 1962 seminal book, *Silent Spring*, which is often credited with starting the environmental movement in the west. As many of this book's chapters were prepared in 2007, a century after Carson's birth, it is an appropriate time to reflect on her legacy. In the first 16 chapters of *Silent Spring*, she focused on the long-term effects on human health and the environment caused by misusing pesticides. In the book's last chapter, *The Other Road*, she advocates for increased use of alternatives to broad-spectrum insecticides. Chief among these is her endorsement of *Bacillus thuringiensis* (*Bt*), a microbial insecticide that was and continues to constitute less than 1% of the total insecticide market (Shelton et al., 2002a). At nearly the same time her book was published, Stern et al. (1959) formally introduced the concept of integrated pest management (IPM), which advocated using multiple tactics to control pest populations and reduced the emphasis on synthetic pesticides. IPM was a paradigm shift in crop protection, but was brought about because of outbreaks of secondary pests; resurgence of target pest populations following destruction of beneficial arthropods and/or insecticide resistance; and environmental and human health concerns (Kennedy, chapter 1). The concept of IPM was linked to the legacy of Carson with the publication of *Beyond Silent Spring: Integrated Pest Management* (van Emden and Peakall, 1996). It is fitting that this was also the year in which GM plants expressing insecticidal proteins (Cry toxins) from *Bt* were first commercialized. Thus, what was once a minor insecticide (*Bt*) soon became a major player when its genes were incorporated into major crop plants (Shelton et al., 2002a). By 2007, varieties of *Bt* cotton and *Bt* maize were grown on a total of 42.1 million hectares worldwide (Fig. 14.1).

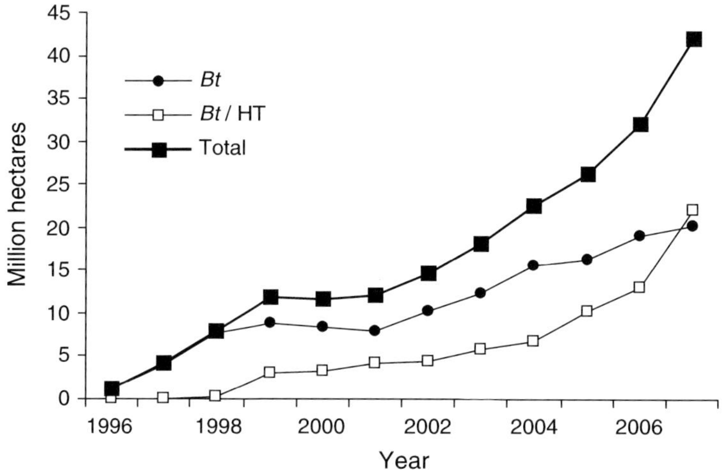

Fig. 14.1 Global adoption of GM crops that express Cry toxins from *Bt* either alone (*Bt*) or in combination with a herbicide-tolerance trait (*Bt*/HT) since the commercial release of the first *Bt* maize variety in 1996 (Data compiled from ISAAA Briefs; http://www.isaaa.org/)

Solutions to biological problems in agriculture, such as those advocated by Borlaug, Carson and Stern, must be viewed in a social as well as a biological context. It is appropriate that this last chapter attempts to address not only the biological aspects of insect-resistant GM plants but also how their adoption is viewed in a wider social context.

14.2 Agronomic and Environmental Advances with Insect-Resistant GM Plants

Bt-transgenic plants have become important components of maize and cotton IPM programs worldwide (Hellmich et al., chapter 5; Naranjo et al., chapter 6). They have provided simultaneous opportunities and challenges to IPM production practices. A similar situation is expected to occur for other important commodities including rice (Cohen et al., chapter 8), potatoes (Grafius and Douches, chapter 7), and vegetables and fruits (Shelton et al., chapter 9). *Bt* plants are only the first wave of insect-resistant GM plants (Malone et al., chapter 13), and continue the history of host plant resistance, as described by Kennedy (chapter 1). These insect-resistant plants pose the same risks (e.g., gene flow and the ability of an insect population to overcome the resistance) as the more "traditional" host plant resistance described in the earlier writings of Painter (1951). The incorporation of *Bt* crops into IPM programs has caused a paradigm shift in the way many of the key pests are being controlled, using much more specific toxins rather than broader spectrum insecticides. Since *Bt* plants were first commercialized in 1996, considerable data has been gathered from many parts of the world on their impact. Although these are crop and regional specific, some general trends have emerged.

Use of *Bt* plants has resulted in changes in insecticide practices, farm income and environmental impact. Worldwide, Brookes and Barfoot (2006a, b) estimated that between 1996 and 2005 the deployment of *Bt* cotton reduced the volume of insecticide active ingredient applied by 94.5 million kilograms (a 19.4% reduction), provided an economic benefit of $7.5 billion, and reduced the environmental impact by 24.3% (as determined by the environmental impact quotient (EIQ), which integrates the various environmental impacts of individual pesticides into a single "field value per hectare" [Kovach et al., 1992]). During this same period, the use of *Bt* maize resulted in a $2.4 billion economic benefit, a reduction of 4.1% of insecticide a.i. used and a 4.6% reduction in the environmental impact. More recent reports have documented similar benefits (Fitt, chapter 11; Qaim et al., chapter 12). These results are impressive and indicate that within a 12-year span no technology previously developed has had such a far reaching and positive impact on the economics and reduced use of insecticides in insect pest management. *Bt* plants have also had a tremendous impact on the biology and ecology of insects within these agricultural systems.

Commercialized *Bt* plants have effectively controlled key species of Lepidoptera affecting maize (Hellmich et al., chapter 5) and cotton (Naranjo et al., chapter 6), and control has also been documented with key species of Coleoptera on potato (Grafius and Douches, chapter 7) and maize (Hellmich et al., chapter 5). However, most cropping systems have insect complexes and, as one order of insects is controlled by one toxin, other pests that normally had been controlled by broader spectrum tactics may become problematic. Such pest shifts have mainly been reported from cotton, which is attacked by a very diverse group of herbivores. The most prominent case involves reported outbreaks of plant bugs in *Bt* cotton in different parts of the world caused by the significant reduction in the use of broad-spectrum insecticides previously applied to control pest Lepidoptera (Naranjo et al., chapter 6).

Within the context of agroecosystems, extensive research has been published on the impact of *Bt* plants on non-target arthropods, especially natural enemies that are essential in helping suppress insect pest populations and are thus an important component of IPM systems (O'Callaghan et al., 2005; Romeis et al., 2006, chapter 4). Additional reviews on non-targets are also available in the context of cotton (Naranjo et al., chapter 6), maize (Hellmich et al., chapter 5), rice (Cohen et al., chapter 8), potatoes (Grafius and Douches, chapter 7) and some vegetables (Shelton et al., chapter 9). Long-term, large-scale field studies have indicated no meaningful impacts of *Bt* cotton on predator populations, even when the predator has acquired the toxin by feeding on intoxicated prey (Naranjo et al., chapter 6). These results are in agreement with those obtained in maize (Hellmich et al., chapter 5). Even studies that have indicated predators can acquire the *Bt* protein have shown no direct negative effects, and studies in which *Bt* crops were compared to conventional crops treated with insecticides have demonstrated the latter to be far more harmful to predators. The situation appears to be more complex for parasitoids. While an insect predator is characterized by feeding on multiple and various hosts during its lifetime, a parasitoid usually completes its entire lifetime within a single host and derives all its nutritional requirements by feeding on the host tissues.

This suggests that the more intimate relationship of a parasitioid with its host would put it at more risk to any hazard its host encounters. Thus, it is not surprising that negative effects of *Bt* plants on mortality, development, weight or longevity of parasitoids were observed in most published studies (Romeis et al., 2006). However, such negative effects on the parasitoids could be due to poor quality of its host (sick or dying hosts) that had fed on the *Bt* plant or to direct toxicity. More recently, a meta-analysis of field studies examining the effects of *Bt* cotton and maize on non-target invertebrates revealed reduced abundance of hymenopteran parasitoids in *Bt* crops compared to non-*Bt*, insecticide-free controls (Marvier et al., 2007). The authors state that it is unclear whether the reduced abundance was due to direct toxicity or reduced availability of prey. This is an important question since parasitoids are essential components of a sustainable IPM program. Using populations of the diamondback moth, *Plutella xylostella* (Lepidoptera: Plutellidae) that were resistant to Cry1C or the commonly-used insecticides λ-cyhalothrin, cypermethrin, indoxacarb or spinosad, Chen et al. (2008) showed that Cry1C *Bt* broccoli plants and the purified Cry1C toxin had no direct toxicity to *Diadegma insulare* (Hymenoptera: Ichneumonidae), *P. xylostella's* major parasitoid in North America, when it fed inside its host after the host had consumed either *Bt* plants or leaves dipped in Cry1C. On the contrary, λ-cyhalothrin, cypermethrin, indoxacarb and spinosad significantly reduced parasitism rates of *D. insulare* on strains of *P. xylostella* resistant to these insecticides. This study was the first to make such direct comparisons between a *Bt* plant and conventional insecticides and to suggest that the previously reported negative impacts of *Bt* plants on parasitoids were likely due to the poor host quality as the result of ingestion and susceptibility to the *Bt* toxin, rather than direct toxicity to the parasitoid.

Overall, the data that are available now show that the *Bt* toxins expressed in today's GM crops do not cause any direct toxic effects on natural enemies. When the impact of *Bt* crops is compared to that of insecticide-treated controls, predators and parasitoids (and the biological control function they provide) benefited from the reduction in insecticides applied to the GM crop. Thus, this indicates that *Bt* plants may allow biological control to become a more important component in IPM than is possible with most insecticides.

A second area of concern about whether *Bt* plants could fit into sustainable IPM programs is whether insects will develop resistance to *Bt* plants. When the US Environmental Protection Agency (USEPA) reviewed the first registration for *Bt* plants, there was considerable concern in some sectors that resistance to the plants would rapidly occur and that not only would this be a concern to growers of *Bt* plants but also to organic farmers who relied on *Bt* as a foliar spray. In the USA, there has been substantial policy interest in maintaining the effectiveness of *Bt* as an important public resource to agricultural production systems, so EPA mandated specific insecticide resistance management (IRM) requirements (Matten et al., chapter 2). Currently the only commercially available strategy in the USA is use of a high dose of one or more toxins, combined with a refuge of non-*Bt* plants, which can serve as a reservoir for susceptible alleles in the population. The size and location of refuges vary depending on the crop and region in the USA. Registrations of *Bt* crops have been tied to IRM strategies and these strategies are refined as more information is made available. After 12 years of *Bt* crops, there have not been any verified cases of field failures due to resistance (Tabashnik et al., 2003; Ferré et al., chapter 3, but see Matten et al., chapter 2 for a recent purported case of field resistance to *Bt* maize). This is in stark contrast to conventional insecticides in which resistance often occurs at soon as 3 years after a product is introduced (Bates et al., 2005). This success of the *Bt* crop IRM strategy has been credited to the proactive IRM strategy composed of the mandated high dose/refuge strategy combined with the monitoring program. Additional factors may include fitness costs associated with resistance alleles and "natural" refuges of non-*Bt* crop and non-crop hosts (Bates et al., 2005). However, it is expected that resistance will eventually evolve (see Matten et al., chapter 2 for a recent example) and strategies should be developed to help delay resistance evolution. In this context, it is also important to remember that, prior to the advent of resistance, substantial benefits (economic, environmental, human health) would have accrued (Fitt, chapter 11; Qiam et al., chapter 12). Also, it should be remembered that if resistance occurs to one *Bt* toxin, others may still be effective when engineered into plants, since many Cry toxins are not cross-resistant (Ferré et al., chapter 3). This is much different than most other insecticide classes where resistance to one insecticide generally confers resistance to other insecticides in that class.

Other biological questions about the use of *Bt* plants, besides those relating to effects on non-target organisms and IRM, have and will continue to arise, but it is important to compare such effects to the use of alternative management tactics. Such an approach has not generally been the framework in which *Bt* plants have been evaluated, thus resulting in a more "precautionary" approach than a risk-benefit analysis. This is unfortunate since it may result in continuing with practices that are more harmful (Goklany, 2002; Shelton et al., 2002a; Sanvido et al., 2007). To ensure that a policy is truly precautionary, one should compare the risk of adopting a technology (such as GM) against the risk of not adopting it. It is apparent that in the current global regulatory climate GM crops are held to higher standards. This is clearly demonstrated by two examples: *Bt* crops are the only insect control strategy that requires a detailed IRM component (Matten et al., chapter 2), and herbicide-tolerant crops that are created through genetic engineering are more highly regulated than herbicide-tolerant crops developed through other methods, although both share the same risks and benefits to the environment (Kennedy, chapter 1). This situation suggests that regulations do not always follow scientific principles. This situation has led to a far slower development and commercialization of GM crops than traditional crops.

14.3 The Social Context

While this book has largely focused on the biological, agricultural and regulatory framework in which GM insecticidal plants operate, it should be recognized that the adoption of GM plants will be strongly influenced by social issues that operate outside these areas. This is most clearly seen in the sociological and political literature.

James (2007) documents that GM crops (all traits) were grown on 114.3 million hectares by 12 million farmers in 23 countries in 2007. While this is an impressive figure, especially considering GM crops were released in 1996, it has a flip side. The majority of the world's 194 countries don't grow GM crops. It is important to recognize the reasons for this. In many cases, maize, cotton, soya and canola (the main GM crops) may not be produced in the country of interest, regardless of whether they are GM or not. Or the GM traits that would help solve pest problems of importance in the region may not yet be available in local varieties for various reasons, including the fact that such varieties represent very small markets for biotech companies. In many cases, however, it is the lack of a biosafety law that restricts the legal commercialization of GM crops. A biosafety law is required under the Cartagena Protocol on Biosafety (CBD Secretariat, 2000) and such laws and their regulations may be written differently. Thus, countries that wish to import GM organisms may write their laws one way while other countries that wish to restrict the importation may write their laws and regulations in such a manner that it will bring about this result. Many countries, especially developing countries, may not have the administrative or scientific resources needed to develop biosafety laws and accompanying regulations (Matten et al., chapter 2). Generally, this has prevented GM crops, including *Bt* plants and GM virus-resistant plants, from being grown commercially. However, important exceptions have occurred, such as in India.

Despite pressure from activists against GM technology, Indian farmers obtained, tested and grew *Bt* cotton before it was approved by Indian regulators. This story has been documented in an aptly named article, *Stealth Seeds: Bioproperty, Biosafety and Biopolitics* (Herring, 2007a). Briefly, one *Bt* cotton line had been smuggled into Gujarat, perhaps in 1998, but no one noticed until the massive bollworm (*Helicoverpa armigera*; Lepidoptera: Noctuidae) outbreak of 2001. In November, 2001, 460 acres of cotton seed farms were found to be producing transgenic cotton in the Kurnool and Mahabubnagar districts of Andhra Pradesh. The government initiated action against the framers but, on March 25, 2002, farmers threatened to launch a civil-disobedience movement if *Bt* cotton was not approved by Delhi. Farmers rallied for immediate approval and threatened to cultivate *Bt* cotton whether or not the government approved. On March 26, 2002 the government approved three varieties of *Bt* cotton. India has doubled its production of cotton in the last 5 years and surpassed the USA to become the world's second largest producer of cotton behind China (Herring, 2007a).

What drove this adoption was the farmers who saw benefits to the technology, despite what Herring (2007b) claims as the "proxy war" between the metropolitan middle classes on the terrain of relatively poor farmers. Herring suggests that "opposition to biotechnology in India has been largely an urban phenomenon, a creature of media and various websites. Opponents are backed by international NGOs and aid projects brokered through claims of indigenous authenticity. Farmers were largely absent, though everyone speaks in their name". Regardless of urban discourses on biotechnology, farmers are adopting GM plants if they see an advantage in using them. This is now happening in Vietnam and Brazil. And if GM seed are too expensive or bureaucratically restricted, farmers make, trade, and save their own stealth seeds (Herring, 2007a).

The Indian example is interesting because it illustrates the long-standing power of Indian farmers. No regime in Delhi, or in many other countries, can ignore the farmers as a voting block, nor can regulations from Delhi that are opposed by rural people be enforced in the villages. The *Bt* cotton episode in Gujarat illustrates this phenomenon: the genetic engineering approval committee (GEAC) ordered the destruction of the stealth-seed cotton crop, but had no power to enforce its edict on either the state government or the farmers. Social and cultural contexts are critical in explaining differential spread of biotechnology, both above ground-through official channels- and underground through farmers' stealthful agency. Unfortunately, for the adoption of GM plants an alternative outcome is also possible. Paarlberg (2007) makes the argument that in all countries in Africa (except for South Africa) no GM crops can be planted because of post-colonial European influence over governmental policies in Africa. He argues that this influence extends to international commodity markets, financial and technical assistance policies, European dominance with the special agencies of the United Nations, and advocacy campaigns by European-based NGOs. Thus, he concludes that Europe, "where farmers can be highly productive and consumers well-fed, has exported its rejection of GMOs to Africa" where the opposite occurs. Unlike in India, African states are not faced with highly mobilized and assertive farmers, nor does an established democratic machinery allow African farmers to present their interests and punish at the polls those who work against their interests. However, the situation in Africa may change with recent pressure in several countries by farmers who wish to produce *Bt* cotton (e.g., Burkina Faso, Kenya, Uganda) and *Bt* maize (Kenya, Uganda) and other crops. Farmers' visits to South Africa and India to see insect-resistant GM crops have spurred their interests (Shelton, unpublished), and studies have shown the benefits of GM plant technology for Africa (Thomson, 2008). However, for legal production, biosafety laws must be passed. But if their biosafety regulations are too restrictive or not balanced (i.e. do not compare risks and benefits of GM technology to other technologies), then GM crops, including insect-resistant varieties, may not be available to farmers…at least legally.

14.4 Conclusions

Controversies in biology are not new. Controversy about evolution continues to exist in many countries, although most biologists accept evolution as a unifying principle of biology. Likewise, controversies in agriculture continue to garner public attention. In the early 1900s consumption of raw milk accounted for high infant mortality rates in the USA. Pasteurization of milk, a major food safety innovation that resulted from basic scientific studies, virtually eliminated this problem but was highly controversial for many of the same reasons some critics use against biotechnology (Shelton et al., 2002b). In genetic engineering, one type of biotechnology, controversy followed the first recombinant DNA experiments (1972) by Boyer and Cohen when they manufactured insulin (Kelves, 2001). Controversy about the use of *Bt* plants for insect management has focused largely on their potential to affect non-target organisms such as the monarch butterfly (Shelton and Sears, 2001) and natural enemies (Romeis et al., chapter 4), or cause insect pests to develop resistance to the plants (Ferré et al., chapter 3). But as documented in the chapters of this book, an abundance of studies have indicated that, compared to other insect-management technologies, use of *Bt* plants has provided definite benefits not only to farmers but also to the environment and overall global economy (Qaim et al., chapter 12; Fitt, chapter 11).

The increasing demands that are being placed on the global food, energy and fresh water supplies by population growth and economic development will necessitate increased efficiency in agricultural production. This is especially evident in developing countries where 83% of the world's population presently resides, where population growth rates are highest and where agricultural problems are most severe. Crops with improved input and output traits derived by genetic engineering will certainly be able play a vital role in meeting this need. For example, insect-resistance GM plants provide the ability to significantly reduce the energy and labor required to manage insect pests, as well as the ecological impact of insect management when used within the context of IPM programs. It is also important to remember that the present genetically engineered insect-resistant crops of cotton and maize are primarily used for fabric and processed food ingredients, respectively. However, insect-resistant plants can and should play an increased role in providing food crops, such as potatoes, rice, fruits and vegetable, to meet the growing needs of consumers. Worldwide, there is increased demand for foods with fewer residues of potentially harmful pesticides and, in fact, this is one of the drivers of the increased marketing of organic foods. We suggest that the organic community reevaluate their opposition to using genetically engineered insect-resistant crops if, in fact, a major goal of their approach to agriculture is to reduce the risk of agriculture to human health and the environment. As the scientific evidence on *Bt* plants has accumulated since 1996, it is clear that they have provided substantial benefits to human health and the environment.

It appears fortunate that the first insect-resistant GM plants produced *Bt* proteins, since they have a long history of safe use. Indeed, *Bt* plants can be considered another delivery method of such proteins that were so strongly advocated by Carson (1962) in *Silent Spring*. As new insecticide molecules are being developed for expression in plants (Malone et al., chapter 13) or plant genes are altered to affect biochemical pathways to elicit insect resistance, they must also be evaluated to ensure their safety. But risk assessments need to be done in context and be conducted in a scientifically rigorous manner with testable hypotheses and formal decision guidelines (Raybould, 2007; Johnson et al., 2007; Romeis et al., 2008). When done properly, there should be a high degree of transparency of the risk assessment process and transportability of results from risk assessment studies across crops and countries. This will ensure that countries with limited resources will have access to vital information with which to make regulatory decisions. Countries that delay in developing workable, biosafety regulations, for whatever reason, will be challenged by farmers if they see benefits to the technology. In today's world, information and products move more freely across political boundaries than ever before.

References

Bates, S.L., Zhao, J.-Z., Roush, R.T., and Shelton, A.M., 2005. Insect resistance management in GM crops: Past, present and future. Nature Biotechnology 23: 57–62.

Brookes, G., and Barfoot, P., 2006a. Global impact of biotech crops: Socio-economic and environmental effects in the first ten years of commercial use. AgBioForum 9: 139–151.

Brookes, G., and Barfoot, P., 2006b. GM Crops: The First Ten Years – Global Socio-Economic and Environmental Impacts. ISAAA Brief No. 36, International Service for the Acquisition of Agri-Biotech Applications, Ithaca, NY, USA.

Carson, R., 1962. Silent Spring. Houghton Mifflin, Boston, MA, USA.

CBD Secretariat, 2000. Secretariat of the Convention on Biological Diversity. Cartagena Protocol on Biosafety to the Convention on Biological Diversity: Text and Annexes. Secretariat of the Convention on Biological Diversity, Montreal, Canada.

Chen, M., Zhao, J.-Z., Colins, H.L., Earle, E.D., Cao, J., and Shelton, A.M., 2008. A critical assessment of the effects of Bt transgenic plants on parasitoids. PLoS ONE 3(5): e2284. doi:10.1371/journal.pone.0002284.

Easterbrook, G., 1997. Forgotten benefactor of humanity. The Atlantic Monthly 279: 75–82.

Goklany, I., 2002. From precautionary principle to risk-risk analysis. Nature Biotechnology 20: 1075.

Herring, R.J., 2007a. Stealth seeds: Bioproperty, biosafety and biopolitics. Journal of Development Studies 43: 130–157.

Herring, R.J., 2007b. Suicide seeds? Biotechnology meets the development state. http://casi.ssc.upenn.edu/india/index.html (accessed 11 January 2008).

Hesser, L., 2006. The man who fed the world: Nobel Peace Prize Laureate Norman Borlaug and his battle to end world hunger. Durban House Publishing, Dallas, TX, USA.

James, C., 2007. Global Status of Commercialized Biotech/GM Crops: 2007. ISAAA Brief No. 37, International Service for the Acquisition of Agri-Biotech Applications, Ithaca, NY, USA.

Johnson, K.L., Raybould, A.F., Hudson, M.D., and Poppy, G.M., 2007. How does scientific risk assessment of GM crops fit within the wider risk analysis? Trends in Plant Sciences 12: 1–5.

Kelves, D.J., 2001. The battle over biotechnology. In: Days of Destiny, Crossroads in American History, D. Rubel, ed., DK Publishing, New York, USA, pp. 453–463.

Kovach, J., Petzoldt, C., Degni, J., and Tette, J., 1992. A method to measure the environmental impact of pesticides. New York's Food and Life Sciences Bulletin. NYS Agricultural Experiment Station, Cornell University, Geneva, NY, USA. http://www.nysipm.cornell.edu/publications/eiq/ (accessed 6 January 2008).

Marvier, M., McCreedy, C., Regetz, J., and Kareiva, P., 2007. A meta-analysis of effects of Bt cotton and maize on nontarget invertebrates. Science 316: 1475–1477.

O'Callaghan, M., Glare, T.R., Burgess, E.P.J., and Malone, L.A., 2005. Effects of plants genetically modified for insect resistance on nontarget organisms. Annual Review of Entomology 50: 271–292.

Paarlberg, R., 2007. Keeping genetically engineered crops out of Africa (Abstract). Workshop on contentious knowledge and the diffusion of social protest. http://www.socialsciences.cornell.edu/0609/Diffusion.html#Agenda/ (accessed 6 January 2008).

Painter, R.H., 1951. Insect Resistance in Crop Plants. University of Kansas Press, Lawrence, KS, USA.

Raybould, A., 2007. Ecological versus ecotoxicological methods for assessing the environmental risks of transgenic crops. Plant Science 173: 589–602.

Romeis, J., Meissle M., and Bigler, F., 2006. Transgenic crops expressing *Bacillus thuringiensis* toxins and biological control. Nature Biotechnology 24: 63–71.

Romeis, J., Bartsch, D., Bigler, F., Candolfi, M.P., Gielkens, M.M.C., Hartley, S.E., Hellmich, R.L., Huesing, J.E., Jepson, P.C., Layton, R., Quemada, H., Raybould, A., Rose, R.I., Schiemann, J., Sears, M.K., Shelton, A.M., Sweet, J., Vaituzis, Z., and Wolt, J.D., 2008. Assessment of risk of insect-resistant transgenic crops to nontarget arthropods. Nature Biotechnology 26: 203–208.

Sanvido, O., Romeis, J., and Bigler, F., 2007. Ecological impacts of genetically modified crops: Ten years of field research and commercial cultivation. Advances in Biochemical Engineering and Biotechnology 107: 235–278.

Shelton, A.M., and Sears, M.K., 2001. The monarch butterfly controversy: Scientific interpretations of a phenomenon. The Plant Journal 27: 483–488.

Shelton, A.M., Zhao, J.-Z., and Roush, R.T., 2002a. Economic, ecological, food safety, and social consequences of the deployment of Bt transgenic plants. Annual Review of Entomology 47: 845–881.

Shelton, A.M., McCandless, L., Lewenstein, B., Hawkes, J., Lyson, T., Bauman, T., and Aldwinckle, H. 2002b. Agricultural biotechnology: Informing the dialogue, 2002. Available online at http://www.nysaes.cornell.edu/comm/gmo/ (accessed 6 January 2008).

Stern, V.M., Smith, R.F., van den Bosch, R., and Hagen, K., 1959. The integrated control concept. Hilgardia 29: 81–101.

Tabashnik, B.E., Carrière, Y., Dennehy, T.J., Morin, S., Sisterson, M.S., Roush, R.T., Shelton, A.M., and Zhao, J.Z., 2003. Insect resistance to transgenic Bt crops: Lessons from the laboratory and field. Journal of Economic Entomology 96: 1031–1038.

Thomson, J.A., 2008. The role of biotechnology for agricultural sustainability in Africa. Philosophical Transactions of the Royal Society B 363: 905–913.

Van Emden, H.F., and Peakall, D.B., eds., 1996. Beyond Silent Spring: Integrated Pest Management and Chemical Safety. Chapman & Hall, London, UK.

Index

R